Laboratory Manual

Concepts of Biology

Third Edition

Sylvia S. Mader

CONCEPTS OF BIOLOGY LABORATORY MANUAL,
THIRD EDITION

This book is printed on acid-free paper.

3 4 5 6 7 8 9 LMN 19 18 17 16

ISBN 978-0-07-751158-6
MHID 0-07-751158-1

Senior Vice President, Products & Markets: Kurt L. Strand
Vice President, General Manager, Products & Markets: Marty Lange
Vice President, Content Production & Technology Services: Kimberly Meriwether David
Managing Director: Michael Hackett
Director, Biology: Lynn Breithaupt
Brand Manager: Eric Weber
Director of Development: Rose Koos
Development Editor: Anne Winch
Marketing Manager: Chris Loewenberg
Director, Content Production: Terri Schiesl
Content Project Manager: Sherry Kane
Senior Buyer: Sandy Ludovissy
Senior Designer: Laurie B. Janssen
Cover Image Credit: FLPA/Alamy
Senior Content Licensing Specialist: Lori Hancock
Compositor: Electronic Publishing Services Inc., NYC
Typeface: 10/12 Utopia Std
Printer: LSC Communications

Some of the laboratory experiments included in this text may be hazardous if materials are handled improperly or if procedures are conducted incorrectly. Safety precautions are necessary when you are working with chemicals, glass test tubes, hot water baths, sharp instruments, and the like, or for any procedures that generally require caution. Your school may have set regulations regarding safety procedures that your instructor will explain to you. Should you have any problems with materials or procedures, please ask your instructor for help.

www.mhhe.com

Contents

Preface

To the Instructor

The 26 laboratory sessions in this manual have been designed to introduce beginning students to the major concepts of biology, while keeping in mind minimal preparation for sequential laboratory use. The laboratories are coordinated with *Concepts of Biology*, a general biology text that covers all fields of biology. In addition, this Laboratory Manual can be adapted to a variety of course orientations and designs. There are a sufficient number of laboratories and exercises within each lab to tailor the laboratory experience as desired. Then, too, many exercises may be performed as demonstrations rather than as student activities, thereby shortening the time required to cover a particular concept.

Laboratory Resource Guide

The Laboratory Resource Guide, an essential aid for instructors and laboratory assistants, free to adopters of the *Concepts of Biology*, is online at **www.mhhe.com/maderconcepts3.** The answers to the Laboratory Review questions are in the Laboratory Resource Guide.

Customized Editions

With McGraw-Hill Create™ you can easily rearrange the labs in this manual and combine them with content from other sources, including your own syllabus or teaching notes. Go to www.mcgrawhillcreate.com to learn how to create your own version of the *Concepts of Biology Laboratory Manual.*

LearnSmart Labs™

McGraw-Hill's LearnSmart Labs™ are available where noted to accompany many labs in this manual. These outcomes-based lab simulations assess a student's knowledge and adaptively correct deficiencies, allowing the student to learn faster and retain more knowledge with greater sucess. A demonstration of the labs can be found at www.LearnSmartAdvantage.com.

The Exercises

All exercises have been tested for student interest, preparation time, estimated time of completion, and feasibility. The following features are particularly appreciated by adopters:

Integrated opening: Each laboratory begins with a list of Learning Outcomes organized according to the major sections of the laboratory. The major sections of the laboratory are numbered on the opening page and in the laboratory text material. This organization will help students better understand the goals of each laboratory session.

Self-contained content: Each laboratory contains all the background information necessary to understand the concepts being studied and to answer the questions asked. This feature will reduce student frustration and increase learning.

Scientific process: All laboratories stress the scientific process, and many opportunities are given for students to gain an appreciation of the scientific method. The first laboratory of this edition explicitly explains the steps of the scientific method and gives students an opportunity to use them.

Student activities: Sequentially numbered steps guide students as they perform each activity. Some student exercises are Observations (designated by a tan bar) and some are Experimental Procedures (designated by a blue bar). A time icon appears whenever a procedure requires a period of time before results can be viewed.

Live materials: Although students work with living material during some part of almost all laboratories, the exercises are designed to be completed within one laboratory session. This facilitates the use of the manual in multiple-session courses.

Virtual labs: New to this edition, virtual labs currently available on the *Concepts of Biology* website, **www.mhhe.com/maderconcepts3,** are announced and described whenever the announcement seems appropriate. Instructors can use the virtual labs as separate assignments or integrate them into the laboratory experience. McGraw-Hill's LearnSmart Labs™ are available where noted to accompany many labs in this manual. These outcomes-based lab simulations assess a student's knowledge and adaptively correct deficiencies, allowing the student to learn faster and retain more knowledge with greater success.

Laboratory safety: Laboratory safety, a listing on the website www.mhhe.com/maderconcepts3, will assist instructors in making the laboratory experience a safe one. Throughout the laboratories, safety precautions specific to an activity are highlighted and identified by a caution symbol.

Improvements This Edition

In general, laboratories have been revised to (1) improve the Introduction so that it becomes an integral part of the laboratory experience; (2) improve the Laboratory Review so that questions better reflect the Learning Outcomes; and (3) use inexpensive plastic transfer pipets when doing exercises that require the use of chemical solutions and test tubes.

Laboratory 1 Scientific Method The laboratory was rewritten to improve the flow and better emphasize the scientific method.

Laboratory 2 Measuring with Metric At adopters' request, study of, and practice with, the metric system has been separated from microscopy and now is presented in a separate laboratory.

Laboratory 7 Photosynthesis This laboratory has been streamlined to better stress the basics of photosynthesis and create a laboratory that is easier to set up and use, so that the emphasis is on the concepts of photosynthesis.

Laboratory 9 Mitosis: Cellular Reproduction By following the suggestions of adopters, a laboratory has been created that gives students an exercise to do as they learn the phases of mitosis.

Laboratory 10 Meoisis: Sexual Reproduction In separating meiosis from mitosis, students will engage in an exercise that allows them to see how meiosis introduces genetic variations in the gametes.

Laboratory 11 Mendelian Genetics The laboratory was rewritten to streamline the experiments to provide students with the opportunity to do hands-on genetics experiments.

Laboratory 12 Human Genetics The laboratory was reorganized and revised to include problems concerning genetic disorders and a genetic counseling section that considers chromosomal anomalies and pedigree analysis.

Laboratory 14 Evidence of Evolution This laboratory now offers an engaging exposure to plant, animal, and primate fossils. Molecular data to support evolutionary principle are now emphasized.

Laboratory 17 Early Invertebrate Evolution This new laboratory stresses a hands-on examination of living hydras and planarians, vinegar eels, and rotifers.

Laboratory 18 Later Invertebrate and Vertebrate Evolution This lab, also new, includes a dissection of the clam, and also gives students an opportunity to examine the earthworm, crayfish, grasshopper, and sea star anatomy.

Laboratory 23 Homeostasis This laboratory was rewritten to introduce more hands-on activities, including blood pressure and lung volume measurements and comparative urine analysis to diagnose particular illnesses.

Laboratory 24 Nervous System and Senses This laboratory was reorganized by changing the first section to the Central Nervous System. This new section now includes both the brain and the spinal cord. The Peripheral Nervous System section, which reviews the reflex arc, is followed by an examination of the eye, ear, and cutaneous receptors.

Laboratory 25 Animal Development The laboratory now utilizes the sea star for cellular development, the frog for tissue development, and the chick for organ development before the extraembryonic membranes and human fetal development are discussed.

To the Student

Special care has been taken in preparing the *Concepts of Biology Laboratory Manual* to enable you to **enjoy** the laboratory experience as you **learn** from it. The instructions and discussions are written clearly so that you can understand the material while working through it. Student aids are designed to help you focus on important aspects of each exercise. Student learning aids are carefully integrated throughout this manual:

The Learning Outcomes set the goals of each laboratory session and help you review the material for a laboratory practical or any other kind of exam. The major sections of each laboratory are numbered, and the Learning Outcomes are grouped according to these topics. This system allows you to study the chapter in terms of the outcomes presented.

The Introduction to the laboratory reviews necessary background information required for comprehending the work you will be doing during the laboratory session. Also, specific information provided before each of the Observations and Experimental Procedures will assist you in being engaged while you do these activities.

The Observations and Experimental Procedures require your active participation. Space is provided in tables for you to record the results and conclusions of observations and experiments. Space is also provided for you to answer questions pertaining to these activities.

Observation: An activity in which models, slides, and preserved or live organisms are observed to achieve a learning outcome.

Experimental Procedure: An activity in which a series of steps uses laboratory equipment to gather data and come to a conclusion.

At the end of an Observation or Experimental Procedure if you are asked to formulate explanations or conclusions, you should be sure you are truly writing an explanation or conclusion and not just restating the observations made. To do so, you will need to synthesize information from a variety of sources including:

1. Your experimental results and/or the results of other groups in the class. If your data are different from those of other groups in your class, do not erase your answer; add the other groups' answers in parentheses.

2. Your knowledge of underlying principles. Obtain this information from the laboratory Introduction or the appropriate section of the laboratory and from the corresponding chapter of your text.
3. Your understanding of how the experiment was conducted and/or the materials used. (If by chance the results seem inappropriate, it's possible the ingredients were contaminated or you misunderstood the directions. If this occurs, consult with other students and your instructor to see if you should repeat the experiment.)

Color Bars, Time Icon, and Safety Boxes

Observations are identified by the color bar shown to the left. Whenever you see this color, you know that the activity will require you to make careful observations and answer questions about these activities.

Experimental Procedures are identified by the color bar shown to the left. Whenever you see this color, you know that the activity will require you to use laboratory equipment to perform an experiment and answer questions about this experiment.

A time icon is used to designate when time is needed for a reaction to occur. You may be asked to start these activities at the beginning of the laboratory, proceed to other activities, and return to these when the designated time is up.

A safety icon throughout the manual alerts you to any specific activity that requires a cautionary approach. Read these boxes, and follow the advice given in the box and/or your instructor when performing the activity.

The Laboratory Review is a set of questions covering the day's work. Do all the review questions as an aid to understanding the laboratory. Your instructor may require you to hand in these questions for credit.

Laboratory Preparation

It will be very helpful to you to read the entire laboratory chapter before coming to lab. **Study** the introductory material and the Observations and Experimental Procedures so you know ahead of time what you will be doing that week and how it correlates with the lecture material. If necessary, to obtain a better understanding, read the corresponding chapter in your text. If your text is *Concepts of Biology* by Sylvia S. Mader, the "text chapter reference" column in the table of contents at the beginning of the *Concepts of Biology Laboratory Manual* lists the corresponding chapter in the text.

Student Feedback

If you have any suggestions for how this Laboratory Manual could be improved, you can send your comments to:

The McGraw-Hill Companies
Product Development—General Biology
501 Bell St.
Dubuque, Iowa 52001

Acknowledgments

We gratefully acknowledge the following reviewers for their assistance in the development of this Laboratory Manual.

Cynthia Anderson
Georgia Military College

Jack Brook
Mt. Hood Community College

Consuella A. Davis
Holmes Community College

Vivian Elder
Ozarks Technical Community College

Maria Gomez
Nicholls State University

Melissa Greene
Northwest Mississippi Community College

Quentin Hays
Eastern New Mexico University, Ruidoso

Holley Langille
Northwest Florida State College

Kimberly Noice
Richmond Community College

Chris Perry
College of the Albemarle

Carol Phillips
Pamlico Community College

Mary Leigh Poole
Holmes Community College

Adele Register
Rogers State University

Lisa Strong
Northwest Mississippi Community College

Shervia Taylor
Southern University, Baton Rouge

LABORATORY

1

Scientific Method

Learning Outcomes

Introduction

- In general, describe pillbug external anatomy and lifestyle. 1–2

1.1 Using the Scientific Method

- Outline the steps of the scientific method. 2–3
- Distinguish among observations, hypotheses, conclusions, and theories. 2–3

1.2 Observing a Pillbug

- Observe and describe the external anatomy of a pillbug, *Armadillidium vulgare.** 4
- Observe and describe how a pillbug moves. 5

1.3 Formulating Hypotheses

- Formulate a hypothesis based on appropriate observations. 6

1.4 Performing the Experiment and Coming to a Conclusion

- Design an experiment that can be repeated by others. 6
- Reach a conclusion based on observation and experimentation. 6–7

Introduction

This laboratory will provide you with an opportunity to use the scientific method in the same manner as scientists. Today your subject will be the pillbug, *Armadillidium vulgare*, a type of crustacean that lives on land.

Pillbugs have an exoskeleton consisting of overlapping "armored" plates that make them look like little armadillos. As pillbugs grow, they molt (shed the exoskeleton) four or five times during a lifetime. A pillbug can roll up into such a tight ball that its legs and head are no longer visible, earning it the nickname "roly-poly." They have three body parts: head, thorax, and abdomen. The head bears compound eyes and two pairs of antennae. The thorax bears pairs of walking legs; gills are located at the top of the first five pairs. The gills must be kept slightly moist, which explains why pillbugs are usually found in damp places. The final pair of appendages, the uropods, which are sensory and defensive in function, project from the abdomen of the animal.

Pillbugs on leaf

Pillbugs are commonly found in damp leaf litter, under rocks, and in basements or crawl spaces under houses. Following an inactive winter, pillbugs mate in the spring. Several weeks later, the eggs hatch and remain for six weeks in a brood pouch on the underside of the female's body. Once they leave the pouch,

*The garden snail, *Helix aspersa,* or the earthworm, *Lumbricus terrestris,* can be substituted as desired.

they eat primarily dead organic matter, including decaying leaves. Therefore, they are easy to find and to maintain in a moist terrarium with leaf litter, rocks, and wood chips. You are encouraged to collect some for your experiment. Since they live in the same locations as snakes, be careful when collecting them.

1.1 Using the Scientific Method

Some scientists work alone, but often scientists belong to a community of scientists who are working together to study some aspect of the natural world. For example, many scientists from different institutions work together to study the AIDS virus (Fig. 1.1).

Figure 1.1 Scientists work together. Robert Gallo and his colleagues do research on how viruses, such as the AIDS virus, invade humans.

> You will share your study of pillbugs with the other members of the class.

Even though the methodology can vary, scientists often use the **scientific method** (Fig. 1.2) when doing research. The scientific method involves these steps:

Making observations. Observations help scientists begin their study of a particular topic.

> To learn about pillbugs you will visually observe one. You could also do a Google search of the Web or talk to someone who has worked with pillbugs for a long time.

Why does the scientific method begin with observations? recieve knowledge through our senses, or using instruments.

Formulating a hypothesis. Based on their observations, scientists come to a tentative decision, called a hypothesis, about their topic. Formulating hypotheses helps scientists decide how an experiment will be conducted.

> Based on your observations you might hypothesize that a pillbug will be attracted to juices.

Now you know what you will actually do. What is the benefit of formulating a hypothesis? tells what is to be tested by experiment or further observation.

Virtual Lab Mealworm Behavior A virtual lab, called Mealworm Behavior, is available on the *Concepts of Biology* website **www.mhhe.com/maderconcepts3.** This virtual lab demonstrates how investigators conducted an experiment similar to the one you will be doing.

Testing the hypothesis involves deciding on an **experimental variable,** that part of the experiment that changes. The dependent variable changes as the experimental variable changes.

> You could decide to expose the pillbug to a variety of juices such as apple juice, orange juice, and pineapple juice.

A well-designed experiment must have a **negative control**—that is, a sample or event—that is not exposed to the testing procedure. If the negative control and the test sample produce the same results, either the procedure is flawed or the hypothesis is false.

> Water can substitute for fruit juice and be the control in your experiment.

Scientists call the results of their experiments the **data.** It is very important for scientists to keep accurate records of all their data.

> You will record your data in a table that can be easily examined by another person.

When another person repeats the same experiment, and the data is the same, both experiments have merit. Why must a scientist keep a complete record of an experiment? So others can repeat the experiment, and check if the data is valid

Coming to a conclusion. Scientists come to a conclusion as to whether their data support or do not support the hypothesis.

> If a pillbug is attracted to fruit juice, your hypothesis is supported. If the pillbug is not attracted to fruit juice, your hypothesis is not supported.

A scientist never says that a hypothesis has been proven true because, after all, some future knowledge might have a bearing on the experiment. What is the purpose of the conclusion? tells what has been learned from the experiment

Developing a scientific theory. A *theory* in science is an encompassing conclusion based on many individual conclusions in the same field. For example, the gene theory states that organisms inherit coded information that controls their anatomy, physiology, and behavior. It takes many years for scientists to develop a theory and, therefore, we will not be developing any theories today. How is a scientific theory different from a conclusion? Scientific theory has many conclusions from various experiments.

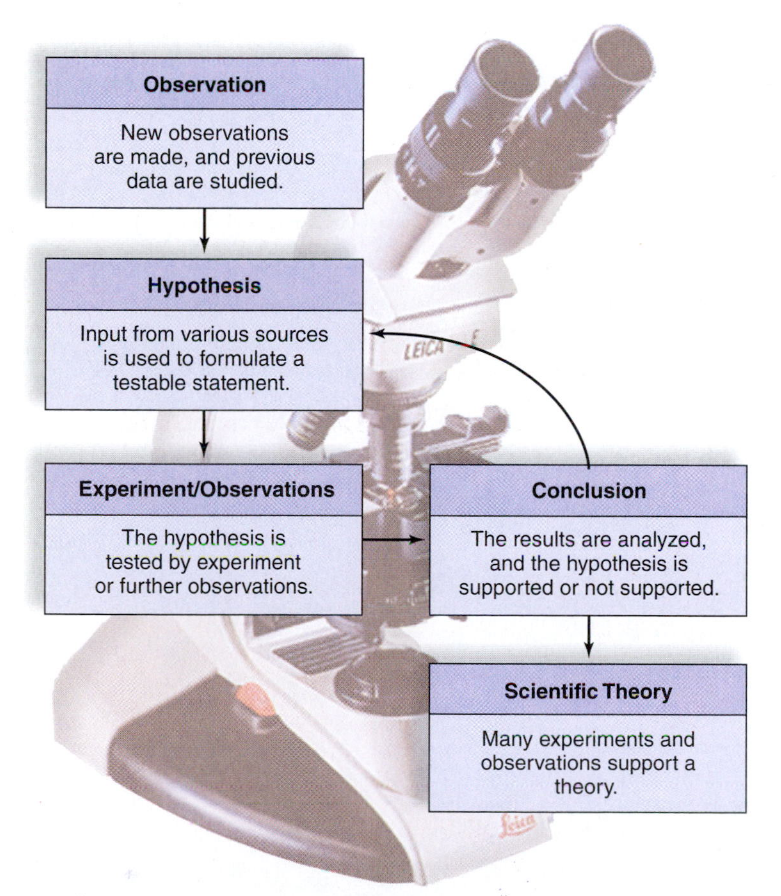

Figure 1.2 Flow diagram for the scientific method.
Often, scientists use this methodology to come to conclusions and develop theories about the natural world. The return arrow shows that scientists often choose to retest the same hypothesis, or test a related hypothesis, before arriving at a conclusion.

1.2 Observing a Pillbug

Wash your hands before and after handling pillbugs. Please handle them carefully so they are not crushed. When touched, they roll up into a ball or "pill" shape as a defense mechanism. They will soon recover if left alone.

Observation: Pillbug's External Anatomy

Obtain a pillbug that has been numbered with white correction fluid or tape tags. Put the pillbug in a small glass or plastic dish to keep it contained.

1. Examine the exterior of the pillbug with the unaided eye and with a magnifying lens or dissecting microscope.
   - How can you recognize the head end of a pillbug? ______________________________
   - How many segments and pairs of walking legs are in the thorax? ______________________
   - The abdomen ends in uropods, appendages with a sensory and defense function. (Females have leaflike growths at the base of some legs where developing eggs and embryos are held in pouches.)
2. In the following space, draw an outline of your pillbug (at least 7 cm long). Label the head, thorax, abdomen, antennae, eyes, uropods, and one of the seven pairs of legs.

3. Draw a pillbug rolled into a ball.

Observation: Pillbug's Motion

1. Watch a pillbug's underside as the pillbug moves up a transparent surface, such as the side of a graduated cylinder or beaker.
   - **a.** Describe the action of the feet and any other motion you see. ______________________
   - **b.** Allow a pillbug to crawl on your hand. Describe how it feels. ______________________
   - **c.** Does a pillbug have the ability to move directly forward? ______________________
   - **d.** Do you see evidence of mouthparts on the underside of the pillbug? ______________________
2. As you watch the pillbug, identify
   - **a.** the anatomical parts that allow a pillbug to identify and take in food. ______________________
   - **b.** behaviors that will help the pillbug acquire food. For example, is the ability of the pillbug to move directly forward a help in acquiring food? How or how not? ______________________

     What other behaviors allow a pillbug to acquire food? ______________________
   - **c.** a behavior that helps a pillbug avoid dangerous situations. ______________________

     If a pillbug rolls up into a ball, wait a few minutes and it may uncurl itself.
3. Measure the speed of three pillbugs.
   - **a.** Place each pillbug on a metric ruler, and use a stopwatch to measure the number of seconds (sec) it takes for the pillbug to move several centimeters (cm). Quickly record here the number of cm moved and the time in sec.

     pillbug 1 ______________________

     pillbug 2 ______________________

     pillbug 3 ______________________
   - **b.** Knowing that 10 millimeters (mm) are in a cm, convert the number of cm traveled to mm and record this in the first column of Table 1.1. Record the total time taken in the second column.
   - **c.** Use the space above to calculate the speed of each pillbug in mm/sec and record the speed for each pillbug in the last column of Table 1.1.
   - **d.** Average the speed for your three pillbugs. (Since you have already calculated the mm/sec for each pillbug, it is only necessary to take an average of the mm moved.) Record the average speed of pillbug motion in Table 1.1.

     When you conduct the experiment in Section 1.4 you will have to be patient with your pillbug as it moves toward or away from a substance.

Table 1.1 Pillbug Speed

Pillbug	Millimeters (mm) Traveled	Time (sec)	Speed (mm/sec)
1			
2			
3			
		Average speed:	

1.3 Formulating Hypotheses

You will be testing whether pillbugs are attracted to (move toward and eat), repelled by (move away from), or unresponsive to (don't move away from and do not move toward and eat) the particular substances, which are potential foods. If a pillbug simply rolls into a ball, nothing can be concluded, and you may wish to choose another pillbug or wait a minute or two to check for further response.

1. Choose
   - **a.** two dry substances, such as flour, cornstarch, coffee creamer, or baking soda. Fine sand will serve as a control for dry substances. Record your "dry" choices as 1, 2, and 3 in the first column of Table 1.2.
   - **b.** two liquids, such as milk, orange juice, ketchup, applesauce, or carbonated beverage. Water will serve as a control for liquid substances. Record your "wet" choices as 4, 5, and 6 in the first column of Table 1.2.
2. In the second column of Table 1.2, hypothesize how you expect the pillbug to respond to each substance. Use a plus (+) sign if you hypothesize that the pillbug will move toward and eat the substance; a minus (−) sign if you hypothesize that the pillbug will be repelled by the substance; and a zero (0) if you expect the pillbug to show neither behavior.
3. In the third column of Table 1.2 offer a reason for your hypothesis based on your knowledge of pillbugs from the introduction and your examination of the animal.

Table 1.2 Hypotheses About Pillbug's Response to Potential Foods

Substance		Hypothesis About Pillbug's Response	Reason for Hypothesis
1			
2			
3	(control)		
4			
5			
6	(control)		

1.4 Performing the Experiment and Coming to a Conclusion

A good experimental design would be to keep your pillbug in a petri dish to test its reaction to the chosen substances. During your experiment, no substance must be put directly on the pillbug, nor can the pillbug be placed directly onto the substances.

Experimental Procedure: Pillbug's Response to Potential Foods

1. Before testing the pillbug's reaction, fill in the first column of Table 1.3. It will look exactly like the first column of Table 1.2.
2. Since pillbugs tend to walk around the edge of a petri dish, you could put the wet or dry substance there; or for the wet substance you could put liquid-soaked cotton in the pillbug's path.
3. Rinse your pillbug between procedures by spritzing it with distilled water from a spray bottle. Then put it on a paper towel to dry it off.
4. Watch the pillbug's response to each substance, and record it in Table 1.3, using +, −, or 0 as before.

Table 1.3 Pillbug's Response to Potential Foods			
Substance		**Pillbug's Response**	**Hypothesis Supported?**
1			
2			
3	(control)		
4			
5			
6	(control)		

5. Do your results support your hypotheses? Answer yes or no in the last column of Table 1.3.
6. Are there any hypotheses that were not supported by the experimental results (data)? How do such data give you more insight into pillbug behavior? ______________________________

____

7. **Class Results.** Compare your results with those of other students who tested the same substance. Calculate the proportional response to each potential food (%+, %−, %0) and record your calculations in Table 1.4. As a group, your class can decide what proportion is needed to designate this response as typical. For example, if the pillbugs as a whole were attracted to a substance 70% or more of the time, you can call that response the "typical response."

Table 1.4 Pillbug's Response to Potential Foods: Class Results					
Substance		**Pillbug's Response**			**Hypothesis Supported?**
1		%+	%−	%0	
2		%+	%−	%0	
3	(control)	%+	%−	%0	
4		%+	%−	%0	
5		%+	%−	%0	
6	(control)	%+	%−	%0	

8. On the basis of the class data, do you need to revise your conclusion for any particular pillbug response? ______________ Scientists prefer to come to conclusions on the basis of many trials. Why is this the best methodology? ______________________________

____

9. Did the pillbugs respond as expected to the controls (i.e., did not eat them)? ______________ If they did not respond as expected, what can you conclude about your experimental results? ______________

____

Laboratory Review 1

1. What are the essential steps of the scientific method? observation, hypothesis, conduct an experiment, conclusion

2. What is a hypothesis? helps scientists decide how an experiment will be conducted

3. Is it sufficient to do a single experiment to test a hypothesis—why or why not? NO, you need more data to prove your hypothesis

4. What do you call a sample that goes through all the steps of an experiment but does not contain the factor being tested? Control group

5. What part of a pillbug is for protection, and what does a pillbug do to protect itself? it's their exoskeleton, they roll up into a ball.

6. State the type of data you used to formulate your hypotheses regarding pillbug reactions toward various substances. ______

7. Why is it important to test one substance at a time when doing an experiment? ______

Indicate whether statements 8–10 are hypotheses, conclusions, or theories.

8. The data show that vaccines protect people from disease. conclusion

9. All organisms are made of cells. theory

10. The breastbone of a chicken is proportionately larger than that of any other bird. hypotheses

success

***Concepts of Biology* Website**

Instructors can find lab prep information and answers to all of the laboratory questions in the Laboratory Resource Guide. *Students* can practice their knowledge with quizzes, animations, flashcards, and much more.

www.mhhe.com/maderconcepts3

McGraw-Hill Access Science Website

An online encyclopedia of science and technology that provides information, including videos, that can enhance the laboratory experience.

www.accessscience.com

Scientific Method

LABORATORY

2

Measuring with Metric

Learning Outcomes

2.1 Length

- Compare and contrast the metric units for length: meter (m), centimeter (cm), millimeter (mm), micrometer (μm), and nanometer (nm). 10–12
- Know the abbreviations for each of these units. 10–12
- Convert the metric units for length from one type of unit to another. 10–12

2.2 Weight

- Compare and contrast the metric units for weight: kilogram (kg), gram (g), and milligram (mg). 13–14
- Know the abbreviations for each of these units. 13–14
- Convert the metric units for weight from one type of unit to another. 13–14

2.3 Volume

- Compare and contrast the metric units for volume: liter (l) and milliliter (ml). 14–15
- Know the abbreviations for each of these units. 14–15
- Convert the metric units for volume from one type of unit to another. 14–15

2.4 Temperature

- Compare and contrast the Fahrenheit (F) and Celsius (C) temperature scales. 16
- Know the abbreviations for each of these units. 16
- Convert one type temperature into another using a provided equation. 16

2.5 Summary

- Use the size relationship between metric units in order to carry out conversions. 17

Introduction

The metric system is the standard system of measurement in the sciences, including biology, chemistry, and physics (Fig. 2.1). It has tremendous advantages because all conversions, whether for volume, mass (weight), or length, are in units of ten. This base-ten system is similar to our monetary system, in which 10 cents equals a dime, 10 dimes equals a dollar, and so on. In this laboratory, you will get experience making measurements of length, volume, mass, and also temperature.

Figure 2.1
The metric system is the system of measurement used in scientific laboratories.

2.1 Length

Metric units of length measurement include the **meter (m), centimeter (cm), millimeter (mm), micrometer (μm),** and **nanometer (nm)** (Table 2.1).

Table 2.1 Metric Units of Length Measurement

Unit	Meters	Centimeters	Millimeters	Relative Size
Meter (m)	1 m	100 cm	1,000 mm	Largest
Centimeter (cm)	0.01 (10^{-2}) m	1 cm	10 mm	
Millimeter (mm)	0.001 (10^{-3}) m	0.1 cm	1.0 mm	
Micrometer (μm)	0.000001 (10^{-6}) m	0.0001 (10^{-4}) cm	0.001 (10^{-3}) mm	
Nanometer (nm)	0.000000001 (10^{-9}) m	0.0000001 (10^{-7}) cm	0.000001 (10^{-6}) mm	Smallest

You will want to know what these abbreviations stand for, so write them out here:

m = Meter μm = Micrometer

cm = Centimeter nm = nanometer

mm = Millimeter

How many cm are in a meter? 100 How many mm are in a centimeter? 0.1

How many μm are in a millimeter? 1000 How many nm are in a micrometer? 1000

Meter, Centimeter, and Millimeter

Observation: A Meterstick

1. Obtain a meterstick. On one side, find the numbers 1 through 39, which denote inches. One meter equals 39.37 inches; therefore, 1 meter is roughly equivalent to 1 yard.

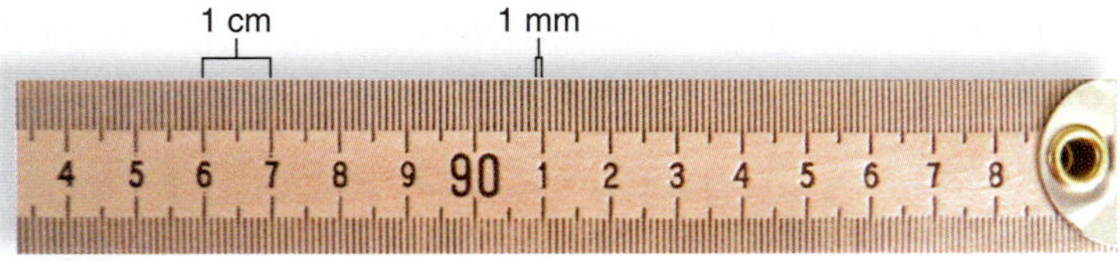

Figure 2.2 The end of a meterstick.

2. Turn the meterstick over and observe the metric subdivisions (Fig. 2.2). How many centimeters are in a meter? 100 The prefix *centi-* means 100. For example, how many cents are in a dollar? 100
3. How many millimeters are in a centimeter? 10 But the prefix *milli-* means 1,000. How many millimeters are in a meter? 1000 Obtain a penny and measure its width in terms of mm. 19.05 Why does it seem preferable to measure a penny in terms of mm? You get a more visible and accurate measurement
4. Use the meterstick and the method shown in Figure 2.3 to measure the length of two long bones from a disarticulated human skeleton. Lay the meterstick flat on the lab table. Place a long bone next to the meterstick between two pieces of cardboard (each about 10 cm × 30 cm), held upright at right angles to the stick. The narrow end of each piece of cardboard should touch the meterstick. The length between the cards is the length of the bone in centimeters. For example, if the bone

measures from the 22 cm mark to the 50 cm mark, the length of the bone is __________ cm. If the bone measures from the 22 cm mark to midway between the 50 cm and 51 cm marks, its length is __________ cm = __________ mm.

5. Record the length of two bones. First bone: __________ cm = __________ mm.
   Second bone: __________ cm = __________ mm.

Figure 2.3 Measurement of a long bone.
How to measure a long bone using a meterstick.

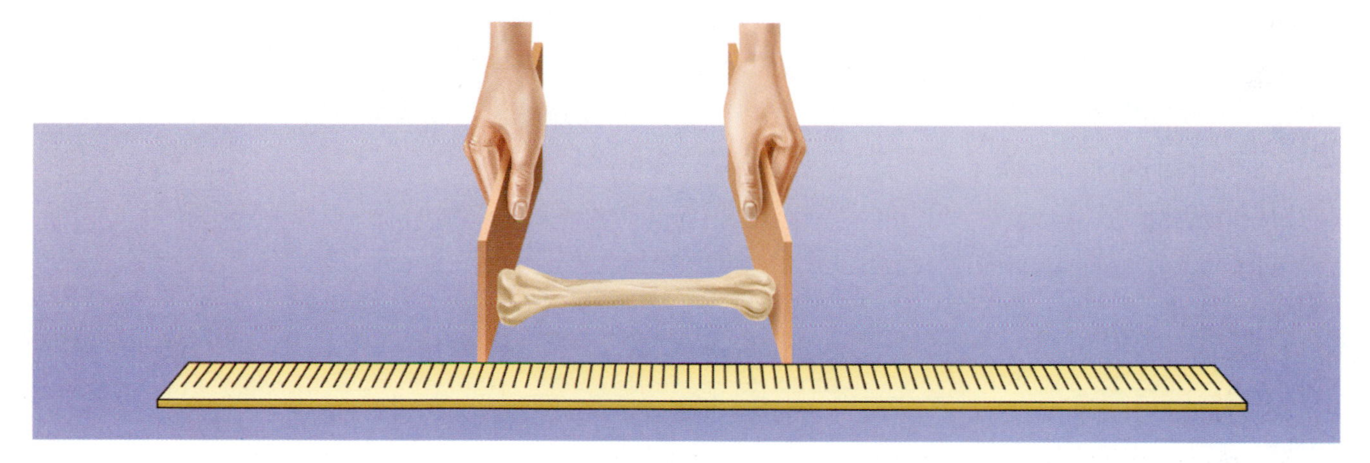

Millimeter, Micrometers, and Nanometer

As you will discover in the next laboratory, the units micrometer (μm) and nanometer (nm) are useful in microscopy because the objects being viewed under the compound light microscope are cells or the contents of cells. Figure 2.4 shows that cells are generally smaller than a millimeter (mm).

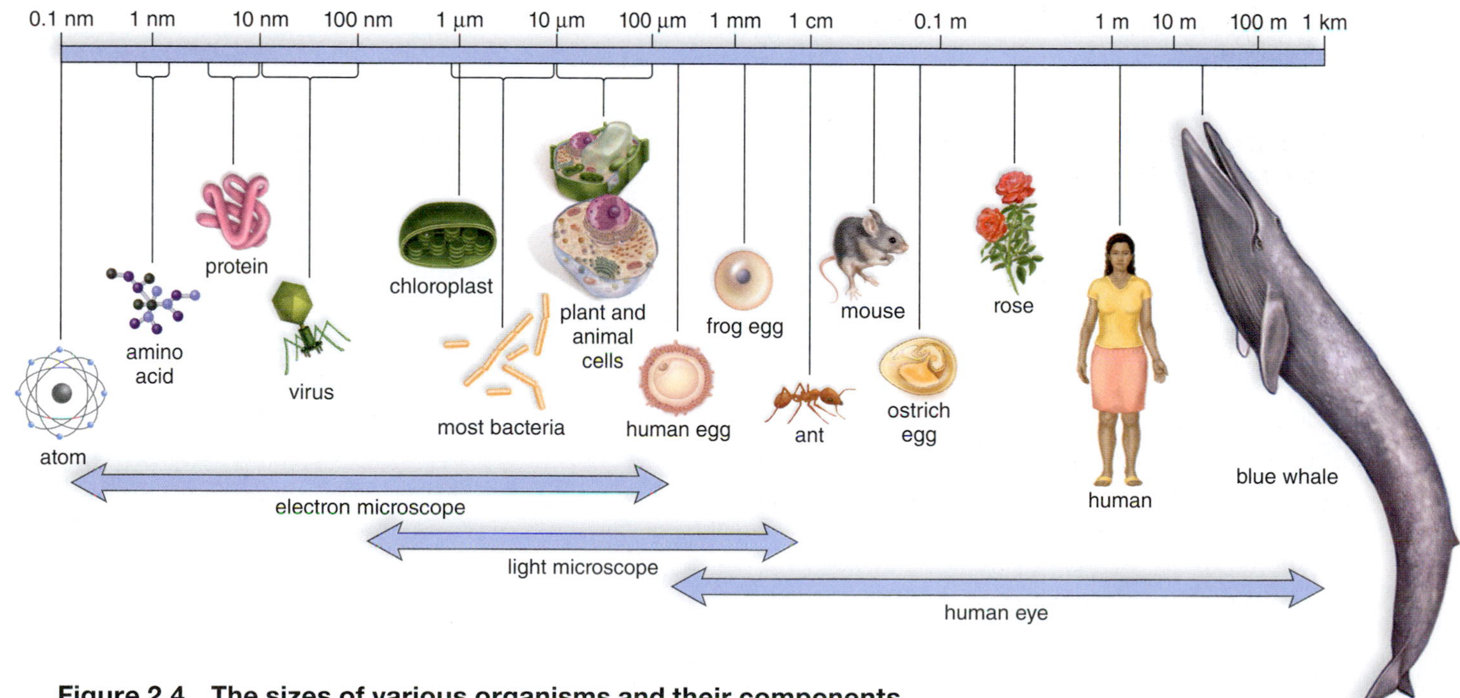

Figure 2.4 The sizes of various organisms and their components.

Observation: Small Metric Ruler

1. Obtain a small metric ruler marked in centimeters (cm) and millimeters (mm). Micrometers (μm) are not on the ruler, and it is necessary to remember that there are 1,000 μm in a mm. Use the ruler to measure the diameter of the circle shown below to the nearest mm.

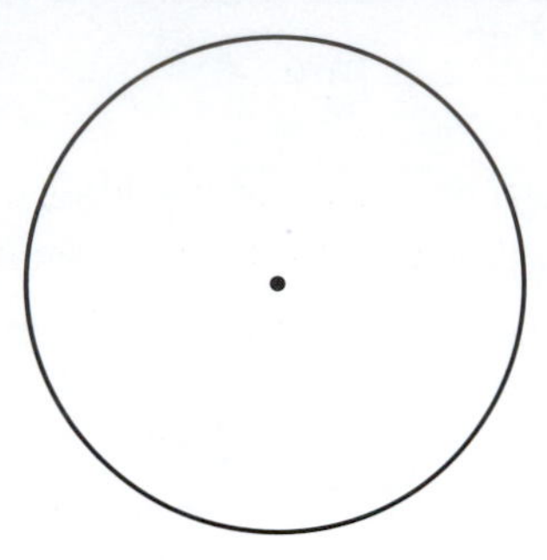

Diameter of circle in mm: __________ mm

2. Use this formula to convert this measurement to μm. Do you expect the answer to be a smaller number or a larger number? ____________________

$$\_\_\_\_\_\_ \text{ mm} \times \frac{1{,}000 \ \mu\text{m}}{\text{mm}} = \_\_\_\_\_\_ \ \mu\text{m}$$

Size of circle in μm: __________ μm

3. Knowing that there are 1,000 nm in a μm, change the formula to convert these μm into nm.

Diameter of circle in nm: __________ nm

4. You have shown that the diameter of this circle is ______ mm = ________ μm = ______ nm. It is expected when doing microscopy that you will make such conversions.

2.2 Weight

For our purposes we will equate mass with weight. Just as a meter is the basic unit of length, the **kilogram** is the basic unit of mass.

1 kilogram (kg) = 1,000 **grams** (g)
1 g = 1,000 **milligrams** (mg)

The weight of domestic animals or a human would be given in kilograms but usually biologists work with the units of grams or milligrams. Using a revision of the formula on page 12 if necessary, do these conversions: 2 g = 2000 mg and 0.2 g = 200 mg.

Experimental Procedure: Weight

Examine the balance scales available in your laboratory. They are called balance scales because the object to be weighed is placed in one pan and weights are placed on the other pan. When the pans contain exactly the same mass the beam between them will be in balance. Figure 2.5 shows a typical balance scale you may be using.

Figure 2.5 A balance scale.

1. Use a balance scale to measure the weight of a wooden block small enough to hold in the palm of your hand. Weigh the block to a tenth of a gram. The weight of the wooden block is
   __________ g = __________ mg.
2. Measure the weight of an item small enough to fit inside the opening of a 50 ml graduated cylinder.
   The item, a(n) __________, is __________ g = __________ mg.
3. A triple beam balance gets its name from its three horizontal beams. If you are going to use a triple beam balance, study it closely (Fig. 2.6). Each beam has a moveable weight. Move the weights in order to see that the closest beam weighs to 0.1 g; the middle beam has 100 g graduations, and the farthest beam has 10 g graduations.

   Before making any measurements, clean the weighing pan and move all the weights to the far left. The weights will line up to indicate zero grams; if they do not, turn the adjustment knob until they do.

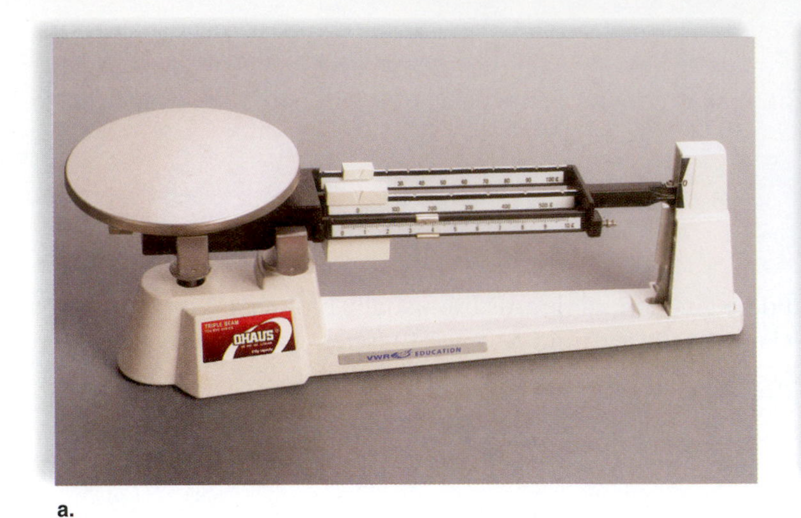

a.

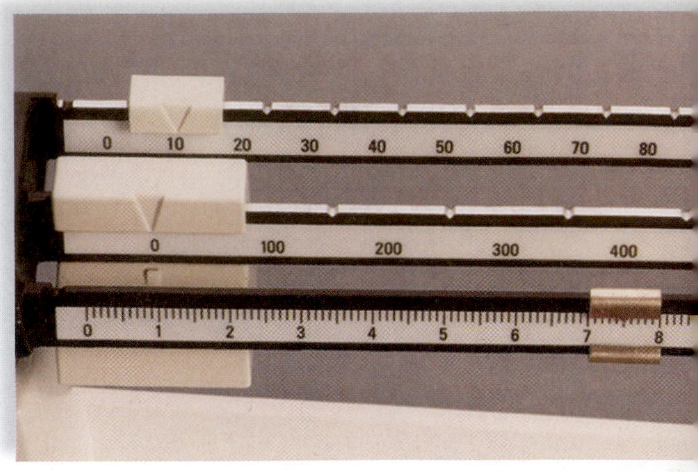

b.

Figure 2.6 A triple beam balance.
a. The entire balance. **b.** Enlargement of triple (3) beams.

Measure the weight of an object by placing it in the center of the weighing pan and moving the weights until the beams balance (are straight across). The weight of the object is the sum of the weights on the three beams.

If so directed by your instructor, use a triple beam balance to take the weight of one or more of these objects to a tenth of a gram:

Penny ____________

Paper clip ____________

Quarter ____________

2.3 Volume

Two metric units of volume are the **liter (l)** and the **milliliter (ml).** One liter = 1,000 ml.

Experimental Procedure: Volume

1. Volume measurements can be related to those of length. For example, use a millimeter ruler to measure the wooden block used in the previous Experimental Procedure to get its length, width, and depth.

   length = __________ cm; width = __________ cm; depth = __________ cm

   The volume, or space, occupied by the wooden block can be expressed in cubic centimeters (cc or cm^3) by multiplying: length $\times$ width $\times$ depth = __________ cm^3. For purposes of this Experimental Procedure, 1 cubic centimeter equals 1 milliliter; therefore, the wooden block you just measured has a volume of = __________ ml.

2. In the biology laboratory, liquid volume is usually measured directly in liters or milliliters with appropriate measuring devices. For example, use a 50 ml graduated cylinder to add 20 ml of water to a test tube. First, fill the graduated cylinder to the 20 ml mark. To do this properly, you have to make sure that the lowest margin of the water level, or the **meniscus** (Fig. 2.7), is at the 20 ml mark.

Place your eye directly parallel to the level of the meniscus, and add water until the meniscus is at the 20 ml mark. (Having a dropper bottle filled with water on hand can help you do this.) A large, blank, white index card held behind the cylinder can also help you see the scale more clearly.

3. Pour the 20 ml of water from the graduated cylinder into a test tube. Hypothesize how you could find the total volume of the test tube. ______________________________

____

____

What is the test tube's total volume? ____________

____

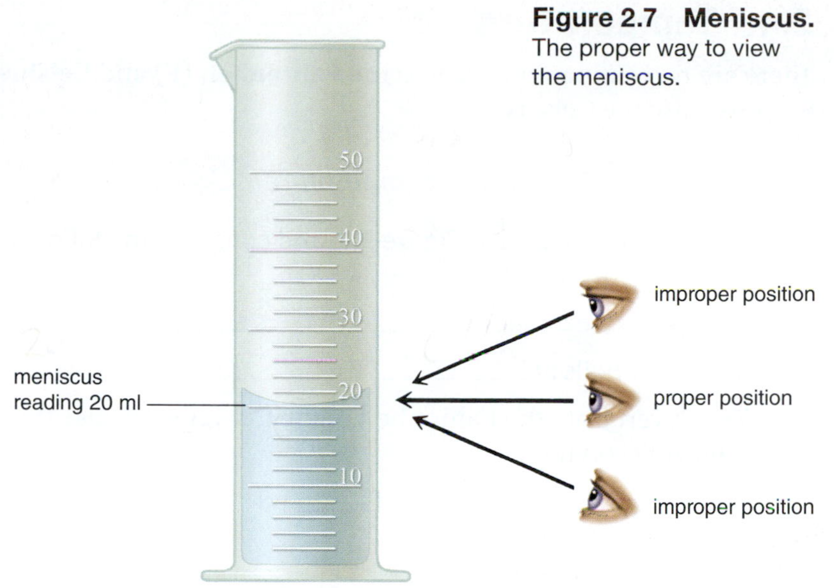

Figure 2.7 Meniscus. The proper way to view the meniscus.

4. Fill a 50 ml graduated cylinder with water to about the 20 ml mark. Hypothesize how you could use this setup to calculate the volume of the small object you weighed previously (see step 2, page 13). ______________________________

____

Now perform the operation you suggested.

The object, __________, has a volume of __________ ml.

5. Hypothesize how you could determine how many drops from the pipet of the dropper bottle equal 1 ml. ______________________________

How many drops from the pipet of the dropper bottle equal 1 ml? ______________

6. Some pipets are graduated (Fig. 2.8) and can be filled to a certain level as a way to measure volume directly. Your instructor will demonstrate this. Are pipets customarily used to measure large or small volumes? ______________________________

____

____

transfer pipet

Figure 2.8 Use of a transfer pipet. Transfer pipets easily measure and transfer small quantities from stock solutions to test tubes.

2.4 Temperature

There are two temperature scales: the **Fahrenheit (F)** and **Celsius (centigrade, C)** scales (Fig. 2.9). Scientists use the Celsius scale.

Experimental Procedure: Temperature

1. Study the two scales in Figure 2.9, and complete the following information:
   a. Water freezes at __________ °F = __________ °C.
   b. Water boils at __________ °F = __________ °C.
2. To convert from the Fahrenheit to the Celsius scale, use the following equation:

$$°C = (°F - 32°)/1.8$$

or

$$°F = (1.8°C) + 32$$

   Human body temperature of 98°F is what temperature on the Celsius scale? __________
3. Record any two of the following temperatures in your lab environment. In each case, allow the Celsius thermometer to remain in or on the sample for one minute.

   Room temperature = __________ °C

   Surface of your skin = __________ °C

   Cold tap water in a 50 ml beaker = __________ °C

   Hot tap water in a 50 ml beaker = __________ °C

   Ice water = __________ °C

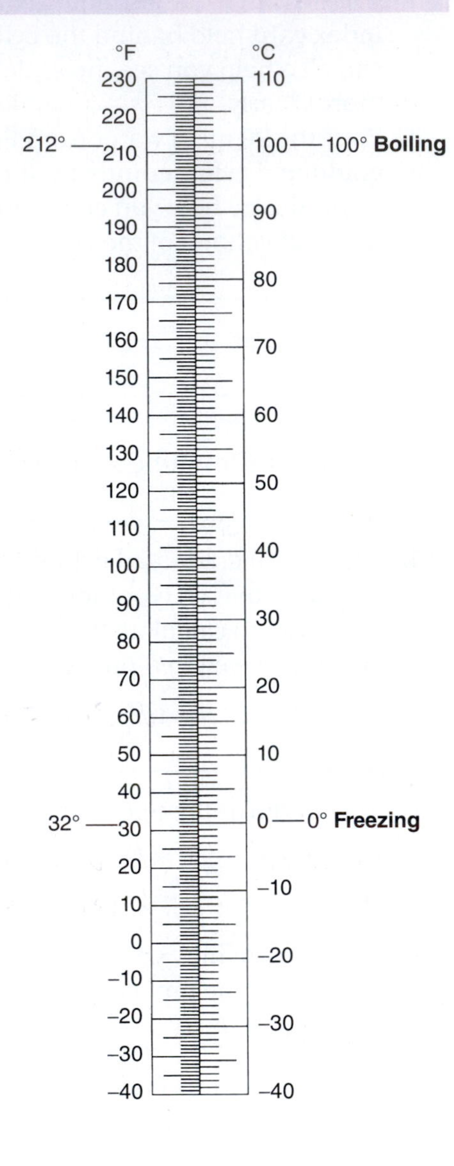

Figure 2.9 Temperature scales.
The Fahrenheit (°F) scale is on the left, and the Celsius (°C) scale is on the right.

2.5 Summary

Table 2.2 will serve as a summary for this laboratory. In Table 2.2, except for centimeter, the units are all ________ × larger or smaller than the next unit. To know whether to multiply or divide when going from one unit to the next, consider this example:

When you cut an apple into smaller pieces, you end up with more pieces. So, when you convert a gram to milligrams, you multiply the gram by __________. On the other hand, if you cut an apple into 4 pieces, then each piece is .25 of the apple. In the same manner, 1 milligram has to be a portion of a gram. To determine what portion, you divide by __________.

Table 2.2 Units of Metric Measurement

Length	
Nanometer (nm)	$= 10^{-6}$ m (10^{-3} µm)
Micrometer (µm)	$= 10^{-6}$ m (10^{-3} mm)
Millimeter (mm)	= 0.001 (10^{-3}) m
Centimeter (cm)	= 0.01 (10^{-2}) m
Meter (m)	= 100 (10^{2}) cm
	= 1,000 mm
Kilometer (km)	= 1,000 (10^{3}) m
Weight (mass)	
Nanogram (ng)	$= 10^{-9}$ g
Microgram (µg)	$= 10^{-6}$ g
Milligram (mg)	$= 10^{-3}$ g
Gram (g)	= 1,000 mg
Kilogram (kg)	= 1,000 (10^{3}) g
Metric ton (t)	= 1,000 kg
Volume	
Microliter (µl)	$= 10^{-6}$ l (10^{-3} ml)
Milliliter (ml)	$= 10^{-3}$ l
	= 1 cm^{3} (cc)
	= 1,000 mm^{3}
Liter (l)	= 1,000 ml
Kiloliter (kl)	= 1,000 l
Temperature	
Degree	= degree Celsius (°C)

Laboratory Review 2

1. What type of measurement is signified by kg? weight ml? volume cm? length
degrees? temperature µm? length

2. What type of measurement would utilize a meterstick? length a graduated cylinder? volume
a balance scale? weight

3. If a triple beam balance shows a weight of 100 g plus 10 g plus 1 g, what is the weight of the object?
111 g.

4. An object is added to a graduated cylinder that holds 250 ml, and the water rises to 300 ml. What is the volume of the object in ml? 50 mL in cm^3? 50 cm^3

5. Name two units of measurement you expect to use in the next laboratory, which concerns microscopy.
micrometer , nanometer

6. How many micrometers are in a millimeter? 1,000
Convert 1.1 mm to µm. 1,100 µm

7. How many milliliters are in a liter? 1,000
Convert 500 ml to liters. 0.50 L

8. How many milligrams are in a gram? 1,000
Convert 5 g to mg. 5,000 mg

9. Convert 1.5 cm to µm. Show your work. 1.5 cm = 15 mm = 15,000 µm

10. A student looking for a shortcut drops an object in a graduated cylinder that contained water to find its weight. What's wrong? This will measure volume and not weight. To measure use a scale.

***Concepts of Biology* Website**

Instructors can find lab prep information and answers to all of the laboratory questions in the Laboratory Resource Guide. *Students* can practice their knowledge with quizzes, animations, flashcards, and much more.

www.mhhe.com/maderconcepts3

McGraw-Hill Access Science Website

An online encyclopedia of science and technology that provides information, including videos, that can enhance the laboratory experience.

www.accessscience.com

Metric Measurements

LABORATORY

3
Light Microscopy

Learning Outcomes

3.1 Light Microscopes Versus Electron Microscopes
- Describe three differences between the compound light microscope and the transmission electron microscope. 20–21

3.2 Stereomicroscope (Binocular Dissecting Microscope)
- Identify the parts and tell how to focus the stereomicroscope. 22–23

3.3 Use of the Compound Light Microscope
- Name and give the function of the basic parts of the compound light microscope. 24–25
- Discuss how to properly bring an object into focus with the compound light microscope. 25–26
- Describe how an image is inverted using a compound light microscope. 26
- Calculate the total magnification for both low- and high-power lens systems. 27

3.4 Microscopic Observations
- Identify and describe the three types of cells studied in this exercise. 28–29
- State two differences between onion epidermal cells and human epithelial cells. 28–29

Introduction

This laboratory examines the features, functions, and use of the compound light microscope and the stereomicroscope. Transmission and scanning electron microscopes are explained, and micrographs produced using these microscopes appear throughout this lab manual. The stereomicroscope and the scanning electron microscope view the surface and/or the three-dimensional structure of an object. The compound light microscope and the transmission electron microscope can view only extremely thin sections of a specimen. If a subject was sectioned lengthwise for viewing, the interior of the projections at the top of the cell, called cilia, would appear in the micrograph (Fig. 3.1). A lengthwise cut through any type of specimen is called a **longitudinal section (l.s.).** On the other hand, if the subject in Figure 3.1 was sectioned crosswise below the area of the cilia, you would see other portions of the interior of the subject. A crosswise cut through any type of specimen is called a **cross section (c.s.).**

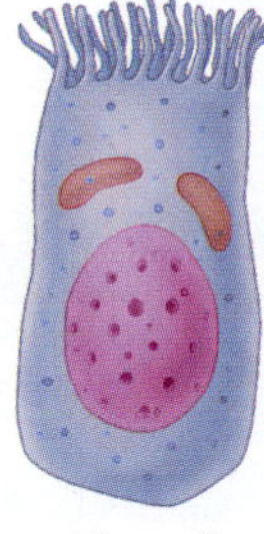
a. The cell

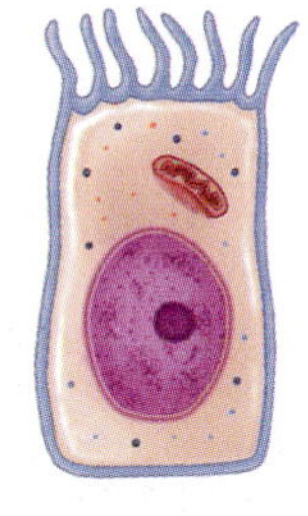
b. Longitudinal section (l.s.)

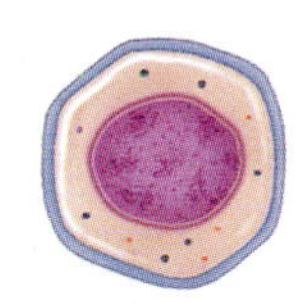
c. Cross section (c.s.)

Figure 3.1 Longitudinal and cross sections. **a.** Transparent view of a cell. **b.** A longitudinal section would show the cilia at the top of the cell. **c.** A cross section shows only the interior where the cut is made.

3.1 Light Microscopes Versus Electron Microscopes

Because biological objects can be very small, we often use a microscope to view them. Many kinds of instruments, ranging from the hand lens to the electron microscope, are effective magnifying devices. A short description of two kinds of light microscopes and two kinds of electron microscopes follows.

Light Microscopes

Light microscopes use light rays passing through lenses to magnify the object. The **stereomicroscope** (binocular dissecting microscope) is designed to study entire objects in three dimensions at low magnification. The **compound light microscope** is used for examining small or thinly sliced sections of objects under higher magnification than that of the stereomicroscope. The term **compound** refers to the use of two sets of lenses: the ocular lenses located near the eyes and the objective lenses located near the object. Illumination is from below, and visible light passes through clear portions but does not pass through opaque portions. To improve contrast, the microscopist uses stains or dyes that bind to cellular structures and absorb light. Photomicrographs, also called light micrographs, are images produced by a compound light microscope (Fig. 3.2*a*).

Figure 3.2 Comparative micrographs of a lymphocyte.
Micrographs of a lymphocyte, a type of white blood cell. **a.** A photomicrograph (light micrograph) shows less detail than a **(b)** transmission electron micrograph (TEM). **c.** A scanning electron micrograph (SEM) shows the cell surface in three dimensions.

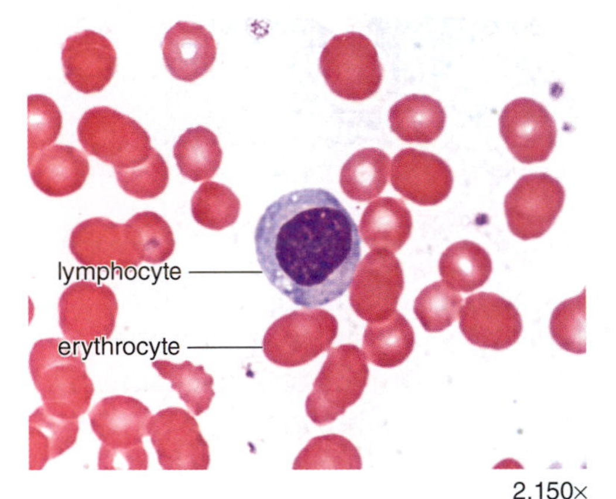

2,150×
a. Photomicrograph or light micrograph (LM)

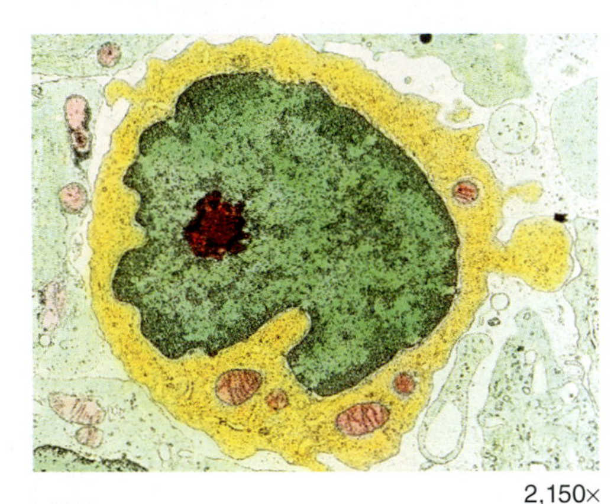

2,150×
b. Transmission electron micrograph (TEM)

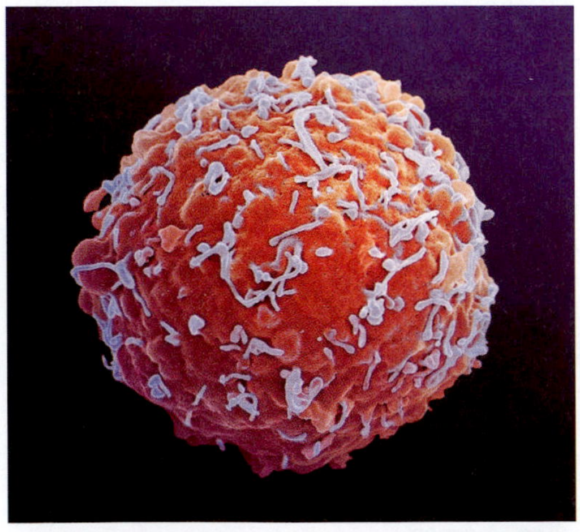

5,000×
c. Scanning electron micrograph (SEM)

Electron Microscopes

Electron microscopes use beams of electrons to magnify the object. The beams are focused on a photographic plate by means of electromagnets. The **transmission electron microscope** is analogous to the compound light microscope. The object is ultra-thinly sliced and treated with heavy metal salts to improve contrast. Figure 3.2*b* is a micrograph produced by this type of microscope. The **scanning electron microscope** is analogous to the stereomicroscope. It gives an image of the surface and dimensions of an object, as is apparent from the scanning electron micrograph in Figure 3.2*c*.

The micrographs in Figure 3.2 demonstrate that an object is magnified more with an electron microscope than with a compound light microscope. The difference between these two types of microscopes, however, is not simply a matter of magnification; it is also the electron microscope's ability to show detail. The electron microscope has greater resolving power. **Resolution** is the minimum distance between two objects at which they can still be seen, or resolved, as two separate objects. The use of high-energy electrons rather than light gives electron microscopes a much greater resolving power since two objects that are much closer together can still be distinguished as separate points. Table 3.1 lists several other differences between the compound light microscope and the transmission electron microscope.

Table 3.1 Comparison of the Compound Light Microscope and the Transmission Electron Microscope

Compound Light Microscope	Transmission Electron Microscope
1. Glass lenses	1. Electromagnetic lenses
2. Illumination by visible light	2. Illumination due to beam of electrons
3. Resolution $\cong$ 200 nm	3. Resolution $\cong$ 0.1 nm
4. Magnifies to 1,500×	4. Magnifies to 100,000×
5. Costs up to tens of thousands of dollars	5. Costs up to hundreds of thousands of dollars

Answer These Questions

- Which two types of microscopes view the surface of an object? ______________________

  ____

- Which two types of microscopes view objects that have been sliced and treated to improve contrast? ______________________
- Of the microscopes just mentioned, which one resolves the greater amount of detail? ______________

  ____

Rules for Microscope Use

Observe the following rules for using a microscope:

1. The lowest power objective (scanning or low) should be in position, both at the beginning and end of microscope use.
2. Use only lens paper for cleaning lenses.
3. Do not tilt the microscope as the eyepieces could fall out, or wet mounts could be ruined.
4. Keep the stage clean and dry to prevent rust and corrosion.
5. Do not remove parts of the microscope.
6. Keep the microscope dust-free by covering it after use.
7. Report any malfunctions.

3.2 Stereomicroscope (Binocular Dissecting Microscope)

The **stereomicroscope** (binocular dissecting microscope) allows you to view objects in three dimensions at low magnifications. It is used to study entire small organisms, any object requiring lower magnification, and opaque objects that can be viewed only by reflected light. It is also called a stereomicroscope because it produces a three-dimensional image.

Identifying the Parts

After your instructor has explained how to carry a microscope, obtain a stereomicroscope and a separate illuminator, if necessary, from the storage area. Place it securely on the table. Plug in the power cord, and turn on the illuminator. There is a wide variety of stereomicroscope styles, and your instructor will discuss the specific style(s) available to you. Regardless of style, the following features should be present:

Figure 3.3 Stereomicroscope (binocular dissecting microscope).
Label this microscope with the help of the text material.

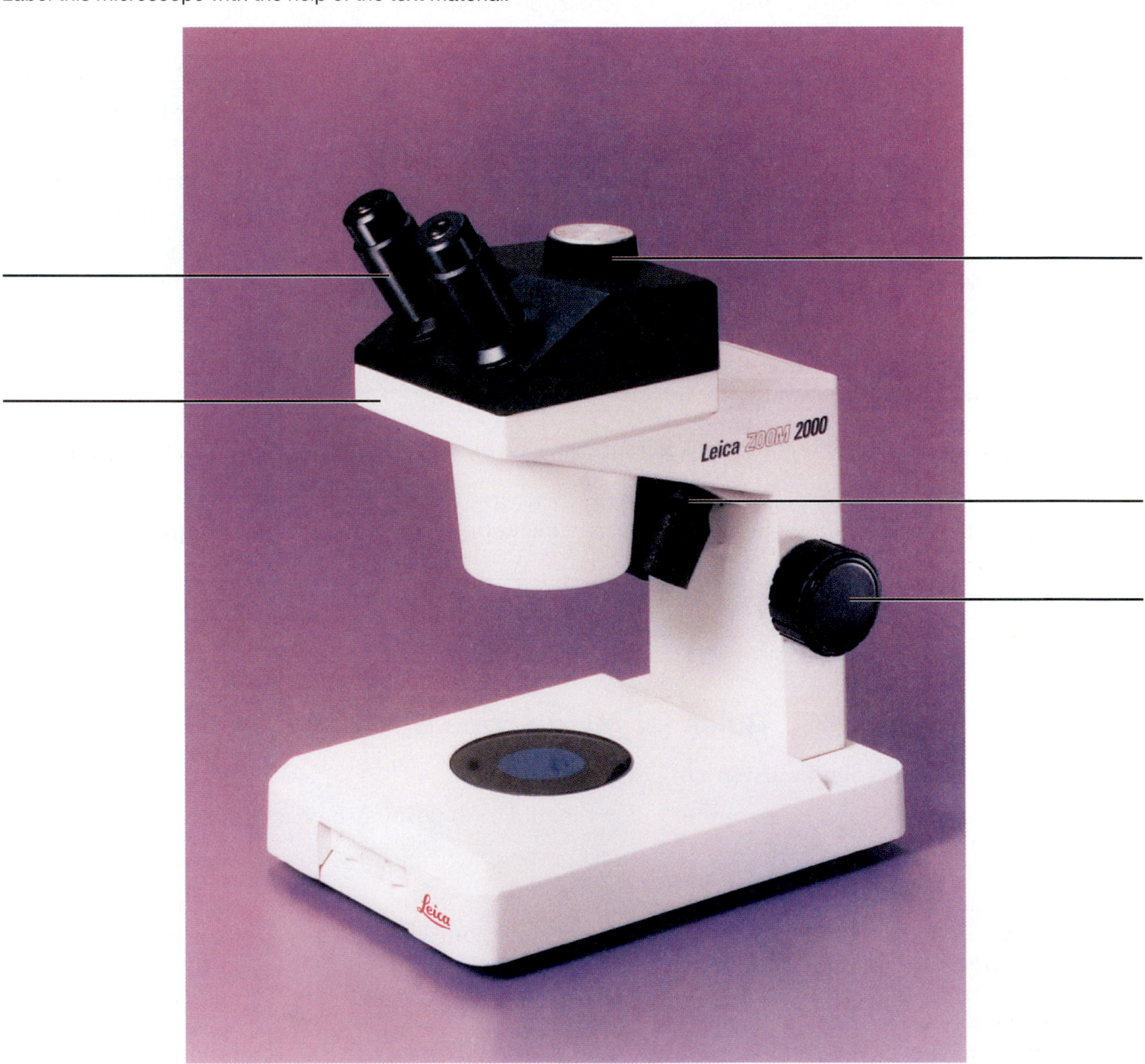

1. **Binocular head:** Holds two eyepiece lenses that move to accommodate for the various distances between different individuals' eyes.
2. **Eyepiece lenses:** The two lenses located on the binocular head. What is the magnification of your eyepieces? __________ Some models have one **independent focusing eyepiece** with a knob to allow independent adjustment of each eye. The nonadjustable eyepiece is called the **fixed eyepiece.**
3. **Focusing knob:** A large, black or gray knob located on the arm; used for changing the focus of both eyepieces together.
4. **Magnification changing knob:** A knob, often built into the binocular head, used to change magnification in both eyepieces simultaneously. This may be a **zoom** mechanism or a **rotating lens** mechanism of different powers that clicks into place.
5. **Illuminator:** Used to illuminate an object from above; may be built into the microscope or separate.

Locate each of these parts on your stereomicroscope, and label them on Figure 3.3.

Focusing the Stereomicroscope

1. Place a plastomount that contains small organisms in the center of the stage.
2. Adjust the distance between the eyepieces on the binocular head so that they comfortably fit the distance between your eyes. You should be able to see the object with both eyes as one three-dimensional image.
3. Use the focusing knob to bring an organism in the plastomount into focus.
4. Does your microscope have an independent focusing eyepiece? __________ If so, use the focusing knob to bring the image in the fixed eyepiece into focus, while keeping the eye at the independent focusing eyepiece closed. Then adjust the independent focusing eyepiece so that the image is clear, while keeping the other eye closed. Is the image inverted? ______________________________
5. Turn the magnification changing knob, and determine the kind of mechanism on your microscope. A zoom mechanism allows continuous viewing while changing the magnification. A rotating lens mechanism blocks the view of the object as the new lenses are rotated. Be sure to click each lens firmly into place. If you do not, the field will be only partially visible. What kind of mechanism is on your microscope? ______________________________
6. Set the magnification changing knob on the lowest magnification. *Sketch an organism from the plastomount in the following circle as though this represents your entire field of view:*

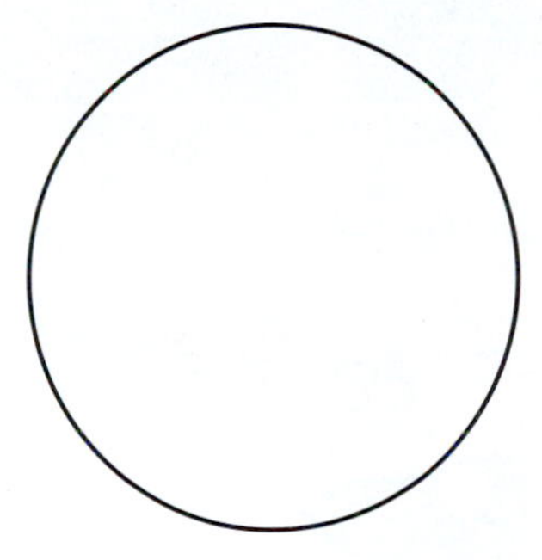

7. Rotate the magnification changing knob to the highest magnification. *Draw another circle within the one provided to indicate the reduction of the field of view.*
8. Experiment with various objects at various magnifications until you are comfortable with using the stereomicroscope.
9. When you are finished, return your stereomicroscope and illuminator to their correct storage areas.

3.3 Use of the Compound Light Microscope

As mentioned, the name "compound light microscope" indicates that it uses two sets of lenses and light to view an object. The two sets of lenses are the ocular lenses located near the eyes and the objective lenses located near the object. Illumination is from below, and the light passes through clear portions but does not pass through opaque portions of the object. This microscope is used to examine small or thinly sliced sections of objects under higher magnification than would be possible with the stereomicroscope.

Identifying the Parts

Obtain a compound light microscope from the storage area, and place it securely on the table. *Identify the following parts on your microscope, and label them in Figure 3.4.*

Figure 3.4 Compound light microscope.
Compound light microscope with binocular head and mechanical stage. Label this microscope with the help of the text material.

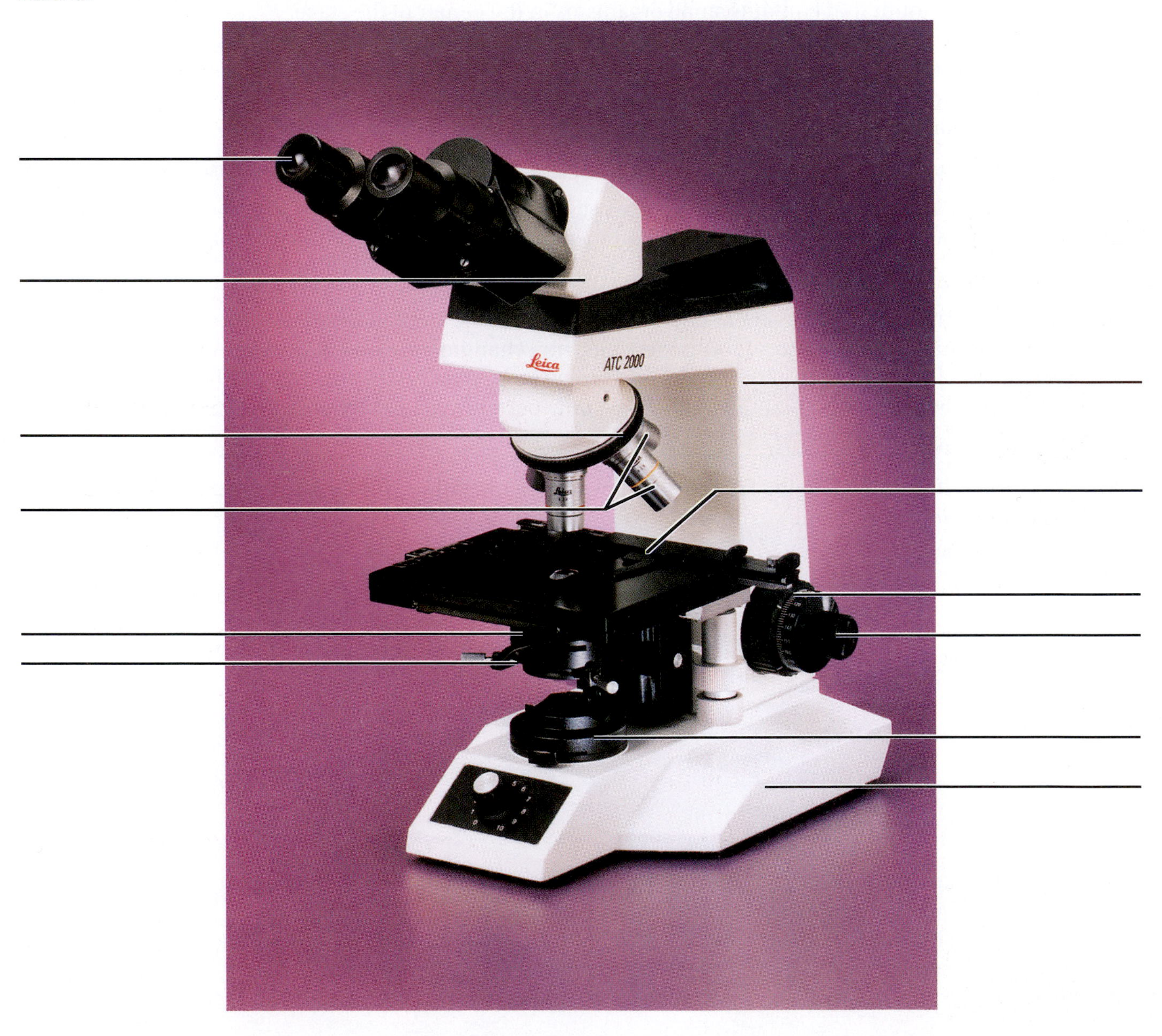

1. **Ocular lenses (eyepieces):** What is the magnifying power of the ocular lenses on your microscope? ____________________
2. **Body tube:** Holds nosepiece at one end and eyepiece at the other end; conducts light rays.
3. **Arm:** Supports upper parts and provides carrying handle. When carrying the microscope, hold the arm with one hand and place the other hand beneath the base.
4. **Nosepiece:** Revolving device that holds objectives.
5. **Objective lenses (objectives):**
   a. **Scanning objective:** This is the shortest of the objective lenses and is used to scan the whole slide. The magnification is stamped on the housing of the lens. It is a number followed by an ×. What is the magnifying power of the scanning objective lens on your microscope? ____________________
   b. **Low-power objective:** This lens is longer than the scanning objective lens and is used to view objects in greater detail. What is the magnifying power of the low-power objective lens on your microscope? ____________________
   c. **High-power objective:** If your microscope has three objective lenses, this lens will be the longest. It is used to view an object in even greater detail. What is the magnifying power of the high-power objective lens on your microscope? ____________________
   d. **Oil immersion objective:** (on microscopes with four objective lenses): Holds a 95× (to 100×) lens and is used in conjunction with immersion oil to view objects with the greatest magnification. Does your microscope have an oil immersion objective? __________ If this lens is available, your instructor will discuss its use when the lens is needed.
6. **Coarse-adjustment knob:** Knob used to bring object into approximate focus; used only with low-power objective.
7. **Fine-adjustment knob:** Knob used to bring object into final focus.
8. **Condenser:** Lens system below the stage used to focus the beam of light on the object being viewed.
9. **Diaphragm** or **diaphragm control lever:** Controls amount of illumination used to view the object.
10. **Light source:** An attached lamp that directs a beam of light up through the object.
11. **Base:** The flat surface of the microscope that rests on the table.
12. **Stage:** Holds and supports microscope slides.
13. **Stage clips:** Hold slides in place on the stage.
14. **Mechanical stage (optional):** A movable stage that aids in the accurate positioning of the slide. Does your microscope have a mechanical stage? ____________________
15. **Mechanical stage control knobs (optional):** Two knobs usually located below the stage. One knob controls forward/reverse movement, and the other controls right/left movement.

Focusing the Compound Light Microscope—Lowest Power

1. Turn the nosepiece so that the lowest power objective on your microscope is in straight alignment over the stage.
2. Always begin focusing with the lowest power objective on your microscope (4× [scanning] or 10× [low power]).
3. With the coarse-adjustment knob, lower the stage (or raise the objectives) until it stops.
4. Place a slide of the letter *e* on the stage, and stabilize it with the clips. (If your microscope has a mechanical stage, pinch the spring of the slide arms on the stage, and insert the slide.) Center the *e* as best you can on the stage, or use the two control knobs located below the stage (if your microscope has a mechanical stage) to center the *e*.

5. Again, be sure that the lowest-power objective is in place. Then, as you look from the side, decrease the distance between the stage and the tip of the objective lens until the lens comes to an automatic stop or is no closer than 3 mm above the slide.
6. While looking into the eyepiece, rotate the diaphragm (or diaphragm control lever) to give the maximum amount of light.
7. Using the coarse-adjustment knob, slowly increase the distance between the stage and the objective lens until the object—in this case, the letter *e*—comes into view, or focus.
8. Once the object is seen, you may need to adjust the amount of light. To increase or decrease the contrast, rotate the diaphragm slightly.
9. Use the fine-adjustment knob to sharpen the focus if necessary.
10. Practice having both eyes open when looking through the eyepiece, as this greatly reduces eyestrain.

Inversion

Inversion refers to the fact that a microscopic image is upside down and reversed.

Observation: Inversion

1. In space 1 provided here, *draw the letter* e *as it appears on the slide (with the unaided eye, not looking through the eyepiece).*

1.	2.

2. In space 2, *draw the letter* e *as it appears when you look through the eyepiece.*
3. What differences do you notice? ____________________
4. Move the slide to the right. Which way does the image appear to move? ____________________
5. Move the slide down. Which way did the image move? ____________________

Focusing the Compound Light Microscope—Higher Powers

Compound light microscopes are **parfocal;** that is, once the object is in focus with the lowest power, it should also be almost in focus with the higher power.

1. Bring the object into focus under the lowest power by following the instructions in the previous section.
2. Make sure that the letter *e* is centered in the field of the lowest objective.
3. Move to the next higher objective (low power [10×] or high power [40×]) by turning the nosepiece until you hear it click into place. Do not change the focus; parfocal microscope objectives will not hit normal slides when changing the focus if the lowest objective is initially in focus. (If you are on low power [10×], proceed to high power [40×] before going on to step 4.)
4. If any adjustment is needed, use only the *fine*-adjustment knob. (*Note:* Always use only the fine-adjustment knob with high power.) On your drawing of the letter *e, draw a circle around the portion of the letter that you are now seeing with high-power magnification.*
5. When you have finished your observations of this slide (or any slide), rotate the nosepiece until the lowest-power objective clicks into place, and then remove the slide.

Total Magnification

Total magnification is calculated by multiplying the magnification of the ocular lens (eyepiece) by the magnification of the objective lens. (The ocular lenses are each 10×.) The magnification of objective lenses is imprinted on the lens casing.

Observation: Total Magnification

In Table 3.2 calculate total magnifications for your microscope according to the ocular lens and the objective lens you are using.

Table 3.2 Total Magnification

Objective	Ocular Lens	Objective Lens	Total Magnification
Scanning power (if present)			
Low power			
High power			
Oil immersion (if present)			

3.4 Microscopic Observations

When a specimen is prepared for observation, the object should always be viewed as a **wet mount**. A wet mount is prepared by placing a drop of liquid on a slide or, if the material is dry, by placing it directly on the slide and adding a drop of water or stain. The mount is then covered with a coverslip, as illustrated in Figure 3.5. Dry the bottom of your slide before placing it on the stage.

Figure 3.5 Preparation of a wet mount.

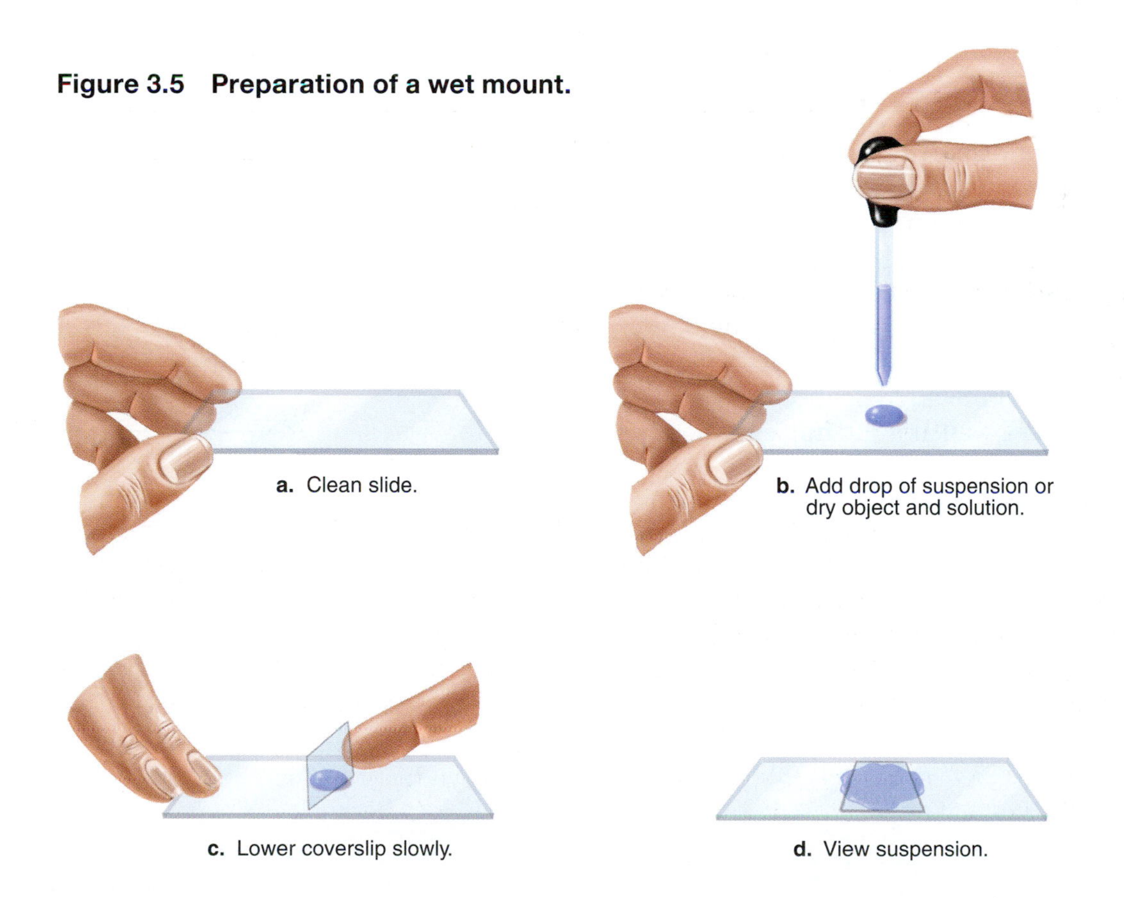

Human Epithelial Cells

Epithelial cells cover the body's surface and line its cavities.

Methylene blue Avoid ingestion, inhalation, and contact with skin, eyes, and mucous membranes. Exercise care in using this chemical. If any should spill on your skin, wash the area with mild soap and water. Methylene blue will also stain clothing.

Observation: Human Epithelial Cells

1. Obtain a prepared slide, or make your own as follows:
   a. Obtain a prepackaged flat toothpick (or sanitize one with alcohol or alcohol swabs).
   b. Gently scrape the inside of your cheek with the toothpick, and place the scrapings on a clean, dry slide. Discard used toothpicks in the biohazard waste container provided.
   c. Add a drop of very weak methylene blue or iodine solution, and cover with a coverslip.
2. Observe under the microscope.
3. Locate the nucleus (the central, round body), the cytoplasm, and the plasma membrane (outer cell boundary). *Label Figure 3.6.*
4. Because your epithelial slides are biohazardous, they must be disposed of as indicated by your instructor.

Figure 3.6 Cheek epithelial cells.
Label the nucleus, the cytoplasm, and the plasma membrane.

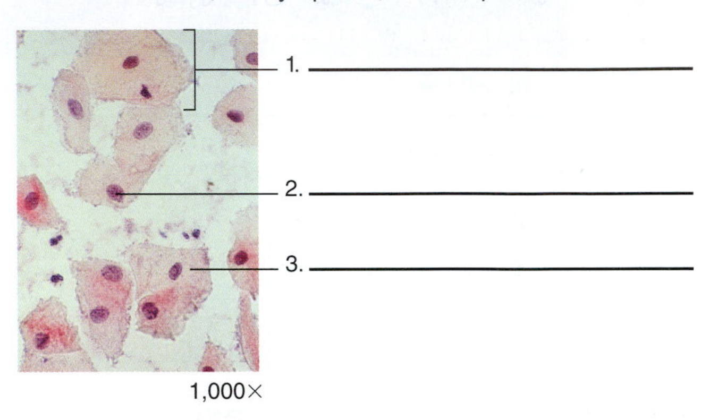

Onion Epidermal Cells

Epidermal cells cover the surfaces of plant organs, such as leaves. The bulb of an onion is made up of fleshy leaves.

Observation: Onion Epidermal Cells

1. With a scalpel, strip a small, thin, transparent layer of cells from the inside of a fresh onion leaf.

2. Place it gently on a clean, dry slide, and add a drop of iodine solution (or methylene blue). Cover with a coverslip.
3. Observe under the microscope.
4. Locate the cell wall and the nucleus. *Label Figure 3.7.*
5. Note some obvious differences between the human cheek cells and the onion cells, and list them in Table 3.3.

Figure 3.7 Onion epidermal cells.
Label the cell wall and the nucleus.

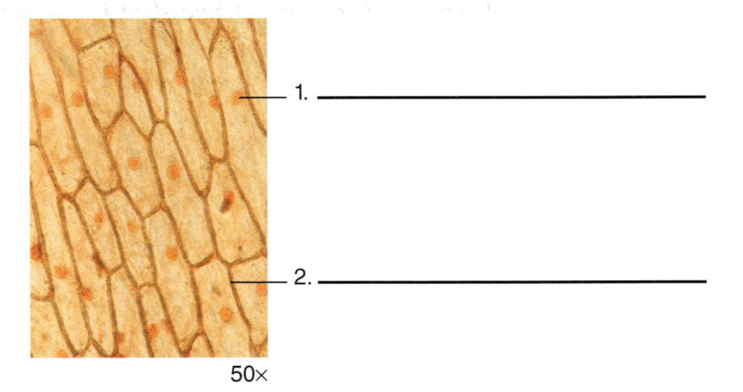

Table 3.3 Differences Between Human Epithelial and Onion Epidermal Cells		
Differences	**Human Epithelial Cells (Cheek)**	**Onion Epidermal Cells**
Shape (sketch)		

Euglena

Examination of *Euglena* (a unicellular organism with a flagellum to facilitate movement) will test your ability to observe objects with the microscope and to control illumination to heighten contrast.

Observation: Euglena

1. Make a wet mount of *Euglena* by using a drop of a *Euglena* culture and adding a drop of Protoslo (methyl cellulose solution) onto a slide. The Protoslo slows the organism's swimming.
2. Mix thoroughly with a toothpick, and add a coverslip.
3. Scan the slide for *Euglena:* Start at the upper left-hand corner, and move the slide forward and back as you work across the slide from left to right. The *Euglena* may be at the edge of the slide because they show an aversion to Protoslo.
4. Experiment by using scanning, low-power, and high-power objective lenses; by focusing up and down with the fine-adjustment knob; and by adjusting the light so that it is not too bright.
5. Compare your *Euglena* specimens with Figure 3.8. How do your specimens compare to Figure 3.8? ____________________

____

Figure 3.8 ***Euglena.*** *Euglena* is a unicellular, flagellated organism.

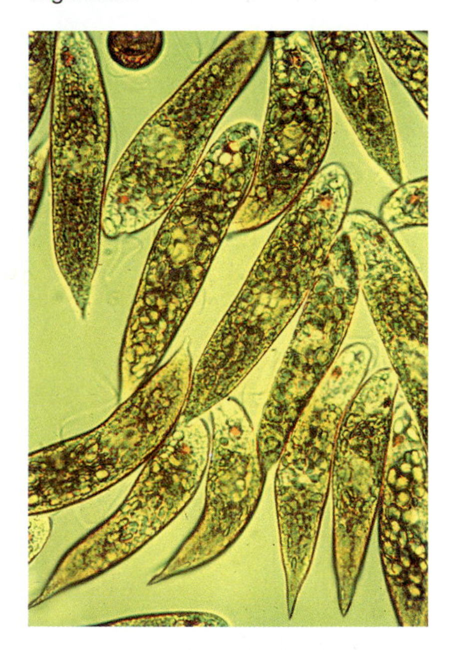

Laboratory Review 3

1. Of the three types of microscopes studied, which one best shows the surface of an object? Scanning electron microscope

2. Explain the designation "compound light" microscope:
   a. compound there are two sets of lens- objective and ocular
   b. light is used to view the object

3. What function is performed by the diaphragm of a microscope? controls the amount of illumination used to view the object

4. Briefly describe the necessary steps for observing a slide at low power under the compound light microscope. Center it, look, move object lens, look through lesses use course adjustment knob and adjust light.

5. Why is it helpful for a microscope to be parfocal? not much adjustment is needed

6. Why is locating an object more difficult if you start with the high-power objective than with the low-power objective? It's smaller in high power than it is in low.

7. How much larger than normal does an object appear with a low-power objective? 100X

8. A virus is 50 nm in size. Would you recommend using a stereomicroscope, compound light microscope, or an electron microscope to see it? electron Why? because it's small

9. What type of microscope, aside from the compound light microscope, might you use to observe the organisms found in pond water? Stereomicroscope

10. State two differences between onion epidermal cells and human epithelial cells. Onion are square and move in rows Human: are round and have random orientation.

Concepts of Biology Website

Instructors can find lab prep information and answers to all of the laboratory questions in the Laboratory Resource Guide. *Students* can practice their knowledge with quizzes, animations, flashcards, and much more.

www.mhhe.com/maderconcepts3

McGraw-Hill Access Science Website

An online encyclopedia of science and technology that provides information, including videos, that can enhance the laboratory experience.

www.accessscience.com

Microscopy Biology

LABORATORY

4

Simpson-brianna Lab 4

Chemical Composition of Cells

Learning Outcomes

Introduction

All organisms consist of basic units of matter called **atoms.** Molecules form when atoms bond with one another. Inorganic molecules are often associated with nonliving things, and biomolecules are associated with organisms. In this laboratory, you will be studying the biomolecules of cells: **proteins, carbohydrates** (monosaccharides, disaccharides, polysaccharides), and **lipids** (i.e., fat).

Planning Ahead To save time, your instructor may have you start the boiling water bath needed for the experiment on page 37 at the beginning of the laboratory session.

Large biomolecules, sometimes called macromolecules, form during *dehydration reactions* when smaller molecules bond as water is given off. During *hydrolysis reactions,* bonds are broken as water is added.

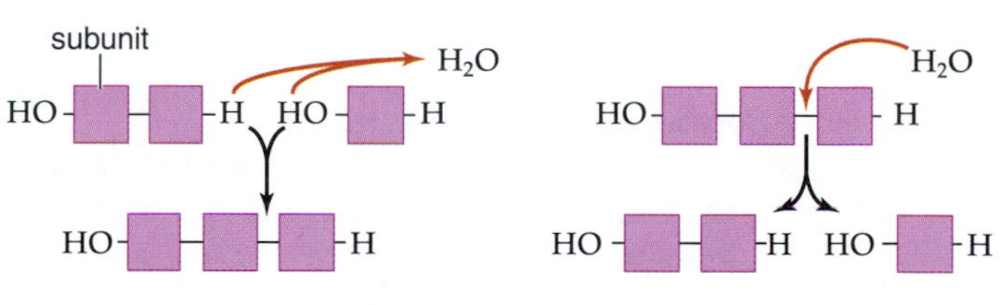

A fat contains one glycerol and three fatty acids. Proteins and some carbohydrates (called polysaccharides) are **polymers** because they are made up of smaller molecules called subunits. Proteins contain a large number of amino acids joined together by a peptide bond. A polysaccharide, such as starch, contains a large number of glucose molecules joined together. Various chemicals will be used in this laboratory to test for the presence of cellular biomolecules. If a color change is observed, the test is said to be *positive* because it indicates that the molecule is present. If the color change is not observed, the test is said to be *negative* because it indicates that the molecule is not present.

What Is a Control?

The experiments in today's laboratory have both a positive control and a negative control, *which should be saved for comparison purposes until the experiment is complete.* The **positive control** goes through all the steps of the experiment and does contain the substance being tested. Therefore, positive results are expected. The **negative control** goes through all the steps of the experiment except it does not contain the substance being tested. Therefore, negative results are expected.

For example, if a test tube contains glucose (the substance being tested) and Benedict's reagent (blue) is added, a red color develops upon heating. This test tube is the positive control; it tests positive for glucose. If a test tube does not contain glucose and Benedict's reagent is added, Benedict's is expected to remain blue. This test tube is the negative control; it tests negative for glucose.

What benefit is a positive control? Positive controls give you a standard by which to tell if the substance being tested is present (or acting properly) in an unknown sample. Negative controls ensure that the experiment is giving reliable results; after all, if a negative control should happen to give a positive result, then the entire experiment may be faulty and unreliable.

4.1 Proteins

Proteins have numerous functions in cells. Antibodies are proteins that combine with pathogens so that the pathogens are destroyed by the body. Transport proteins combine with and move substances from place to place. Hemoglobin transports oxygen throughout the body. Albumin is another transport protein in our blood. Regulatory proteins control cellular metabolism in some way. For example, the hormone insulin regulates the amount of glucose in blood so that cells have a ready supply. Structural proteins include keratin, found in hair, and myosin, found in muscle. **Enzymes** are proteins that speed chemical reactions. A reaction that could take days or weeks to complete can happen within an instant if the correct enzyme is present. Amylase is an enzyme that speeds the breakdown of starch in the mouth and small intestine.

Proteins are made up of **amino acids** joined together. About 20 different common amino acids are found in cells. All amino acids have an acidic group (—COOH) and an amino group (—H_2N). They differ by the ***R* group** (remainder group) attached to a carbon atom, as shown in Figure 4.1. The *R* groups have varying sizes, shapes, and chemical activities.

A chain of two or more amino acids is called a **peptide,** and the bond between the amino acids is called a **peptide bond.** A **polypeptide** is a very long chain of amino acids. A protein can contain one or more polypeptide chains. Insulin contains a single chain, while hemoglobin contains four polypeptides. A protein has a particular shape, which is important to its function. The shape comes about because the *R* groups of the polypeptide chain(s) can interact with one another in various ways.

Figure 4.1 Formation of a dipeptide.
During a dehydration reaction, a dipeptide forms when an amino acid joins with an amino acid as a water molecule is removed. The bond between amino acids is called a peptide bond. During a hydrolysis reaction, water is added and the peptide bond is broken.

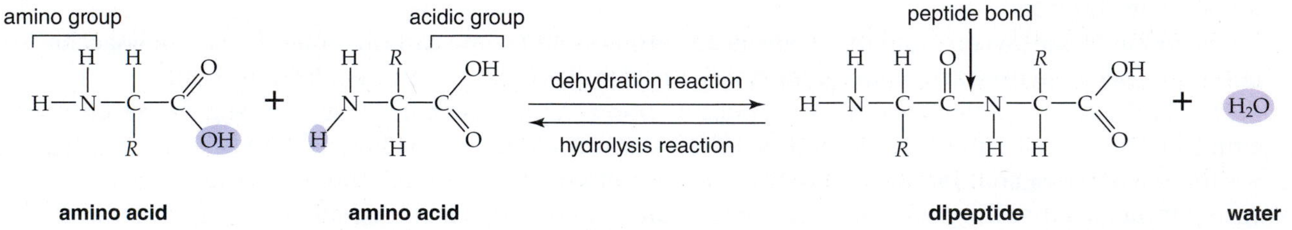

amino group
acidic group
peptide bond
dehydration reaction
hydrolysis reaction
amino acid
amino acid
dipeptide
water

Test for Proteins

Biuret reagent (blue color) contains a strong solution of sodium or potassium hydroxide (NaOH or KOH) and a small amount of dilute copper sulfate ($CuSO_4$) solution. The reagent changes color in the presence of proteins or peptides because the peptide bonds of the protein or peptide chemically combine with the copper ions in biuret reagent (Table 4.1). *Label these test results.*

Biuret test for protein and peptides

Table 4.1 Biuret Test for Protein and Peptides

	Protein	Peptides
Biuret reagent (blue)	Purple	Pinkish-purple

Experimental Procedure: Test for Proteins

1. Label four clean test tubes (1 to 4).
2. Using the designated graduated transfer pipets, add 1 ml of the experimental solutions listed in Table 4.2 to the test tubes according to their numbers.
3. Then add five drops of Biuret reagent to the tubes, with swirling to mix.
4. The reaction is almost immediate. Record your observations in Table 4.2.

> **Biuret reagent** Biuret reagent is highly corrosive. Exercise care in using this chemical. If any should spill on your skin, wash the area with mild soap and water. Follow your instructor's directions for its disposal.

Table 4.2 Biuret Test for Protein

Tube	Contents	Final Color	Conclusion (+ or −)
1	Distilled water		
2	Albumin		
3	Pepsin		
4	Starch		

Conclusions: Proteins

- From your test results, conclude if a protein is present (+) or absent (−) in each tube. Enter your conclusions in Table 4.2.
- Pepsin is an enzyme. Enzymes are composed of what type of biomolecules? proteins
- According to your results, is starch a protein? NO
- Which of the four tubes is the negative control sample? 1, 4 Why? did not have a color change, so it was negative
- Why do experimental procedures include control samples? tell you if the substance being tested is carried throughout the experiment to be tested or if it was not carried throughout the experiment.

4.2 Carbohydrates

Carbohydrates include sugars and molecules that are chains of sugars. **Glucose,** which has only one sugar unit, is a monosaccharide; **maltose,** which has two sugar units, is a disaccharide (Fig. 4.2). Glycogen, starch, and cellulose are polysaccharides, made up of chains of glucose units (Fig. 4.3).

Glucose is used by all organisms as an energy source. Energy is released when glucose is broken down to carbon dioxide and water. This energy is used by the organism to do work. Animals store glucose as glycogen and plants store glucose as starch. Plant cell walls are composed of cellulose.

Figure 4.2 Formation of a disaccharide.
During a dehydration reaction, a disaccharide, such as maltose, forms when a glucose joins with a glucose as a water molecule is removed. During a hydrolysis reaction, the components of water are added, and the bond is broken.

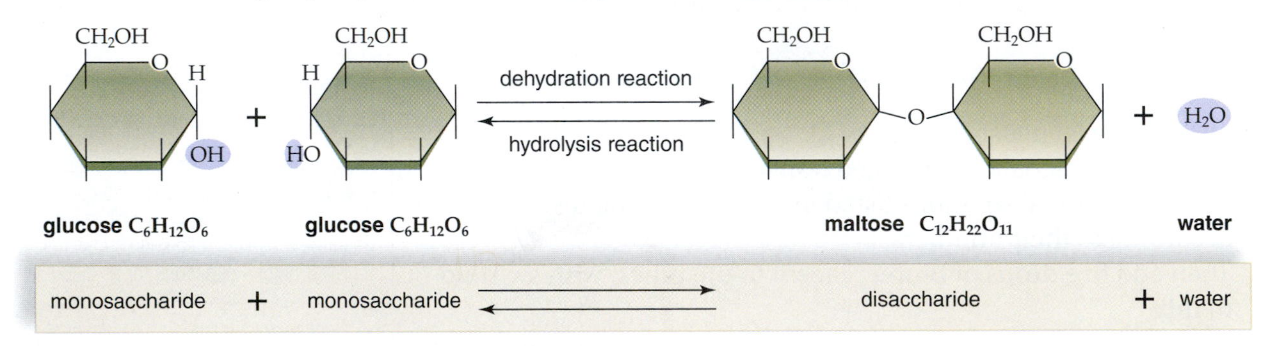

Figure 4.3 Starch.
Starch is a polysaccharide composed of many glucose units. **a.** Photomicrograph of starch granules in cells of a potato. **b.** Structure of starch. Starch consists of amylose that is nonbranched and amylopectin that is branched.

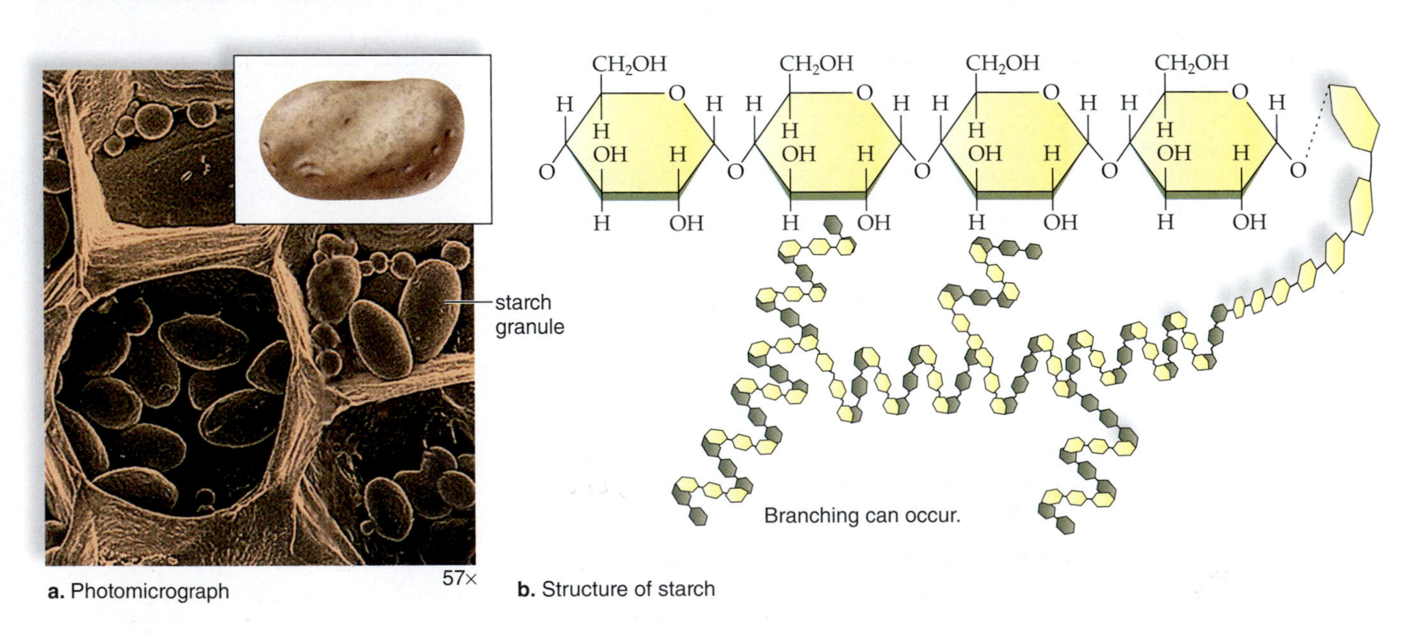

a. Photomicrograph

b. Structure of starch

Test for Starch

In the presence of starch, iodine solution (yellowish-brown) reacts chemically with starch to form a blue-black color (Table 4.3).

Table 4.3 Iodine Test for Starch

	Starch
Iodine solution	Blue-black

Experimental Procedure: Test for Starch

1. Label five clean test tubes (1 to 5).
2. Using the designated graduated transfer pipets, add 1 ml of the experimental solutions listed in Table 4.4 to the test tubes according to their numbers.
3. Then add five drops of iodine solution to the tubes at the same time.
4. Note the final color changes and record your observations in Table 4.4.

Iodine test for starch

Table 4.4 Iodine (IKI) Test for Starch

Tube	Contents	Final Color	Conclusion (+ or −)
1	Water		
2	Starch suspension		
3	Onion juice		
4	Potato juice		
5	Glucose solution		

Conclusions: Starch

- From your test results, conclude if starch is present (+) or absent (−) in each tube. Enter your conclusions in Table 4.4.
- Does the potato or the onion store glucose as starch? potato How do you know? the Starch was originally black, and the potato Solution turned black
- If your results are not as expected, offer an explanation. Then inform your instructor, who will advise you how to proceed.

Experimental Procedure: Microscopic Study

Potato

1. With a scalpel, slice a very thin piece of potato. Place it on a microscope slide, add a drop of water and a coverslip, and observe under low power with your compound light microscope. Compare your slide with the photomicrograph of starch granules (see Fig. 4.3a). Find the cell wall (large, geometric compartments) and the starch grains (numerous clear, oval-shaped objects).
2. Without removing the coverslip, place two drops of iodine solution onto the microscope slide so that the iodine touches the coverslip. Draw the iodine under the coverslip by placing a small piece of paper towel in contact with the water on the *opposite* side of the coverslip.
3. Microscopically examine the potato again on the side closest to where the iodine solution was applied.

   What is the color of the small, oval bodies? black

   What is the chemical composition of these oval bodies? Starch

Onion

1. Peel a single layer of onion from the bulb. On the inside surface, you will find a thin, transparent layer of onion skin. Peel off a small section of this layer and place on a slide.
2. Add a large drop of iodine solution.
3. Does onion contain starch? NO
4. Are these results consistent with those you recorded for onion juice in Table 4.4? yes

Test for Sugars

Monosaccharides and some disaccharides will react with **Benedict's reagent** after being heated in a boiling water bath. In this reaction, copper ion (Cu^{2+}) in the Benedict's reagent reacts with part of the sugar molecule, causing a distinctive color change. The color change can range from green to red, and increasing concentrations of sugar will give a continuum of colored products as shown in Table 4.5.

> **Benedict's reagent** Benedict's reagent is highly corrosive. Exercise care in using this chemical. If any should spill on your skin, wash the area with mild soap and water. Follow your instructor's directions for disposal of this chemical.

Table 4.5 Benedict's Test for Sugars (Some Typical Reactions)

Chemical	Chemical Category	Benedict's Reagent (After Heating)
Water	Inorganic	Blue (no change)
Glucose	Monosaccharide (carbohydrate)	Varies with concentration: very low—green low—yellow moderate—yellow-orange high—orange very high—orange-red
Maltose	Disaccharide (carbohydrate)	Varies with concentration—see "Glucose"
Starch	Polysaccharide (carbohydrate)	Blue (no change)

Benedict's test for sugars

Experimental Procedure: Test for Sugars

1. Prepare a boiling water bath and label five clean test tubes (1 to 5).
2. Using the designated graduated transfer pipets, add 1 ml of the experimental solutions listed in Table 4.6 to the test tubes according to their numbers.
3. Then add five drops of Benedict's reagent to all the tubes at this time.
4. Place all the tubes into the boiling water bath at the same time.
5. When, after a few minutes, you see a change of colors, remove all the tubes from the water bath and record your observations in Table 4.6.
6. Save your tubes for comparison purposes when you do Section 4.4.

Table 4.6 Benedict's Test for Sugars

Tube	Contents	Color (After Heating)	Conclusions
1	Water		
2	Glucose solution		
3	Starch suspension		
4	Onion juice		
5	Potato juice		

Conclusions: Sugars

- With the help of Table 4.5, conclude whether glucose, maltose, starch, or none of those is present in each test tube. Enter your conclusion in Table 4.6.
- Which tube served as a negative and which as a positive control? Starch was negative onion and potato juice was positive
- Compare Table 4.4 with Table 4.6. Sugars are an immediate energy source in cells. In plant cells, glucose (a primary energy molecule) is often stored in the form of starch. Is glucose stored as starch in the potato? ________ Is glucose stored as starch in the onion? ________ Does this explain your results in Table 4.6? ________ Why? ______________________________

4.3 Lipids

Lipids are compounds that are insoluble in water and soluble in solvents, such as alcohol and ether. Lipids include fats, oils, phospholipids, steroids, and cholesterol. Typically, **fat,** such as in the adipose tissue of animals, and **oils,** such as the vegetable oils from plants, are composed of three molecules of fatty acids bonded to one molecule of glycerol (Fig. 4.4). **Phospholipids** have the same structure as fats, except that in place of the third fatty acid there is a phosphate group (a grouping that contains phosphate). **Steroids** are derived from **cholesterol** and, like this molecule, have skeletons of four fused rings of carbon atoms, but they differ by functional groups (attached side chains). Fat, as we know, is long-term stored energy in the human body. Phospholipids are found in the plasma membrane of cells. Cholesterol, a molecule transported in the blood, has been implicated in causing cardiovascular disease. Regardless, steroids are very important compounds in the body; for example, the sex hormones are steroids.

Figure 4.4 Formation of a fat.
During a dehydration reaction, a fat molecule in animals and a vegetable oil in plants forms when glycerol joins with three fatty acids as three water molecules are removed. During a hydrolysis reaction, water is added, and the bonds are broken between glycerol and the three fatty acids. Note the double bond between the carbons in third fatty acid.

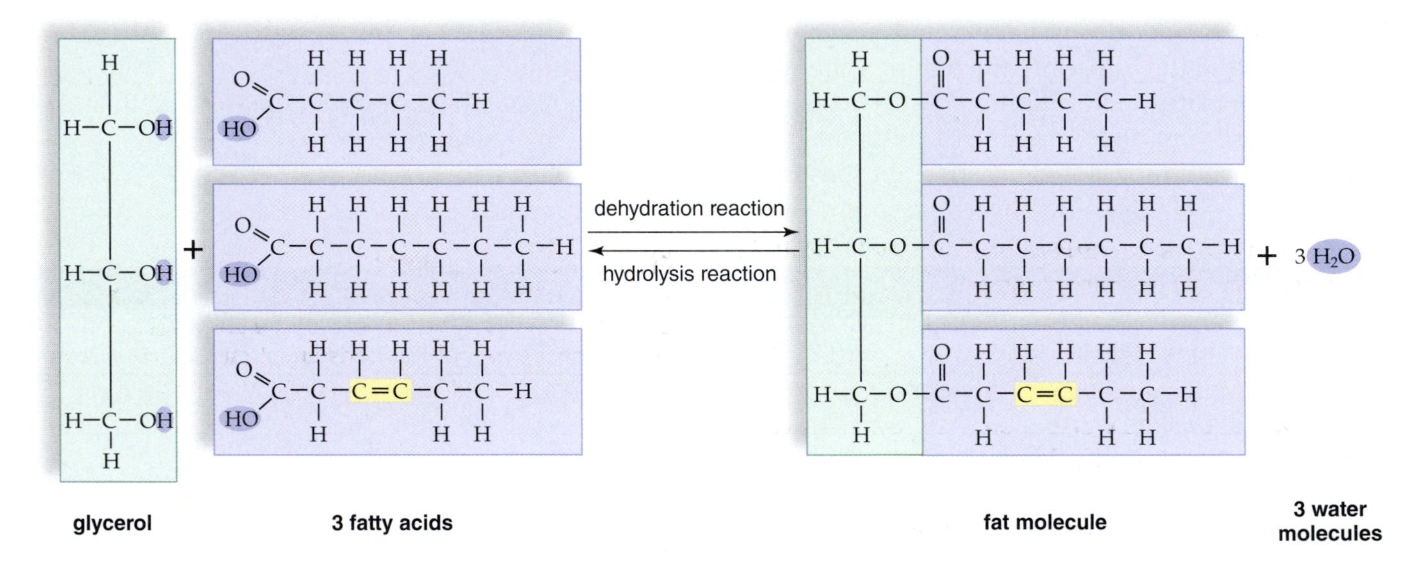

Fats Versus Vegetable Oils

In this laboratory you will have an opportunity to observe that vegetable oil does not disperse in water. What's the difference between a fat molecule and a vegetable oil? Both of these molecules are formed when glycerol and three fatty acids, which are polar molecules, combine to form a nonpolar molecule (Fig. 4.4). However, the fatty acids in a vegetable oil tend to have more double bonds between the carbons than do those in a fat. The double bonds introduce "kinks" in the molecule that cause the vegetable oil to be a liquid while a fat with fewer double bonds is solid at room temperature.

Test for Fat

Fats and oils do not evaporate from brown paper or loose-leaf paper; instead, they leave an oily spot.

Experimental Procedure: Paper Test for Fat

1. Place a small drop of water on a square of brown paper or loose-leaf paper. Describe the immediate effect. ______________________
2. Place a small drop of vegetable oil on a square of the paper. Describe the immediate effect. ______________________
3. Wait at least 15 minutes for the paper to dry. Evaluate which substance penetrates the paper and which is subject to evaporation. Record your observations and conclusions in Table 4.7. Save the paper for comparison use with Section 4.4.

Table 4.7 Paper Test for Fat

Sample	Observations	Conclusions
Water spot		
Oil spot		

Emulsification of Oil

Some molecules are **polar,** meaning that they have charged groups or atoms, and some are **nonpolar,** meaning that they have no charged groups or atoms. A water molecule is polar, and therefore, water is a good solvent for other polar molecules. When the charged ends of water molecules interact with the charged groups of polar molecules, these polar molecules disperse in water.

Water is not a good solvent for nonpolar molecules, such as fats. A fat has no polar groups to interact with water molecules. An **emulsifier,** however, can cause a fat to disperse in water. An emulsifier contains molecules with both polar and nonpolar ends. When the nonpolar ends interact with the fat and the polar ends interact with the water molecules, the fat disperses in water, and an **emulsion** results (Fig. 4.5).

Figure 4.5 Emulsification. An emulsifier contains molecules with both a polar and a nonpolar end. The nonpolar ends are attracted to the nonpolar fat, and the polar ends are attracted to the water. This causes droplets of fat molecules to disperse.

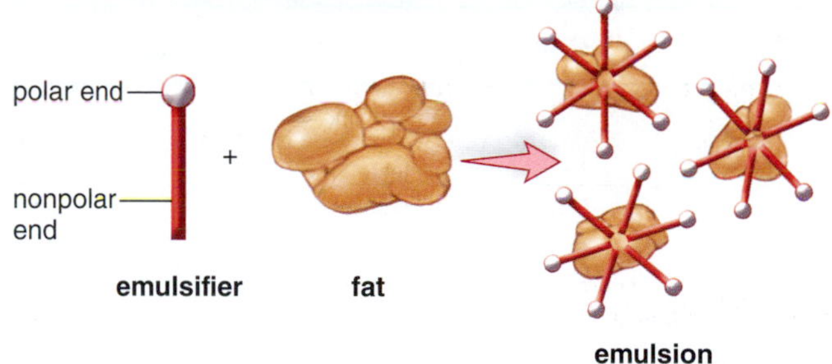

Bile salts (emulsifiers found in bile produced by the liver) are used in the digestive tract. Today milk, such as 1% milk, has been homogenized so that fat droplets do not congregate and rise to the top of the container. Homogenization requires the addition of natural emulsifiers such as phospholipids—the phosphate part of the molecule is polar and the lipid portion is nonpolar.

Experimental Procedure: Emulsification of Lipids

Label three clean test tubes (1 to 3), and use the appropriate graduated transfer pipet to add solutions to the test tubes as follows:

Tube 1

1. Add 3 ml of water and 1 ml of vegetable oil. Shake.
2. Observe for the initial dispersal of oil, followed by rapid separation into two layers. Is vegetable oil soluble in water? ____________
3. Let the tube settle for 5 minutes. Label a microscope slide as 1.
4. Use a dropper to remove a sample of the solution that is just below the layer of oil. Place the drop on the slide, add a coverslip, and examine with the low power of your compound light microscope.
5. Record your observations in Table 4.8.

Tube 2

1. Add 2 ml of water, 3 ml of vegetable oil, and 1 ml of the available emulsifier (Tween or bile salts). Shake.
2. Describe how the distribution of oil in tube 2 compares with the distribution in tube 1. ____________
3. Let the tube settle for 5 minutes. Label a microscope slide as 2.
4. Use a different dropper to remove a sample of the solution that is just below the layer of oil. Place the drop on the slide, add a coverslip, and examine with the low power of your compound light microscope.
5. Record your observations in Table 4.8.

Tube 3

1. Add 1 ml of milk and 2 ml of water. Shake well.
2. Use a different dropper to remove a sample of the solution. Place a drop on a slide, add a coverslip, and examine with the low power of your compound light microscope.
3. Record your observations in Table 4.8.

Table 4.8 Emulsification

Tube	Contents	Observations	Conclusions
1	Oil Water		
2	Oil Water Emulsifier		
3	Milk Water		

Conclusions: Emulsification

- From your observations, conclude why the contents of each tube appear as they do under the microscope. Record your conclusions in Table 4.8.
- Explain the correlation between your macroscopic observations (how the tubes look to your unaided eye) and your microscopic observations. ____________

4.4 Testing Foods and Unknowns

It is common for us to associate the term *organic* with the foods we eat, including carbohydrate foods (Fig. 4.6), protein foods (Fig. 4.7), and lipid foods (Fig. 4.8). Though we may recognize foods as being organic, often we are not aware of what specific types of compounds are found in what we eat. In the following Experimental Procedure, you will use the same tests you used previously to determine the composition of everyday foods and unknowns.

Figure 4.6 Carbohydrate foods.

Figure 4.7 Protein foods.

Figure 4.8 Lipid foods.

Experimental Procedure: Testing Foods and Unknowns

Your instructor will provide you with several everyday foods including unknowns, and your task is to:

1. With the help of your instructor decide how you will test substances for protein (page 33), carbohydrates (pages 35 and 37), and fat (page 39).
2. Record your results as positive (+) or negative (–) in Table 4.9.

Table 4.9 Testing Foods and Unknowns

Sample Name	Protein (Biuret)	Starch (Iodine)	Sugar (Benedict's)	Fat (Brown or loose-leaf paper)
Unknown A				
Unknown B				

Conclusions: Testing Foods and Unknowns

- What foods tested positive for more than one of the organic compounds? ______________________
- What does more than one positive test tell you about these foods? ______________________

____

Laboratory Review 4

1. What biomolecules studied today are present in cells? Proteins, Carbohydrates, lipids

2. You have been assigned the task of constructing a protein. What type of subunit would you use?
Amino acids

3. A digestive enzyme such as amylase breaks down starch to what disaccharide studied in this laboratory?
maltose

4. Why is it necessary to shake an oil and vinegar salad dressing before adding it to a salad?
to mix oil and water

5. How would you test for each of the following substances?
   - **a.** Sugars benedicts reagent
   - **b.** Fat brown paper bag test
   - **c.** Starch Iodine
   - **d.** Protein biuret reagent

6. Assume that you have tested an unknown sample with both Biuret solution and Benedict's solution and that both tests result in a blue color. What have you learned? that there are no Sugars and no proteins in the sample

7. What purpose is served when a test is done using water instead of a sample substance?
water is a negative test expected for negative results. If you get a Positive, experiment is invalid

8. A test tube contains albumin. The test for protein is positive and the test for starch is negative. Explain.
albumin is a protein

9. A test tube contains starch and the enzyme amylase. After 30 minutes, the test for starch is negative and the test for sugar is positive. Explain. the Starch is broken down into sugars

10. Which type of lipid contains four fused rings of carbon atoms? Steriods

Concepts of Biology Website

Instructors can find lab prep information and answers to all of the laboratory questions in the Laboratory Resource Guide. *Students* can practice their knowledge with quizzes, animations, flashcards, and much more.

www.mhhe.com/maderconcepts3

McGraw-Hill Access Science Website

An online encyclopedia of science and technology that provides information, including videos, that can enhance the laboratory experience.

www.accessscience.com

Chemical Composition of Cells

LABORATORY

5

Cell Structure and Function

Learning Outcomes

5.1 Animal Cell and Plant Cell Structure
- Label an animal cell diagram, and state a function for the structures labeled. 44–45
- Label a plant cell diagram, and state a function for the structures labeled. 46
- Use microscopic techniques to observe plant cell structure. 47

5.2 Diffusion
- Define and describe the process of diffusion as affected by the medium. 48
- Predict and observe which substances will or will not diffuse across a plasma membrane. 48–49

5.3 Osmosis: Diffusion of Water Across Plasma Membrane
- Explain an osmosis experiment based on a knowledge of diffusion principles. 50
- Define isotonic, hypertonic, and hypotonic solutions, and give examples in terms of NaCl concentrations. 51
- Predict the effect of different tonicities on animal (e.g., red blood) cells and on plant (e.g., *Elodea*) cells. 51–53

5.4 pH and Cells
- Predict the change in pH before and after the addition of an acid to nonbuffered and buffered solutions. 54–55
- Suggest a method by which it is possible to test the effectiveness of antacid medications. 55

Introduction

The molecules we studied in Laboratory 3 are not alive—the basic units of life are cells. The **cell theory** states that all organisms are composed of cells and that cells come only from other cells. While we are accustomed to considering the heart, the liver, or the intestines as enabling the human body to function, it is actually cells that do the work of these organs.

Planning Ahead To save time, your instructor may have you start a boiling water bath (page 49) and the potato strip experiment (page 53) at the beginning of the laboratory.

Figure 3.6 shows human cheek epithelial cells as viewed by an ordinary compound light microscope available in general biology laboratories. It shows that the content of a cell, called the **cytoplasm,** is bounded by a **plasma membrane.** The plasma membrane regulates the movement of molecules into and out of the cytoplasm. In this lab, we will study how the passage of water into a cell depends on the difference in concentration of solutes (particles) between the cytoplasm and the surrounding medium or solution. The well-being of cells also depends upon the pH of the solution surrounding them. We will see how a buffer can maintain the pH within a narrow range and how buffers within cells can protect them against damaging pH changes.

Because a photomicrograph shows only a minimal amount of detail, it is necessary to turn to the electron microscope to study the contents of a cell in greater depth. The models of plant and animal cells available in the laboratory today are based on electron micrographs.

5.1 Animal Cell and Plant Cell Structure

Table 5.1 lists the structures found in animal and plant cells. The **nucleus** in a eukaryotic cell is bounded by a **nuclear envelope** and contains **nucleoplasm.** The cytoplasm, found between the plasma membrane and the nucleus, consists of a background fluid and the organelles, such as the nucleolus, endoplasmic reticulum, Golgi apparatus, vacuoles and vesicles, lysosomes, peroxisome, mitochondrion, and chloroplast.

Table 5.1 Eukaryotic Structures in Animal Cells and Plant Cells

Name	Composition	Function
Cell wall*	Contains cellulose fibrils	Provides support and protection
Plasma membrane	Phospholipid bilayer with embedded proteins	Outer cell surface that regulates entrance and exit of molecules
Nucleus	Enclosed by nuclear envelope; contains chromatin (threads of DNA and protein)	Storage of genetic information; synthesis of DNA and RNA
Nucleolus	Concentrated area of chromatin	Produces subunits of ribosomes
Ribosome	Protein and RNA in two subunits	Carries out protein synthesis
Endoplasmic reticulum (ER)	Membranous, flattened channels and tubular canals; rough ER and smooth ER	Synthesis and/or modification of proteins and other substances; transport by vesicle formation
Rough ER	Studded with ribosomes	Protein synthesis
Smooth ER	Lacks ribosomes	Synthesis of lipid molecules
Golgi apparatus	Stack of membranous saccules	Processes, packages, and distributes proteins and lipids
Vesicle/vacuole	Membrane-bounded sac; large central vacuole in plant cells*	Stores and transports substances
Lysosome	Vesicle containing hydrolytic enzymes	Digests macromolecules and cell parts
Peroxisome	Vesicle containing specific enzymes	Breaks down fatty acids and converts resulting hydrogen peroxide to water
Mitochondrion	Bounded by double membrane; inner membrane is cristae	Cellular respiration, producing ATP molecules
Chloroplast*	Membranous grana bounded by double membrane	Photosynthesis, producing sugars
Cytoskeleton	Microtubules, intermediate filaments, actin filaments	Maintains cell shape and assists movement of cell parts
Cilia and flagella	Attachments supported by microtubules	Movement of cell
Centrioles** in centrosome	Microtubule-containing, cylindrically shaped organelle in a structure of complex composition.	Centrioles organize microtubules in cilia and flagella; centrosome organizes microtubules in cell

*Plant cells only

**Animal cells only

Study Table 5.1 to determine structures that are unique to plant cells and unique to animal cells, and write them below the examples given.

	Plant Cells	Animal Cells
Unique structures:	1. Large central vacuole	1. Small vacuoles
	2. ____________	2. ____________
	3. ____________	

Animal Cell Structure

Label Figure 5.1. With the help of Table 5.1, give a function for each labeled structure.

Structure	Function
Plasma membrane	
Nucleus	
Nucleolus	
Endoplasmic reticulum	
Rough ER	
Smooth ER	
Golgi apparatus	
Vesicle	
Lysosome	
Mitochondrion	
Centrioles in centrosome	
Cytoskeleton	

Figure 5.1 Animal cell structure.

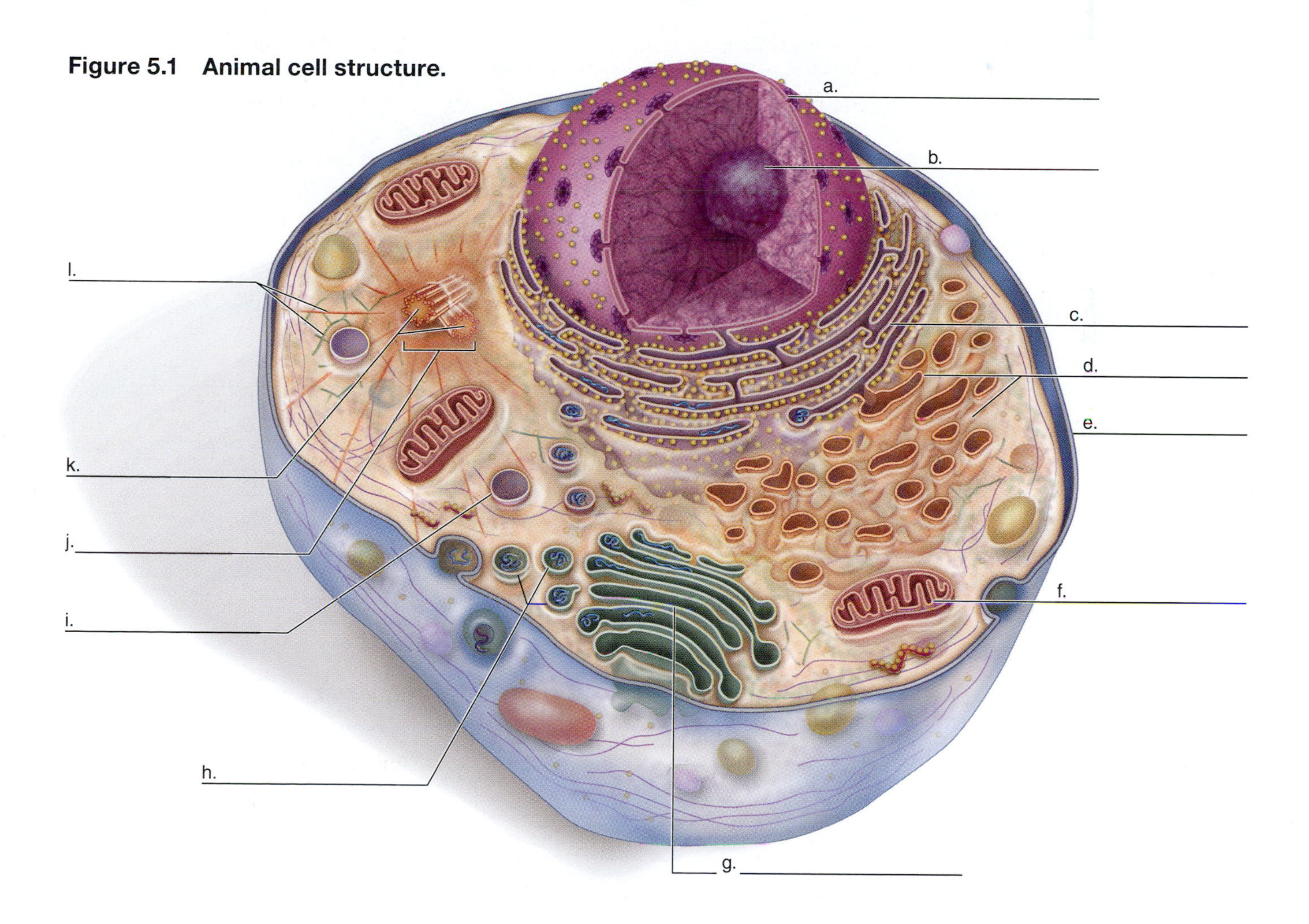

Plant Cell Structure

Label Figure 5.2. With the help of Table 5.1, give a function for each labeled structure unique to plant cells.

Structure	Function
Cell wall	
Central vacuole, large	
Chloroplast	

Figure 5.2 Plant cell structure.

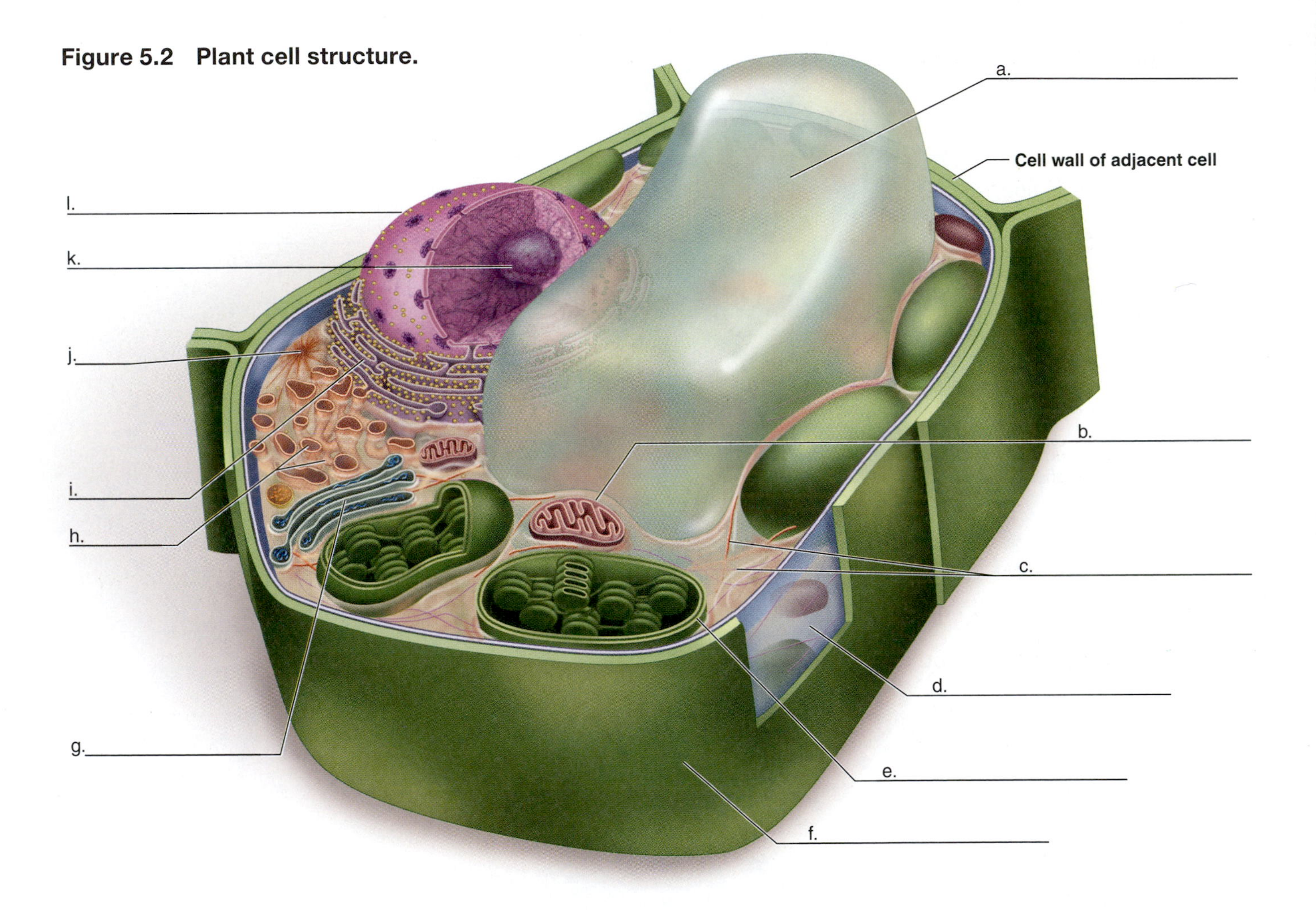

Observation: Plant Cell Structure

1. Prepare a wet mount of a small piece of young *Elodea* leaf in fresh water. *Elodea* is a multicellular, eukaryotic plant found in freshwater ponds and lakes.
2. Have the drop of water ready on your slide so that the leaf does not dry out, even for a few seconds. Take care that the leaf is mounted with its top side up.
3. Using low power bring the leaf surface into focus.
4. Select a cell with numerous chloroplasts for further study, and switch to high power.
5. Carefully focus on the sides of the cell. The chloroplasts appear to be only along the sides of the cell because the large, fluid-filled, membrane-bounded central vacuole pushes the cytoplasm against the cell walls (Fig. 5.3*a*). Then focus on the surface and notice an even distribution of chloroplasts (Fig. 5.3*b*).

Figure 5.3 ***Elodea*** **cell structure.**

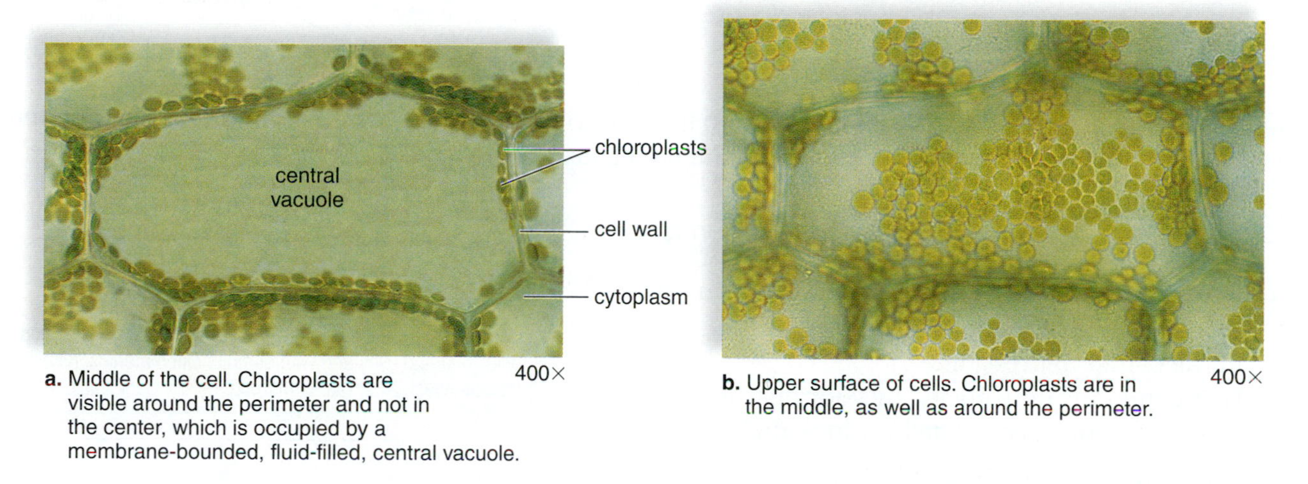

a. Middle of the cell. Chloroplasts are visible around the perimeter and not in the center, which is occupied by a membrane-bounded, fluid-filled, central vacuole.

b. Upper surface of cells. Chloroplasts are in the middle, as well as around the perimeter.

6. Can you locate the cell nucleus? __________ It may be hidden by the chloroplasts, but when visible, it appears as a faint gray lump on one side of the cell.
7. Why can't you see the other organelles featured in Figure 5.2? ______________________________

   ____
8. Can you detect movement of chloroplasts in this cell or any other cell? __________ The chloroplasts are not moving under their own power but are being carried by a streaming of the nearly invisible cytoplasm.
9. Save your slide for use later in this laboratory.

5.2 Diffusion

Diffusion is the movement of molecules from a higher to a lower concentration until equilibrium is achieved and the molecules are distributed equally (Fig. 5.4). At equilibrium, molecules may still be moving back and forth, but there is no net movement in any one direction.

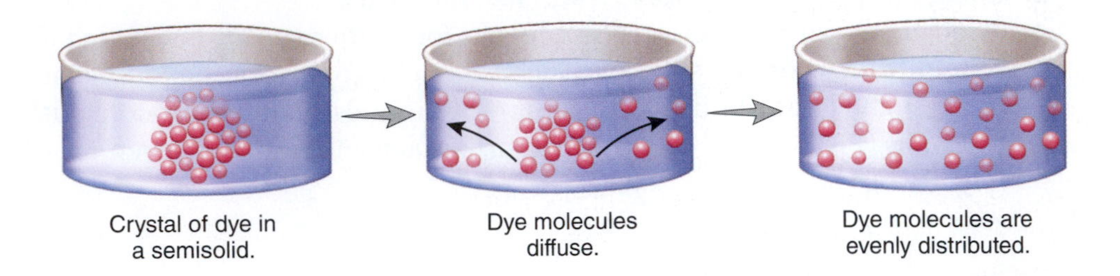

Figure 5.4 Process of diffusion. Diffusion is apparent when dye molecules have equally dispersed.

Diffusion Through a Semisolid

Diffusion of molecules is a general phenomenon in the environment. The speed of diffusion is dependent on such factors as the temperature, the size of the molecule, and the type of medium.

Experimental Procedure: Diffusion in a Semisolid

> **Potassium permanganate ($KMnO_4$)** $KMnO_4$ is highly poisonous and is a strong oxidizer. Avoid contact with skin and eyes and with combustible materials. If spillage occurs, wash all surfaces thoroughly. $KMnO_4$ will also stain clothing.

1. Observe a petri dish containing 1.5% gelatin (or agar) to which potassium permanganate ($KMnO_4$) was added in the center depression.
2. Using the start time posted by your instructor and the time of day now, calculate how long diffusion has been occurring. Length of time has been ____________.
3. Using a ruler placed over the petri dish, measure (in mm) the movement of color from the center of the depression outward in one direction: __________ mm.
4. Calculate the speed of diffusion: __________ mm/60 min = mm/hr.

Diffusion Across the Plasma Membrane

Some molecules can diffuse across a plasma membrane, and some cannot. In general, small, noncharged molecules can cross a membrane by simple diffusion, but large molecules cannot diffuse across a membrane. The dialysis tube membrane in the Experimental Procedure simulates a plasma membrane.

Experimental Procedure: Diffusion Across Plasma Membrane

At the start of the experiment,

1. Cut a piece of dialysis tubing approximately 40 cm (about 16 in) long. Soak the tubing in water until it is soft and pliable.
2. Close one end of the dialysis tubing with two knots.
3. Fill the bag halfway with glucose solution.
4. Add four full droppers of starch solution to the bag.

5. Hold the open end while you mix the contents of the dialysis bag. Rinse off the outside of the bag with distilled water.
6. Fill a beaker ⅔ full with distilled water.
7. Add droppers of iodine solution (IKI) to the water in the beaker until an amber (tealike) color is apparent.
8. Record the color of the solution in the beaker in Table 5.2.
9. Place the bag in the beaker with the open end hanging over the edge. Secure the open end of the bag to the beaker with a rubber band as shown (Fig. 5.5). Make sure the contents do not spill into the beaker.

Figure 5.5 Placement of dialysis bag in water containing iodine.

rubber band
open end of dialysis bag
dialysis membrane (simulates plasma membrane)
water and iodine solution
glucose and starch
closed end of dialysis bag

After about 5 minutes, at the end of the experiment,

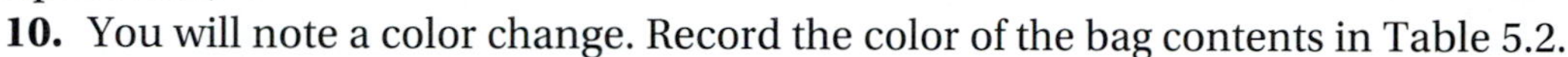

10. You will note a color change. Record the color of the bag contents in Table 5.2.
11. Mark off a test tube at 1 cm and 3 cm.
12. Draw solution from near the bag and at the bottom of the beaker for testing with Benedict's reagent. Fill the test tube to the first mark with this solution. Add Benedict's reagent to the 3 cm mark. Heat in a boiling water bath for 5 to 10 minutes, observe any color change, and record your results as + or – in Table 5.2. (Optional use of glucose test strip: Dip glucose test strip into beaker. Compare stick with chart provided by instructor.)

> **Benedict's reagent** Exercise care in using this chemical. It is highly corrosive. If any should spill on your skin, wash the area with mild soap and water. Follow your instructor's directions for its disposal.

13. Remove the dialysis bag from the beaker. Dispose of it and the used Benedict's reagent solution in the manner directed by your instructor.

Table 5.2 Solute Diffusion Across Plasma Membrane

At Start of Experiment			At End of Experiment		
	Contents	Color	Color	Benedict's Test	Conclusion
Bag	Glucose Starch			______	
Beaker	Water Iodine				

Conclusions: Solute Diffusion Across the Plasma Membrane

- Based on the color change noted in the bag, conclude what solute diffused across the dialysis membrane from the beaker to the bag, and record your conclusion in Table 5.2.
- From the results of the Benedict's test on the beaker contents, conclude what solute diffused across the dialysis membrane from the bag to the beaker, and record your conclusion in Table 5.2.
- Which solute did not diffuse across the dialysis membrane from the bag to the beaker? ______

  How do you know? ____

5.3 Osmosis: Diffusion of Water Across Plasma Membrane

Osmosis is the diffusion of water across the plasma membrane of a cell. Just like any other molecule, water follows its concentration gradient and moves from the area of higher concentration to the area of lower concentration.

Experimental Procedure: Osmosis

To demonstrate osmosis, a thistle tube is covered with a membrane at its lower opening and partially filled with 50% corn syrup (starch solution) or a similar substance. The whole apparatus is placed in a beaker containing distilled water (Fig. 5.6). The water concentration in the beaker is 100%. Water molecules can move freely between the thistle tube and the beaker.

1. Note the level of liquid in the thistle tube, and measure how far it travels in 10 minutes:

   ___________ mm

2. Calculate the speed of osmosis under these conditions: ___________ mm/hr

Figure 5.6 Osmosis demonstration.
a. A thistle tube, covered at the broad end by a differentially permeable membrane, contains a corn syrup solution. The beaker contains distilled water. **b.** The solute is unable to pass through the membrane, but the water (arrows) passes through in both directions. There is a net movement of water toward the inside of the thistle tube, where there is a lower percentage of water molecules. **c.** Due to the incoming water molecules, the level of the solution rises in the thistle tube.

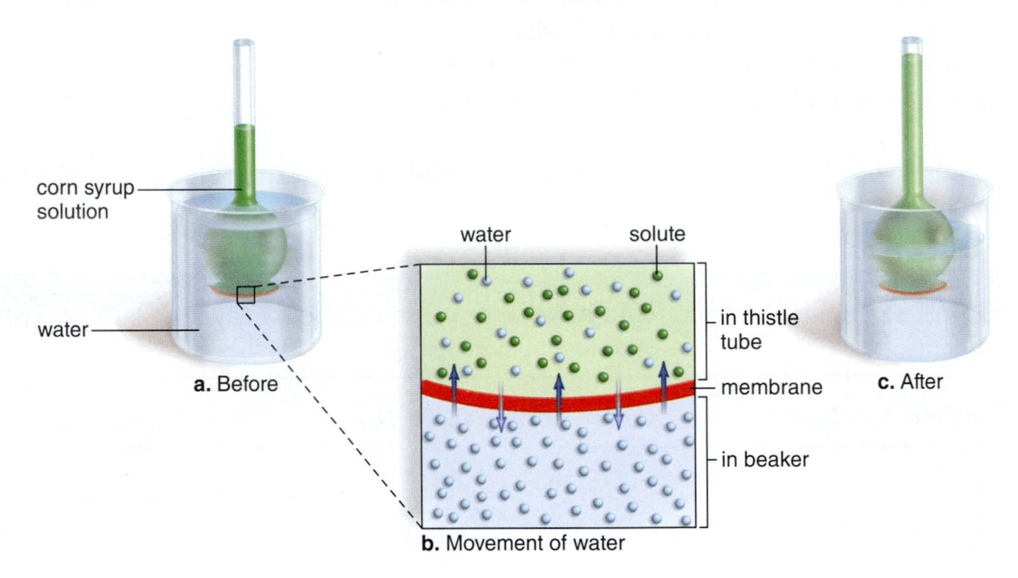

Conclusions: Osmosis

- In which direction was there a net movement of water? ___________________________

  Explain what is meant by "net movement" after examining the arrows in Figure 5.6*b*.

  ____

- If the starch molecules in corn syrup moved from the thistle tube to the beaker, would there have been a net movement of water into the thistle tube? ___________ Why wouldn't large starch molecules be able to move across the membrane from the thistle tube to the beaker? ___________________

  ____

- Explain why the water level in the thistle tube rose: In terms of solvent concentration, water moved from the area of __________ water concentration to the area of __________ water concentration across a differentially permeable membrane.

Tonicity in Cells

Tonicity is the relative concentration of solute (particles), and therefore also of solvent (water), outside the cell compared with inside the cell.

- An **isotonic solution** has the same concentration of solute (and therefore of water) as the cell. When cells are placed in an isotonic solution, there is no net movement of water (see Fig. 5.7*a*).
- A **hypertonic solution** has a higher solute (therefore, lower water) concentration than the cell. When cells are placed in a hypertonic solution, water moves out of the cell into the solution (see Fig. 5.7*b*).
- A **hypotonic solution** has a lower solute (therefore, higher water) concentration than the cell. When cells are placed in a hypotonic solution, water moves from the solution into the cell (see Fig.5.7*c*).

Animal Cells (Red Blood Cells)

A solution of 0.9% NaCl is isotonic to red blood cells (Fig. 5.7*a*). A solution greater than 0.9% NaCl is hypertonic to red blood cells. In such a solution, the cells shrivel up, a process called **crenation** (Fig. 5.7*b*). A solution of less than 0.9% NaCl is hypotonic to red blood cells. In such a solution, the cells swell to bursting, a process called **hemolysis** (Fig. 5.7*c*).

Figure 5.7 Tonicity and red blood cells.

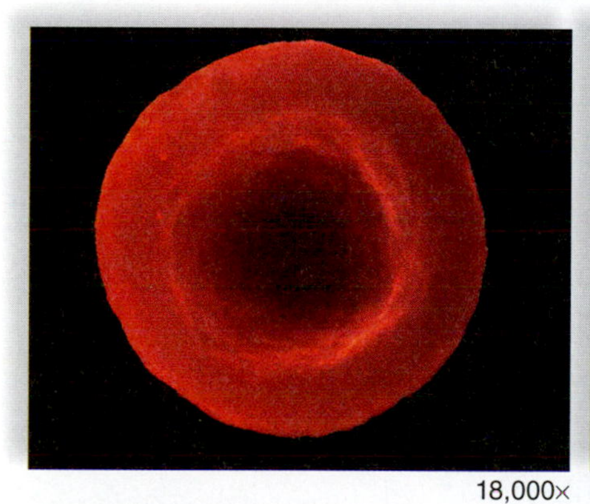

18,000×

a. Isotonic solution. Red blood cell has normal appearance due to no net gain or loss of water.

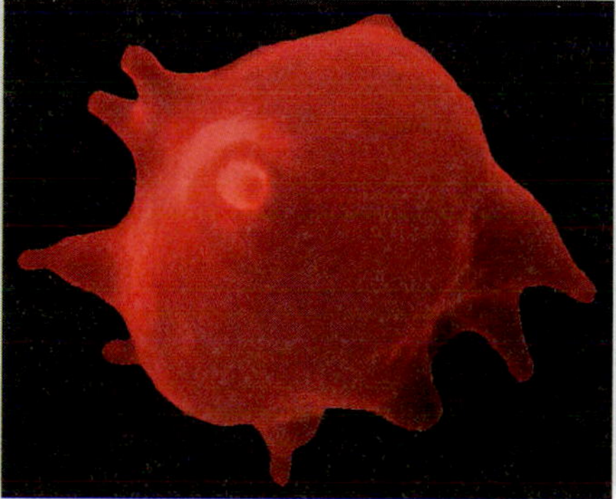

18,000×

b. Hypertonic solution. Red blood cell shrivels due to loss of water.

18,000×

c. Hypotonic solution. Red blood cell fills to bursting due to gain of water.

Experimental Procedure: Demonstration of Tonicity in Red Blood Cells

Do not remove the stoppers of test tubes during this procedure.

Three stoppered test tubes on display have the following contents:

Tube 1: 0.9% NaCl plus a few drops of whole sheep blood
Tube 2: 10% NaCl plus a few drops of whole sheep blood
Tube 3: 0.9% NaCl plus distilled water and a few drops of whole sheep blood

1. In the second column of Table 5.3, record the tonicity of each tube in relation to red blood cells.
2. Hold each tube in front of one of the pages of your Lab Manual. Determine whether you can see the print on the page through the tube. Record your findings in the fourth column of Table 5.3.
3. In the fifth column of Table 5.3, relate print visibility to effect of tonicity on the cells.

Table 5.3 Effect of Tonicity on Red Blood Cells

Tube	Tonicity	Effect on Cells	Print Visibility	Explanation
1				
2				
3				

Plant Cells

When plant cells are in a hypotonic solution, such as fresh water, the large central vacuole gains water and exerts pressure, called **turgor pressure.** The cytoplasm, including the chloroplasts, is pushed up against the cell wall. You observed turgor pressure in Figure 5.3.

When plant cells are in a hypertonic solution, such as 10% NaCl, the central vacuole loses water, and the cytoplasm, including the chloroplasts, pulls away from the cell wall. This is called **plasmolysis.** You will observe plasmolysis in the following experimental procedure.

Experimental Procedure: Tonicity in Elodea *Cells*

1. If possible, use the *Elodea* slide you prepared earlier in this laboratory. If not, prepare a new wet mount of a small *Elodea* leaf using fresh water. Your slide should look like Figure 5.3.
2. Complete the portion of Table 5.4 that pertains to a hypotonic solution.
3. Prepare a new wet mount of a small *Elodea* leaf using a 10% NaCl solution.
4. After several minutes, focus on the surface of the cells. Note that plasmolysis has occurred and the cell contents are now in the center because the cytoplasm has pulled away from the cell wall due to loss of water from the large central vacuole.
5. Complete the portion of Table 5.4 that pertains to a hypertonic solution.

plasmolysis

Table 5.4 Effect of Tonicity on *Elodea* Cells

Tonicity	Appearance of Cells	Due to (Scientific Term)
Hypotonic		
Hypertonic		

Experimental Procedure: Tonicity in Potato Strips

(This Experimental Procedure runs for one hour. Prior setup can maximize time efficiency.)

1. Cut two strips of potato, each about 7 cm long and 1.5 cm wide.
2. *Label two test tubes 1 and 2.* Place one potato strip in each tube.
3. Fill tube 1 with water to cover the potato strip.
4. Fill tube 2 with 10% sodium chloride (NaCl) to cover the potato strip.
5. After one hour, remove the potato strips from the test tubes and place them on a paper towel. Observe each strip for limpness (water loss) or stiffness (water gain). Which tube has the limp potato strip? ________ Use tonicity to explain why water diffused out of the potato strip in this tube. ____

   Which tube has the stiff potato strip? ________ Use tonicity to explain why water diffused into the potato strip in this tube. ____

   ____
6. Use this space to create a table to display your results. Give your table a title and columns for tube number and contents, tonicity, results, and explanation.

Conclusions: Tonicity

- In a hypotonic solution, animal cells ________________. In red blood cells, this is called ________________. In a hypertonic solution, animal cells ________________. In red blood cells, this is called ________________.
- In a hypotonic solution, the central vacuole of *Elodea* cells exerts ________________ pressure, and chloroplasts are seen ________________. In a hypertonic solution, the central vacuole loses water and ________________ occurs. The cytoplasm plus the chloroplasts are seen ________________.
- In a hypotonic solution, potato strips ________________ water; in a hypertonic solution, potato strips ________________ water and become ________________.

5.4 pH and Cells

The pH of a solution tells its hydrogen ion concentration [H^+]. The **pH scale** ranges from 0 to 14. A pH of 7 is neutral (Fig. 5.8). A pH lower than 7 indicates that the solution is acidic (has more hydrogen ions than hydroxide ions), whereas a pH greater than 7 indicates that the solution is basic (has more hydroxide ions than hydrogen ions). A **buffer** is a system of chemicals that takes up excess hydrogen ions or hydroxide ions, as appropriate.

The concept of pH is important in biology because organisms are very sensitive to hydrogen ion concentration. For example, in humans the pH of the blood must be maintained at about 7.4 or we become ill. All organisms need to maintain the hydrogen ion concentration, or pH, at a constant level.

Why are cells and organisms buffered? ____

____

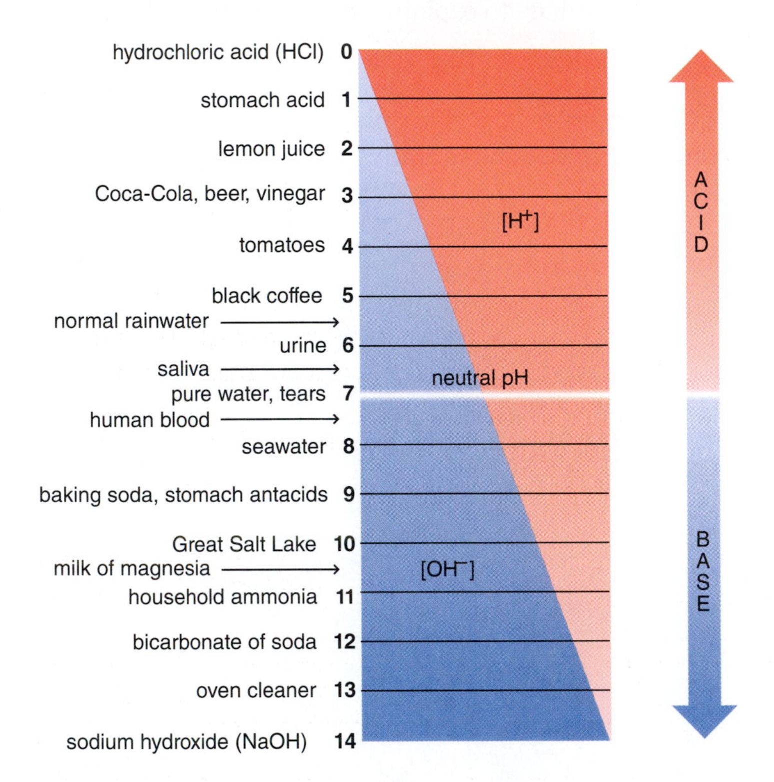

Figure 5.8 The pH scale. The proportionate amount of hydrogen ions (H^+) to hydroxide ions (OH^-) is indicated by the diagonal line.

Experimental Procedure: pH and Cells

> **Hydrochloric acid (HCl)** used to produce an acid pH is a strong, caustic acid. Exercise care in using this chemical. If any HCl spills on your skin, rinse immediately with clear water. Follow your instructor's directions for disposal of tubes that contain HCl.

1. Label three test tubes, and fill them to the halfway mark as follows: tube 1: water; tube 2: buffer (inorganic) solution; and tube 3: simulated cytoplasm (buffered protein solution).
2. Use pH paper to determine the pH of each tube. Dip the end of a stirring rod into the solution, and then touch the stirring rod to a 5 cm strip of pH paper. Read the current pH by matching the color observed with the color code on the pH paper package. Record your results in the "pH Before Acid" column in Table 5.5.
3. Add 0.1 N hydrochloric acid (HCl) dropwise to each tube until you have added five drops—shake or swirl after each drop. Use pH paper as in step 2 to determine the new pH of each solution. Record your results in the "pH After Acid" column in Table 5.5.

Table 5.5 pH and Cells				
Tube	**Contents**	**pH Before Acid**	**pH After Acid**	**Explanation**
1	Water			
2	Buffer			
3	Cytoplasm			

Conclusions: pH and Cells

- Enter your explanations in the last column of Table 5.5.
- Why would you expect cytoplasm to be as effective as the buffer in maintaining pH? ______________

____

Experimental Procedure: Effectiveness of Antacids

This procedure tests the ability of commercial antacid tablets such as Alka-Seltzer, Rolaids, or Tums to absorb excess H^+.

1. Use a mortar and pestle to grind up the amount of antacid that is listed as one dose.
2. For each antacid tested, use a 100 ml of phenol red solution diluted to a faint pink to wash the antacid into a 250 ml beaker. Phenol red solution is a pH indicator that turns yellow in an acid and red in a base. Use a stirring rod to get the powder to dissolve.
3. Add and count the number of 0.1 N HCl drops it takes for the solution to turn light yellow.
4. Record your results in Table 5.6.

Table 5.6 Effectiveness of Antacids		
Antacid	**Drops of Acid Needed to Reach End Point**	**Evaluation**
1		
2		
3		

Conclusions: Effectiveness of Antacids

- Participate with others in concluding which of the antacids tested neutralizes the most acid.

  ____
- Did dosage in mg have any effect on the results? ______________________________

  ____
- Which of the substances on the label could be a buffer? ______________________________

Laboratory Review 5

1. What characteristic do all eukaryotic cells have in common? ______________________

____

2. Which organelle digests macromolecules and cell parts? ______________________
3. Why would you predict that an animal cell, but not a plant cell, might burst when placed in a hypotonic solution? ______________________

____

4. Which of the cellular organelles would be included in the following categories?
   a. Membranous canals and vacuoles ______________________

   ____

   b. Energy-related organelles ______________________
5. How do you distinguish between rough endoplasmic reticulum and smooth endoplasmic reticulum?
   a. Structure ______________________
   b. Function ______________________
6. If a dialysis bag filled with water is placed in a molasses solution, what do you predict will happen to the weight of the bag over time? ______________________

   Why? ______________________

   ____

7. What is the relationship between plant cell structure and the ability of plants to stand upright?

   ____

8. The police are trying to determine if material removed from the scene of a crime was plant matter. What would you suggest they look for? ______________________

   ____

9. A test tube contains red blood cells and a salt solution. When the tube is held up to a page, you cannot see the print. With reference to a concentration of 0.9% sodium chloride (NaCl), how concentrated is the salt solution? ______________________
10. Predict the microscopic appearance of cells in the leaf tissue of a wilted plant. ______________________

    ____

Concepts of Biology Website

Instructors can find lab prep information and answers to all of the laboratory questions in the Laboratory Resource Guide. *Students* can practice their knowledge with quizzes, animations, flashcards, and much more.

www.mhhe.com/maderconcepts3

McGraw-Hill Access Science Website

An online encyclopedia of science and technology that provides information, including videos, that can enhance the laboratory experience.

www.accessscience.com

LEARNSMART LABS™

Cell Anatomy
Diffusion
Osmosis
pH & Cells

LABORATORY

6 How Enzymes Function

Learning Outcomes

Introduction

The cell carries out many chemical reactions. All the chemical reactions that occur in a cell are collectively called **metabolism.** A possible chemical reaction can be indicated like this:

$$\underset{\text{reactants}}{A + B} \longrightarrow \underset{\text{products}}{C + D}$$

In all chemical reactions, the **reactants** are molecules that undergo a change, which results in the **products.** The arrow means "produces," as in A + B produces C + D. The number of reactants and products can vary; in the one you are studying today, a single reactant breaks down to two products. All the reactions that occur in a cell have an enzyme. **Enzymes** are organic catalysts that speed metabolic reactions. Because enzymes are specific and speed only one type of reaction, they are given names. In today's laboratory, you will be studying the action of the enzyme **catalase.** The reactants in an enzymatic chemical reaction are called **substrate(s)** (Fig. 6.1).

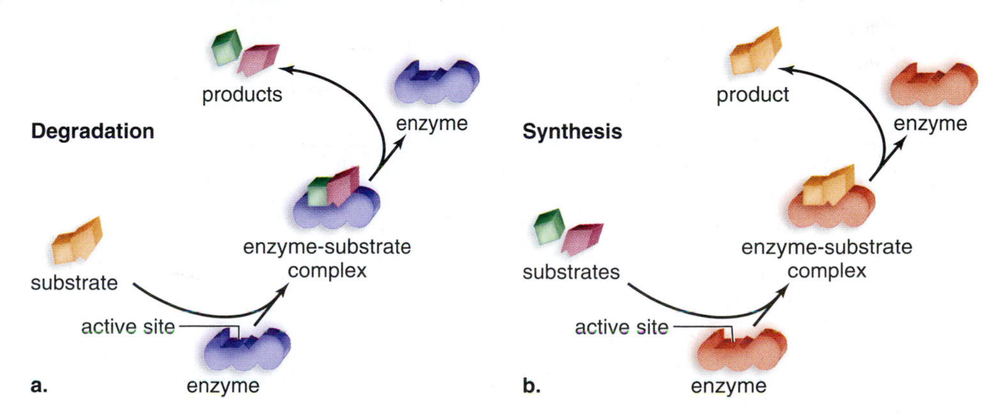

Figure 6.1 Enzymatic action. The reaction occurs on the surface of the enzyme at the active site. The enzyme is reusable. **a.** Degradation: substrate is broken down. **b.** Synthesis: substrates are combined.

Enzymes are specific because they have a shape that accommodates the shape of their substrates. Enzymatic reactions can be indicated like this:

$$E + S \longrightarrow ES \longrightarrow E + P$$

In this reaction, E = enzyme, ES = enzyme-substrate complex, and P = product.

Two types of enzymatic reactions in cells are shown in Figure 6.1. During degradation reactions, the substrate is broken down to the product(s), and during synthesis reactions, the substrates are joined to form a product. A number of other types of reactions also occur in cells. The location where the enzyme and substrate form an enzyme-substrate complex is called the **active site** because the reaction occurs here. At the end of the reaction, the product is released, and the enzyme can then combine with its substrate again. A cell needs only a small amount of an enzyme because enzymes are used over and over. Some enzymes have turnover rates well in excess of a million product molecules per minute.

> **Planning Ahead** To save time, your instructor may have you start a boiling water bath (page 60) at the beginning of the laboratory.

6.1 Catalase Activity

Catalase is involved in a degradation reaction: Catalase speeds the breakdown of hydrogen peroxide (H_2O_2) in nearly all organisms including bacteria, plants, and animals. A cellular organelle called a peroxisome, which contains catalase, is present in every plant and animal organ. This means that we could use any plant or animal organ as our source of catalase today. Commonly, school laboratories use the potato as a source of catalase because potatoes are easily obtained and cut up.

Catalase performs a useful function in organisms because hydrogen peroxide is harmful to cells. Hydrogen peroxide is a powerful oxidizer that can attack and denature cellular molecules like DNA! Knowing its harmful nature, humans use hydrogen peroxide as a commercial antiseptic to kill germs (Fig. 6.2). In reduced concentration, hydrogen peroxide is a whitening agent used to bleach hair and teeth. Skillful technicians use it to provide oxygen to aquatic plants and fish, but it is also used industrially to clean most anything from tubs to sewage. It's even put in glow sticks, where it reacts with a dye that then emits light.

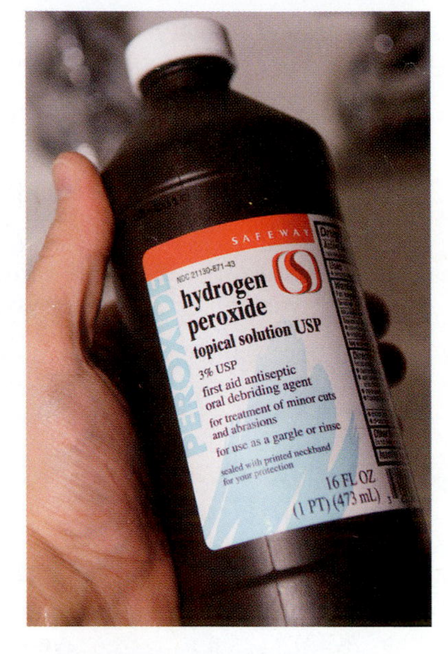

Figure 6.2 Hydrogen peroxide.
Bubbling occurs when you apply hydrogen peroxide to a cut because oxygen is being released when catalase, an enzyme present in the body's cells, degrades hydrogen peroxide.

When catalase speeds the breakdown of hydrogen peroxide, water and oxygen are released.

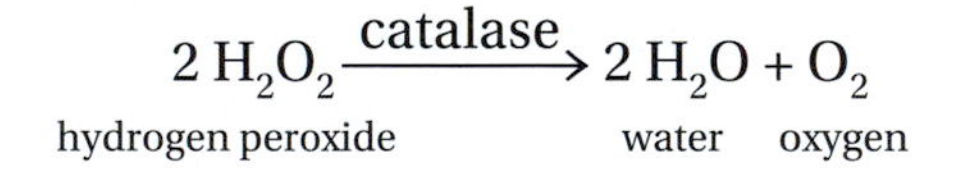

$$\underset{\text{hydrogen peroxide}}{2\,H_2O_2} \xrightarrow{\text{catalase}} \underset{\text{water}}{2\,H_2O} + \underset{\text{oxygen}}{O_2}$$

What is the reactant in this reaction? ______________________ What is the substrate for catalase? ______________________ What are the products in this reaction? ______________________ and ______________________

Bubbling occurs as the reaction proceeds. Why? ______________________

In the experimental procedure that follows, you will use bubble height to indicate the amount of product per unit time and therefore enzyme activity. Examine Table 6.1 and hypothesize which tube (1, 2, or 3) will have a greater bubble column height. Include a complete explanation in your hypothesis. ______________________

____

____

Experimental Procedure: Catalase Activity

Label three clean test tubes (1 to 3), and use the appropriate graduated transfer pipet to add solutions to the test tubes as follows:

Tube 1
1. Add 1 ml of catalase buffered at pH 7.0, the optimum pH for catalase.
2. Add 4 ml of hydrogen peroxide. Swirl well to mix, and wait at least 20 seconds for bubbling to develop.
3. Measure the height of the bubble column (in mm), and record your results in Table 6.1.

Tube 2
1. Add 1 ml of water.
2. Add 4 ml of hydrogen peroxide. Swirl well to mix, and wait at least 20 seconds.
3. Measure the height of the bubble column and record your results in Table 6.1.

Tube 3
1. Add 1 ml of catalase.
2. Add 4 ml of sucrose solution. Swirl well to mix; wait 20 seconds.
3. Measure the height of the bubble column, and record your results in Table 6.1.

Table 6.1 Catalase Activity

Tube	Contents	Bubble Column Height (mm)	Explanation
1	Catalase Hydrogen peroxide	Mito: alpers syndrome Perox: ALD Lysos: wolfman disease	: farber : hurler
2	Water Hydrogen peroxide	: Tay Sachs disease	plasma: Alzheimer Schindlers
3	Catalase Sucrose solution	: hunter : krabbe	gaucher

: Pompe dise
: Niemann-Pick

Conclusions: Catalase Activity

- Which tube showed the amount of bubbling you expected? ___________ Record your explanation in Table 6.1.
- Which tube is a negative control? ___________ If this tube showed bubbling, what could you conclude about your procedure? ___________________________
  Record your explanation in Table 6.1.
- Enzymes are specific; they speed only a reaction that contains their substrate. Which tube exemplifies this characteristic of an enzyme? ___________ Record your explanation in Table 6.1.

6.2 Effect of Temperature on Enzyme Activity

The active sites of enzymes increase the likelihood that substrate molecules will find each other and interact. Therefore, enzymes lower the energy of activation (the temperature needed for a reaction to occur). Still, increasing the temperature is expected to increase the likelihood that active sites will be occupied because molecules move about more rapidly as the temperature rises. In this way, a warm temperature increases enzyme activity.

The shape of an enzyme and its active site must be maintained or else they will no longer be functional. A very high temperature, such as the one that causes water to boil, is likely to cause weak bonds of a protein to break; and if this occurs, the enzyme **denatures**—it loses its original shape and the active site will no longer function to bring reactants together. Now enzyme activity plummets.

With this information in mind, examine Table 6.2 and hypothesize which tube (1, 2, or 3) will have more product per unit time as judged by bubble column height. Include a complete explanation in your hypothesis. ____

____

Experimental Procedure: Effect of Temperature

Label three clean test tubes (1 to 3), and use the appropriate graduated transfer pipet to add solutions to the test tubes as follows:

1. To each tube add 1 ml of catalase buffered at pH 7.0, the optimum pH for catalase.
2. Place tube 1 in a refrigerator or cold water bath, tube 2 in an incubator or warm water bath, and tube 3 in a boiling water bath. Complete the second column in Table 6.2. Wait 15 minutes.
3. As soon as you remove the tubes one at a time from the refrigerator, incubator, and boiling water, add 4 ml of hydrogen peroxide.
4. Swirl well to mix, and wait 20 seconds.
5. Measure the height of the bubble column in each tube, and record your results in Table 6.2. Plot your results in Figure 6.3.

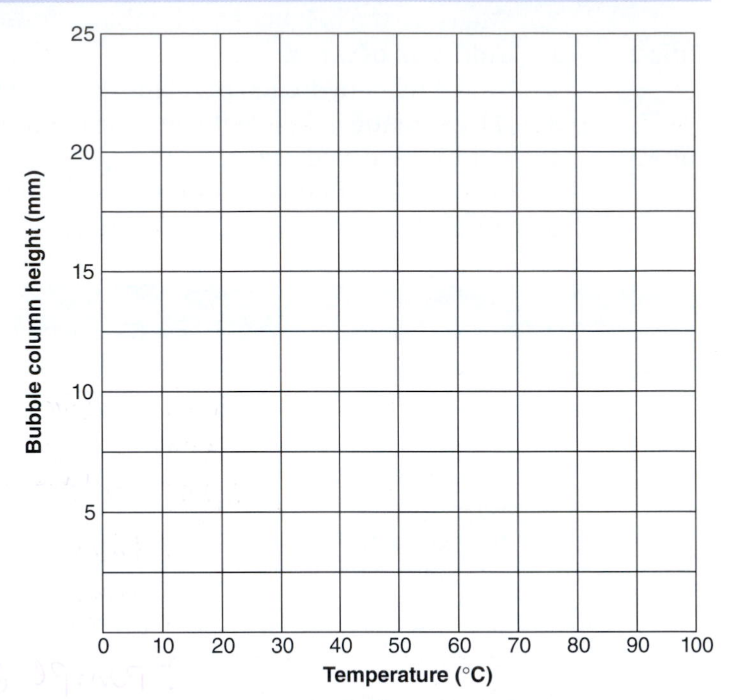

Figure 6.3 Effect of temperature on enzyme activity.

Table 6.2 Effect of Temperature

Tube	Temperature °C	Bubble Column Height (mm)	Explanation
1 Refrigerator			
2 Incubator			
3 Boiling water			

Conclusions: Effect of Temperature

- The bubble column height indicates the degree of enzyme activity. Was your hypothesis supported? __________ Explain in Table 6.2 the degree of enzyme activity per tube.
- What is your conclusion concerning the effect of temperature on enzyme activity? ____________________

____

6.3 Effect of Concentration on Enzyme Activity

Consider that if you increase the number of caretakers per number of children, it is more likely that each child will have quality time with a caretaker. So it is if you increase the amount of enzyme per amount of substrate—it is more likely that substrates will find the active site of an enzyme and a reaction will take place. With this in mind, examine Table 6.3 and hypothesize which tube (1, 2, or 3) will have more product per unit time as judged by bubble column height. Include a complete explanation in your hypothesis. ____

____

Experimental Procedure: Effect of Enzyme Concentration

Label three clean test tubes (1 to 3), and use the appropriate graduated transfer pipet to add solutions to the test tubes as follows:

Tube 1
1. Add 1 ml of water and 4 ml of hydrogen peroxide.
2. Swirl well to mix, and wait 20 seconds.
3. Measure the height of the bubble column, and record your results in Table 6.3.

Tube 2
1. Add 1 ml of buffered catalase and 4 ml of hydrogen peroxide.
2. Swirl well to mix, and wait 20 seconds.
3. Measure the height of the bubble column, and record your results in Table 6.3.

Tube 3
1. Add 3 ml of buffered catalase and 4 ml of hydrogen peroxide.
2. Swirl well to mix, and wait 20 seconds.
3. Measure the height of the bubble column, and record your results in Table 6.3.

Table 6.3 Effect of Enzyme Concentration

Tube	Amount of Enzyme	Bubble Column Height (mm)	Explanation
1	none		
2	1 cm		
3	3 cm		

Conclusions: Effect of Concentration

- The bubble column height indicates the degree of enzyme activity. Was your hypothesis supported? ____________ Explain in Table 6.3 the degree of enzyme activity per tube.
- If unlimited time was allotted, would the results be the same in all tubes? ____________ Explain why or why not. ____
- Would you expect similar results if the substrate concentration were varied in the same manner as the enzyme concentration? __________ Why or why not? ______________________
- What is your conclusion concerning the effect of concentration on enzyme activity? __________

____

6.4 Effect of pH on Enzyme Activity

> ⚠ **Hydrochloric acid (HCl)** used to produce an acid pH is a strong, caustic acid, and sodium hydroxide (NaOH) used to produce a basic pH is a strong, caustic base. Exercise care in using these chemicals, and follow your instructor's directions for disposal of tubes that contain these chemicals. If any acidic or basic solutions spill on your skin, rinse immediately with clear water.

Each enzyme has a pH at which the speed of the reaction is optimum (occurs best). Any higher or lower pH affects hydrogen bonding and the structure of the enzyme, leading to reduced activity.

Catalase is an enzyme found in cells where the pH is near 7 (called neutral pH). Other enzymes prefer different pHs. The pancreas secretes a slightly basic (below pH 7) juice into the digestive tract and the stomach wall releases a very acidic digestive juice which can be as low as pH 2. With this information about catalase in mind, examine Table 6.4 and hypothesize which tube (1, 2, or 3) will have more product per unit time as judged by the bubble column height. Include a complete explanation in your hypothesis. ____

____

Experimental Procedure: Effect of pH

Label three clean test tubes (1 to 3), and use the appropriate graduated transfer pipet to add solutions to the test tubes as follows:

Tube 1
1. Add 1 ml of nonbuffered catalase and 2 ml of water adjusted to pH 3 by the addition of HCl. Wait one minute.
2. Add 4 ml of hydrogen peroxide.
3. Swirl to mix, and wait 20 seconds.
4. Measure the height of the bubble column, and record your results in Table 6.4.

Tube 2
1. Add 1 ml of nonbuffered catalase and 2 ml of water adjusted to pH 7. Wait one minute.
2. Add 4 ml of hydrogen peroxide.
3. Swirl to mix, and wait 20 seconds.
4. Measure the height of the bubble column, and record your results in Table 6.4.

Tube 3
1. Add 1 ml of nonbuffered catalase and 2 ml of water adjusted to pH 11 by the addition of NaOH. Wait one minute.
2. Add 4 ml of hydrogen peroxide.
3. Swirl to mix, and wait 20 seconds.
4. Measure the height of the bubble column, and record your results in Table 6.4.

Plot your results from Table 6.4 here (Fig. 6.4).

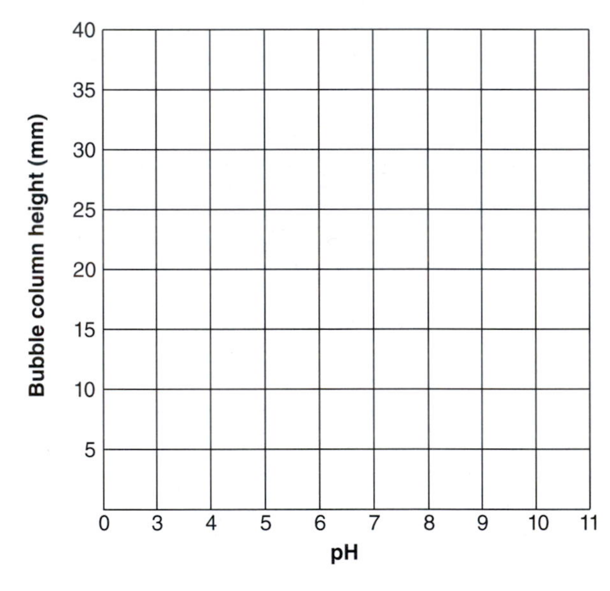

Figure 6.4 Effect of pH on enzyme activity.

Table 6.4 Effect of pH

Tube	pH	Bubble Column Height (mm)	Explanation
1	3		
2	7		
3	11		

Conclusions: Effect of pH

- The amount of bubble column height indicates the degree of enzyme activity. Was your hypothesis supported? ____________ Explain in Table 6.4 the degree of enzyme activity per tube.
- What is your conclusion concerning the effect of pH on enzyme activity? ____________

6.5 Factors That Affect Enzyme Activity

In Table 6.5, summarize what you have learned about factors that affect the speed of an enzymatic reaction. For example, in general, what type of temperature promotes enzyme activity, and what type inhibits enzyme activity? Answer similarly for enzyme or substrate concentration and pH.

Table 6.5 Factors That Affect Enzyme Activity

Factors	Promote Enzyme Activity	Inhibit Enzyme Activity
Enzyme specificity		
Temperature		
Enzyme or substrate concentration		
pH		

Conclusions: Factors That Affect Enzyme Activity

- Why does enzyme specificity promote enzyme activity? ____________
- Why does a warm temperature promote enzyme activity? ____________
- Why does increasing enzyme concentration promote enzyme activity? ____________
- Why does optimum pH promote enzyme activity? ____________

Virtual Lab

Enzyme-Controlled Reactions A virtual laboratory called Enzyme-Controlled Reactions is available on the *Concepts of Biology* website **www.mhhe.com/maderconcepts3**. After opening this virtual lab, click on the TV screen to watch a video about enzymes.

Experiment 1 Determining an Enzyme's Optimum pH

Hypothesis

Considering that the pH of the medium affects the shape of protein and that the shape of an enzyme is necessary to its enzymatic function, formulate a general hypothesis about the relationship between pH and enzyme function.

Hypothesis: ____.

Steps of the Experiment

In this experiment, it is possible to vary both the amount of substrate and the pH. Which of these factors should you hold constant during this experiment for determining an enzyme's optimum pH? ______________ Knowing how enzymes work, why might you decide to use the maximum concentration of substrate available? ______________

1. Sequentially change the pH to pH 3, pH 5, pH 7, pH 9, and pH 11 by using the down or up arrows beneath the test tubes.
2. Click and drag 8 g of substrate (far right only) to each test tube.

Results

3. Click the computer monitor to see the number of product molecules per minute formed for each test tube. In Table 1, sequentially enter your results per pH when the substrate is held constant at 8 g.

Table 1 Product/min/pH at Substrate Concentration of 8 g

	pH 3	pH 5	pH 7	pH 9	pH 11
Product/min					

Did the enzyme perform best at a particular pH? ______________ What pH? ______________

Use this graph to show your results:

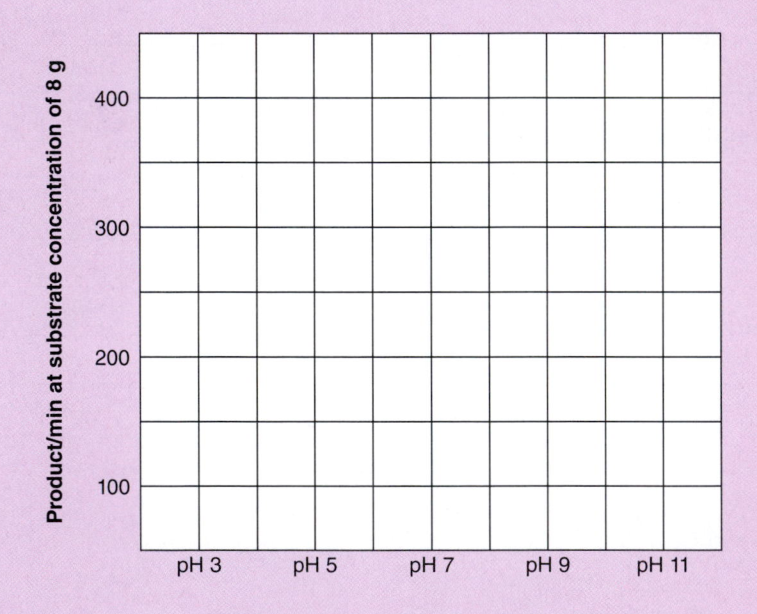

Conclusion

Results of this experiment (support or do not support) the hypothesis that an enzyme performs best at a particular pH. ______________ This particular pH is called the **optimum pH.**

Experiment 2 Determining the Optimum Substrate Concentration

Hypothesis

Knowing that enzymes perform best when their active site is always filled with substrate, hypothesize how increasing the substrate concentration could affect the product/min:

Hypothesis: ______________________________

Steps of the Experiment

In this experiment what experimental factor will you hold constant? ______________ Which pH would you use and why? ______________________________

1. Click the reset button and this time hold the pH constant. Note that all test tubes already indicate pH 7.

2. Click and drag substrate concentrations from 0.5 to 8.0 g to the test tubes.

Results

3. Click the computer monitor to see the number of product molecules formed per minute for each test tube. In Table 2, sequentially enter your results per amount of substrate when the pH is held constant at pH 7.

Table 2 Product/min at pH 7

Amount of Substrate	0.5 g	1.0 g	2.0 g	4.0 g	8.0 g
Product/min					

Use this graph to show your results:

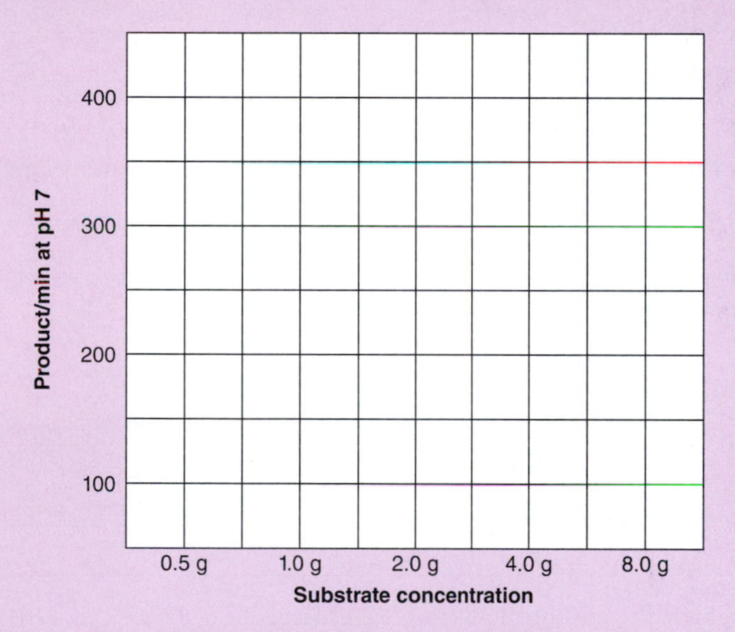

Did the amount of product/min continue to increase as substrate concentration increased to 8.0 g? Explain. ______________________________

Why do scientists favor the use of graphs to present their results? ______________________________

Conclusion

Results of this experiment (support or do not support) the hypothesis? ______________________________

Explain your results on the basis of the active site model. ______________________________

Laboratory Review 6

1. What happens at the active site of an enzyme? ______
2. On the basis of the active site, explain why the following conditions speed a chemical reaction:
   a. More enzyme ______
   b. More substrate ______
3. Name two other conditions (other than the ones mentioned in question 2) that maximize enzymatic reactions.
   a. ______
   b. ______
4. Explain the necessity for each of the two conditions you listed in question 3.
   a. ______
   b. ______
5. Lipase is a digestive enzyme that digests fat droplets in the basic conditions ($NaHCO_3$ is present) of the small intestine. Indicate which of the following test tubes would show digestion following incubation at 37°C, and explain why the others would not.

   Tube 1: Water, fat droplets ______

   Tube 2: Water, fat droplets, lipase ______

   Tube 3: Water, fat droplets, lipase, $NaHCO_3$ ______

   Tube 4: Water, lipase, $NaHCO_3$ ______
6. Fats are digested to fatty acids and glycerol. As the reaction described in question 5 proceeds, the solution will become what type pH? ______ Why? ______

   ______
7. Given the following reaction:

$$\underset{\text{hydrogen peroxide}}{2\,H_2O_2} \xrightarrow{\text{catalase}} \underset{\text{water}}{2\,H_2O} + \underset{\text{oxygen}}{O_2}$$

   a. Which substance is the substrate? ______
   b. Which substance is the enzyme? ______
   c. Which substances are the end products? ______
   d. Is this a synthetic or degradative reaction? ______

   How do you know? ______

Concepts of Biology Website

Instructors can find lab prep information and answers to all of the laboratory questions in the Laboratory Resource Guide. *Students* can practice their knowledge with quizzes, animations, flashcards, and much more.

www.mhhe.com/maderconcepts3

McGraw-Hill Access Science Website

An online encyclopedia of science and technology that provides information, including videos, that can enhance the laboratory experience.

www.accessscience.com

How Enzymes Function

LABORATORY

7

Photosynthesis

Learning Outcomes

7.1 Photosynthetic Pigments

- Describe the separation of plant pigments by paper chromatography. 68–69

7.2 Solar Energy

- Describe how white light is composed of many colors of light and how it benefits a plant to utilize several photosynthetic pigments. 70
- Explain an experiment that shows how solar energy promotes photosynthesis. 70–71
- Explain how you would use the same experimental setup to show that a plant carries on cellular respiration. 71

7.3 Carbon Dioxide Uptake

- Describe an experiment that indicates carbon dioxide is utilized during photosynthesis. 72

7.4 The Light Reactions and the Calvin Cycle Reactions

- Use the overall equation for photosynthesis to relate the light reactions to the Calvin cycle reactions. 73

Introduction

Photosynthesis involves the use of solar energy to produce a carbohydrate:

$$6\,CO_2 + 6\,H_2O \xrightarrow{\text{solar energy}} (C_6H_{12}O_6) + 6\,O_2$$

In this equation, glucose ($C_6H_{12}O_6$) appears as an end product of photosynthesis.

Photosynthesis takes place in chloroplasts (Fig. 7.1). Here membranous thylakoids are stacked in grana surrounded by the stroma. During the light reactions, pigments within the thylakoid membranes absorb solar energy, water is split, and oxygen is released. The Calvin cycle reactions occur within the stroma. During these reactions, carbon dioxide (CO_2) is reduced and solar energy is now stored in a carbohydrate (CH_2O).

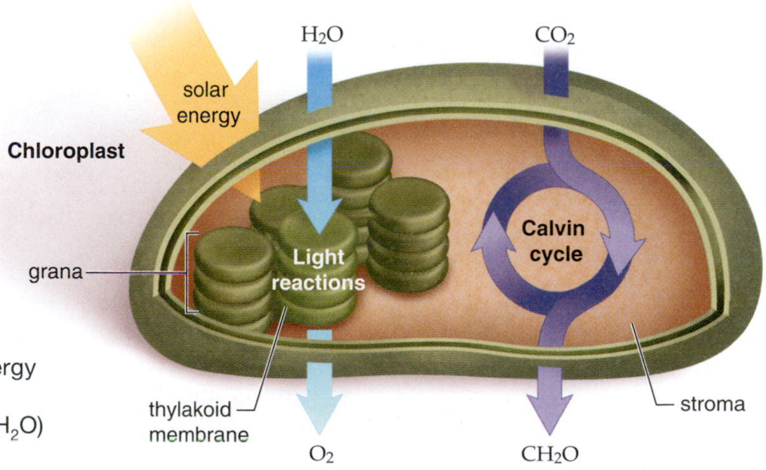

Figure 7.1 Overview of photosynthesis. Photosynthesis includes the light reactions when energy is collected and O_2 is released and the Calvin cycle reactions when CO_2 is reduced and carbohydrate (CH_2O) is formed.

7.1 Photosynthetic Pigments

Later in this laboratory, we will learn that sunlight is composed of different colors of light (see Fig. 7.3). We can hypothesize that leaves contain various pigments that absorb the solar energy of these different colors. Restate this hypothesis here:

Hypothesis: ____

To test this hypothesis, we will use a technique called chromatography to separate the pigments located in leaves. Chromatography separates molecules from each other on the basis of their solubility in particular solvents. The solvents used in the following Experimental Procedure are petroleum ether and acetone, which have no charged groups and are therefore nonpolar. As a nonpolar solvent moves up the chromatography paper, the pigment moves along with it. The more nonpolar a pigment, the more soluble it is in a nonpolar solvent and the faster and farther it proceeds up the chromatography paper.

Experimental Procedure: Photosynthetic Pigments

1. Assemble a **chromatography apparatus** (Fig. 7.2*a*):
   - Obtain a large, dry test tube and a cork with a hook.
   - Attach a strip of precut **chromatography paper** (hold from the top) to the hook and test for fit. The paper should hang straight down and barely touch the bottom of the test tube; trim if necessary.
   - Measure 2 cm from the bottom of the paper; place here a small dot with a pencil (not a pen).
   - With the stopper in place, mark the test tube with a wax pencil 1 cm below where the dot is on the paper.
   - Set the chromatography apparatus in a test tube rack.

Figure 7.2 Paper chromatography.
a. For a chromatography apparatus, the paper must be cut to size and arranged to hang down without touching the sides of a dry tube. **b.** The pigment (chlorophyll) solution is applied to a designated spot. **c.** The chromatogram, which develops after the spotted paper is suspended in the chromatography solution, will show these pigments.

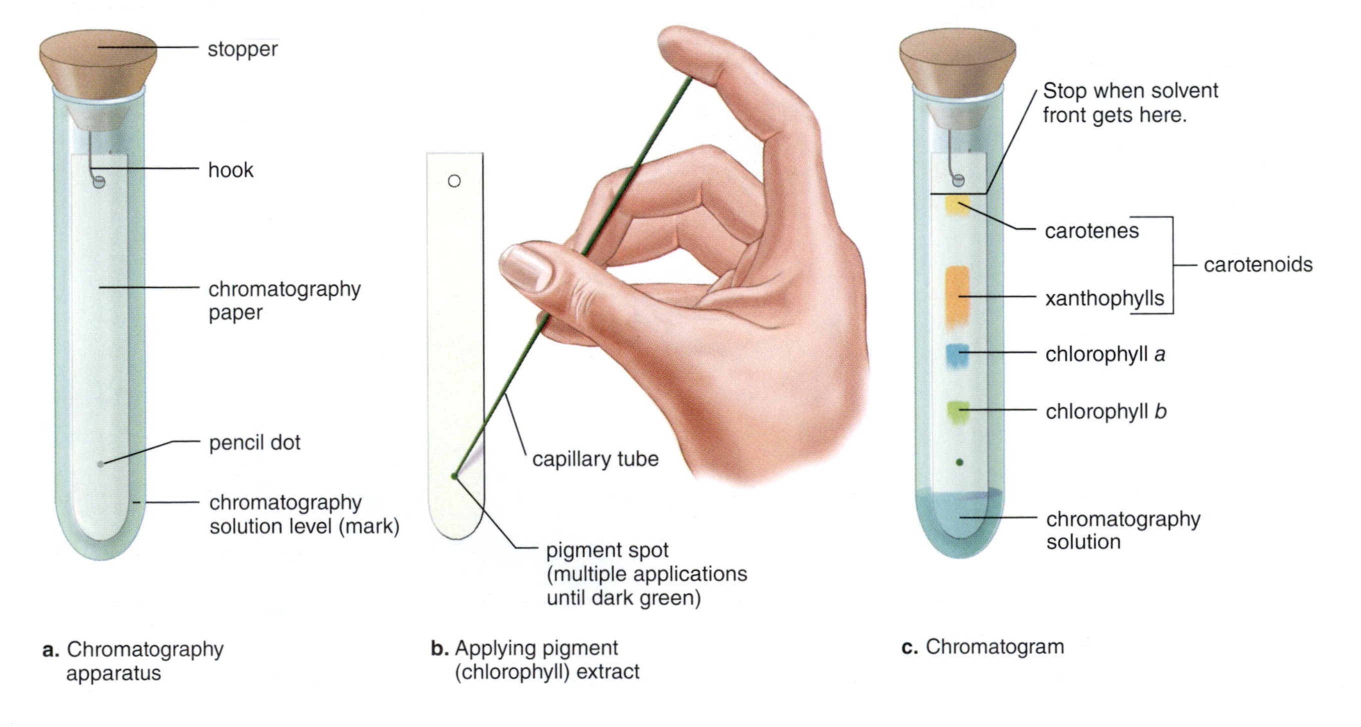

2. Prepare the chromatography paper (Fig. 7.2*b*):
   - Remove the chromatography paper from the test tube; place on a paper towel and apply **plant pigments** to the dot on the paper as directed by your instructor. Figure 7.2*b* shows how to apply pigment extract using a capillary tube.
   - In the **fume hood**, add **chromatography solution** up to the mark you made on the test tube. Place the chromatography paper attached to the hook back in the test tube. The pigment spot should remain above the chromatography solution, but close the chromatography apparatus tightly.

> ⚠ The chromatography solution is toxic and extremely flammable. Do not breathe the fumes, and do not place the chromatography solution near any source of heat. A fume (ventilation) hood is recommended.

3. Develop the chromatogram (Fig. 7.2*c*):
   - Place the reassembled chromatography apparatus in the test tube rack and allow 10 minutes for the chromatogram to develop, but check frequently so that the solution does not reach the top of the paper.
   - When the solvent has moved to within 1 cm of the upper edge of the paper, remove the paper. Close the empty apparatus tightly. With a pencil, lightly mark the location of the solvent front (where the solvent stopped on the paper) and allow the chromatogram to dry in the fume hood.
4. Read the chromatogram:
   - Compare your chromatogram to that shown in Figure 7.2*c*. Measure the distance in mm from the dark green pigment spot to the top of each individual pigment band, and record these values in Table 7.1.
   - Measure the distance the solvent moved from the dark green pigment spot to the solvent front and add this value to Table 7.1.
   - Use this formula to calculate the R_f (ratio-factor) values for each pigment, and record these values in Table 7.1:

$$R_f = \frac{\text{distance moved by pigment}}{\text{distance moved by solvent}}$$

Table 7.1 R_f (Ratio-Factor) Values for Each Pigment

Pigments	Distance Moved (mm)	R_f Values
Carotenes		
Xanthophylls		
Chlorophyll *a*		
Chlorophyll *b*		
Solvent		———

Conclusions: Photosynthetic Pigments

- Do your results support the hypothesis that plant leaves contain various photosynthetic pigments? ______________ Explain your answer. ______________________________
  ____

Figure 7.3 shows sunlight is made up of various colors. The chlorophylls absorb predominantly violet-blue and orange-red light. The carotenoids (carotenes and xanthophylls) absorb mostly blue-green light.

7.2 Solar Energy

During photosynthesis, solar energy is absorbed by the photosynthetic pigments and is transformed into the chemical energy of a carbohydrate (CH_2O). Without solar energy, photosynthesis would be impossible. Verify that photosynthesis releases oxygen by reviewing the overall equation for photosynthesis on page 67. Release of oxygen from a plant also indicates that the light reactions of photosynthesis are occurring. The next experiment on page 72 will utilize the same equipment and will illustrate that carbon dioxide is used during photosynthesis.

Role of White Light

White (sun) light contains different colors of light, as is demonstrated when white light passes through a prism (Fig. 7.3). White light is the best for photosynthesis because it contains all the colors of light. The various plant pigments absorb different portions of solar energy.

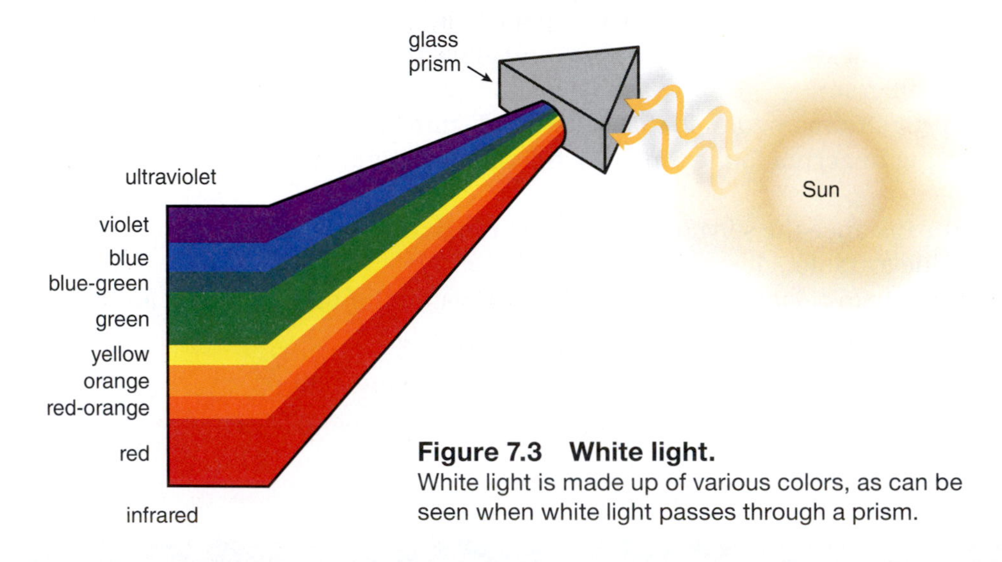

Figure 7.3 White light.
White light is made up of various colors, as can be seen when white light passes through a prism.

Experimental Procedure: White Light

1. Place a generous quantity of duckweed or *Elodea* with the cut end up (make sure the cuts are fresh) in a test tube with a rubber stopper containing a piece of glass tubing, as illustrated in Figure 7.3. When assembled, this is your volumeter for studying the need for light in photosynthesis. (Do not hold the volumeter in your hand, as body heat will also drive the reaction forward.) Your instructor will show you how to fix the volumeter in an upright position.
2. Before stoppering the test tube, add sufficient 3% sodium bicarbonate ($NaHCO_3$)—a source of carbon dioxide—solution so that, when the rubber stopper is inserted into the tube, the solution comes to rest at about ¼ the length of the upright glass tubing. Mark this location on the glass tubing with a wax pencil.
3. Place a beaker of plain water next to the *Elodea* tube to serve as a heat absorber. Place a lamp (150 watt) next to the beaker. The tube, beaker, and lamp should be as close to one another as possible.
4. Turn on the lamp. As soon as the edge of the solution in the tubing begins to move, time the reaction for 10 minutes. Be careful not to bump the tubing or to readjust the stopper, or your readings will be altered. After 10 minutes, mark the edge of the solution, and measure in millimeters the distance the edge moved upward: __________ mm/10 min. This is **net photosynthesis,** a measurement that does not take into account the oxygen that was used up for cellular respiration. Record your results in Table 7.2. Why did the edge move upward? ____

____

Figure 7.4 Volumeter. A volumeter apparatus is used to study the role of light in photosynthesis.

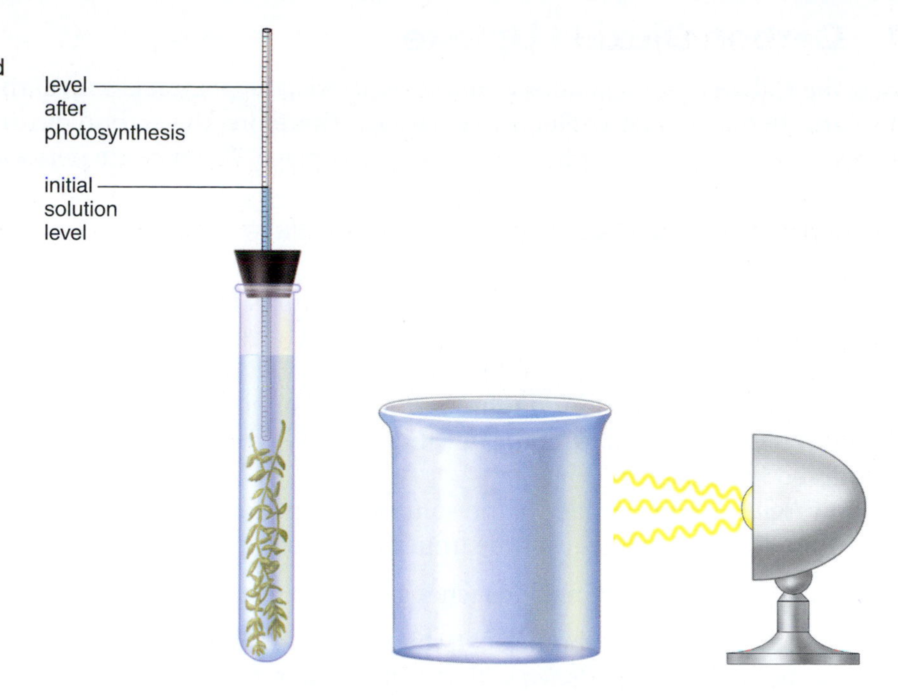

5. Turn off the lamp. Carefully wrap the tube containing *Elodea* in aluminum foil, and record here the length of time it takes for the edge of the solution in the tubing to move downward 1 mm: ________. Convert your measurement to ________ mm/10 min, and record this value for **cellular respiration** in Table 7.2. (Do not use a minus sign, even though the edge moved downward.) Why does cellular respiration, which occurs in a plant whether it is light or dark, cause the edge to move downward?

____

What organelle takes in oxygen and carries on cellular respiration? ________________

6. If the *Elodea* had *not* been respiring in step 4, how far would the edge have moved upward? ________ mm/10 min. This is **gross photosynthesis** (net photosynthesis + cellular respiration). Record this number in Table 7.2.

7. Calculate the **rate of photosynthesis** (mm/hr) by multiplying gross photosynthesis (mm/10 min) by 6 (that is, 10 min × 6 = 60 min = 1 hr): ________ mm/hr. Record this value in Table 7.2.

Table 7.2 Rate of Photosynthesis (White Light)

	Movement of Edge (mm/10 min)	Rate of Photosynthesis (mm/hr)
Net photosynthesis (white light)		________
Cellular respiration (no light)		________
Gross photosynthesis (net + cellular respiration)		

Conclusion

- This experiment showed that solar energy must be present for a plant to photosynthesize. How so?

____

7.3 Carbon Dioxide Uptake

During the Calvin cycle reactions of photosynthesis, the plant takes up carbon dioxide (CO_2) and reduces it to a carbohydrate, such as glucose ($C_6H_{12}O_6$). Therefore, the carbon dioxide in the solution surrounding *Elodea* should disappear as photosynthesis takes place. To illustrate perform the following experiment.

Experimental Procedure: Carbon Dioxide Uptake

> ⚠ **Phenol red** Avoid ingestion, inhalation, and contact with skin, eyes, and mucous membranes. Follow your instructor's directions for disposal of this chemical. Use protective eyewear when performing this experiment.

1. Temporarily remove the *Elodea* from the test tube. Empty the sodium bicarbonate ($NaHCO_3$) solution from the test tube, rinse the test tube thoroughly, and fill with a phenol red solution diluted to a faint pink. (Add more water if the solution is too dark.) Phenol red is a pH indicator that turns yellow in an acid and red in a base.
2. Blow *lightly* on the surface of the solution. Stop blowing as soon as the surface color changes to yellow. Then shake the test tube until the rest of the solution turns yellow.

   Blowing onto the solution adds what gas to the test tube? __________ When carbon dioxide combines with water, it forms carbonic acid; therefore, the solution appears yellow.
3. Thoroughly rinse the *Elodea* with distilled water, return it to the test tube, which now contains a yellow solution, and assemble your volumeter as before.
4. The water in the beaker used to absorb heat should be clear.
5. Turn on the lamp, and wait until the edge of the solution just begins to move upward. Note the time.

   Observe until you note a change in color. How long did it take for the color to change? ______________
6. Hypothesize why the solution in the test tube eventually turns red. ______________________________

   ____
7. The carbon cycle includes all the many ways that organisms exchange carbon dioxide with the atmosphere. Figure 7.5 notes the relationship between cellular respiration and photosynthesis. Animals produce carbon dioxide used by plants to carry out photosynthesis. Plants produce the food (and oxygen) that they and animals require to carry out cellular respiration. Therefore, the same carbon atoms pass between animals and plants and between plants and animals (Fig. 7.5).

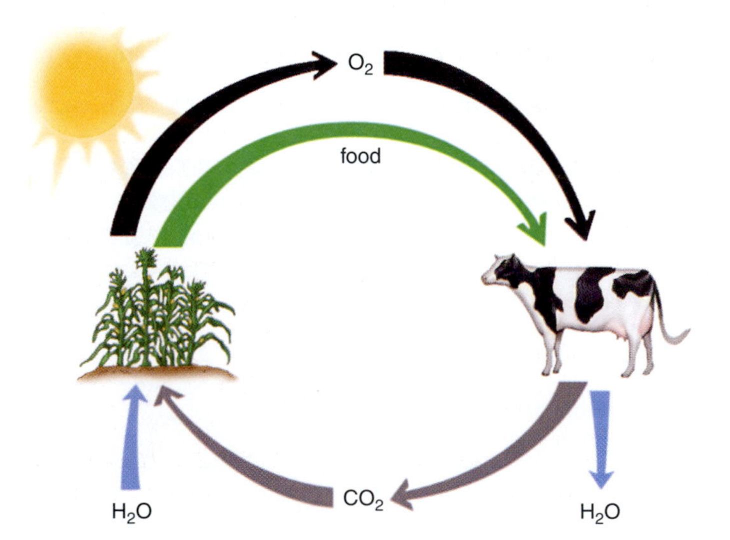

Figure 7.5 Photosynthesis and cellular respiration.
Animals are dependent on plants for a supply of oxygen, and plants are dependent on animals for a supply of carbon dioxide.

7.4 The Light Reactions and the Calvin Cycle Reactions

Review the introduction to this laboratory. Note the overall equation for photosynthesis and that photosynthesis consists of the light reactions and the Calvin cycle reactions. Solar energy is absorbed by photosynthetic pigments during the light reactions and energy is used during the Calvin cycle reactions to reduce carbon dioxide to a carbohydrate.

Light Reactions

1. Photosynthetic pigments
   a. What is the function of the photosynthetic pigments in photosynthesis? ______________________

   ____
   b. How does it benefit a plant to have a variety of photosynthetic pigments? ______________________

   ____
2. Water and oxygen
   a. What happens to water during the light reactions? ______________________
   b. What happens to the released oxygen? ______________________
3. Location of the light reactions

   Fill in the blank: The light reactions take place in the ______________________ membranes.
4. In your own words, summarize the light reactions based on this laboratory.

Calvin Cycle Reactions

1. Carbon dioxide and carbohydrate

   What happens to carbon dioxide after it is taken up during the Calvin cycle reactions? ______________________

   ____
2. Location of the Calvin cycle reaction in a chloroplast

   Fill in the blank: The Calvin cycle reactions take place in the ______________________ of a chloroplast.
3. In your own words, summarize the Calvin cycle reactions based on this laboratory.

Light Reactions and the Calvin Cycle Reactions

1. Examine the overall equation for photosynthesis and show that there is a relationship between the light reactions and the Calvin cycle reactions by drawing an arrow between the hydrogen atoms in water and the hydrogen atoms in the carbohydrate.

$$CO_2 + H_2O \longrightarrow (CH_2O) + O_2$$

2. Only because solar energy splits water can hydrogen atoms be used to reduce carbon dioxide. In this sense, solar energy is now stored in the carbohydrate. This energy sustains all the organisms in the biosphere.

Laboratory Review 7

1. How are plant pigments involved in photosynthesis? ______________________________

____

2. Why is it beneficial to have several different plant pigments involved in photosynthesis? ______________________________

____

3. On what basis does chromatography separate substances? ______________________________

____

4. Consider the following reaction:

$$CO_2 + H_2O \rightleftarrows \underset{\text{carbonic acid}}{H_2CO_3} \rightleftarrows HCO_3^-$$

5. Phenol red, a pH indicator, turns yellow (indicating acid) when you breathe into a solution. How does the reaction explain why the solution turned acidic? ______________________________

____

6. Phenol red turns back to red when light is present and a plant is added to the solution. In terms of the reaction, why does this occur? ______________________________

____

7. Gas exchange occurs in both photosynthesis and cellular respiration. Contrast these two processes by completing the following table:

	Organelle	Gas Given Off	Gas Taken Off
Photosynthesis			
Cellular respiration			

8. What experimental conditions were used in this laboratory to test for cellular respiration in plant cells? ______________________________

____

9. Suppose you replaced *Elodea* with animal cells in the experimental test tube from #7. Would the results differ according to the use of a white light or no light? __________ Explain. ______________________________

____

***Concepts of Biology* Website**

Instructors can find lab prep information and answers to all of the laboratory questions in the Laboratory Resource Guide. *Students* can practice their knowledge with quizzes, animations, flashcards, and much more.

www.mhhe.com/maderconcepts3

McGraw-Hill Access Science Website

An online encyclopedia of science and technology that provides information, including videos, that can enhance the laboratory experience.

www.accessscience.com

Photosynthesis

LABORATORY

8

Organization of Flowering Plants

Learning Outcomes

Introduction

Despite their great diversity in size and shape, all flowering plants have three vegetative organs that have nothing to do with reproduction: the root, the stem, and the leaf (Fig. 8.1). Roots anchor a plant and absorb water and minerals from the soil. A stem transports substances and supports the leaves so that they are exposed to sunlight. Leaves carry on photosynthesis, and thereby produce the nutrients that sustain a plant and allow it to grow.

Figure 8.1 Organization of plant body. Roots, stems, and leaves—the vegetative organs of a plant—are shown in this onion plant.

8.1 External Anatomy of a Flowering Plant

Figure 8.2 shows that a plant has a root system and a shoot system. The **root system** consists of the roots. The **shoot system** consists of the stem and the leaves.

Observation: A Living Plant

Shoot System

What structures are in the shoot system?

____

____

____

The Leaves

Leaves carry on photosynthesis. Which part of a leaf (blade or petiole) is the most expansive part?

____

____

The Stem

1. In the **stem**, locate a **node** and an **internode**.
2. Measure the length of the internode in the middle of the stem. Does the internode get larger or smaller toward the apex of the stem? ____________ Toward the roots? _______

   Based on the fact that a stem elongates as it grows, explain your observation. ____________

   ____

   ____

   ____

3. Where is the **terminal bud** (also called the shoot tip) of a stem? ____________________

   ____

   Where are axillary buds? ____________________

   ____

Root System

Observe the root system of a living plant if the root system is exposed. Does this plant have a strong primary root or many roots of the same size?

____

What other structures are in the root system?

____

____

Where is the root tip? ____________________

Figure 8.2 Organization of a plant.
Roots, stems, and leaves are vegetative organs.

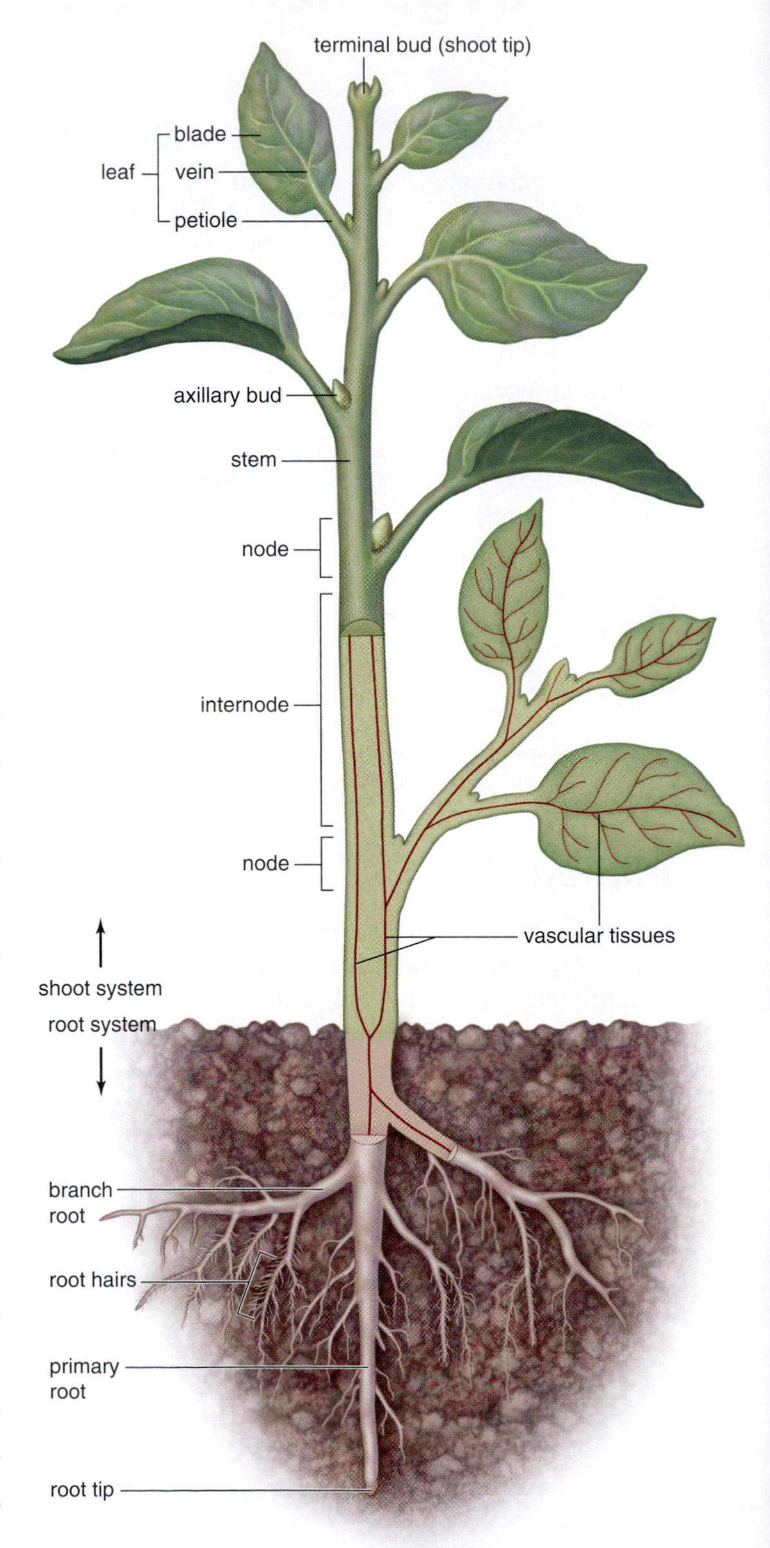

8.2 Major Tissues of Roots, Stems, and Leaves

Unlike humans, flowering plants grow in size their entire life because they have an immature tissue called **meristematic tissue** composed of cells that divide. **Apical meristem** is located at the terminal end of the stem, the branches, and at the root tip and the root branches. When apical meristem cells divide, some of the cells differentiate into the mature tissues of a plant:

Dermal tissue: Forms the outer protective covering of a plant organ

Ground tissue: Fills the interior of a plant organ; photosynthesizes and stores the products of photosynthesis

Vascular tissue: Transports water and sugar, the product of photosynthesis, in a plant and provides support

Note in Table 8.1 that roots, stems, and leaves have all three tissues but they may be given different specific names.

Table 8.1 Mature Tissues of Vegetative Organs

Tissue Type	Roots	Stems	Leaves
1. Dermal tissue (epidermis)	Protect inner tissues Root hairs absorb water and minerals.	Protect inner tissues	Protect inner tissues Cuticle prevents H_2O loss. Stomata carry on gas exchange.
2. Ground tissue	Cortex: Store products of photosynthesis Pith: Store products of photosynthesis	Cortex: Carry on photosynthesis, if green Pith: Store products of photosynthesis	Mesophyll: Photosynthesis
3. Vascular tissue (xylem and phloem)	Vascular cylinder: Transport water and nutrients	Vascular bundle: Transport water and nutrients	Leaf vein: Transport water and nutrients

Observation: Tissues of Roots, Stems, and Leaves

Meristematic tissue. View a slide on demonstration showing the apical meristem in a shoot tip and another showing the apical meristem in a root tip. Meristematic cells are spherical and stain well when they are dense and have thin cell walls (Fig. 8.3). What is the function of meristematic tissue? ______________________

____

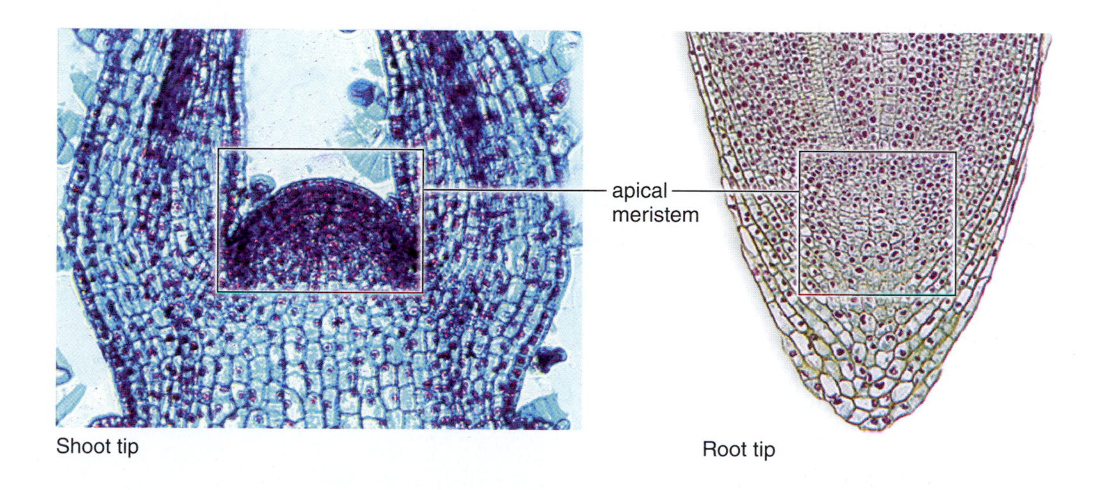

Figure 8.3 Apical meristem. A shoot tip and a root tip contain meristem tissue, which allows them to grow longer the entire life of a plant.

Mature Tissues

1. Dermal tissue. View a cross-section slide of a leaf (Fig. 8.4), and focus only on the upper or lower epidermis (Fig. 8.4). Epidermal cells tend to be square or rectangular in shape. In a leaf, the epidermis is interrupted by openings called **stomata** (sing., stoma). Later you will have an opportunity to see the epidermis in roots, stems, and leaves. What is a function of epidermis in all three organs (see Table 8.1)?

   ____

   ____

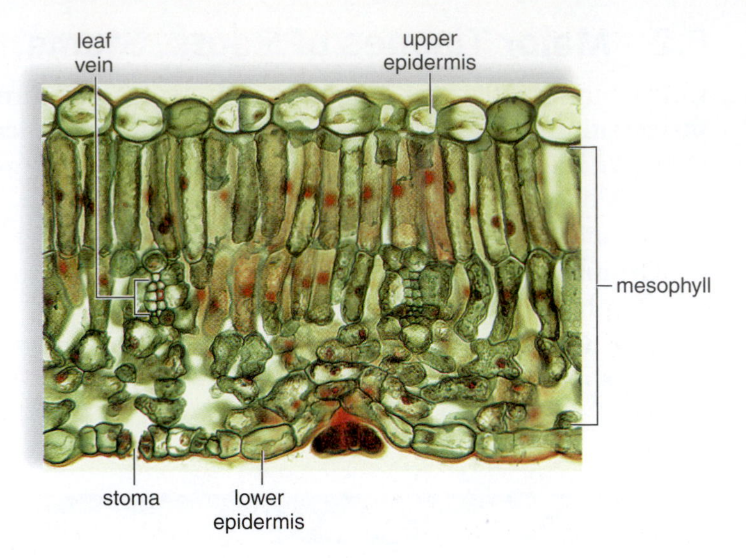

Figure 8.4 Microscopic leaf structure.
Like the stem and root, a leaf contains epidermal tissue, vascular tissue (leaf vein), and ground tissue (mesophyll).

2. Ground tissue. The ground tissue fills the space between epidermis in roots, stems, and leaves. In leaves, for example, ground tissue is called mesophyll (see Fig. 8.4). Ground tissue largely contains parenchyma cells and sclerenchyma cells. **Parenchyma cells** can be of different sizes and vary from fairly circular to oval. Those that contain chloroplasts carry on photosynthesis. Those that contain leucoplasts store starches and oils. **Sclerenchyma cells** are usually elongate in shape and have thick walls impregnated with lignin. These dead cells appear hollow and the presence of lignin means that they stain a red color. Sclerenchyma cells are strong and provide support. Which type cell (parenchyma or sclerenchyma) carries on photosynthesis or stores the products of photosynthesis in a leaf? ______________________ Which one lends strength to ground tissue in roots and stems? ______________________
3. Vascular tissue. In a leaf, strands of vascular tissue are called leaf veins (see Fig. 8.4). There are two types of vascular tissue, called xylem and phloem. **Xylem** contains hollow dead cells that transport water. The presence of lignin makes the cell walls strong, stains red, and makes xylem easy to spot. **Phloem** contains thin-walled, smaller living cells that transport sugars in a plant. Phloem is harder to locate than xylem, but it is always found in association with xylem. Which type tissue (xylem or phloem) transports sugars in a plant? ____

   Which type tissue transports water? ____

Monocots Versus Eudicots

Flowering plants are classified into two major groups: **monocots** and **eudicots.** In this laboratory, you will be studying the differences between monocots and eudicots with regard to the roots, stems, and leaves as noted in Figure 8.5.

Experimental Procedure: Monocot Versus Eudicot

1. Examine the live plant again (see Fig. 8.2). The leaf vein pattern—that is, whether the veins run parallel to one another or whether the veins spread out from a central location (called the net

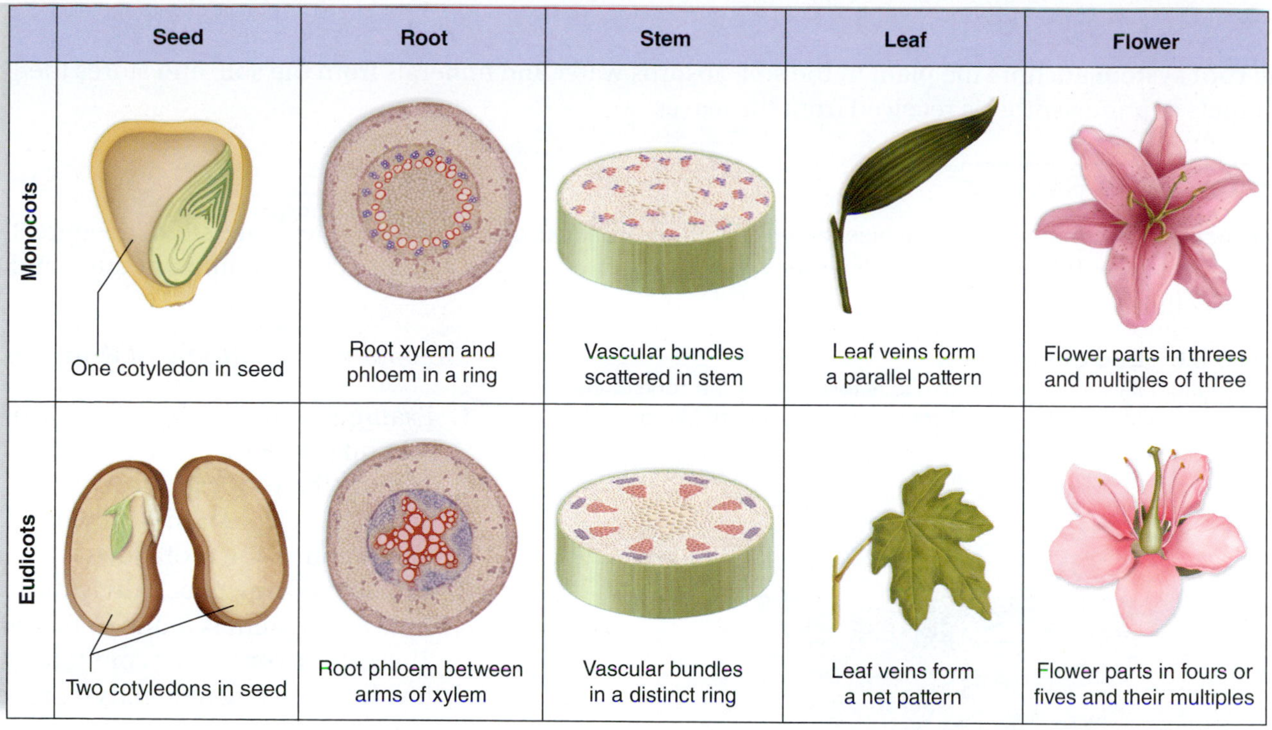

	Seed	Root	Stem	Leaf	Flower
Monocots	One cotyledon in seed	Root xylem and phloem in a ring	Vascular bundles scattered in stem	Leaf veins form a parallel pattern	Flower parts in threes and multiples of three
Eudicots	Two cotyledons in seed	Root phloem between arms of xylem	Vascular bundles in a distinct ring	Leaf veins form a net pattern	Flower parts in fours or fives and their multiples

Figure 8.5 Monocots versus eudicots.
The five features illustrated here are used to distinguish monocots from eudicots.

pattern)—indicates that a plant is either a monocot or a eudicot. Is the plant in Figure 8.2 a monocot or eudicot? ________________ The leaf pattern in Figure 8.4 appears to be parallel. Is this the leaf of a monocot or eudicot? ________________

2. Observe any other available plants, and note in Table 8.2 the name of the plant and whether it is a monocot or a eudicot based on leaf vein pattern.
3. Aside from leaf vein pattern, other external features indicate whether a plant is a monocot or a eudicot. For example, open a peanut if available. The two halves you see are cotyledons. Is the plant that produced the peanut a monocot or eudicot? ________________
   If available, examine a flower. If a flower has three petals or six petals, or any multiple of three, is the plant that produced this flower a monocot or a eudicot? ________________
4. In this laboratory you (pp. 87–89) will have the opportunity to examine the cross sections of roots and stems microscopically; the arrangement of vascular tissue roots and stems also indicates whether a plant is a monocot or a eudicot.

Table 8.2 Monocots Versus Eudicots

Name of Plant	Organization of Leaf Veins	Monocot or Eudicot?

8.3 Root System

The **root system** anchors the plant in the soil, absorbs water and minerals from the soil, and stores the products of photosynthesis received from the leaves.

Anatomy of a Root Tip

Primary growth of a plant increases its length. Note the location of the root apical meristem in Figure 8.6. As primary growth occurs, root cells enter zones that correspond to various stages of differentiation and specialization.

Figure 8.6 Eudicot root tip.
a. In longitudinal section, the root cap is followed by the zone of cell division, zone of elongation, and zone of maturation. **b.** Micrograph.

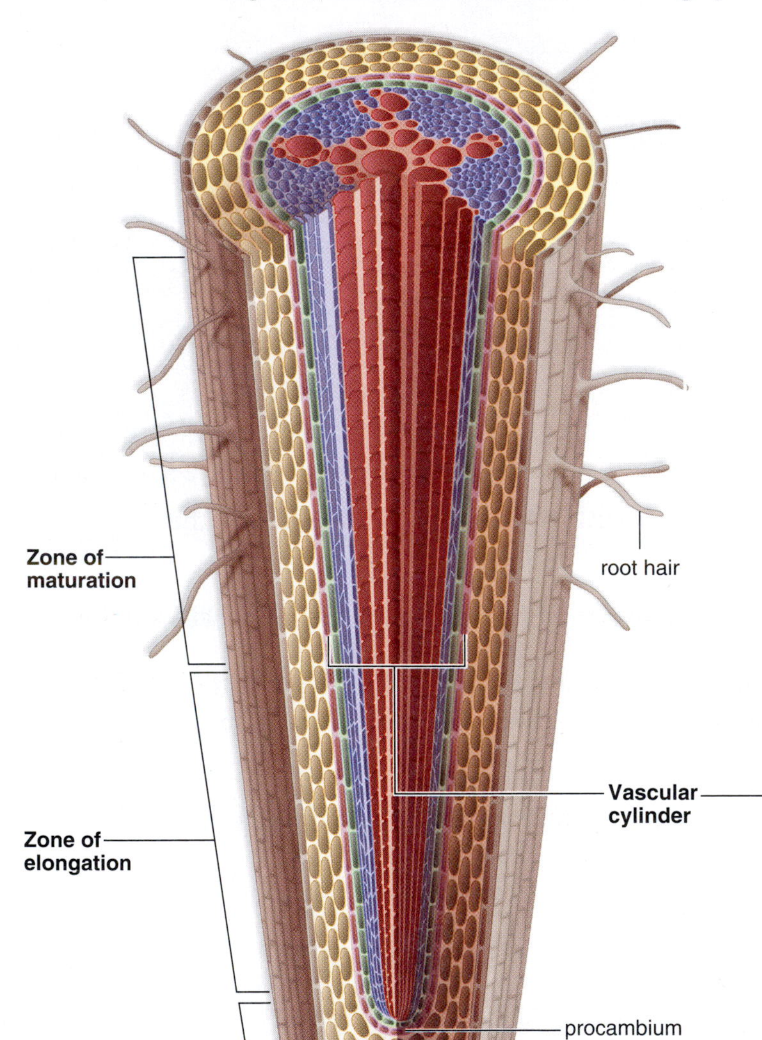

Observation: Anatomy of Root Tip

1. Examine a model and/or a slide of a root tip (Fig. 8.6).
2. Identify the **root cap** (dead cells at the tip of a plant that provide protection as the root grows).
3. Locate the **zone of cell division.** Apical meristem is found in this zone. These embryonic cells continually divide, providing new cells for root growth.
4. Find the **zone of elongation.** In this zone, the newly produced cells elongate as they begin to grow larger.
5. Identify the **zone of maturation.** In this zone, the cells become particular cell types. When epidermal cells differentiate, they produce **root hairs**, small extensions that absorb water and minerals. Also noticeable is the newly formed vascular cylinder.

Anatomy of Eudicot and Monocot Roots

Eudicot and monocot roots differ in the arrangement of their vascular tissue.

Observation: Cross-Section Anatomy of Eudicot and Monocot Roots

Eudicot Root

1. Obtain a prepared cross-section slide of a buttercup *(Ranunculus)* root. Use both low power and high power to identify the **epidermis** (the outermost layer of small cells that gives rise to root hairs). The epidermis protects inner tissues and absorbs water and minerals.
2. Locate the **cortex,** which consists of several layers of thin-walled cells (Fig. 8.7*a, b*). In Figure 8.7*b,* note the many stained starch grains in the cortex cells. The cortex is ground tissue that functions in food storage.
3. Find the **endodermis,** a single layer of cells whose walls are thickened by a layer of waxy material known as the **Casparian strip.** (It is as though these cells are glued together with a waxy glue.) Because of the Casparian strip, the only access to the xylem is through the living endodermal cells. The endodermis regulates what materials that enter a plant through the root? ______________________________

   ____

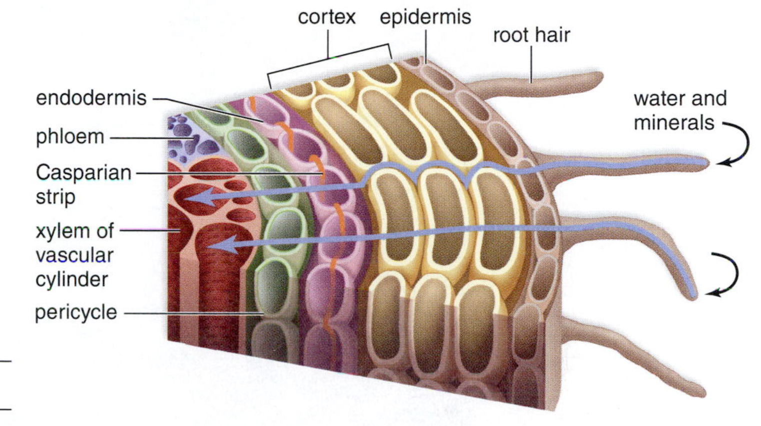

   Use this illustration to trace the path of water and minerals from the root hairs to xylem. ______________________________

4. Identify the **pericycle,** a layer one or two cells thick just inside the endodermis. Branch roots originate from this tissue.
5. Locate the **xylem** in the vascular cylinder of the root. Xylem has several "arms" that extend like the spokes of a wheel. This tissue conducts water and minerals from the roots to the stem.
6. Find the **phloem,** located between the arms of the xylem. Phloem conducts organic nutrients from the leaves to the roots and other parts of the plant.

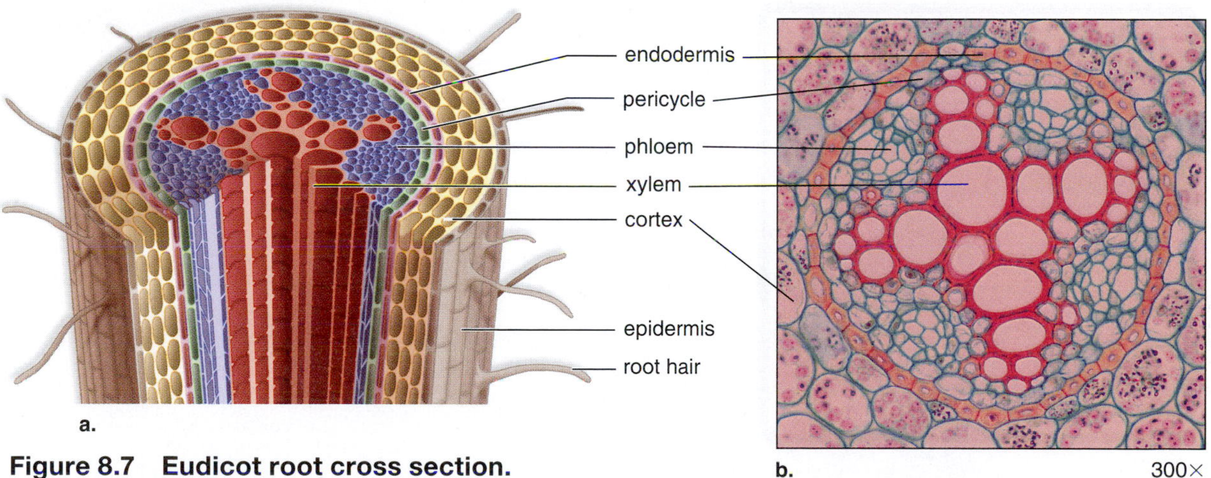

Figure 8.7 Eudicot root cross section. The vascular cylinder of a dicot root contains the vascular tissue. Xylem is typically star-shaped, and phloem lies between the points of the star. **a.** Drawing. **b.** Micrograph.

Monocot Root

1. Obtain a prepared cross-section slide of a corn *(Zea mays)* root (Fig. 8.8*a, b*). Use both low power and high power to identify the six tissues mentioned for the eudicot root.
2. In addition, identify the **pith,** a centrally located ground tissue that functions in food storage.

Figure 8.8 Monocot root cross section.
a. Micrograph of a monocot root cross section. **b.** An enlarged portion.

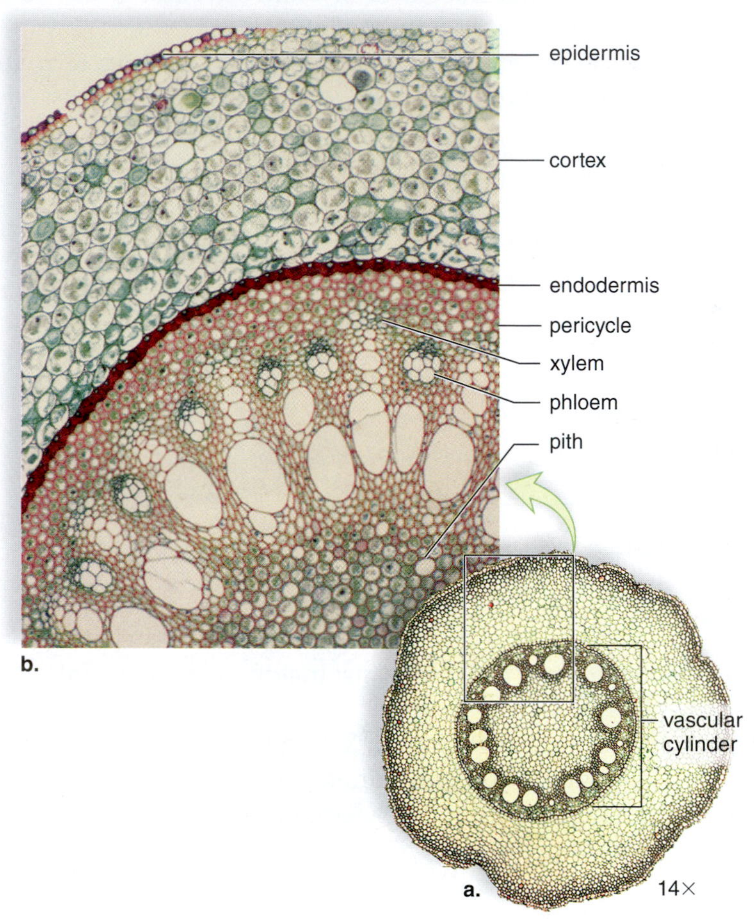

Comparison

Contrast the arrangement of vascular tissue (xylem and phloem) in the vascular cylinder of monocot roots and eudicot roots by writing "monocot" or "eudicot" on the appropriate line.

____________________ Xylem has the appearance of a wheel. Phloem is between the spokes of the wheel.

____________________ Ring of xylem (inside) and phloem (outside) surrounds pith.

Root Diversity

Roots are quite diverse, and we will take this opportunity to become acquainted with only a few select types.

Observation: Root Diversity

1. Taproots and fibrous roots. Most plants have either a taproot or a fibrous root. Note in Figure 8.9*a* that carrots have a **taproot.** The main root is many times larger than the branch roots. Grasses such as in Figure 8.9*b* have a **fibrous root:** All the roots are approximately the same size.

   Examine the taproots on display, and name one or two in which the taproot is enlarged for storage.

   ____

   The dandelion on display has a taproot. Describe. ______________________________

   ____

2. Adventitious roots. Some plants have adventitious roots. Roots that develop from nonroot tissues, such as nodes of stems, are called **adventitious** roots. Examples include the prop roots of corn (Fig. 8.9*c*) and the aerial roots of ivy that attach this plant to structures such as stone walls.

   Which plants on display have adventitious roots? ______________________________

3. Other types of roots. Mangroves and other swamp-dwelling trees have roots called pneumatophores that rise above the water line. Pneumatophores have numerous lenticels, which are openings that allow gas exchange to occur.

   What root modifications not noted here are on display in your laboratory? ______________________________

   ____

Figure 8.9 Root diversity.
a. Carrots have a taproot. **b.** Grass has a fibrous root system. **c.** A corn plant has prop roots, and **d.** black mangroves have pneumatophores to increase their intake of oxygen.

a. Taproot system

b. Fibrous root system

c. Prop roots, a type of adventitious root

d. Pneumatophores of black mangrove trees

8.4 Stems

Stems are usually found aboveground where they provide support for leaves and flowers. Vascular tissue extends from the roots, through the stem and its branches to the leaves. Therefore, what function do botanists assign to stems in addition to support for branches and leaves? ______________________________ ______________________________ Explain why a branch cannot live if severed from the rest of the plant. ____

Stems that do not contain wood are called **herbaceous,** or nonwoody, stems. Usually, monocots remain herbaceous throughout their lives. Some eudicots, such as those that live a single season, are also herbaceous. Other eudicots, namely trees, become woody as they mature.

Anatomy of Herbaceous Stems

Herbaceous stems undergo primary growth. **Primary growth** results in an increase in length due to the activity of the **apical meristem** located in the terminal bud (see Fig. 8.2*a*) of the shoot system.

Observation: Anatomy of Eudicot and Monocot Herbaceous Stems

Eudicot Herbaceous Stem

1. Examine a prepared slide of a eudicot herbaceous stem (Fig. 8.10), and identify the **epidermis** (the outer protective layer). *Label the epidermis in Figure 8.10*a.
2. Locate the **cortex,** which may photosynthesize or store nutrients.
3. Find a **vascular bundle,** which transports water and organic nutrients. The vascular bundles in a eudicot herbaceous stem occur in a ring pattern. *Label the vascular bundle in Figure 8.10*a. Which vascular tissue (xylem or phloem) is closer to the surface? ______________________________
4. *Label the central* ***pith,*** which stores organic nutrients. Both cortex and pith are composed of which tissue type listed in Table 8.1? ______________________________

Figure 8.10 Eudicot herbaceous stem.
The vascular bundles are in a definite ring in this photomicrograph of a eudicot herbaceous stem. Complete the labeling as directed by the Observation.

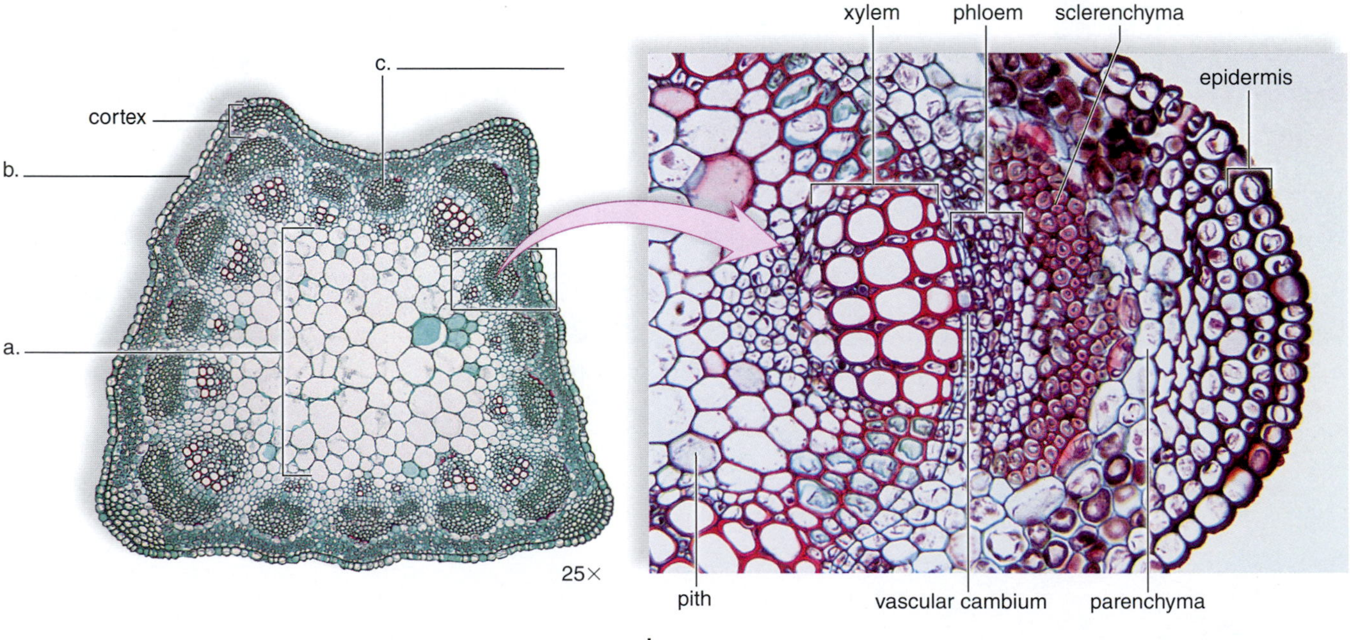

Monocot Herbaceous Stem

1. Examine a prepared slide of a monocot herbaceous stem (Fig. 8.11). Locate the epidermal, ground, and vascular tissues in the stem.
2. The vascular bundles in a monocot herbaceous stem are said to be scattered. Explain. ______________________________

____

Figure 8.11 Monocot stem.
The vascular bundles, one of which is enlarged, are scattered in this photomicrograph of a monocot herbaceous stem.

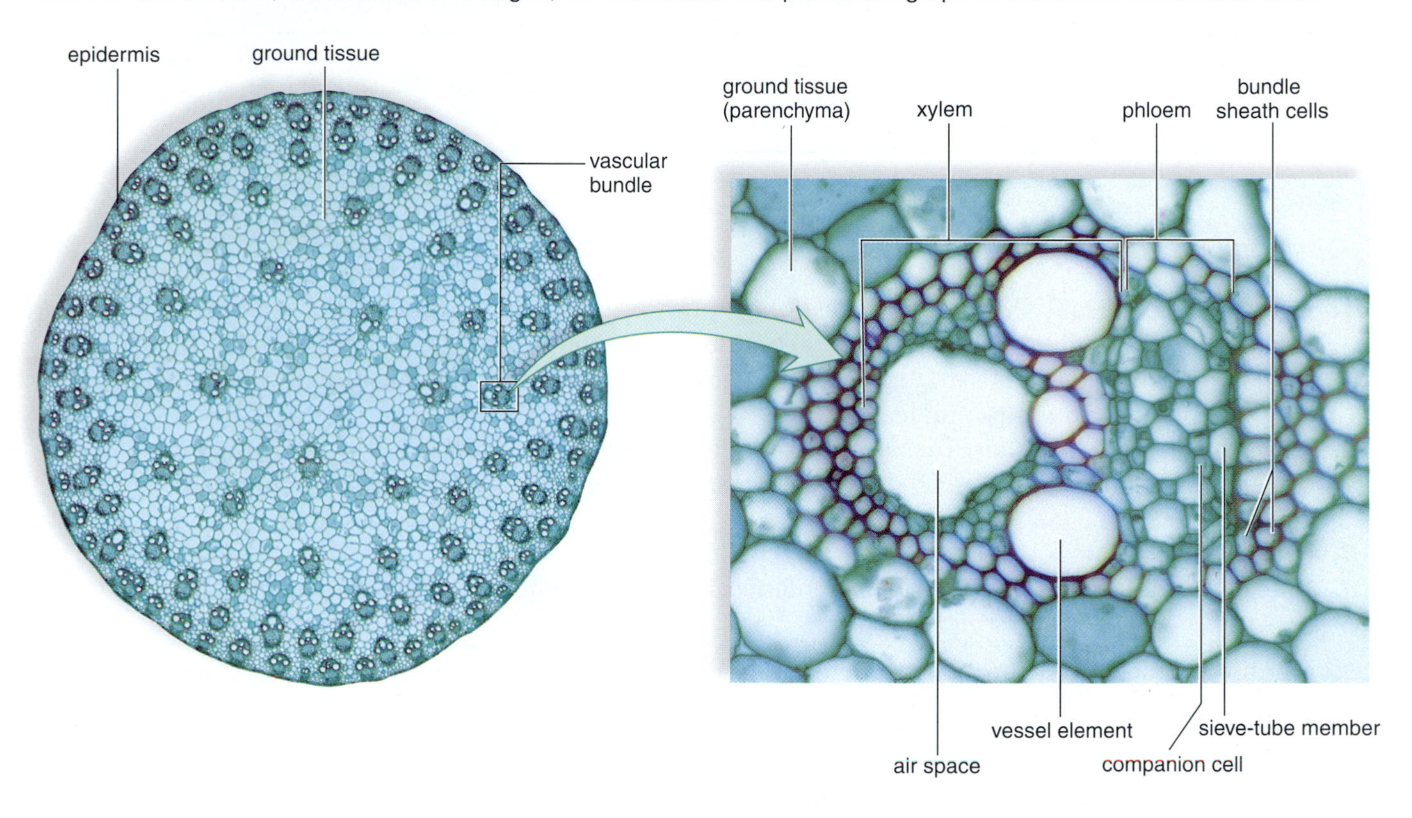

Comparison

1. Compare the arrangement of ground tissue in eudicot and monocot stems. ______________________________

____

2. Compare the arrangement of vascular bundles in the stems of eudicots and monocots. ______________________________

____

Anatomy of Woody Stems

Woody stems undergo both primary growth (increase in length) and secondary growth (increase in girth). When *primary growth* occurs, the apical meristem within a terminal bud is active. When *secondary growth* occurs, the vascular cambium is active. **Vascular cambium** is meristem tissue, which produces new xylem and phloem called **secondary xylem** and **phloem** each year. The buildup of secondary xylem year after year is called **wood.** Complete Table 8.3 to distinguish between primary growth and secondary growth of a stem.

Table 8.3 Primary Growth Versus Secondary Growth

	Primary Growth	Secondary Growth
Active meristem		
Result		

Observation: Anatomy of a Winter Twig

1. A winter twig typically shows several past years' primary growth. Examine several examples of winter twigs (Fig. 8.12), and identify the **terminal bud** located at the tip of the twig. This is where new primary growth will originate. During the next growing season, the terminal bud produces new tissues including vascular bundles and leaves.
2. Locate a **terminal bud scar.** These are marks left on stem from terminal bud scales (modified leaves protecting the bud). The distance between two adjacent terminal bud scars equals one year's primary growth.
3. Find a **leaf scar.** Mark where a leaf was attached to the stem.
4. Note the **bundle scars.** Complete this sentence: Marks left in the leaf scar where the vascular tissue ____________________.
5. Identify a **node.** This is the region where you find leaf scars and bundle scars. The region between nodes is called an ____________________.
6. Locate an **axillary bud.** This is where new branch growth can occur.
7. Note the numerous lenticels, breaks in the outer surface where gas exchange can occur.

Figure 8.12 External structure of a winter twig.
Counting the terminal bud scars tells the age of a particular branch.

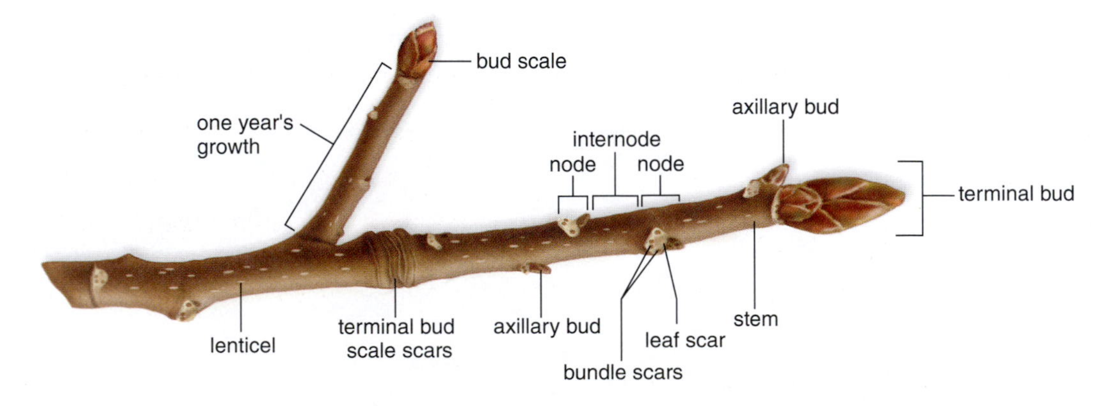

Observation: Anatomy of Woody Stem

1. Examine a prepared slide of a cross section of a woody stem (Fig. 8.13), and identify the **bark** (the dark outer area), which contains **cork,** a protective outer layer; **cortex,** which stores nutrients; and **phloem,** which transports organic nutrients.

Figure 8.13 Woody eudicot stem cross section.
Because xylem builds up year after year, it is possible to count the annual rings to determine the age of a tree. This tree is three years old.

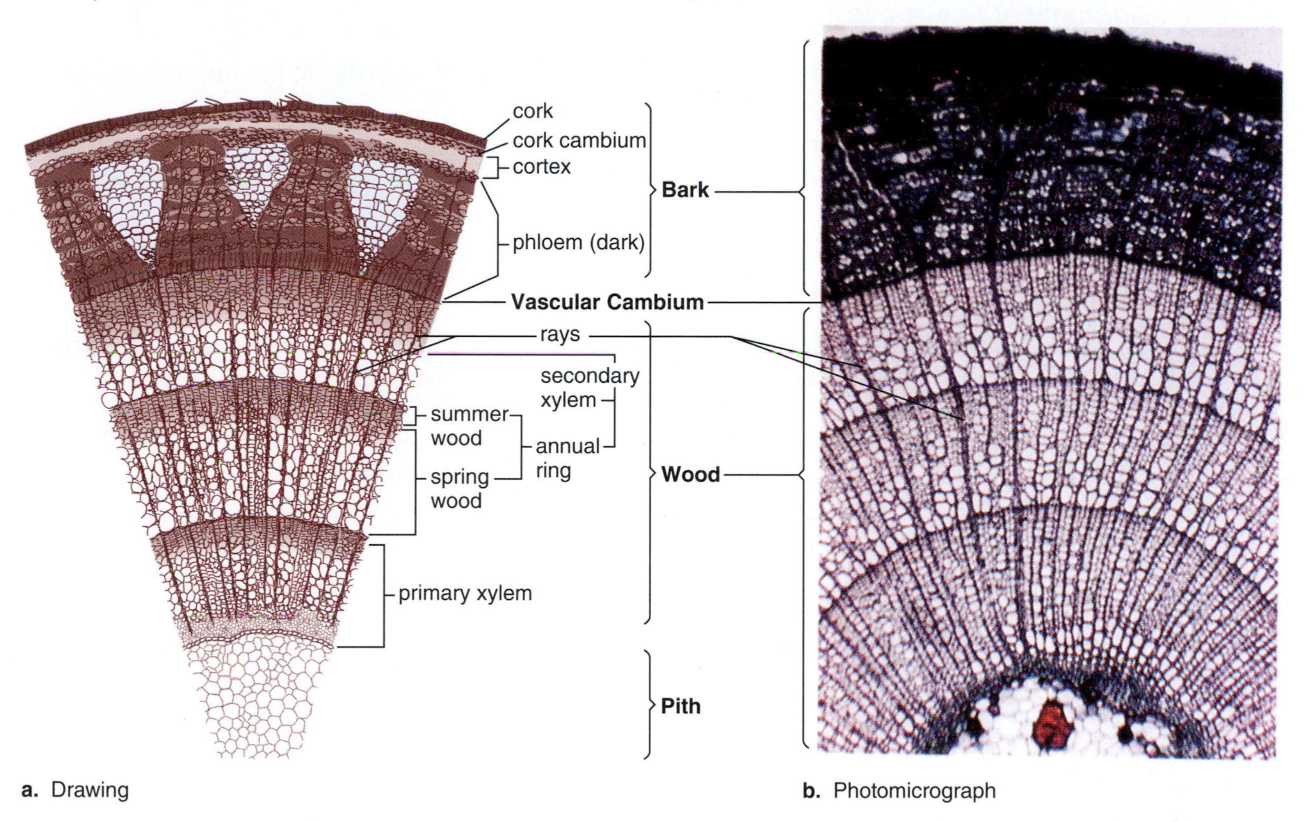

a. Drawing

b. Photomicrograph

2. Locate the **vascular cambium** at the inner edge of the bark, between the bark and the wood. Vascular cambium is meristem tissue whose activity accounts for secondary growth, which causes increased girth of a tree. Secondary phloem (which disappears) and secondary xylem (which builds up) are produced by vascular cambium each growing season.
3. Find the **wood,** which contains annual rings. An **annual ring** is the amount of xylem added to the plant during one growing season. Rings appear to be present because spring wood has large xylem vessels and looks light in color, while summer wood has much smaller vessels and appears much darker. How old is the stem you are observing? ____________ Are all the rings the same width? ____________
4. Identify the **pith,** a ground tissue at the center of a woody stem that stores organic nutrients and may disappear.
5. Locate **rays,** groups of small, almost cuboid cells that extend out from the pith laterally.

8.5 Leaves

A **leaf** is the organ that produces food for the plant by carrying on photosynthesis. Leaves are generally broad and quite thin. An expansive surface facilitates the capture of solar energy and gas exchange. Water and nutrients are transported to the cells of a leaf by leaf veins, extensions of the vascular bundles from the stem.

Anatomy of Leaves

Observation: Anatomy of Leaves

1. Examine a model of a leaf. With the help of Figure 8.14, identify the waxy **cuticle,** the outermost layer that protects the leaf and prevents water loss.
2. Locate the **upper epidermis** and **lower epidermis,** single layers of cells at the upper and lower surfaces. Trichomes are hairs that grow from the upper epidermis and help protect the leaf from insects and water loss.
3. Find the leaf veins in your model. The bundle sheath is the outer boundary of a vein; its cells surround and protect the vascular tissue. If this is a model of a monocot, all the leaf veins will be ________________ ________________________. If this is a model of a eudicot, some leaf veins will be circular and some will be oval. Why? ____
4. Identify the **palisade mesophyll,** located near the upper epidermis. These cells contain chloroplasts and carry on most of the plant's photosynthesis. Locate the **spongy mesophyll,** located near the lower epidermis. These cells have air spaces that facilitate the exchange of gases across the plasma membrane. *Label the layers of mesophyll in Figure 8.14.* Collectively, the mesophyll represents which of the three types of tissue found in all parts of a plant? ________________________
5. *Label the two layers of epidermis in Figure 8.14.* Find a **stoma** (pl., stomata), an opening through which gas exchange occurs. Stomata are more numerous in the lower epidermis. Each stoma has two guard cells that regulate opening and closing of the opening.

Figure 8.14 Leaf anatomy.

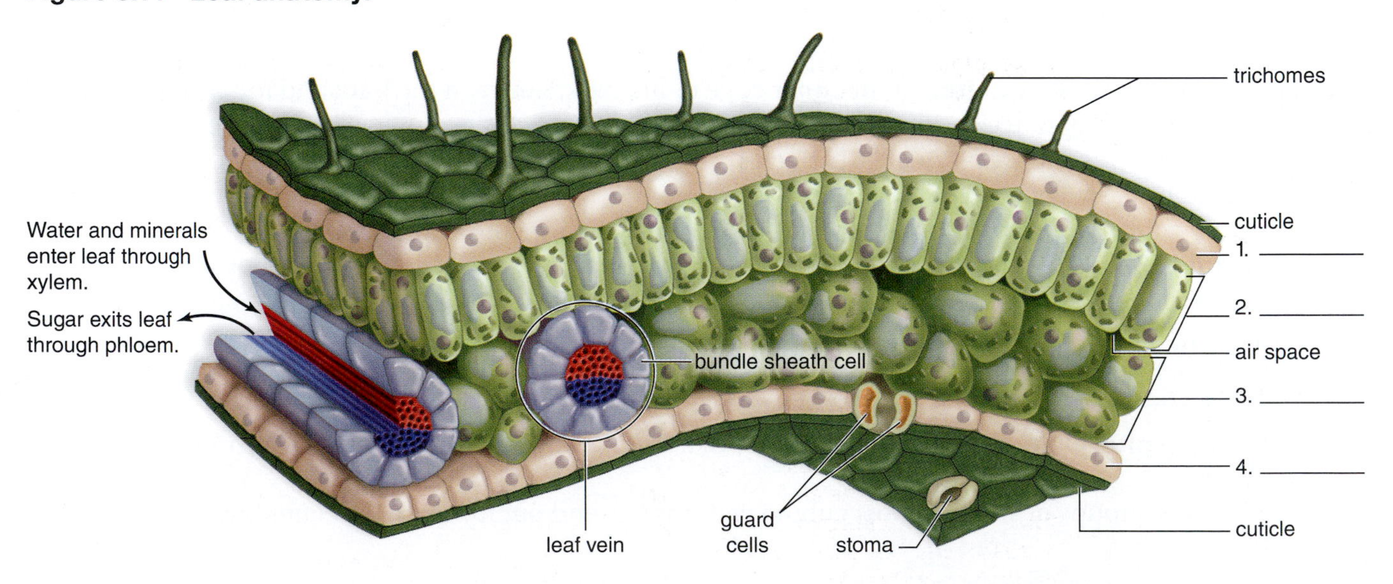

Leaf Diversity

A eudicot leaf consists of a flat blade and a stalk, called the petiole. An axillary bud appears at the point where the petiole attaches a leaf to the stem. In other words, an axillary bud is a tip-off that you are looking at a single leaf.

Observation: Leaf Diversity

Several types of eudicot leaves will be on display in the laboratory. Examine them using these directions.

- A leaf may be **simple,** in which case it consists of a single blade; or a leaf may be **compound,** meaning its single blade is divided into leaflets. *In Figure 8.15*a *write "simple" or "compound" next to the word* leaf *in 1 to 3.* (Among the leaves on display, find one that is simple and one that is compound.)
- A compound leaf can be **palmately** compound, meaning the leaflets are spread out from one point. Which leaf in Figure 8.15*a* is palmately compound? *Add "palmately" in front of "compound" where appropriate.* (See if you can find a palmately compound leaf among those on display.)
- A compound leaf can be **pinnately** compound, meaning the leaflets are attached at intervals along the petiole. Which leaf in Figure 8.15*a* is pinnately compound? *Add "pinnately" in front of "compound" where appropriate.* (See if you can find a pinnately compound leaf among those on display.)
- As shown in Figure 8.15*b*, leaves can be in various positions on a stem. On which stem in Figure 8.15*b*, 4 to 6, are the leaves **opposite** one another? *Write the word "opposite" where appropriate.* On which stem do the leaves **alternate** along the stem? *Write the word "alternate" where appropriate.* On which stem do the leaves **whorl** about a node? *Add the word "whorl" where appropriate in Figure 8.15*b. (See if you can find different arrangements of leaves among the stems on display.)

Figure 8.15 Classification of leaves.

Laboratory Review 8

1. Given the information in this laboratory, how would you distinguish between a monocot plant and a eudicot plant based on their external anatomy? ______________________________

____

2. What is meristem tissue? ______________________________ How is this tissue different from all other types of plant tissue? ______________________________

3. In which zone of a eudicot root would you expect to find vascular tissue? Why? ______________________________

____

____

4. In a eudicot root, what structural feature allows the endodermis to regulate the entrance of water and materials into the vascular cylinder, where xylem and phloem are located? ______________________________

____

____

5. Characterize the root of a carrot. ______________________________

6. How would you microscopically distinguish a eudicot stem from a monocot stem? ______________________________

____

____

7. Distinguish between primary and secondary growth of a woody stem, and explain how each arises.

____

____

8. Contrast how you could determine one year's growth by looking at a winter twig with how you determine one year's growth in a cross section of a tree stem. ______________________________

____

9. Contrast the manner in which water reaches the inside of a leaf with the manner in which carbon dioxide reaches the inside of a leaf. ______________________________

____

10. How would you recognize the epidermis of a root versus the epidermis of a leaf? ______________________________

____

LABORATORY

9

Mitosis: Cellular Reproduction

Learning Outcomes

Introduction

Dividing cells experience nuclear division, cytoplasmic division, and a period of time between divisions called interphase. During **interphase,** the nucleus appears normal, and the cell is performing its usual cellular functions. Also, the cell is increasing all of its components, including such organelles as the mitochondria, ribosomes, and centrioles (in animal cells). DNA replication (making an exact copy of the DNA) occurs toward the end of interphase. Then, during nuclear division, called **mitosis,** the new nuclei receive the same number of chromosomes as the parental nucleus. When the cytoplasm divides, a process called **cytokinesis,** two daughter cells are produced (Fig. 9.1).

In multicellular organisms, mitosis permits growth and repair of tissues. In eukaryotic, unicellular organisms, mitosis is a form of asexual reproduction. Sexually reproducing organisms utilize another form of nuclear division, called **meiosis.** In animals, meiosis is a part of gametogenesis, the production of gametes (sex cells). The gametes are sperm in male animals and eggs in female animals. Meiosis is the topic for Laboratory 10.

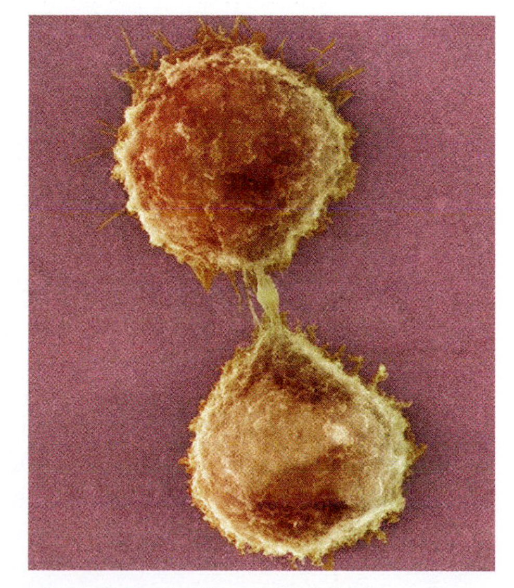

Figure 9.1 Cytokinesis. Following chromosome distribution during mitosis, daughter cells are produced by cytokinesis, a division of the cytoplasm.

9.1 The Cell Cycle

As stated in the Introduction, the period of time between cell divisions is known as interphase. Because early investigators noted little visible activity between cell divisions, they dismissed this period of time as a resting state. But when later investigators discovered that DNA replication and chromosome duplication occur during interphase, the **cell cycle** concept was proposed. The cell cycle is divided into the four stages noted in Figure 9.2.

Figure 9.2 The cell cycle.
Immature cells go through a cycle that consists of four stages: G_1, S (for synthesis), G_2, and M. The cell divides during the M stage, which consists of mitosis and cytokinesis.

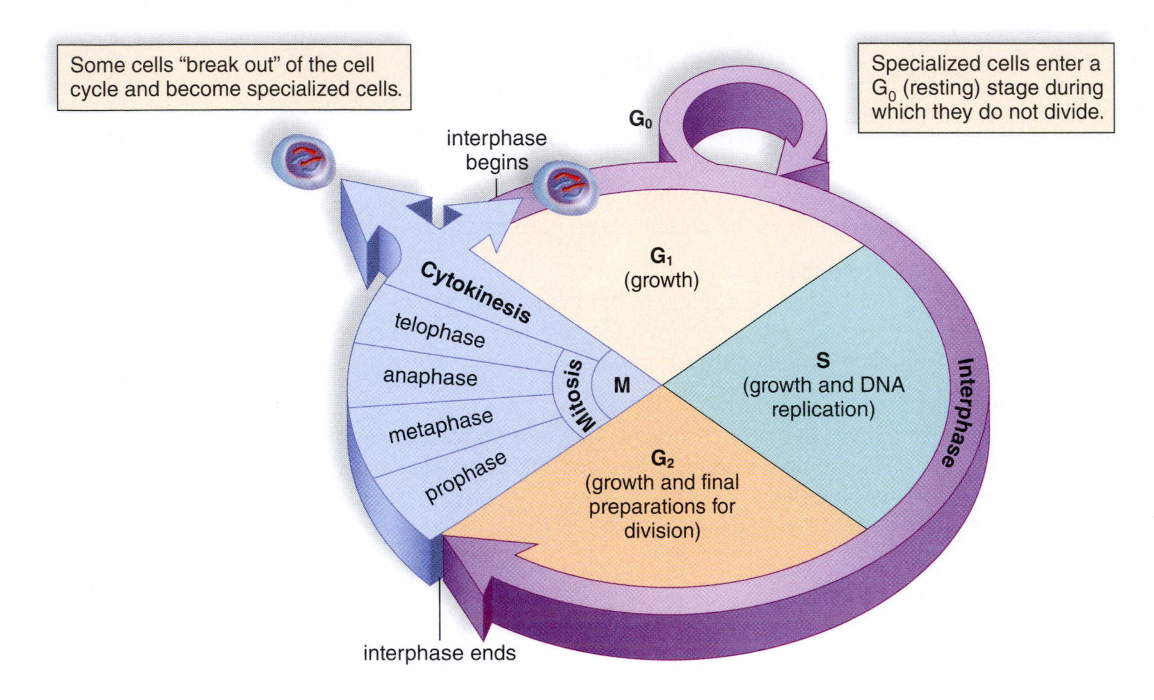

State the events of each stage of the cell cycle:

G_1 ________________________________

S ________________________________

G_2 ________________________________

M ________________________________

Explain why the entire process is called the "cell cycle." ____________________

The Cell Cycle and Cancer

Ordinarily, animal cells require about 18 to 24 hours for one cell cycle. Consult Figure 9.2, and notice that interphase consists of G_1, S, and G_2 stages of the cell cycle. When cancer occurs, cell cycle regulation is disturbed; interphase is severely shortened and very abnormal cells repeatedly undergo the cell cycle. A tumor develops and treatment consists of shrinking or removing the tumor.

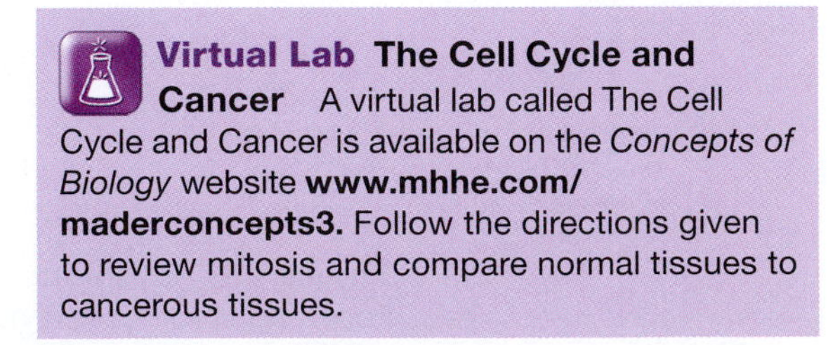

Virtual Lab The Cell Cycle and Cancer A virtual lab called The Cell Cycle and Cancer is available on the *Concepts of Biology* website **www.mhhe.com/maderconcepts3.** Follow the directions given to review mitosis and compare normal tissues to cancerous tissues.

The S Stage of the Cell Cycle

While each stage of the cell cycle is critical to successful mitosis (division of the nucleus), the S stage is particularly important because it is the stage during which DNA replicates. At the completion of replication, there are two double helices and each chromosome is duplicated.

When mitosis is about to occur, we can see each duplicated chromosome because chromatin has condensed and compacted to form two chromatids held together at a centromere. Consult Table 9.1, and *label the sister chromatids and centromere in Figure 9.3.*

What happens to the sister chromatids during mitosis? During mitosis, the sister chromatids separate, and then they are called daughter chromosomes. This is the manner in which DNA replication causes each body (somatic) cell of an organism to contain the same number of chromosomes. In humans, each cell in the body contains 46 chromosomes.

It is customary to call the cell that divides the **parent cell** and the resulting two cells the **daughter cells**. If a parent cell undergoing mitosis has 18 chromosomes, each daughter cell will have ____________ chromosomes.

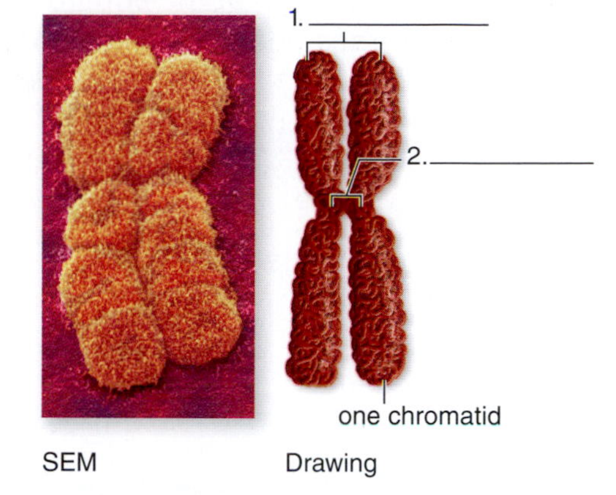

Figure 9.3 Duplicated chromosomes. DNA replication results in duplicated chromosomes that consist of two sister chromatids held together at a centromere.

The M Stage of the Cell Cycle

Study the terms in Table 9.1. These terms all pertain to mitosis. During mitosis sister chromatids (now called daughter chromosomes) move into the daughter nuclei. The process (mitosis) requires several phases. Consult Figure 9.2 and write the phases of mitosis here: ____

____

Table 9.1 Structures Associated with Mitosis

Structure	Description
Nucleus	A large organelle containing the chromosomes and acting as a control center for the cells
Nucleolus	An organelle found inside the nucleus that produces the subunits of ribosomes
Chromosome	Rod-shaped body in the nucleus that is seen during mitosis and meiosis and that contains DNA, and therefore the hereditary units, or genes
Chromatids	The two identical parts of a chromosome following DNA replication
Centromere	A constriction where duplicates (sister chromatids) of a chromosome are held together
Spindle	Microtubule structure that brings about chromosome movement during cell division
Centrioles*	Short, cylindrical organelles at the spindle poles in animal cells
Aster*	Short, radiating fibers surrounding the centrioles in dividing cells

*Animal cells only

9.2 Animal Cell Mitosis and Cytokinesis

When an animal cell divides, it first undergoes mitosis, and then it undergoes cytokinesis, which is division of the cytoplasm. Mitosis is called duplication division because the daughter cells have the same chromosome makeup as the parent cell. As we now know, a spindle, sometimes called the mitotic spindle, occurs during mitosis. A **spindle** is composed of spindle fibers (microtubules) formed by the centrosomes, which are located at the **poles** of the spindle (Fig. 9.4). In animal cells, the poles contain centrioles surrounded by an **aster,** which is an array of fibers. The centromeres of the duplicated chromosomes attach to spindle fibers at the equator of the spindle. When the centromeres split the chromatids (now daughter chromosomes) move toward the poles.

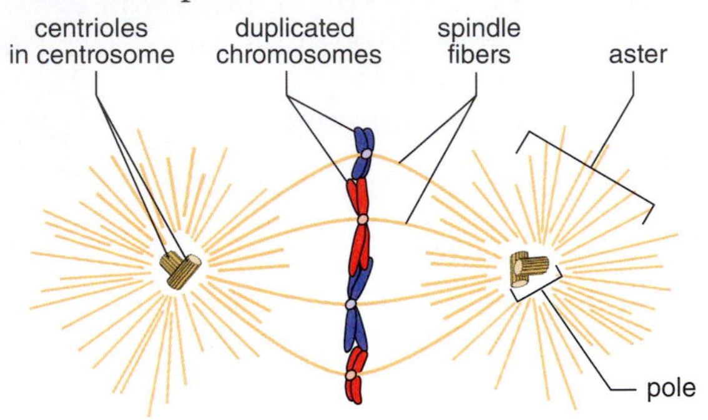

Figure 9.4 The mitotic spindle. Each centromere of a duplicated chromosome is attached to spindle fibers at the equator of the spindle. When the spindle fibers shorten, the centromeres split and the sister chromatids, now called daughter chromosomes, move toward opposite spindle poles.

Animal Cell Mitosis

We will now have the opportunity to observe the stages of mitosis in animal cell models and a prepared slide. Mitosis is the type of nuclear division that occurs when embryos develop, organisms grow larger, and when injuries heal. Without mitosis, none of these important events could occur.

Observation: Animal Cell Mitosis

Animal Mitosis Models

1. Each species has its own chromosome number. Counting the number of centromeres tells you the number of chromosomes in models or slides. What is the number of chromosomes in the parent cell and in the daughter cells in this model series? ______________________________
2. Examine the phases of mitosis as depicted by the models.
3. Do these models show the *spindle,* which is illustrated in Figure 9.4? ____________ In animal cells, the centrioles are surrounded by an *aster,* an array of fibers, at the poles of the spindle.
4. Name two ways you would be able to recognize animal cell mitosis: What is the shape of animal cells?

   ____

   What is the appearance of the spindle pole? ______________________________

   Plant cells do not have centrioles; therefore, their spindle poles lack centrioles and their poles do not have asters.

Whitefish Blastula Slide

1. Examine a prepared slide of whitefish blastula cells undergoing mitosis. The blastula is an early embryonic stage in the development of animals.
2. Try to find a cell in each phase of mitosis using Figure 9.5 as a guide. Have a partner or your instructor check your ability to identify these cells.
3. Match these statements to the correct phase of animal cell mitosis in Figure 9.5 and *write the correct statements on the lines provided.*

   ***Statements*:**
   - Daughter chromosomes are moving to the poles of the spindle.
   - Duplicated chromosomes have no particular arrangement in the cell.
   - Two daughter cells are now forming.
   - Duplicated chromosomes are aligned at the equator of the spindle.
4. The prophase cell in Figure 9.5 has the same number of chromosomes as the telophase nuclei. Explain the different appearance of the chromosomes. ______________________________

____

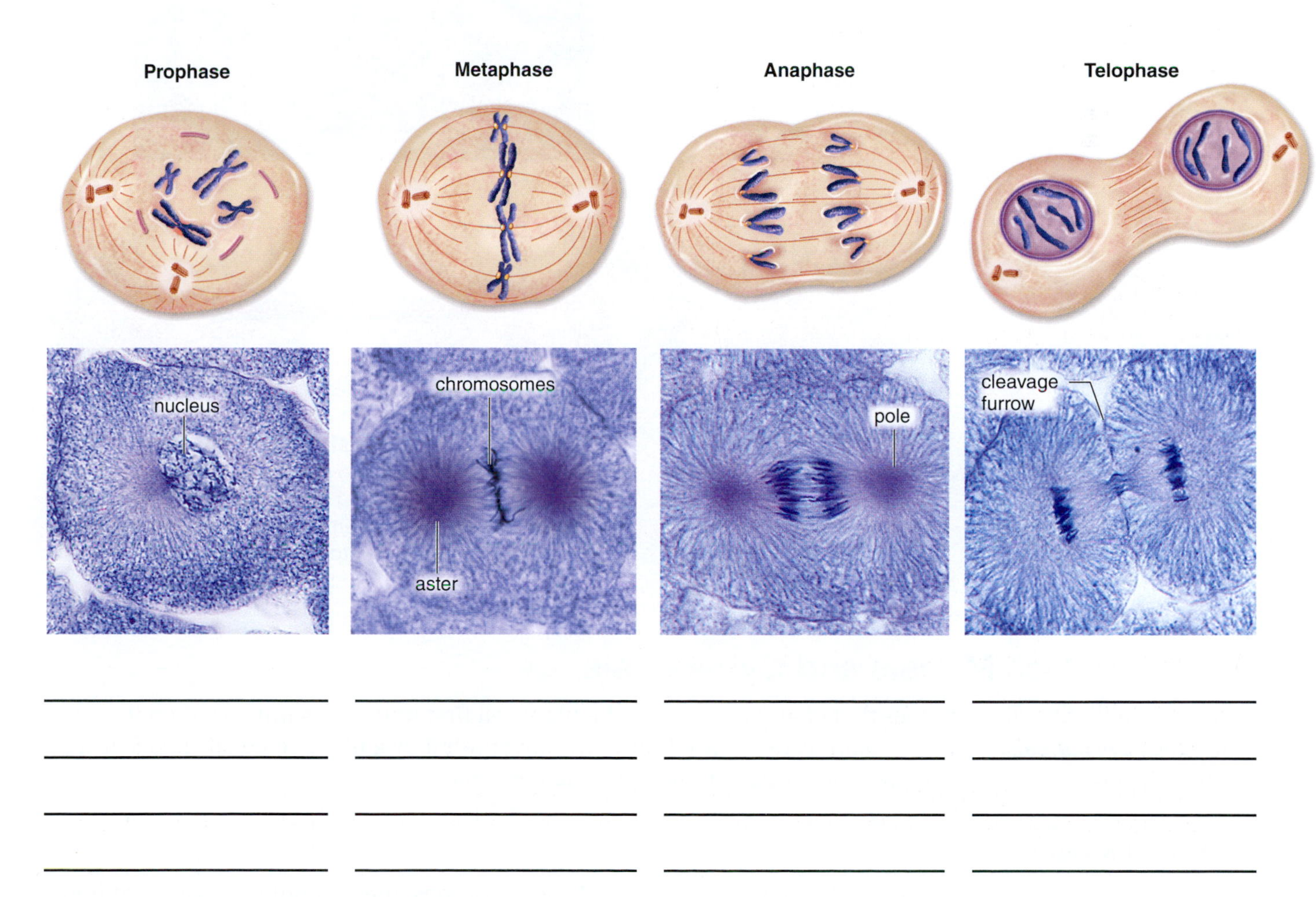

Figure 9.5 The phases of animal cell mitosis.
Mitosis always has these four main phases. Others, designated by the terms "early" or "late" as in early prophase or late anaphase can also be cited.

Cytokinesis in Animal Cells

Cytokinesis, division of the cytoplasm, usually accompanies mitosis. During cytokinesis, each daughter cell receives a share of the organelles that duplicated during interphase. Cytokinesis begins in anaphase, continues in telophase, and reaches completion by the start of the next interphase.

In animal cells, a **cleavage furrow,** an indentation of the membrane between the daughter nuclei, begins as anaphase draws to a close (Fig. 9.6). The cleavage furrow deepens as a band of actin filaments called the contractile ring slowly constricts the cell, forming two daughter cells. Note in Figure 9.5 that cleavage furrow is labeled in telophase. Are any of the cells in your whitefish blastula slide undergoing cytokinesis? ____________________

Do you see any cleavage furrows? ____________

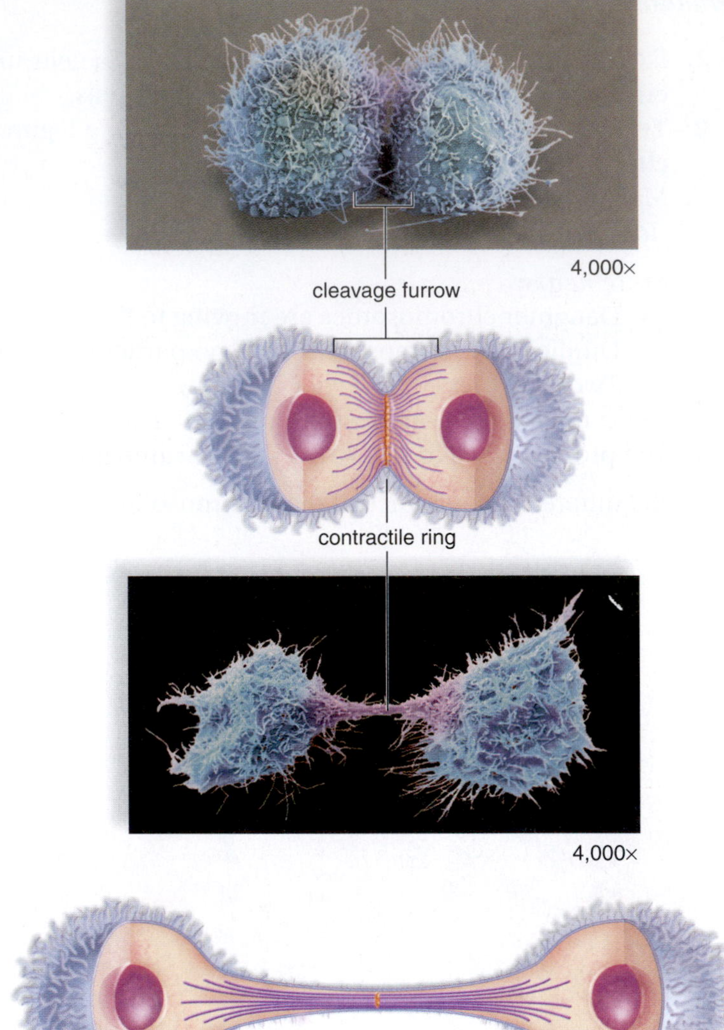

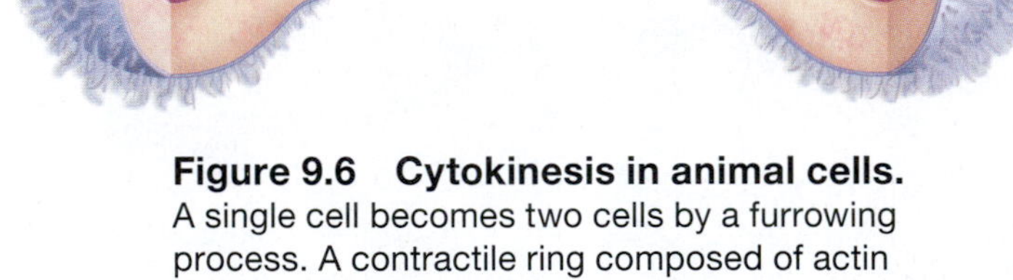

Figure 9.6 Cytokinesis in animal cells. A single cell becomes two cells by a furrowing process. A contractile ring composed of actin filaments gradually gets smaller, and the cleavage furrow pinches the cell into two cells.

9.3 Plant Cell Mitosis and Cytokinesis

Division of a plant cell resembles that of an animal cell. The plant cell first undergoes mitosis, then it undergoes cytokinesis. However, plant cells, as you know, are surrounded by a plant cell wall. This feature has no effect on mitosis but it does affect cytokinesis as we shall observe.

Plant Cell Mitosis

Although plant cells do utilize a spindle to divide duplicated chromosomes, they do not have well-defined spindle poles because they lack centrioles and asters.

Observation: Plant Cell Mitosis

Interphase is not a phase *of* mitosis, but a preliminary stage *to* mitosis.

Plant Mitosis Models

1. Identify interphase and the phases of plant cell mitosis using models of plant cell mitosis and Figure 9.7 as a guide.
2. As mentioned previously, plant cells do not have centrioles and asters, which are short radiating fibers produced by centrioles.
3. What is the number of chromosomes in each of the cells in your model series? ____________________

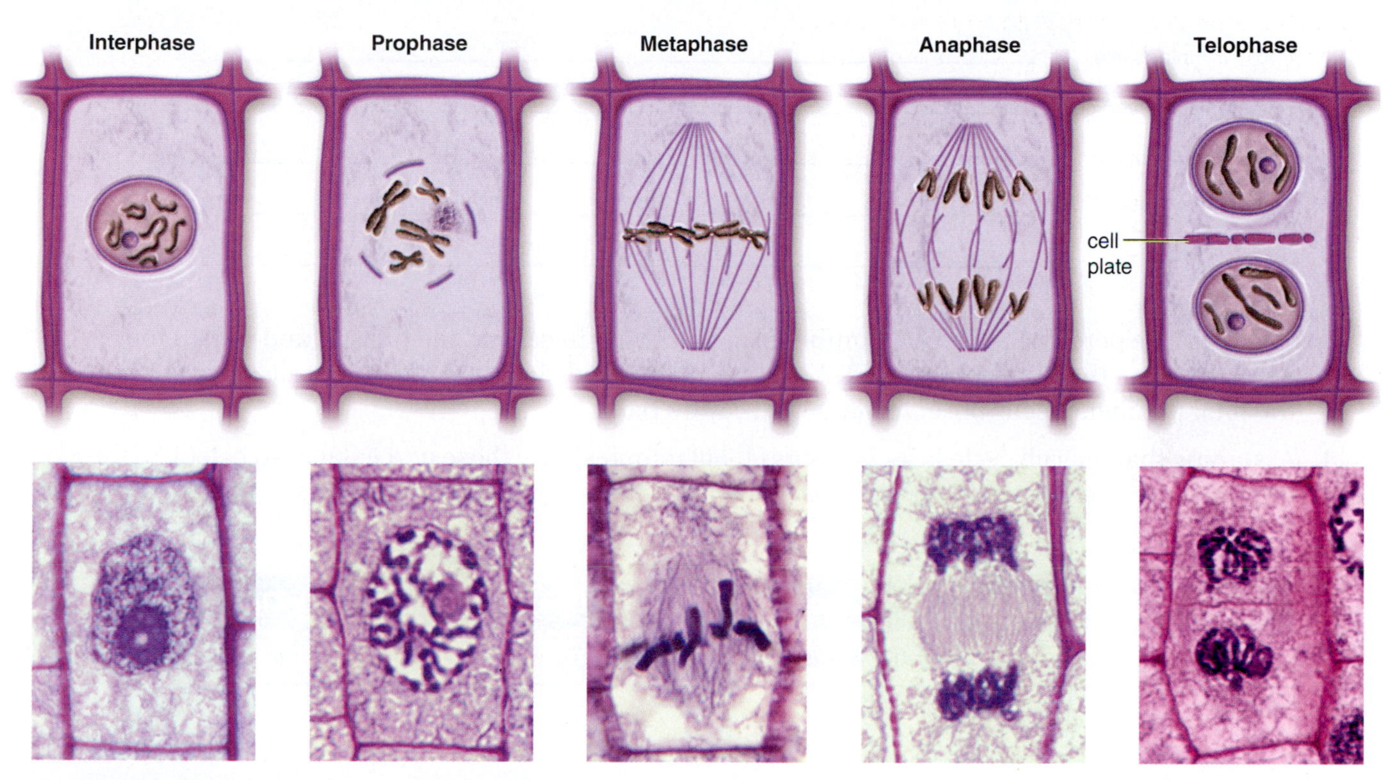

Figure 9.7 Phases of plant cell mitosis.
Mitosis has these phases in all eukaryotic cells. Plant cells are recognizable by the presence of a cell wall and the absence of centrioles and asters.

Onion Root Tip Slide

1. Examine a prepared slide of onion root tip cells (*Allium*) undergoing mitosis. In plants, the root tip contains tissue that is continually dividing and producing new cells (see Fig. 8.6).
2. Focus in low power and then switch to high power. Practice identifying phases that correspond to those shown in Figure 9.7.
3. Using high power, focus up and down on a cell in telophase. You may be able to just make out the cell plate, the region where a plasma membrane is forming between the two prospective daughter cells. Later, cell walls will appear in this region.

Time Span for Phases of the Cell Cycle in the Onion Root Tip

Knowing that the cell cycle consists of interphase plus four phases of mitosis and that the cell cycle typically lasts about 24 hours (1,440 minutes), hypothesize how many minutes the cell spends during each of these phases of the cell cycle.

Interphase ______ Prophase ______ Metaphase ______ Anaphase ______ Telophase ______

Experimental Procedure: Time Span for Phases of the Cell Cycle in the Onion Root Tip

1. Select an area of the onion root tip slide that contains cells in all phases of mitosis and also interphase. Concentrate on examining a confined area of about 20 to 30 cells.
2. Don't stray beyond a confined region, and as you identify phases, put a slash mark on the lines provided in this chart. Preferably work with a partner who will enter the slash marks as you call out observed phases in a confined region of the root tip. Convert your 20 to 30 slashes to Arabic numbers on the lines provided under the title Total. Also record these numbers in the second column of Table 9.2.

Phase	Slash Marks	Total
Interphase	____________________	______
Prophase	____________________	______
Metaphase	____________________	______
Anaphase	____________________	______
Telophase	____________________	______

3. Calculate the percentage of total number of cells that are in each of the phases, and record the percentages in the third column of Table 9.2. (To do this, divide the number of cells in each phase by the total number of cells observed and multiply by 100.)
4. Assuming that the cell cycle lasts 24 hours (1,440 minutes), use these percentages to calculate the time span for each phase of the cell cycle. Enter the time span for each phase in the fourth column of Table 9.2.

Table 9.2 Time Span for Phases of the Cell Cycle in the Onion Root Tip

Phase	Number Seen	% of Total	Time Span (min)
Interphase			
Prophase			
Metaphase			
Anaphase			
Telophase			
Total			

Conclusions: Time Span for Phases of the Cell Cycle in the Onion Root Tip

- Were your hypotheses supported or not supported by your observation of onion root tip cells undergoing the cell cycle? ______________________ Describe any specific discrepancies.
  ____
  ____
- Suggest a possible explanation for the length of time a cell spends on different phases of the cell cycle. ____
  ____

Cytokinesis in Plant Cells

After mitosis, the cytoplasm divides by cytokinesis. In plant cells, membrane vesicles derived from the Golgi apparatus migrate to the center of the cell and form a **cell plate** (Fig. 9.8), which is the location of a new plasma membrane for each daughter cell. Later, individual cell walls appear in this area. Were any of the cells of the onion root tip slide undergoing cytokinesis as shown in Figure 9.7, during:

Telophase? ______________________________

How do you know? ______________________________

Offer an explanation for why Figure 9.8 is so detailed. ______________________________

____

Would you predict that the vesicles of the cell plate lay down the new cell wall inside or outside the vesicles? Explain your answer. ______________________________

____

Figure 9.8 Micrograph showing cytokinesis in plant cells.
During plant cell cytokinesis, vesicles fuse to form a cell plate that separates the daughter nuclei. Later, the cell plate gives rise to a new cell wall.

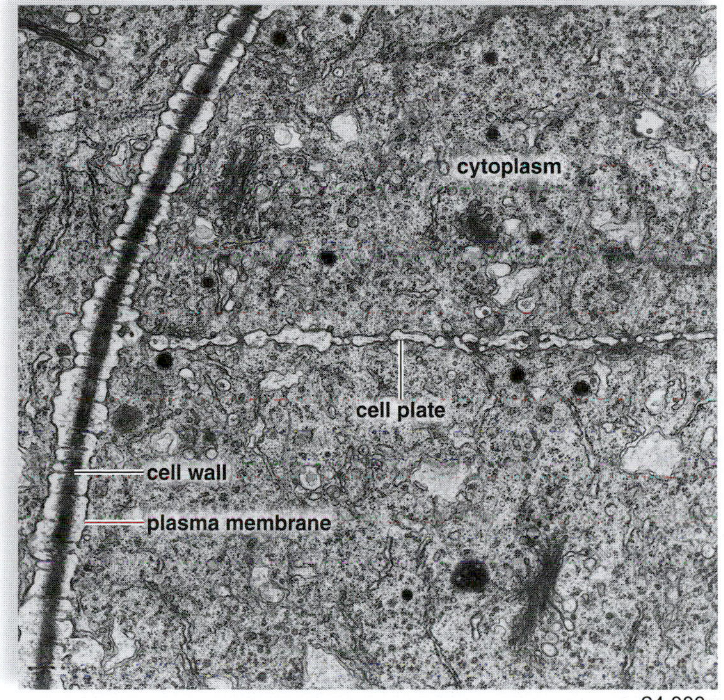

Laboratory Review 9

1. Divide the cell cycle into two main portions, and tell in general what is happening in these two portions.
____
____

2. Most of the time the cell is in which of these portions of the cell cycle? Why is this advantageous? _______________
____
____

3. Name and define the two events that take place when a cell divides. _______________
____
____

4. What is the function of the centromere during mitosis? _______________
____

5. Evolution of the spindle was of central importance to eukaryotic cells. What role is played by the spindle during mitosis? _______________
____

6. Do the chromosomes have sister chromatids during these phases of mitosis (yes or no)?
   a. Prophase _______________
   b. Metaphase _______________
   c. Anaphase _______________
   d. Telophase _______________

7. Contrast the appearance of animal cells and plants during mitosis. _______________
____

8. Explain how it is possible for each phase of mitosis and the daughter cells to have the same number of chromosomes. _______________
____

9. Contrast how cytokinesis occurs in animal cells with how it occurs in plant cells. _______________
____
____

10. What would a tissue look like if cytokinesis did not occur? _______________
____

***Concepts of Biology* Website**

Instructors can find lab prep information and answers to all of the laboratory questions in the Laboratory Resource Guide. *Students* can practice their knowledge with quizzes, animations, flashcards, and much more.

www.mhhe.com/maderconcepts3

McGraw-Hill Access Science Website

An online encyclopedia of science and technology that provides information, including videos, that can enhance the laboratory experience.

www.accessscience.com

Mitosis & Meiosis

LABORATORY

10

Meiosis: Sexual Reproduction

Learning Outcomes

10.1 Meiosis: Reduction Division

- Name and describe the phases of meiosis I and meiosis II. 102–103
- Describe the mechanism for chromosome reduction during meiosis I, and explain why a reduced chromosome number is beneficial in gametes. 102–103
- Recognize the various phases of meiosis when examining slides of meiosis. 102–103
- Describe the chromosome composition in the daughter cells following meiosis II. 103

10.2 Production of Variation During Meiosis

- Using pop bead chromosomes, demonstrate two ways meiosis introduces variations among the daughter cells of meiosis and therefore the gametes. 104–106

10.3 Human Life Cycle

- Draw the human life cycle, showing the occurrence of mitosis and meiosis. 107
- State the function of mitosis and meiosis in the human life cycle. 107
- Contrast the process and the events of mitosis and meiosis. 108–109

Introduction

In sexually reproducing organisms, meiosis is a part of or preparatory to gametogenesis, the production of gametes (sex cells). The gametes are sperm (the smaller gamete) and egg (the larger gamete). Fusion of sperm and egg results in a zygote that develops into a new individual (Fig. 10.1).

Meiosis is nuclear division which reduces the chromosome number so that the gametes have half the species number of chromosomes. At the start of meiosis, the parent cells have the full number of chromosomes and each is duplicated. Following two divisions called **meiosis I** and **meiosis II,** the four daughter cells have only one copy of each type chromosome, and these chromosomes consist of only one chromatid.

Not only does meiosis reduce the chromosomes number, it also introduces variation, and this laboratory will show you exactly how the chromosomes are shuffled during meiosis and how the genetic material is recombined during the process of meiosis.

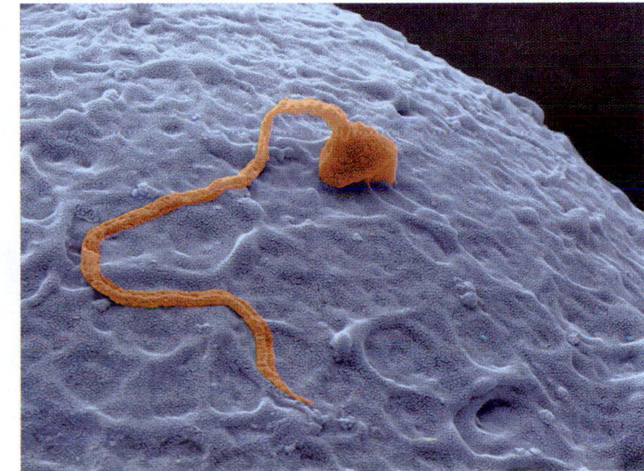
4,200×

Figure 10.1 Zygote formation.
During sexual reproduction, union of the sperm and egg produces a zygote that becomes a new individual.

10.1 Meiosis: Reduction Division

Meiosis is a form of nuclear division that reduces the chromosome number by half. Therefore, when a sperm fertilizes an egg, the zygote will have the normal number of chromosomes for that species.

Before meiosis begins, the parent cell is **diploid (2n);** it contains pairs of chromosomes called **homologues.** Consider the pair of homologues shown in Figure 10.2. A pair of homologues is called a **tetrad** because it contains two pairs of sister chromatids or four chromatids altogether. Following meiosis, the daughter cells are **haploid (n);** they contain only one from each pair of homologues.

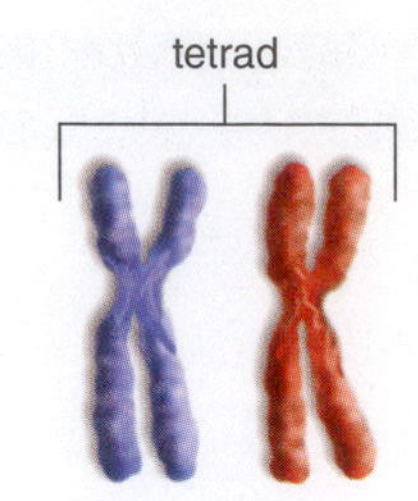

Figure 10.2 Homologues. A pair of homologues have the same shape and size. They form a tetrad because each homologue has two chromatids.

Before proceeding:

a. Distinguish between diploid (2n) and haploid (n): ______________________________

____

b. Distinguish between a homologue and a tetrad: ______________________________

____

Observation: Meiosis in Lily Anther

Almost all eukaryotes practice meiosis during sexual reproduction. It turns out that it's easier to examine meiosis in plants rather than animals because meiosis in plants is not a part of gametogenesis—that happens later. The anther is the male part of the flower where meiosis, consisting of **meiosis I** (Fig. 10.3) and **meiosis II** (Fig. 10.4), occurs preparatory to producing sperm.

Phases of Meiosis I

1. Examine a prepared slide of a lily anther where cells are undergoing meiosis I. Homologues pair up (called **synapsis**) during prophase I. During metaphase I, tetrads are at the equator and during anaphase I, homologues separate. Separation of homologues makes the daughter cells following meiosis I haploid because each daughter cell receives only one chromosome from each pair of homologues.

Figure 10.3 Meiosis I.
Phases of meiosis I in plant cell micrographs and drawings that depict the movement of the chromosomes. (The blue chromosomes were inherited from one parent and the red chromosomes were inherited from the other parent.)

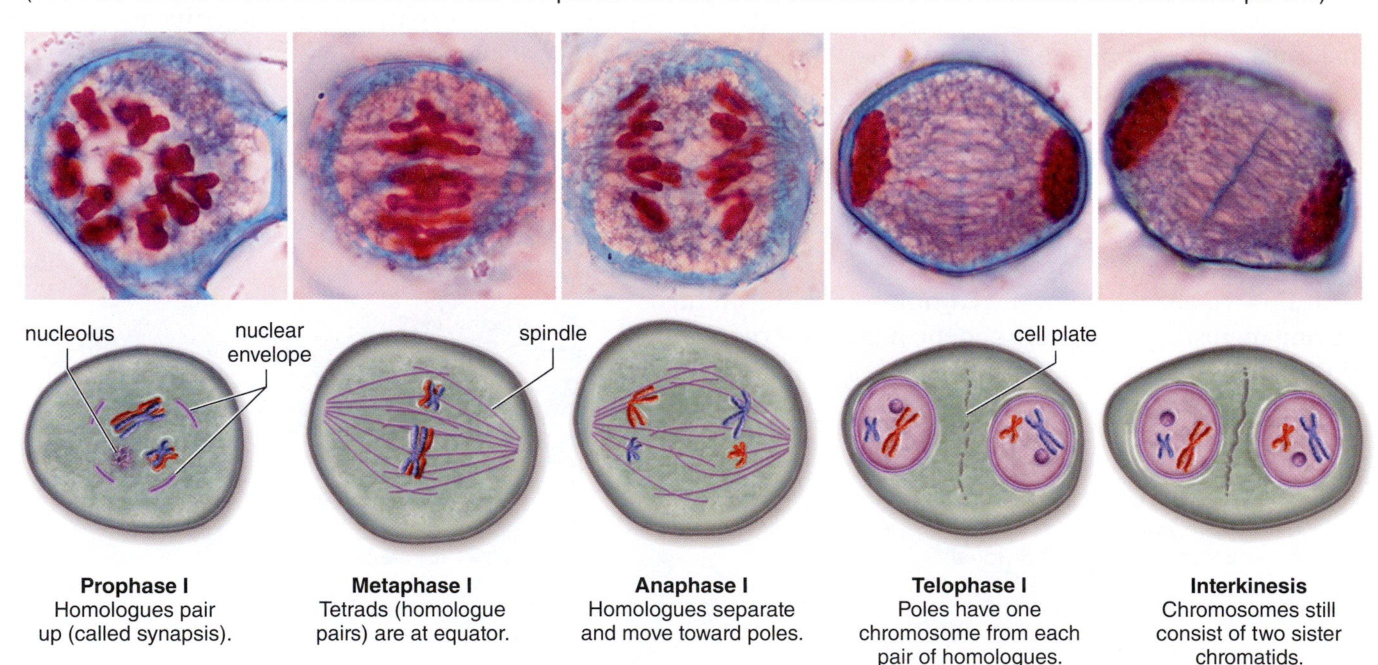

Prophase I Homologues pair up (called synapsis).

Metaphase I Tetrads (homologue pairs) are at equator.

Anaphase I Homologues separate and move toward poles.

Telophase I Poles have one chromosome from each pair of homologues.

Interkinesis Chromosomes still consist of two sister chromatids.

2. Using Figure 10.3 as a guide, try to find a cell in each phase of meiosis I. Tell what is happening in each of these phases:

   a. Prophase I ______________________________

   b. Metaphase I ______________________________

   c. Anaphase I ______________________________

   d. Telophase I ______________________________

3. *In Figure 10.3, place a 2n or n beside each drawing.*

Phases of Meiosis II

1. Examine a prepared slide of a lily anther where cells are undergoing meiosis II. A brief period of time called **interkinesis** occurs between meiosis I and meiosis II. What happens during meiosis II? Separation of the sister chromatids, of course, just as in mitosis (Fig. 10.4).
2. Using Figure 10.4 as a guide, try to find a cell in each phase of meiosis II. Tell what is happening in each of these phases:

   a. Prophase II ______________________________

   b. Metaphase II ______________________________

   c. Anaphase II ______________________________

   d. Telophase II ______________________________

3. *In Figure 10.4, place a 2n or n beside each drawing.* Why is it important for both sperm and egg to have the haploid number of chromosomes? ______________________________

____

Figure 10.4 Meiosis II.
Phases of meiosis II in plant cell micrographs and drawings that depict the movement of the chromosomes.

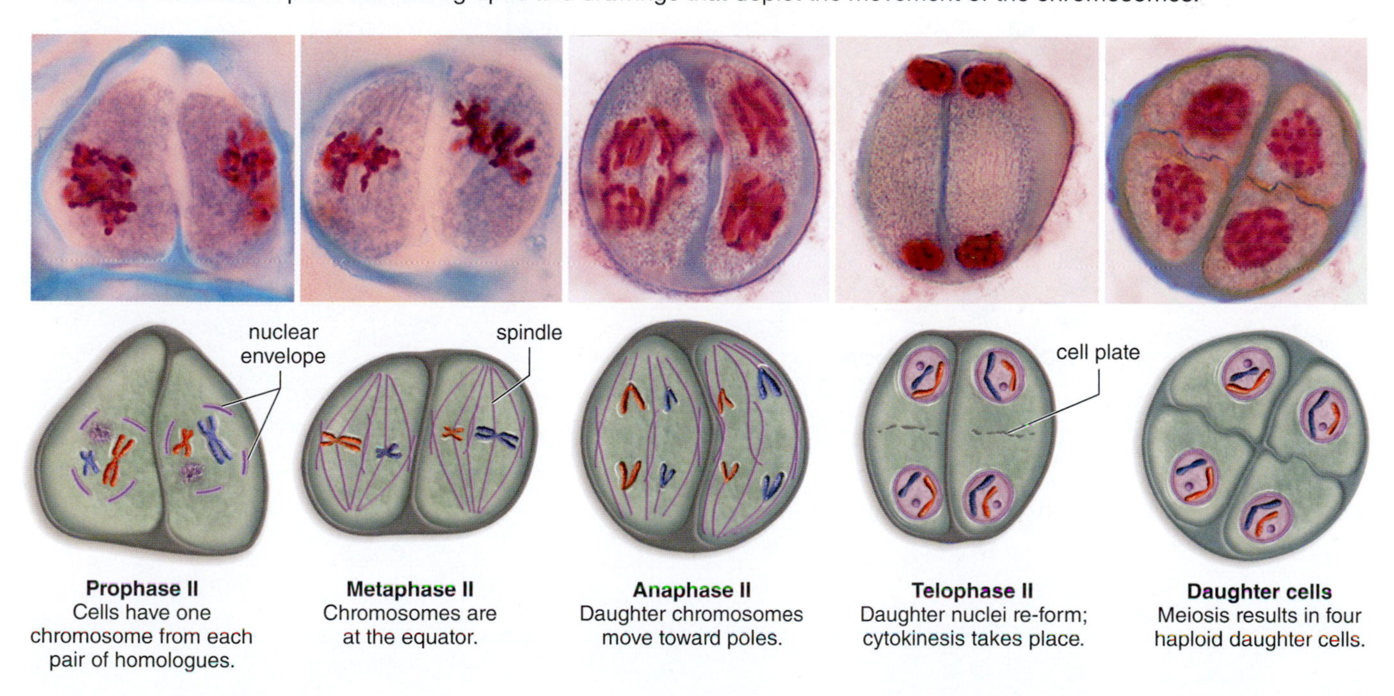

10.2 Production of Variation During Meiosis

The following experimental procedure is designed to show that during meiosis, crossing-over and independent separation of homologues leads to diversity of genetic material in the gametes.

Experimental Procedure: Production of Variation During Meiosis

First, you will build four chromosomes: two pairs of homologues, as in Figure 10.5. In other words the parent cell is 2n = 4.

Building Chromosomes

1. Obtain the following materials: 48 pop beads of one color (e.g., red) and 48 pop beads of another color (e.g., blue) for a total of 96 beads; eight magnetic centromeres.
2. Build a homologue pair of duplicated chromosomes using Figure 10.5*a* as a guide. Each chromatid will have 16 beads. Be sure to bring the centromeres of the same color together so that they form one duplicated chromosome.
3. Build another homologue pair of duplicated chromosomes using Figure 10.5*b* as a guide. Each chromatid will have eight beads.
4. Note that your chromosomes look the same as those in Figure 10.5.

Figure 10.5 Two pairs of homologues.
The red chromosomes were inherited from one parent, and the blue chromosomes were inherited from the other parent. Color does not signify homologues; size and shape signify homologues.

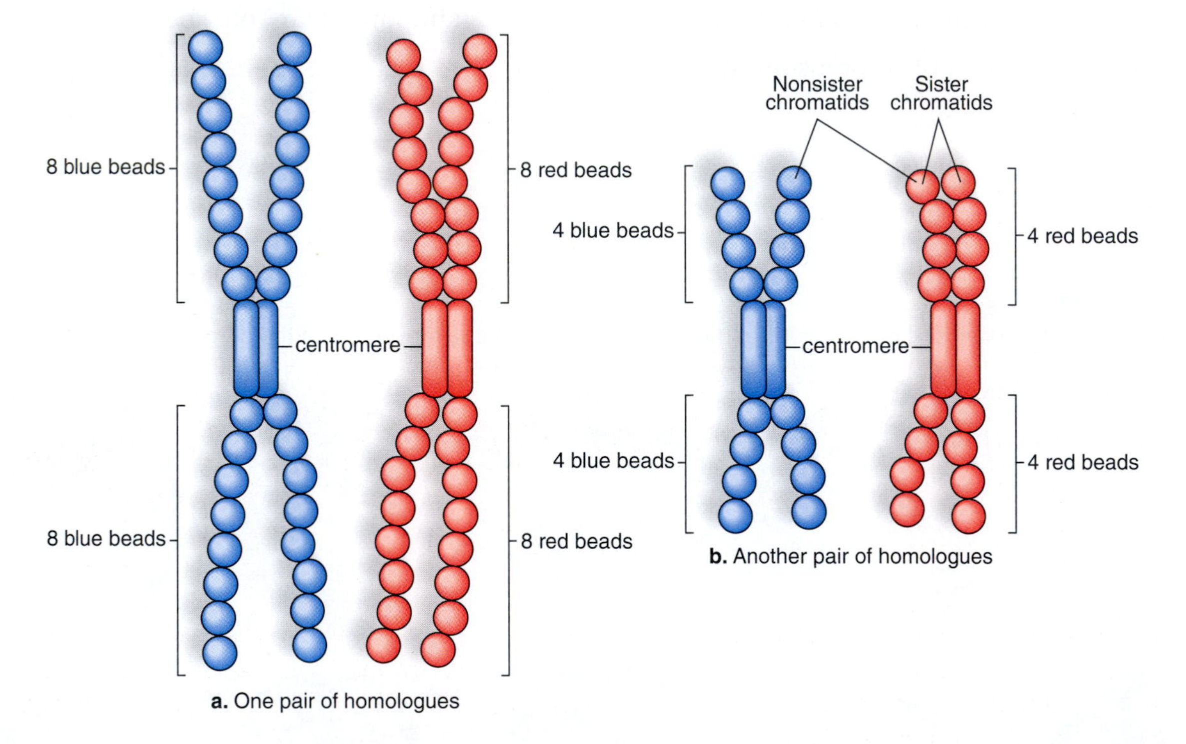

Meiosis I

You will now have your chromosomes undergo meiosis I.

Prophase I

5. Put all four of your chromosomes in the center of your work area: this area represents the nucleus. Synapsis, a very important event, occurs during meiosis. To simulate synapsis, place the long blue chromosome next to the long red chromosome and the short blue chromosome next to the short red chromosome to show that the homologues pair up as in Figure 10.3 (prophase I). Now **crossing-over** occurs. During crossing-over an exchange of genetic material occurs between nonsister chromatids. Perform crossing-over by switching some blue beads for red beads between the inside chromatids of the homologues. The genetic material contains genes and therefore crossing-over recombines genes.

Genetic Variation Due to Meiosis As a result of crossing-over, the genetic material on a chromosome in a gamete can be different from that in the parent cell.

Metaphase I

6. Keep the homologues together and align them at the metaphase plate. Add centrioles to represent the poles of a spindle apparatus. These are animal cells.

Anaphase I and Telophase I

7. Separate the homologues, so that each pole of the spindle receives one chromosome from each pair of homologues. This separation causes each pole to receive the haploid number of chromosomes.

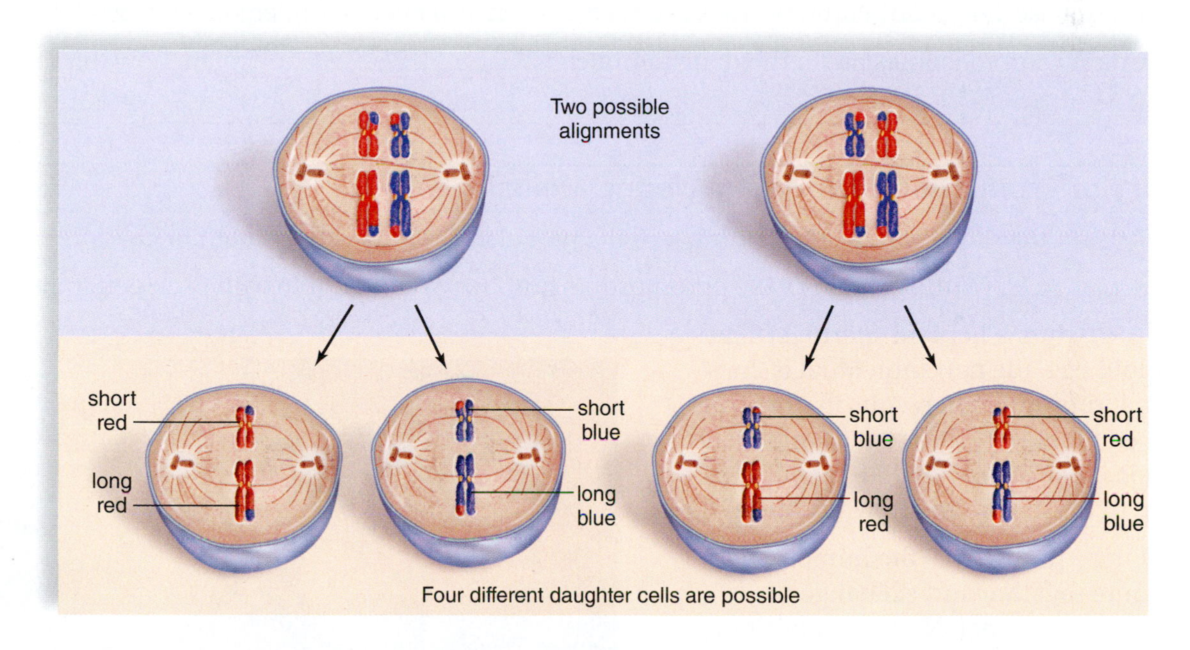

Only two daughter cells result from meiosis I, but two possible alignments of chromosomes can occur at the equator during metaphase I. Therefore, four possible combinations of haploid chromosomes are possible in the daughter cells following meiosis I in a $2n = 4$ parent cell. The possible combinations increase as the chromosome number increases.

Genetic Variation Due to Meiosis As a result of independent homologue separation, all possible combinations of the haploid number of chromosomes can occur among the gametes.

Meiosis II

Follow these directions to simulate meiosis II.

Prophase II

8. Choose one daughter nucleus (see step 7, page 105) to be the parent nucleus undergoing meiosis II.

Metaphase II

9. Move the duplicated chromosomes to the metaphase II metaphase plate, as shown in the art to the right.

Anaphase II

10. Pull the two magnets of each duplicated chromosome apart. What does this action represent? ____

____

Telophase II

11. Put the chromosomes—each having one chromatid—at the poles near the centrioles. At the end of telophase, the daughter nuclei reform.

- You chose only one daughter nucleus from meiosis I to be the nucleus that divides. In reality both daughter nuclei go on to divide again.

  Therefore, how many nuclei are usually present when meiosis II is complete? __________

- In this exercise, how many chromosomes were in the parent cell nucleus undergoing meiosis II? ________
- How many chromosomes are in the daughter nuclei? ________________ Explain how this is possible.

  ____

Summary of Production of Variation During Meiosis

1. Meiosis reduces the chromosome number. If the parent cell is 2n = 4, the daughter cells are n = __________. Without meiosis the chromosome number would double with each generation. Instead, when a haploid sperm fertilizes a haploid egg, the new individual is 2n.

2. Sexual reproduction results in offspring that can look very different as represented in this illustration. Why do the puppies born to these parents show variation?

a. During prophase I, the homologues come together and exchange genetic material. Now the inherited chromosomes will be different from those in the parent cell. This process is called ____________________________.

b. During metaphase I, the homologues align independently and therefore differently. This means that daughter cells following telophase I can have different ________________ of chromosomes.

c. During fertilization variant sperm fertilize variant eggs, further helping to ensure that the new individual inherits different ____________________ of homologues than a parent had.

10.3 Human Life Cycle

The term **life cycle** in sexually reproducing organisms refers to all the reproductive events that occur from one generation to the next. The human life cycle involves both mitosis and meiosis (Fig. 10.6). As you read the following text, *fill in boxes in Figure 10.6 with the terms "mitosis" or "meiosis."*

During development and after birth, mitosis is involved in the continued growth of the child and the repair of tissues at any time. As a result of mitosis, each somatic (body) cell has the diploid number of chromosomes (2n), which is 46 chromosomes.

During gamete formation, meiosis reduces the chromosome number from the diploid to the haploid number (n) in such a way that the gametes (sperm and egg) have one chromosome derived from each pair of homologues. In males, meiosis is a part of **spermatogenesis,** which occurs in the testes and produces sperm. In females, meiosis is a part of **oogenesis,** which occurs in the ovaries and produces eggs. After the sperm and egg join during fertilization, the resulting zygote is 2n. The zygote then undergoes mitosis with differentiation of cells to become a fetus, and eventually a new human being.

Meiosis keeps the number of chromosomes constant between the generations, and it also, as we have seen, causes the gametes to be different from one another. Therefore, due to sexual reproduction, there are more variations among individuals.

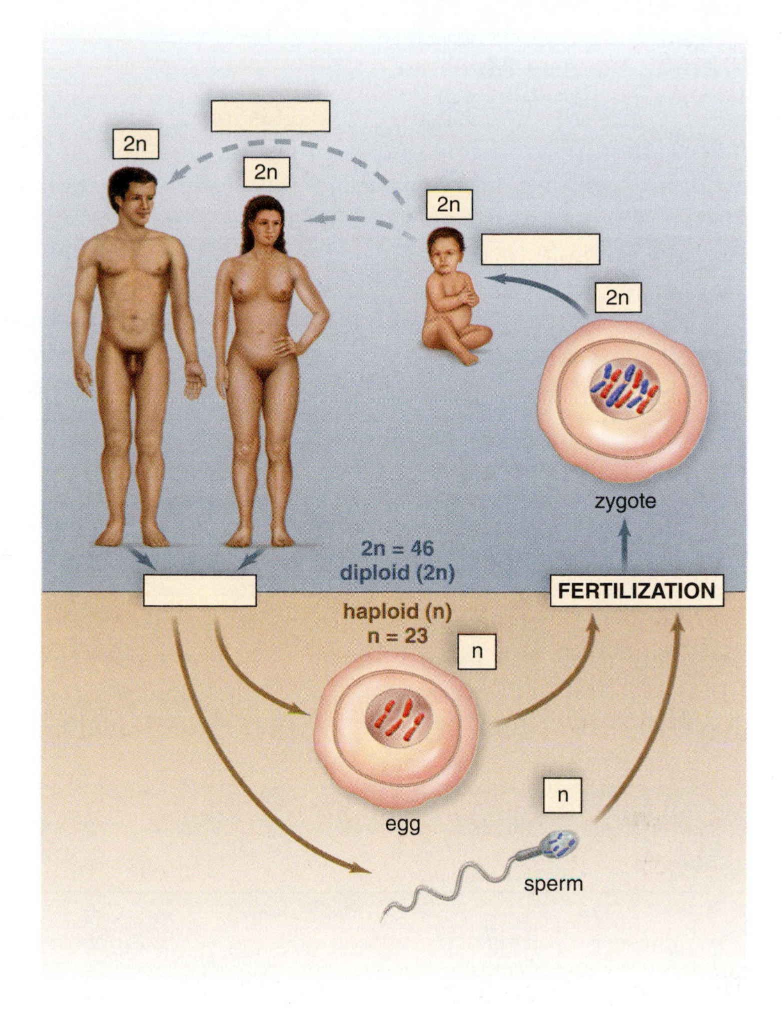

Figure 10.6 Life cycle of humans.

Summary of Human Life Cycle

Fill in the blanks to ensure your understanding of the role of meiosis and mitosis in humans.

1. Name of organ that produces gametes in males ______________ in females ______________
2. Name of process that produces gametes in males ______________ in females ______________
3. Type of cell division involved in process in males ______________ in females ______________
4. Name of gamete in males ______________ in females ______________
5. Number of chromosomes in gamete in males ______________ in females ______________
6. Results of fertilization ______________
7. Number of chromosomes provided ______________

Answer these questions.

Mitosis Versus Meiosis

When comparing mitosis to meiosis it is important to note that meiosis requires two nuclear divisions but mitosis requires only one nuclear division. Therefore, mitosis produces two daughter cells and meiosis produces four daughter cells. Following mitosis, the daughter cells are still diploid but following meiosis, the daughter cells are haploid. Figure 10.7 explains why.

Fill in Table 10.1 to indicate general differences between mitosis and meiosis.

Table 10.1 Differences Between Mitosis and Meiosis

	Mitosis	Meiosis
1. Number of divisions		
2. Chromosome number in daughter cells		
3. Number of daughter cells		

Complete Table 10.2 to indicate specific differences between mitosis and meiosis I. Mitosis need be compared only with meiosis I because the same events occur during both mitosis and meiosis II, except that the cells are diploid during mitosis and haploid during meiosis II.

Table 10.2 Mitosis Compared with Meiosis I

Mitosis	Meiosis I
Prophase: No pairing of chromosomes	Prophase I: ____________
Metaphase: Duplicated chromosomes at metaphase plate	Metaphase I: ____________
Anaphase: Sister chromatids separate	Anaphase I: ____________
Telophase: Chromosomes have one chromatid	Telophase I: ____________

Provide the correct term for each definition.

1. Type of cell division that keeps the chromosome number the same and occurs during growth and repair ____________
2. Type of cell division that reduces the chromosome number and occurs during gamete formation ____________
3. Half the diploid number of chromosomes ____________
4. Male gamete with the n number of chromosomes ____________
5. Female gamete with the n number of chromosomes ____________

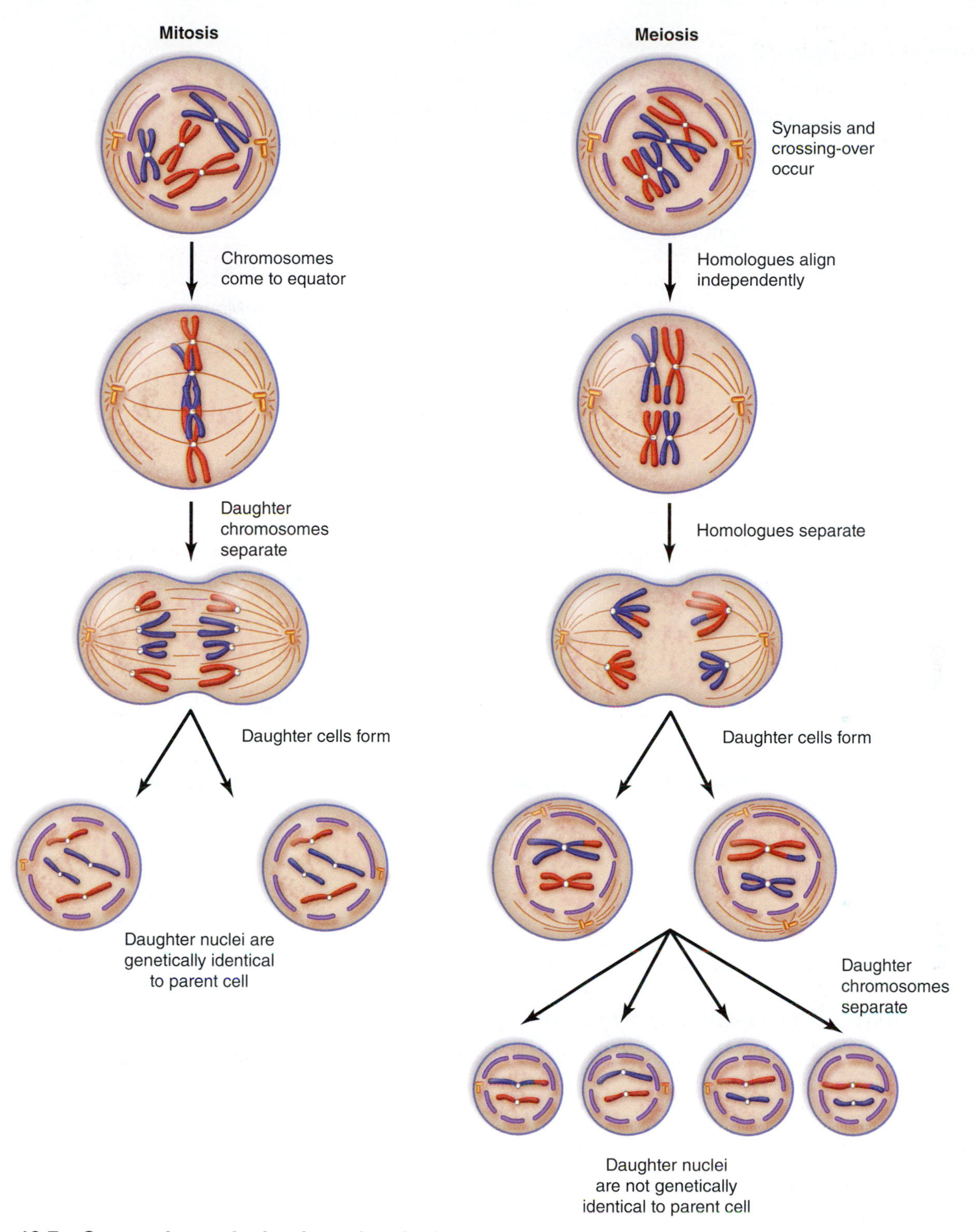

Figure 10.7 Comparison of mitosis and meiosis.
The blue chromosomes were inherited from one parent and the red chromosomes were inherited from the other parent.

Laboratory Review 10

1. Where would you expect to find meiosis taking place in human males? ______________________ in human females? ______________________
2. If there are 13 pairs of homologues in a parent cell, how many chromosomes are there in a sperm? Explain how you arrived at this number. ______________________
3. Account for why each of your body cells contains two of each kind of chromosome. ______________________
4. Does metaphase of mitosis, meiosis I, or meiosis II have the haploid number of chromosomes at the equator of the spindle? ______________________
5. List four differences when comparing mitosis with meiosis. ______________________
6. If the cells of an organism have 12 chromosomes, what is the number chromosomes at the equator during metaphase of mitosis? __________ during metaphase of meiosis II? __________
7. A student is simulating meiosis I with a pair of homologues that are red-long and yellow-long. Why would you not expect to find both red-long and yellow-long in one resulting daughter cell? ______________________
8. With reference to a pair of homologues, describe the change in the two participating nonsister chromatids following crossing-over. ______________________
9. What would be the appearance of a cell that completes mitosis but does not undergo cytokinesis? ______________________

   with a cell that completes meiosis but does not undergo cytokinesis? ______________________
10. In the life cycle of humans, when does mitosis occur? ______________________

***Concepts of Biology* Website**

Instructors can find lab prep information and answers to all of the laboratory questions in the Laboratory Resource Guide. *Students* can practice their knowledge with quizzes, animations, flashcards, and much more.

www.mhhe.com/maderconcepts3

McGraw-Hill Access Science Website

An online encyclopedia of science and technology that provides information, including videos, that can enhance the laboratory experience.

www.accessscience.com

Mitosis & Meiosis

LABORATORY

11 Mendelian Genetics

Learning Outcomes

11.1 One-Trait Crosses

- State Mendel's law of segregation, and relate this law to laboratory exercises. 112–16
- Describe and predict the results of a one-trait cross in both tobacco seedlings and corn. 113–16

11.2 Two-Trait Crosses

- State Mendel's law of independent assortment, and relate this law to laboratory exercises. 116–21
- Explain and predict the results of a two-trait cross in corn plants and *Drosophila*. 117–21

11.3 X-Linked Crosses

- Explain and predict the results of an X-linked cross in *Drosophila*. 121–22

Introduction

Gregor Mendel, sometimes called the "father of genetics," formulated the basic laws of genetics examined in this laboratory. He determined that individuals have two alternate forms of a gene (now called **alleles**) for each trait in their body cells. Today, we know that alleles are on the chromosomes. An individual can be **homozygous dominant** (two dominant alleles, *GG*), **homozygous recessive** (two recessive alleles, *gg*), or **heterozygous** (one dominant and one recessive allele, *Gg*). **Genotype** refers to an individual's genes, while **phenotype** refers to an individual's appearance (Fig. 11.1). Homozygous dominant and heterozygous individuals show the dominant phenotype; homozygous recessive individuals show the recessive phenotype.

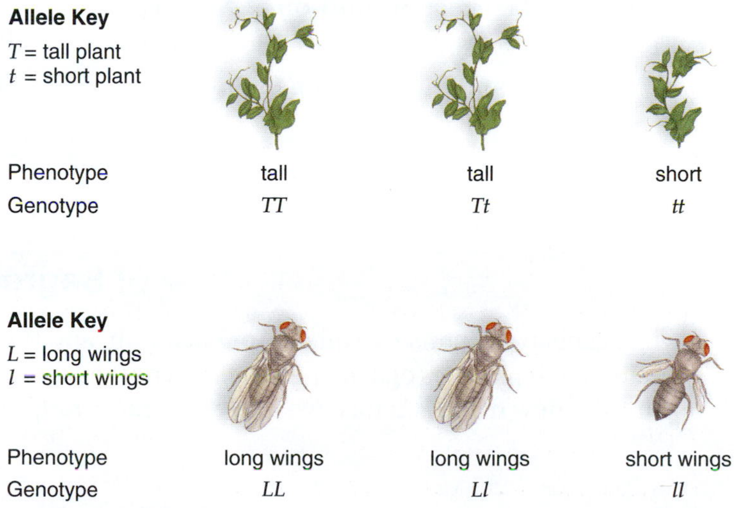

Figure 11.1 Genotype versus phenotype. Only with homozygous recessive do you immediately know the genotype.

Allele Key
T = tall plant
t = short plant

Phenotype	tall	tall	short
Genotype	*TT*	*Tt*	*tt*

Allele Key
L = long wings
l = short wings

Phenotype	long wings	long wings	short wings
Genotype	*LL*	*Ll*	*ll*

Punnett Squares

Punnett squares, named after the man who first used them, allow you to easily determine the results of a cross between individuals whose genotypes are known. Consider that when fertilization occurs, two gametes, such as a sperm and an egg, join together. Whereas individuals have two alleles for every trait, gametes have only one allele because alleles are on the chromosomes and homologues separate during meiosis. Heterozygous parents with the genotype *Aa* produce two types of gametes: 50% of the gametes contain an *A* and 50% contain an *a*. A Punnett square allows you to vertically line up all possible types of sperm and to horizontally line up all possible types of eggs. Every possible combination of gametes occurs within the squares and these combinations indicate the genotypes of the offspring. In Figure 11.2, one offspring is *AA* = homozygous dominant, two are *Aa* = heterozygous, and one is *aa* = homozygous recessive. Therefore, three of the offspring will have the dominant phenotype and one individual will have the recessive phenotype. This is said to be a **phenotypic ratio** of 3:1.

A Punnett square can be used for any cross regardless of the trait(s) and the genotypes of the parents. All you need to do is use the correct letters for the particular trait(s), and make sure you have given the parents the correct genotypes and correct proportion of each type gamete. Then you can determine the genotypes of the offspring and the resulting phenotypic ratio among the offspring.

Figure 11.2 What are the expected results of a cross?
A Punnett square allows you to determine the expected phenotypic ratio for a cross.

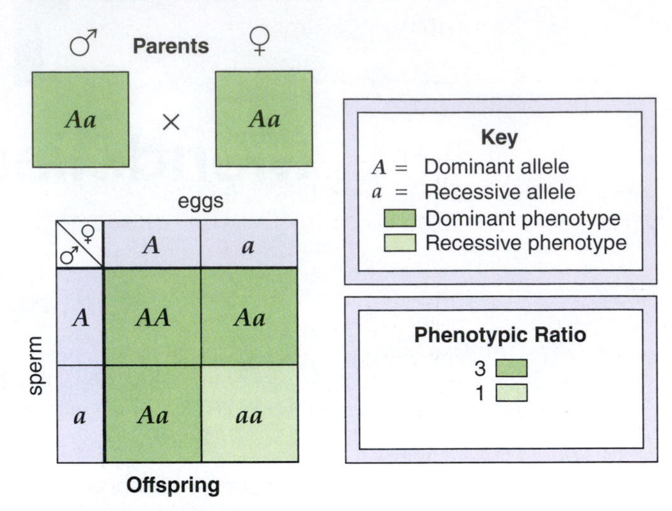

Virtual Lab Punnett Squares A virtual lab called Punnett Squares is available on the *Concepts of Biology* website **www.mhhe.com/maderconcepts3**. It will allow you to practice filling in Punnett squares and determining your results.

11.1 One-Trait Crosses

A single pair of alleles is involved in one-trait crosses. Mendel found that reproduction between two heterozygous individuals *(Aa)*, called a **monohybrid cross**, results in both dominant and recessive phenotypes among the offspring. In Figure 11.2, the expected phenotypic ratio among the offspring is 3:1. Three offspring have the dominant phenotype for every one that has the recessive phenotype.

Mendel realized that these results are obtainable only if the alleles of each parent segregate (separate from each other) during meiosis. Therefore, Mendel formulated his first law of inheritance:

Law of Segregation

Each organism contains two alleles for each trait, and the alleles segregate during the formation of gametes. Each gamete (egg or sperm) then contains only one allele for each trait. When fertilization occurs, the new organism has two alleles for each trait, one from each parent.

Inheritance is a game of chance. Just as there is a 50% probability of heads or tails when tossing a coin, there is a 50% probability that a sperm or egg will have an *A* or an *a* when the parent is *Aa*. The chance of an equal number of heads or tails improves as the number of tosses increases. In the same way, the chance of an equal number of gametes with *A* and *a* improves as the number of gametes increases. Therefore, the 3:1 ratio among offspring is more likely the more offspring you count for the same type cross.

Color of Tobacco Seedlings

In tobacco plants, a dominant allele *(C)* for chlorophyll gives the plants a green color, and a recessive allele *(c)* for chlorophyll causes a plant to appear white. If a tobacco plant is homozygous for the recessive allele *(c)*, it cannot manufacture chlorophyll and thus appears white (Fig. 11.3).

Figure 11.3 Monohybrid cross. These tobacco seedlings are growing on an agar plate. The white plants cannot manufacture chlorophyll.

Experimental Procedure: Color of Tobacco Seedlings

1. Obtain a numbered agar plate on which tobacco seedlings are growing. They are the offspring of a cross between heterozygous parents: the cross *Cc* × *Cc*. Complete the Punnett square to determine the expected phenotypic ratio.

   What is the expected phenotypic ratio? ______________________________

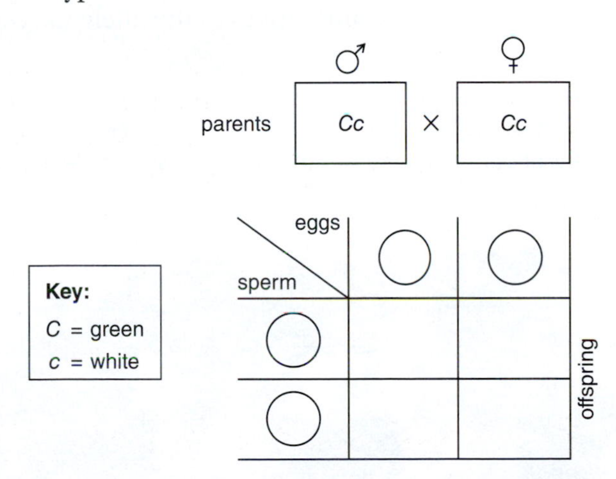

2. Record the plate number and using a stereomicroscope, view the seedlings. Count the number that are green and the number that are white. Record your results in Table 11.1.
3. Repeat this procedure for two additional plates. Total the number that are green and the number that are white.
4. Complete Table 11.1 by recording the class data. Total the number that are green and the number that are white per class.

Table 11.1 Color of Tobacco Seedlings			
	Number of Offspring		
	Green Color	**White Color**	**Phenotypic Ratio**
Plate # ________			
Plate # ________			
Plate # ________			
Totals			
Class data			

Conclusions: Color of Tobacco Seedlings

- In the last column of Table 11.1, record the actual phenotypic ratio per observed plate; per total number of green versus white plants you counted; and per the entire class. To determine the actual phenotypic ratio, divide the number of green color seedlings in a plate by the number of white color seedlings in a plate. Do your results differ from the expected phenotypic ratio? ____________________ If so, explain.

  ____

  ____

- Mendel found that the more plants he counted, the closer he came to the expected phenotypic ratio. Was your class data closer to the expected phenotypic ratio than your individual data? ____________

  This is expected, because the more crosses you observe, the more likely it is that all types of sperm and eggs will have a chance to come together.

Color of Corn Kernels

In corn plants, the allele for purple kernel *(P)* is dominant over the allele for yellow kernel *(p)* (Fig. 11.4).

Figure 11.4 Monohybrid cross.
Two types of kernels are seen on an ear of corn following a monohybrid cross: purple and yellow.

Experimental Procedure: Color of Corn Kernels

1. Obtain an ear of corn from the supply table. You will be examining the results of the cross *Pp* × *pp*. Complete the Punnett square to determine the expected phenotypic ratio. Note that when one parent has only one possible type of gamete, only one column is needed in the Punnett square.

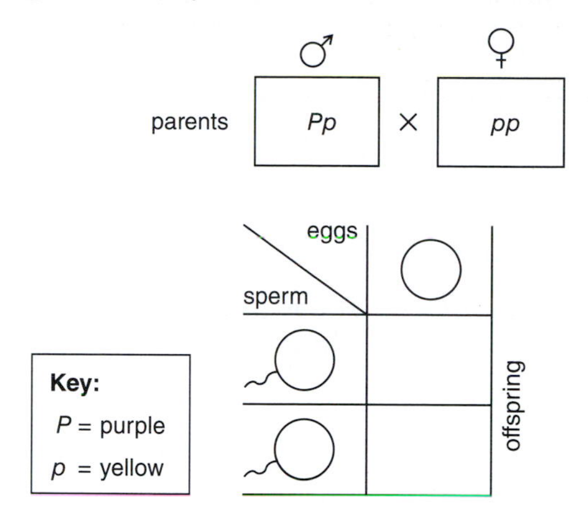

What is the expected phenotypic ratio among the offspring? ______________________

2. Record the sample number and then count the number of kernels that are purple and the number that are yellow. As before, use two more samples, and record your results in Table 11.2.

Table 11.2 Color of Corn Kernels

	Number of Kernels		
	Purple	**Yellow**	**Phenotypic Ratio**
Sample # ________			
Sample # ________			
Sample # ________			
Totals			
Class data			

Conclusions: Color of Corn Kernels

- In the last column of Table 11.2, record the actual phenotypic ratio per observed sample; per total number of purple versus yellow kernels you counted; and per the entire class. Do your results differ from the expected phenotypic ratio? ________________ If so, explain. ____
- Was your class data closer to the expected phenotypic ratio than your individual data? ____________ If so, explain. ____

One-Trait Genetics Problems

1. In pea plants, purple flowers *(P)* is dominant and white flowers *(p)* is recessive. What is the genotype of pure-breeding white plants? Pure-breeding means that they produce plants with only one phenotype. ___________ If pure-breeding purple plants are crossed with these white plants, what phenotype is expected? ___________
2. In pea plants, tall *(T)* is dominant and short *(t)* is recessive. A heterozygous tall plant is crossed with a short plant. What is the expected phenotypic ratio? ___________
3. Unexpectedly to the farmer, two tall plants have some short offspring. What is the genotype of the parent plants and the short offspring? parent ___________ offspring ___________
4. In horses, two trotters are mated to each other and produce only trotters; two pacers are mated to each other and produce only pacers. When one of these trotters is mated to one of the pacers, all the horses are trotters. Create a key and show the cross. key ___________ cross ___________
5. A brown dog is crossed with two different black dogs. The first cross produces only black dogs and the second cross produces equal numbers of black and brown dogs. What is the genotype of the brown dog? ___________ the first black dog? ___________ the second black dog? ___________
6. In pea plants, green pods *(G)* is dominant and yellow pods *(g)* is recessive. When two pea plants with green pods are crossed, 25% of the offspring have yellow pods. What is the genotype of all plants involved? plants with green pods? ___________ plants with yellow pods? ___________
7. A breeder wants to know if a dog is homozygous black or heterozygous black. If the dog is heterozygous, which cross is more likely to produce a brown dog, *Bb* × *bb* or *Bb* × *Bb*? Explain your answer.

   ____

   ____
8. If the cross in question 6 produces 220 plants, how many offspring have green pods and how many have yellow pods? ___________ If the cross in question 2 produces 220 plants, how many offspring are tall and how many are short? ___________

11.2 Two-Trait Crosses

Two-trait crosses involve two pairs of alleles. Mendel found that during a **dihybrid cross,** when two dihybrid individuals *(AaBb)* reproduce, the phenotypic ratio among the offspring is 9:3:3:1, representing four possible phenotypes. He realized that these results could be obtained only if the alleles of the parents segregated independently of one another when the gametes were formed. From this, Mendel formulated his second law of inheritance:

Law of Independent Assortment

Members of an allelic pair segregate (assort) independently of members of another allelic pair. Therefore, all possible combinations of alleles can occur in the gametes.

The FOIL method is a way to determine the gametes. FOIL stands for *F*irst two alleles from each trait; *O*uter two alleles from each trait; *I*nner two alleles from each trait; *L*ast two alleles from each trait. Here is how the FOIL method can help you determine the gametes for the genotype *PpSs:*

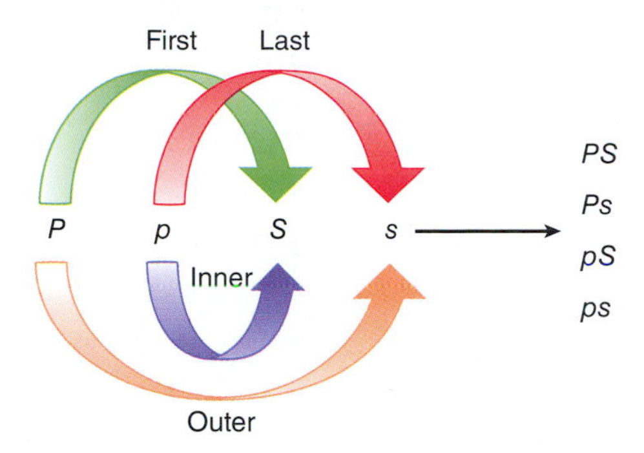

Color and Texture of Corn

In corn plants, the allele for purple kernel *(P)* is dominant over the allele for yellow kernel *(p),* and the allele for smooth kernel *(S)* is dominant over the allele for rough kernel *(s)* (Fig. 11.5).

Figure 11.5 Dihybrid cross. Four types of kernels are seen on an ear of corn following a dihybrid cross: purple smooth, purple rough, yellow smooth, and yellow rough.

Experimental Procedure: Color and Texture of Corn

1. Obtain an ear of corn from the supply table. You will be examining the results of the cross *PpSs* × *PpSs.*
2. Do the Punnett square on page 118 in order to state the expected phenotypic ratio among the offspring. ____________________
3. Count the number of kernels of each possible phenotype listed in Table 11.3. Record the sample number and your results in Table 11.3. Use three samples, and total your results for all samples. Also record the class data (i.e., the number of kernels that are the four phenotypes per class).

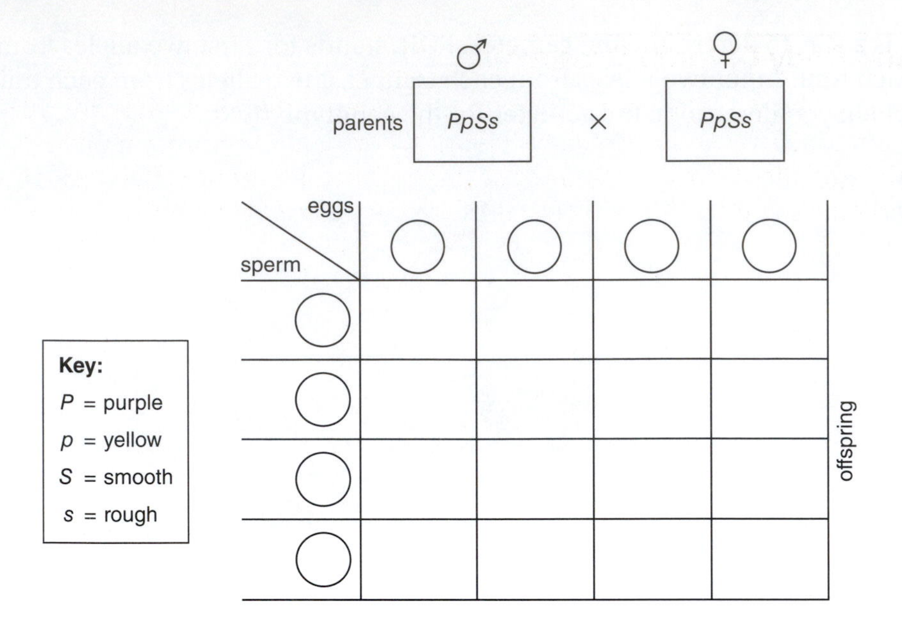

Table 11.3 Color and Texture of Corn					
	Number of Kernels				
	Purple Smooth	**Purple Rough**	**Yellow Smooth**	**Yellow Rough**	**Phenotypic Ratio**
Sample # ________					
Sample # ________					
Sample # ________					
Totals					
Class data					

Conclusions: Color and Texture of Corn

- Calculate the actual phenotypic ratios based on the data and record in Table 11.3. Do the results differ from the expected ratio per your data? ____________ Per class data? ____________ If so, explain.

  ____

  ____

  ____

Wing Length and Body Color in *Drosophila*

Drosophila are the tiny flies you often see flying around ripe fruit; therefore, they are called fruit flies. If a culture bottle of fruit flies is on display, take a look at it. Because so many flies can be grown in a small culture bottle, fruit flies have contributed substantially to our knowledge of genetics. If you were to examine *Drosophila* flies under the stereomicroscope, they would appear like this:

In *Drosophila,* long wings *(L)* are dominant over short (vestigial) wings *(l),* and gray body *(G)* is dominant over black (ebony) body *(g).* Consider the cross *LlGg* × *llgg* and complete this Punnett square:

Key:
L = long wing
l = short (vestigial) wing
G = gray body
g = ebony (black) body

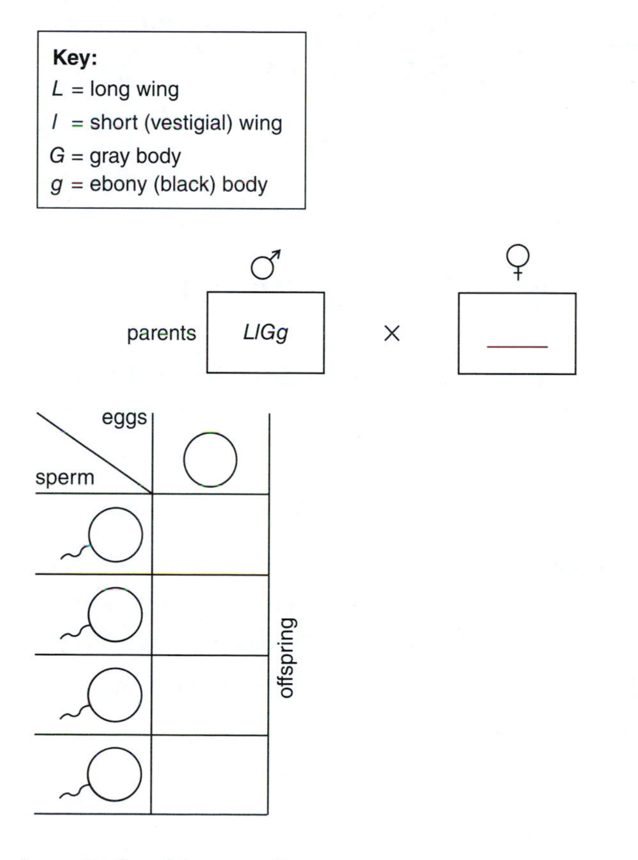

What is the expected phenotypic ratio for this cross? ____________________

Experimental Procedure: Wing Length and Body Color in Drosophila

If your instructor has frozen flies available, cross out the numbers in Table 11.4 and use the stereomicroscope or a hand lens to count the flies of each type given in Table 11.4. Otherwise simply use the data supplied for you in Table 11.4.

Table 11.4 Wing Length and Body Color in *Drosophila**

	Phenotypes				
	Long Gray	**Long Ebony**	**Short Gray**	**Short Ebony**	**Phenotypic Ratio**
Number of offspring	28	32	28	30	
Class data	128	120	120	120	

*Wings and body are understood in this table.

Conclusions: Wing Length and Body Color in Drosophila

- Calculate the actual phenotypic ratio based on the data and record in Table 11.4. Do the results differ from the expected ratio per your data? ___________ Per class data? ___________ If so, explain.

  ____

  ____

Two-Trait Genetics Problems

1. In tomatoes, tall is dominant and short is recessive. Red fruit is dominant and yellow fruit is recessive. Choose a key for height ___________ for color of fruit ___________ What is the genotype of a plant heterozygous for both traits? ___________ What are the possible gametes for this plant?

   ____

2. Using words, what are the likely parental genotypes if the results of a two-trait problem are 1:1:1:1 among the offspring? ____________________ × ____________________
3. In horses, black *(B)* and a trotting gait *(T)* are dominant, while brown *(b)* and a pacing gait *(t)* are recessive. If a black trotter (homozygous for both traits) is mated to a brown pacer, what phenotypic ratio is expected among the offspring? ____________________
4. Two black trotters have a brown pacer offspring. What is the genotype of all horses involved? black trotter parents ___________ brown pacer offspring ___________
5. The phenotypic ratio among the offspring for two corn plants producing purple and smooth kernels is 9:3:3:1. (See page 118 for the key.) What is the genotype of these plants? parental plants ___________ the 9 offspring ___________ 3 of the offspring ___________ the other 3 offspring ___________ the 1 offspring ___________

6. Which matings could produce at least some fruit flies heterozygous in both traits? Write yes or no beside each. (You do not need a key.)

ggLl × *Ggll* ____________ *GGLl* × *ggLl* ____________ *GGLL* × *ggll* ____________

Explain your answer. ____

____

7. State two new crosses that could not produce fruit flies heterozygous in both traits.

__________ × __________ __________ × __________

8. Chimpanzees are not deaf if they inherit both an allele *E* and an allele *G*. A cross between two deaf chimpanzees produces only chimpanzees that can hear. What are the genotypes of all chimpanzees involved? parents __________ × __________ offspring __________

11.3 X-Linked Crosses

In animals such as fruit flies, chromosomes differ between the sexes. All but one pair of chromosomes in males and females are the same; these are called **autosomes** because they do not actively determine sex. The pair that is different is called the **sex chromosomes.** In fruit flies and humans, the sex chromosomes in females are XX and those in males are XY.

Some alleles on the X chromosome have nothing to do with gender, and these genes are said to be X-linked. The Y chromosome does not carry these genes and indeed carries very few genes. Males with a normal chromosome inheritance are never heterozygous for X-linked alleles, and if they inherit a recessive X-linked allele it will be expressed.

Red/White Eye Color in *Drosophila*

In fruit flies, red eyes (X^R) are dominant over white eyes (X^r). You will be examining the results of the cross $X^RY \times X^RX^r$. Complete this Punnett square and state the expected phenotypic ratio for this cross.

females ____________________ males ____________________

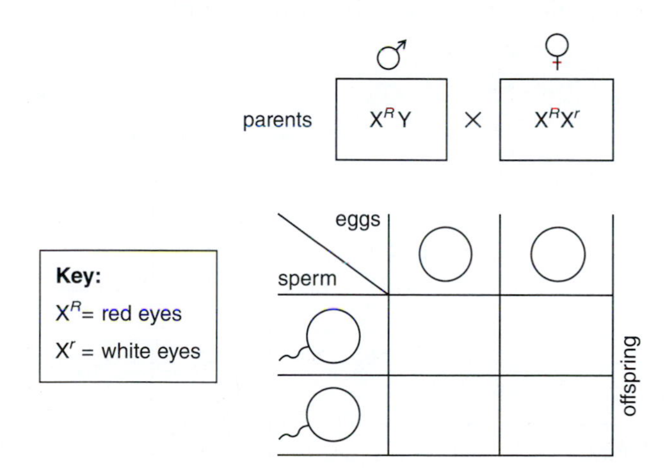

Experimental Procedure: Red/White Eye Color in Drosophila

If your instructor has frozen flies available, cross out the numbers in Table 11.5 and use the stereomicroscope or a hand lens to count the flies of each type given in Table 11.5. Use the art on page 119 to tell males from females and record male and female data separately. If frozen flies are not available, simply use the data supplied for you in Table 11.5.

Table 11.5 Red/White Eye Color in *Drosophila*

		Number of Offspring	
Your Data:	**Red Eyes**	**White Eyes**	**Phenotypic Ratio**
Males	16	17	
Females	63	0	
Class Data:			
Males	45	48	
Females	215	0	

Conclusions: Red/White Eye Color in Drosophila

- Calculate the phenotypic ratios based on the data for males and females separately and record in Table 11.5. Do the results differ from the expected ratio per individual data? ___________

  Per class data? ___________ If so, explain. ____

  ____

  ____

Virtual Lab Sex-linked Traits A virtual lab called Sex-linked Traits is available on the *Concepts of Biology* website **www.mhhe.com/maderconcepts3**. It will allow you to practice filling in Punnett squares and determining your results for X-linked traits.

- Using the Punnett square provided, calculate the expected phenotypic results for the cross $X^RY \times X^rX^r$. What is the expected phenotypic ratio among the offspring? males ___________ females ___________

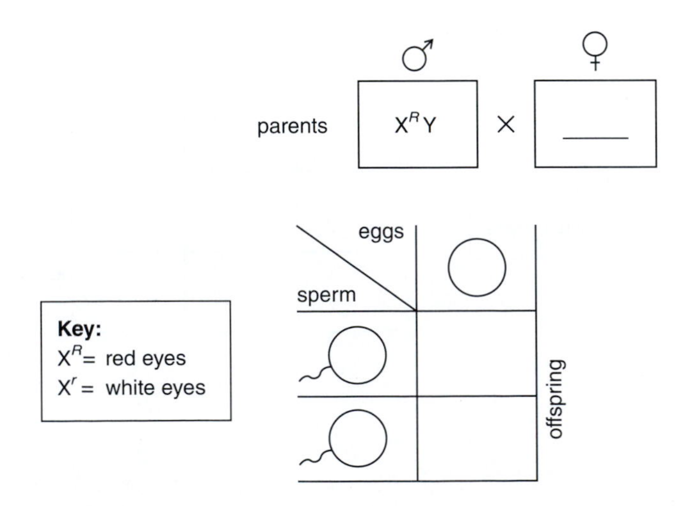

X-Linked Genetics Problems

1. State the genotypes and gametes for each of these fruit flies:

	genotype	gamete(s)
white-eyed male	________	________
white-eyed female	________	________
red-eyed male	________	________
homozygous red-eyed female	________	________
heterozygous red-eyed female	________	________

2. What are the phenotypic ratios if a white-eyed female is crossed with a red-eyed male?

   males ________ females ________

3. Regardless of any type cross, do white-eyed males inherit the allele for white eyes from their father or their mother? ________ Explain your answer. ________

4. In sheep, horns are sex linked; H = horns and h = no horns. Using symbols, what cross do you recommend if a farmer wants to produce hornless males? ________ × ________

5. In *Drosophila,* bar eye is sex linked; B = bar eye and b = no bar eye. What are the phenotypic ratios for these crosses?

   bar-eyed male × non-bar-eyed female ________, ________

   bar-eyed male × heterozygous female ________, ________

   non-bar-eyed male × heterozygous female ________, ________

6. A female fruit fly has white eyes. What is the genotype of the father? ________ What could be the genotype of the mother? ________ or ________

7. In a cross between fruit flies, all the males have white eyes and the females are 1:1. What is the genotype of the parents? female parent ________ male parent ________

8. In a cross between fruit flies, a white-eyed male and a red-eyed female produce no offspring that have white eyes. What is the genotype of the parents? male parent ________ female parent ________

9. Make up a sex-linked genetic cross using words and parental genotypes. Create a key and show a Punnett square and the phenotypic ratios.

Laboratory Review 11

1. If offspring exhibit a 3:1 phenotypic ratio, what are the genotypes of the parents? ____________

2. In fruit flies, which of the characteristics you studied was X-linked? ____________

3. If offspring exhibit a 9:3:3:1 phenotypic ratio, what are the genotypes of the parental *(P)* generation?

   ____

4. If a cross results in 90 long-winged flies to 30 vestigial-winged flies, what are the phenotypes of the parents?

   ____

5. Briefly describe the life cycle of *Drosophila*. ____________

   ____

6. In the cross *AaBb* × *aaBb*, what are the gametes for *AaBb*? ________ For *aaBb*? ________ What are the genotypic (not phenotypic) results for this cross? ____________

   ____

7. What is the genotype of a white-eyed male fruit fly? ____________

8. Suppose you count 40 green tobacco seedlings and 2 white tobacco seedlings in one agar plate. Do your results show that both parent plants were heterozygous for the color allele? ____________

   Explain your answer. ____________

9. Suppose you count tobacco seedlings in six agar plates, and your data are as follows: 125 green plants and 39 white plants. What is the phenotypic ratio? ____________

   ____

10. Suppose that students in the laboratory periods before you removed some of the purple and yellow corn kernels on the ears of corn as they were performing the Experimental Procedure. What specific effects would this have on your results? ____________

    ____

Concepts of Biology Website

Instructors can find lab prep information and answers to all of the laboratory questions in the Laboratory Resource Guide. *Students* can practice their knowledge with quizzes, animations, flashcards, and much more.

www.mhhe.com/maderconcepts3

McGraw-Hill Access Science Website

An online encyclopedia of science and technology that provides information, including videos, that can enhance the laboratory experience.

www.accessscience.com

Mendelian Genetics

LABORATORY

12

Human Genetics

Learning Outcomes

12.1 Determining the Genotype

- Determine the genotype by observation of the person and their relatives. 125–27

12.2 Determining Inheritance

- Do genetic problems involving autosomal dominant, autosomal recessive, and X-linked recessive alleles. 128–31
- Do genetic problems involving multiple allele inheritance and use blood type to help determine paternity. 132–33

12.3 Genetic Counseling

- Analyze a karyotype to determine if a person's chromosomal inheritance is as expected or whether a chromosome anomaly has occurred. 133–34
- Analyze a pedigree to determine if the pattern of inheritance is autosomal dominant, autosomal recessive, or X-linked recessive. 135–36
- Construct a pedigree to determine the chances of inheriting a particular phenotype when provided with generational information. 137

Introduction

In this laboratory, you will discover that the same principles of genetics apply to humans as they do to plants and fruit flies. A gene has two alternate forms, called **alleles,** for any trait, such as hairline, finger length, and so on. One possible allele, designated by a capital letter, is **dominant** over the **recessive** allele, designated by a lowercase letter. An individual can be **homozygous dominant** (two dominant alleles, *EE*), **homozygous recessive** (two recessive alleles, *ee*), or **heterozygous** (one dominant and one recessive allele, *Ee*). **Genotype** refers to an individual's alleles, and **phenotype** refers to an individual's appearance (Fig. 12.1). Homozygous dominant and also heterozygous individuals show the dominant phenotype; homozygous recessive individuals show the recessive phenotype.

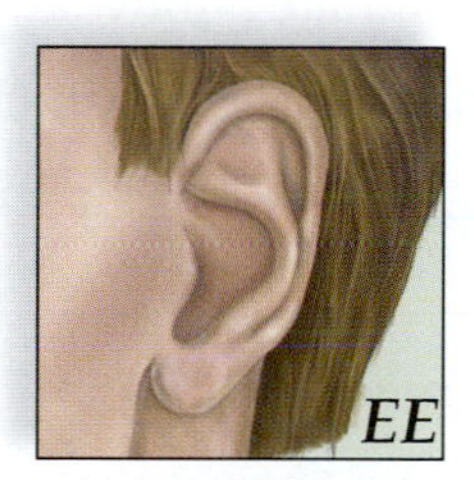

a. Unattached earlobe

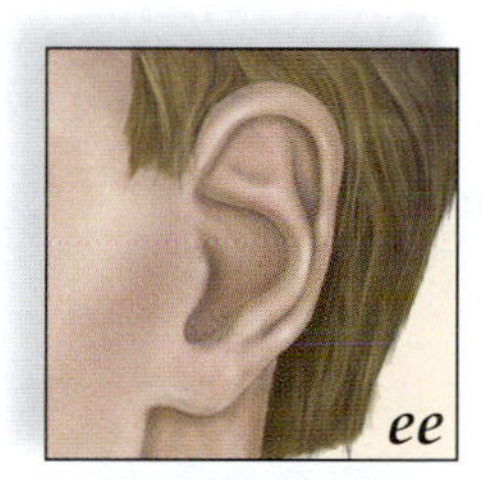

b. Attached earlobe

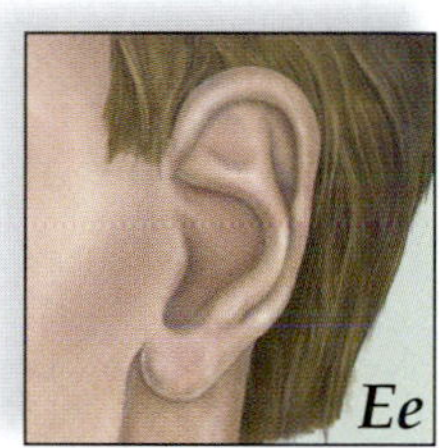

c. Unattached earlobe

Figure 12.1 Genotype versus phenotype. Unattached earlobes *(E)* are dominant over attached earlobes *(e)*. **a.** Homozygous dominant individuals have unattached earlobes. **b.** Homozygous recessive individuals have attached earlobes. **c.** Heterozygous individuals have unattached earlobes.

12.1 Determining the Genotype

Humans inherit 46 chromosomes that occur in 23 pairs. Twenty-two of these pairs are called autosomes and one pair is the sex chromosomes. Autosomal traits are determined by alleles on the autosomal chromosomes.

Autosomal Dominant and Recessive Traits

Figure 12.2 shows a few human traits.

1. What is the homozygous dominant genotype for type of hairline? ______

   What is the phenotype? ______

2. What is the homozygous recessive genotype for finger length? ______

   What is the phenotype? ______

3. Why does the heterozygous individual *Ff* have freckles? ______

Figure 12.2 Commonly inherited traits in humans.
The alleles indicate which traits are dominant and which are recessive.

a. Widow's peak: *WW* or *Ww*

b. Straight hairline: *ww*

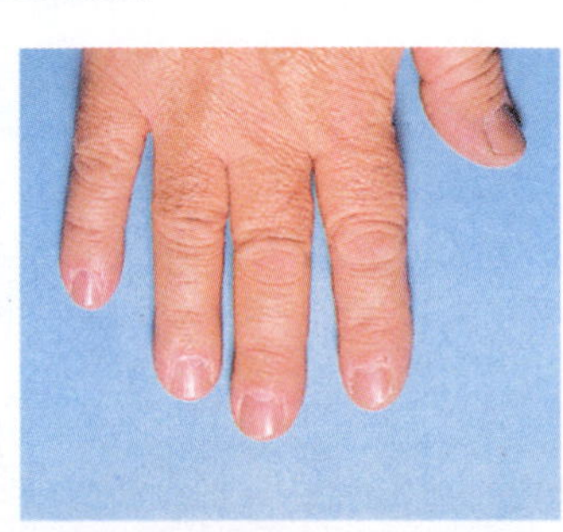

e. Short fingers: *SS* or *Ss*

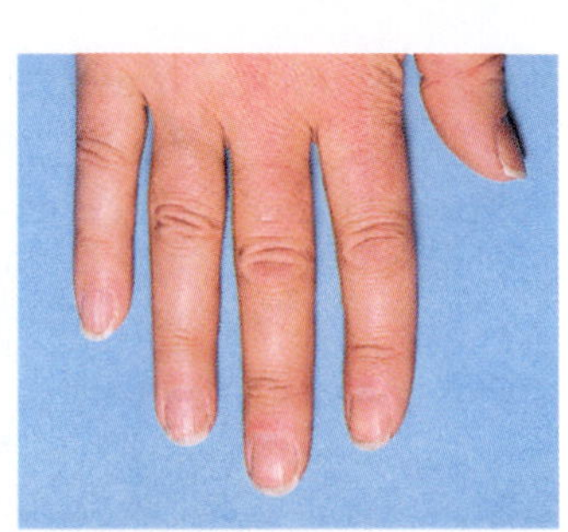

f. Long fingers: *ss*

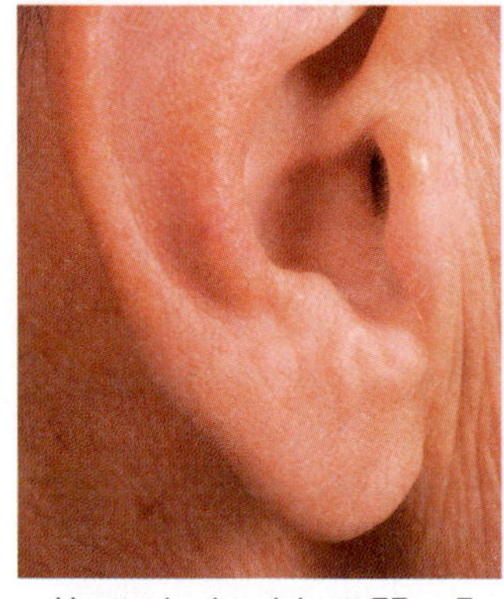

c. Unattached earlobes: *EE* or *Ee*

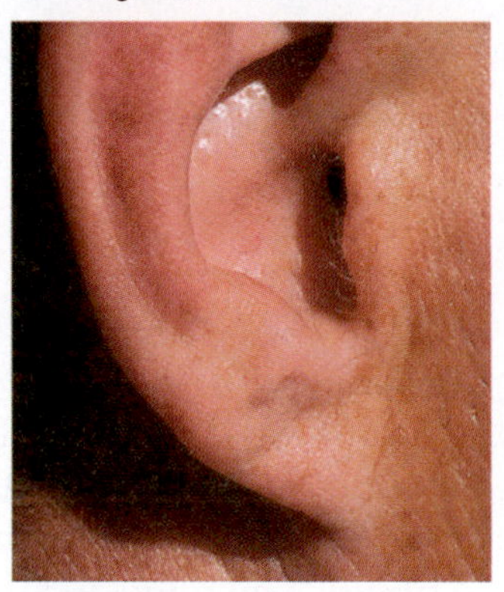

d. Attached earlobes: *ee*

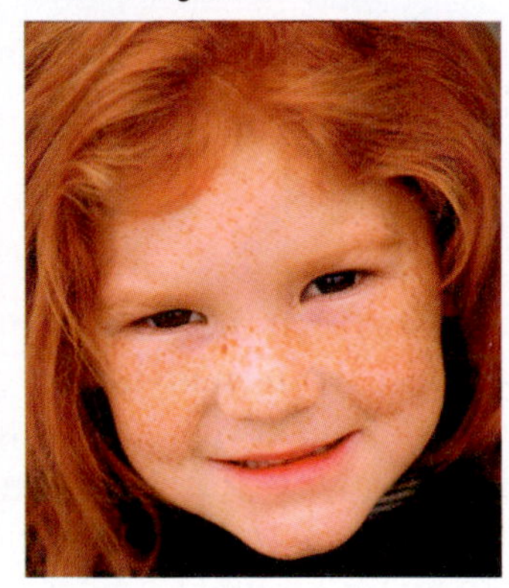

g. Freckles: *FF* or *Ff*

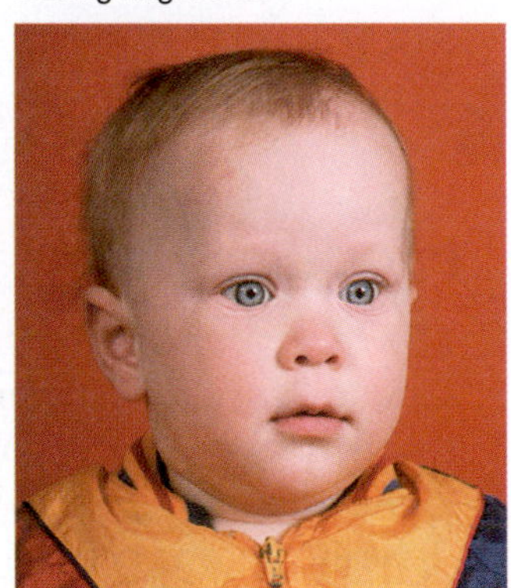

h. No freckles: *ff*

These genetic problems use the alleles from Figure 12.2 and Table 12.1.

4. Maria and the members of her immediate family have attached earlobes. What is Maria's genotype? ______ Her maternal grandfather has unattached earlobes. Deduce the genotype of her maternal grandfather. ______ Explain your answer. ______

   ______

5. Moses does not have a bent little finger, but his parents do. Deduce the genotype of his parents. ______ Deduce the genotype of Moses. ______ Explain your answer. ______

   ______

6. Manny is adopted. He has hair on the back of his hand. Could both of his parents have had hair on the back of the hand? ______ Could both of his parents have had no hair on the back of the hand? ______ Explain your answer. ______

   ______

Experimental Procedure: Human Traits

1. For this Experimental Procedure, you will need a lab partner to help you determine your phenotype for the traits listed in the first column of Table 12.1.
2. Determine your probable genotype. If you have the recessive phenotype, you know your genotype. If you have the dominant phenotype, you may be able to decide whether you are homozygous dominant or heterozygous by recalling the phenotype of your parents, siblings, or children. Circle your probable genotype in the second column of Table 12.1.
3. Your instructor will tally the class's phenotypes for each trait so that you can complete the third column of Table 12.1.
4. Complete Table 12.1 by calculating the percentage of the class with each trait. Are dominant phenotypes always the most common in a population? ____________ Explain your answer. ____________

____

Table 12.1 Autosomal Human Traits

Trait: d · Dominant r · Recessive	Probable Genotypes	Number in Class	Percentage of Class with Trait
Hairline:			
Widow's peak (d)	*WW* or *Ww*	________	________
Straight hairline (r)	*ww*	________	________
Earlobes:			
Unattached (d)	*UU* or *Uu*	________	________
Attached (r)	*uu*	________	________
Skin pigmentation:			
Freckles (d)	*FF* or *Ff*	________	________
No freckles (r)	*ff*	________	________
Hair on back of hand:			
Present (d)	*HH* or *Hh*	________	________
Absent (r)	*hh*	________	________
Thumb hyperextension—"hitchhiker's thumb":			
Last segment cannot be bent backward (d)	*TT* or *Tt*	________	________
Last segment can be bent back to 60° (r)	*tt*	________	________
Bent little finger:			
Little finger bends toward ring finger (d)	*LL* or *Ll*	________	________
Straight little finger (r)	*ll*	________	________
Interlacing of fingers:			
Left thumb over right (d)	*II* or *Ii*	________	________
Right thumb over left (r)	*ii*	________	________

12.2 Determining Inheritance

Recall that a Punnett square is a means to determine the genetic inheritance of offspring if the genotypes of both parents are known. In a **Punnett square**, all possible types of sperm are lined up vertically, and all possible types of eggs are lined up horizontally, or vice versa, so that every possible combination of gametes occurs within the square. Figure 12.3 shows how to construct a Punnett square when autosomal alleles are involved.

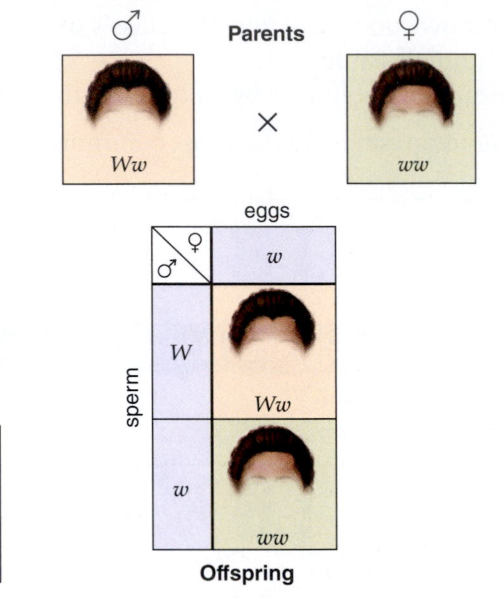

Figure 12.3 Punnett square.
In a Punnett square, all possible sperm are displayed vertically and all possible eggs are displayed horizontally, or vice versa. The genotypes of the offspring (in this case, also the phenotypes) are in the squares.

Results of Cross
Phenotypic ratio 1:1
Chance of widow's peak $\frac{1}{2}$ = 50%
Chance of straight hairline $\frac{1}{2}$ = 50%

Inheritance of Genetic Disorders

Figure 12.4 can be used to learn the chances of a particular phenotype.

In Figure 12.4*a*,

¼ of the offspring have the recessive phenotype = ____________ % chance

¾ of the offspring have the dominant phenotype = ____________ % chance

In Figure 12.4*b*,

½ of the offspring have the recessive or the dominant phenotype = ____________ % chance

In all the following genetic problems, use letters to fill in the parentheses with the genotype of the parents.

1. **a.** With reference to Figure 12.4*a*, if a genetic disorder is recessive and both parents are heterozygous (__________), what are the chances that an offspring will have the disorder? ____________
   b. With reference to Figure 12.4*a*, if a genetic disorder is dominant and both parents are heterozygous (__________), what are the chances that an offspring will have the disorder? ____________
2. **a.** With reference to Figure 12.4*b*, if the parents are heterozygous (__________) by homozygous recessive (__________), and the genetic disorder is recessive, what are the chances that the offspring will have the disorder? ____________
   b. With reference to Figure 12.4*b*, if the parents are heterozygous (__________) by homozygous recessive (__________), and the genetic disorder is dominant, what are the chances that an offspring will have the disorder? ____________

Figure 12.4 Two common patterns of autosomal inheritance in humans.
a. Both parents are heterozygous. **b.** One parent is heterozygous and the other is homozygous recessive. The letter *A* stands for any trait that is dominant and the letter *a* stands for any trait that is recessive. Substitute the correct alleles for the problem you are working on. For example, *C* = normal; *c* = cystic fibrosis.

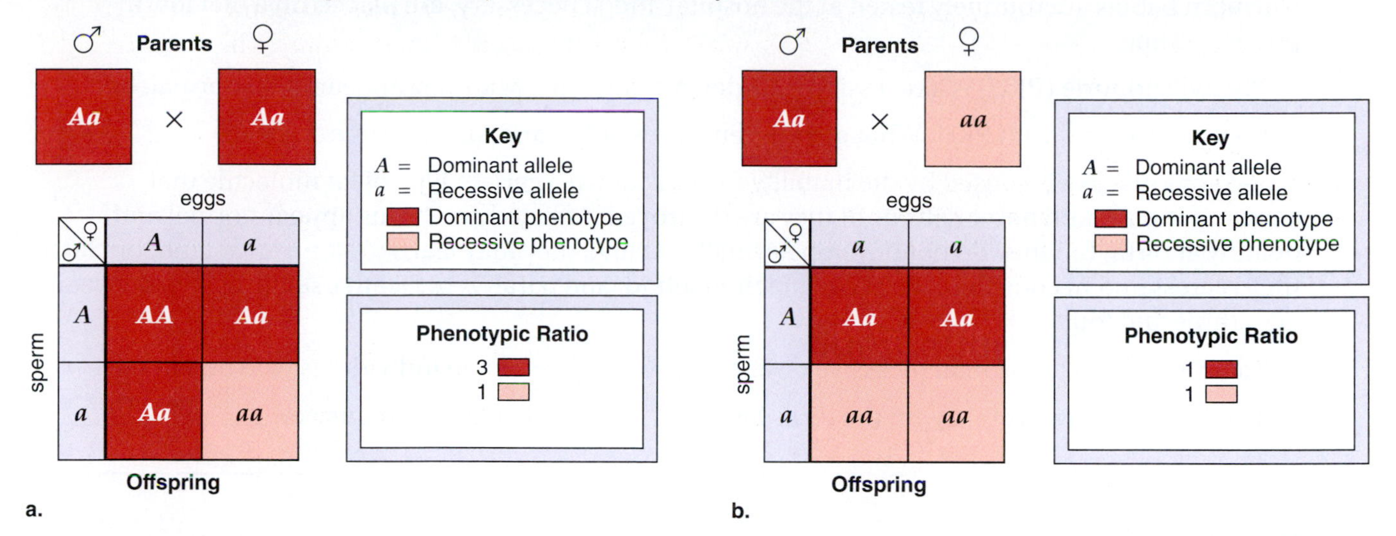

Autosomal Disorders

1. **Neurofibromatosis (NF),** sometimes called von Recklinghausen disease, is one of the most common genetic disorders. It affects roughly 1 in 3,000 people. It is seen equally in every racial and ethnic group throughout the world. At birth or later, the affected individual may have six or more large tan spots on the skin. Such spots may increase in size and number and become darker. Small benign tumors (lumps) called neurofibromas may occur under the skin or in the muscles. Neurofibromas are made up of nerve cells and other cell types.

   Neurofibromatosis is a dominant disorder. If a heterozygous woman reproduces with a homozygous normal man, what are the chances a child will have neurofibromatosis? ____________

2. **Cystic fibrosis** is due to abnormal mucus-secreting tissues. At first, the infant may have difficulty regaining the birth weight despite good appetite and vigor. A cough associated with a rapid respiratory rate but no fever indicates lung involvement. Large, frequent, and foul-smelling stools are due to abnormal pancreatic secretions. Whereas children previously died in infancy due to infections, they now often survive because of antibiotic therapy.

   Cystic fibrosis is a recessive disorder. A **carrier** is an individual that appears to be normal but carries a recessive allele for a genetic disorder. A man and a woman are both carriers (__________) for cystic fibrosis. What are the chances a child will have cystic fibrosis? ____________

3. **Huntington disease** does not appear until the 30s or early 40s. There is a progressive deterioration of the individual's nervous system that eventually leads to constant thrashing and writhing movements until insanity precedes death. Studies suggest that Huntington disease is due to a single faulty gene that has multiple effects, in which case there is now hope for a cure.

   People with Huntington disease seem to be more fertile than others. It is amazing that more than 1,000 of the cases in the United States in the past century can be traced to one man born in 1831.

   Huntington disease is a dominant disorder. Drina is 25 years old and as yet has no signs of Huntington disease. Her mother does have Huntington disease (________), but her father is free (________) of the disorder. What are the chances that Drina will develop Huntington disease?

   ____

4. **Phenylketonuria (PKU)** is characterized by severe intellectual impairment due to an abnormal accumulation of the common amino acid phenylalanine within cells, including neurons. The disorder takes its name from the presence of a breakdown product, phenylketone, in the urine and blood. Newborn babies are routinely tested at the hospital and, if necessary, are placed on a diet low in phenylalanine.

   Phenylketonuria (PKU) is a recessive disorder. Mr. and Mrs. Martinez appear to be normal, but they have a child with PKU. What are the genotypes of Mr. and Mrs. Martinez? ______

5. **Tay–Sachs disease** is caused by the inability to break down a certain type of fat molecule that accumulates around nerve cells until they are destroyed. Afflicted newborns appear normal and healthy at birth, but they do not develop normally. At first, they may learn to sit up and stand, but later they regress and become intellectually impaired, blind, and paralyzed. Death usually occurs between ages three and four.

   Tay–Sachs is an autosomal recessive disorder. Is it possible for two individuals who do not have Tay–Sachs to have a child with the disorder? ____________ Explain your answer. ________________

   ____

X-Linked Disorders

The sex chromosomes designated X and Y carry genes just like the autosomal chromosomes. Some genes, particularly on the X chromosome, have nothing to do with gender inheritance and are said to be X-linked. **X-linked recessive disorders** are due to recessive genes carried on the X chromosomes. Males are more likely to have an X-linked recessive disorder than females because the Y chromosome is blank for this trait. Does a color-blind male give his son a recessive-bearing X or a Y that is blank for the recessive allele? ______________

The possible genotypes and phenotypes for an X-linked recessive disorder are as follows:

Females	**Males**
X^BX^B = normal vision	X^BY = normal vision
X^BX^b = normal vision (carrier)	X^bY = color blindness
X^bX^b = color blindness	

An X-linked recessive disorder in a male is always inherited from his mother. Most likely, his mother is heterozygous and therefore does not show the disorder. She is designated a carrier for the disorder. Figure 12.5 shows how females can become carriers.

1. **a.** What is the genotype for a color-blind female? ________________ How many recessive alleles does a female inherit to be color blind? ________________

   b. What is the genotype for a color-blind male? ______________ How many recessive alleles does a male inherit to be color blind? ____________

2. **a.** With reference to Figure 12.5*a*, if the mother is a carrier (____________) and the father has normal vision (____________), what are the chances that a daughter will be color blind? ________________

   b. A daughter will be a carrier? ______________ **c.** A son will be color blind? ____________

3. **a.** With reference to Figure 12.5*b*, if the mother has normal vision (____________) and the father is color blind (____________), what are the chances that a daughter will be color blind? ____________

   b. A daughter will be a carrier? ______________ **c.** A son will be color blind? ____________

Figure 12.5 Two common patterns of X-linked inheritance in humans.
a. The sons of a carrier mother have a 50% chance of being color blind. **b.** A color-blind father has carrier daughters.

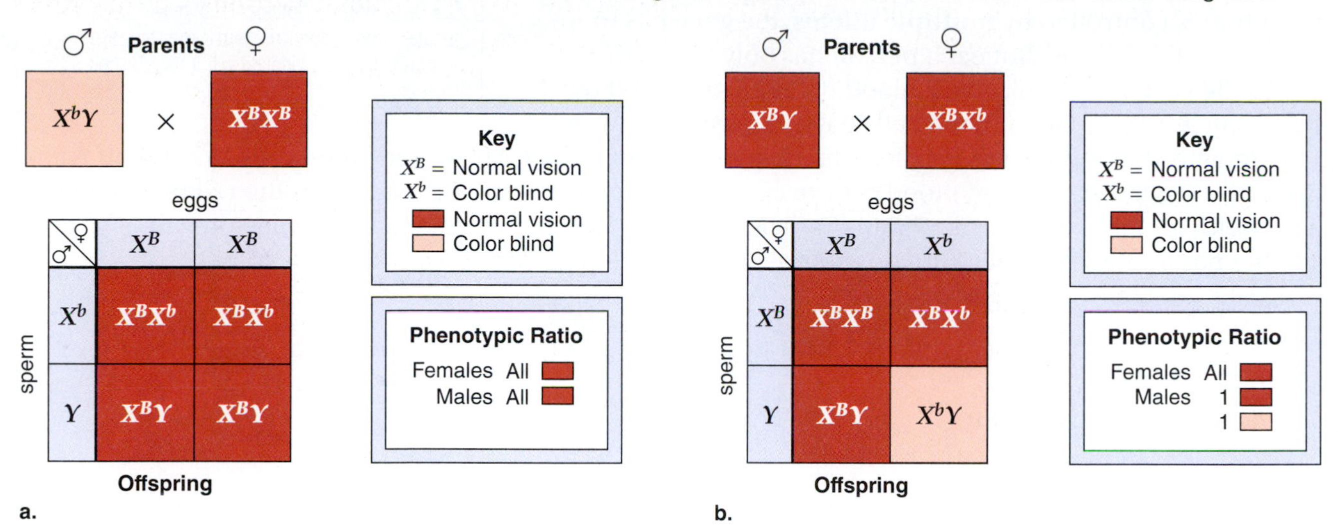

X-Linked Genetics Problems

For **color blindness,** there are two possible X-linked alleles involved. One affects the green-sensitive cones, whereas the other affects the red-sensitive cones. About 6% of men in the United States are color blind due to a mutation involving green perception, and about 2% are color blind due to a mutation involving red perception.

1. A woman with normal color vision (______________), whose father was color blind (__________), marries a man with normal color vision (__________). What genotypes could occur among their offspring? ____

What genotypes could occur if it was the normal-visioned man's father who was color blind?

____

2. Antonio's father is color blind (__________) but his mother is not color blind (__________).

Is Antonio necessarily color blind? ____________ Explain. ______________________________

Could he be color blind? __________ Explain. ______________________________

Hemophilia is called the bleeder's disease because the affected person's blood is unable to clot. Although hemophiliacs do bleed externally after an injury, they also suffer from internal bleeding, particularly around joints. Hemorrhages can be checked with transfusions of fresh blood (or plasma) or concentrates of the clotting protein. The most common type of hemophilia is hemophilia A, due to absence or minimal presence of a particular clotting factor called factor VIII.

3. Make up a cross involving hemophilia that could be answered by a Punnett square, as in Figure 12.5*a* or *b*. ____

____

What is the answer to your genetics problem? ______________________________

Multiple Alleles

> ⚠ **Protective clothing** Wear protective laboratory clothing, latex gloves, and goggles. If the chemicals touch the skin, eyes, or mouth, wash immediately. If inhaled, seek fresh air.

When a trait is controlled by **multiple alleles,** the gene has more than two possible alleles. But each person has only two of the possible alleles. For example, ABO blood type is determined by multiple alleles: I^A, I^B, i. Red blood cells have surface molecules called antigens that indicate they belong to the person. The I^A allele causes red blood cells to carry an A antigen, the I^B allele causes red blood cells to carry a B antigen, and the i allele causes the red blood cells to have neither of these antigens. I^A and I^B are dominant to i. Remembering that each person can have any two of the possible alleles, these are possible genotypes and phenotypes for blood types.

Genotypes	Antigens on Red Cells	Blood Types
I^AI^A, I^Ai	A	A
I^BI^B, I^Bi	B	B
I^AI^B	A and B	AB
ii	none	O

Blood type also indicates whether the person is Rh positve or Rh negative. If the genotype is *DD* or *Dd,* the person is Rh positive and if the genotype is *dd,* the person is Rh negative. It is customary to simply attach a + or − superscript to the ABO blood type, as in A^-.

Experimental Procedure: Using Blood Type to Help Determine Paternity

In this experimental procedure a mother, Wanda, is seeking support for her child, Sophia. We will use blood typing to decide which of three men could possibly be the father.

1. Obtain three testing plates, each of which contains three depressions; vials of blood from possible fathers 1, 2, and 3, respectively; vials of anti-A serum, anti-B serum, and anti-Rh serum. (All of these are synthetic.)
2. Using a wax pencil, number the plates so you know which plate is for possible father 1, 2, or 3. Look carefully at a plate and notice the wells are designated as A, B, or Rh.
3. Being sure to close the cap to each vial in turn, do the following using plate #1:
   Add a drop of father 1 blood to all three wells—close the cap.
   Add a drop of anti-A (blue) to the well designated A—close the cap.
   Add a drop of anti-B (yellow) to the well designated B—close the cap.
   Add a drop of anti-Rh (clear) to the well designated Rh—close the cap.
4. Stir the contents of each well with a mixing stick of the correct color. After a few minutes, examine the wells for agglutination (i.e., granular appearances that indicate the blood type). (Rh^+ takes the longest to react.) If a person had AB^+ blood, which wells would show agglutination? ______________________

   ____
5. Repeat steps 3 and 4 for plates 2 and 3.
6. Record the blood type results for each of the men in Table 12. 2.

Table 12.2 Blood Types of Involved Persons

	Mother*	Child*		Father?	
	Wanda	Sophia	#1	#2	#3
Blood type	B^-	AB^+			

*Your instructor may have you confirm these results.

Conclusion

1. Noting that only father 3 could have given Sophia the Rh antigen, from whom did she receive the I^B allele? ____________ From which parent did she receive the I^A allele? ____________ Is there any other possible interpretation to the results of blood typing? ____

____

Blood Typing Problems

1. A man with type A blood reproduces with a woman who has type B blood. Their child has blood type O. Using I^A, I^B, and i, give the genotype of all persons involved.

   man ____________ woman ____________ child ____________

2. If a child has type AB blood and the father has type B blood, what could the genotype of the mother be?

   ____________ or ____________

3. If both mother and father have type AB blood, they cannot be the parents of a child who has what blood type? ____________

4. What blood types are possible among the children if the parents are $I^A i \times I^B i$? (*Hint:* Do a Punnett square using the possible gametes for each parent.)

   Punnett Square:

12.3 Genetic Counseling

Potential parents are becoming aware that many illnesses are caused by abnormal chromosomal inheritance or by gene mutations. Therefore, they are seeking genetic counseling, which is available in many major hospitals. The counselor helps the couple understand the mode of inheritance for a condition of concern so that the couple can make an informed decision about how to proceed.

Determining Chromosomal Inheritance

If a genetic counselor suspects that a condition is due to a chromosome anomaly, he or she may suggest that the chromosomal inheritance be examined. It is possible to view the chromosomes of an individual because cells can be microscopically examined and photographed just before cell division occurs. A computer is then used to arrange the chromosomes by pairs. The resulting pattern of chromosomes is called a **karyotype.**

A trisomy occurs when the individual has three chromosomes instead of two chromosomes at one karyotype location. **Trisomy 21** (Down syndrome) is the most common autosomal trisomy in humans. Survival to adulthood is common. Characteristic facial features include an eyelid fold, a flat face, and a large fissured tongue. Some degree of intellectual impairment is common as is early-onset Alzheimer disease. Sterility due to sexual underdevelopment may be present.

Observation: Sex Chromosome Anomalies

A female with **Turner syndrome** (XO) has only one sex chromosome, an X chromosome; the O signifies the absence of the second sex chromosome. Because the ovaries never become functional, these females do not undergo puberty or menstruation, and their breasts do not develop. Generally, females with Turner syndrome have a short build, folds of skin on the back of the neck, difficulty recognizing various spatial patterns, and normal intelligence. With hormone supplements, they can lead fairly normal lives.

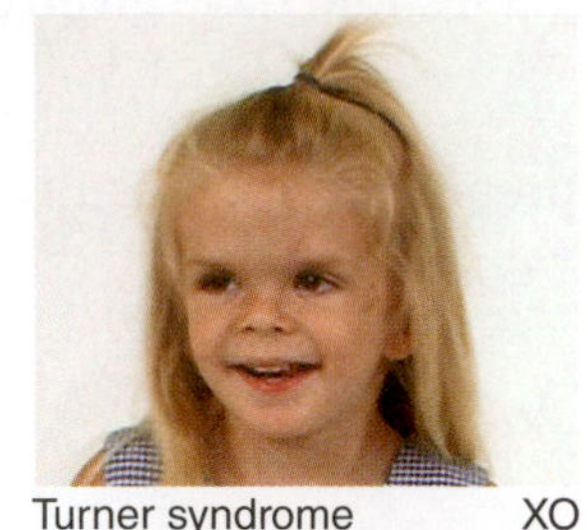

Turner syndrome XO

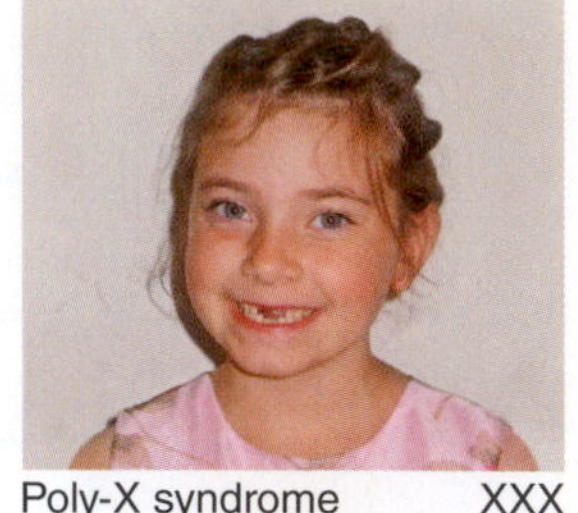

Poly-X syndrome XXX

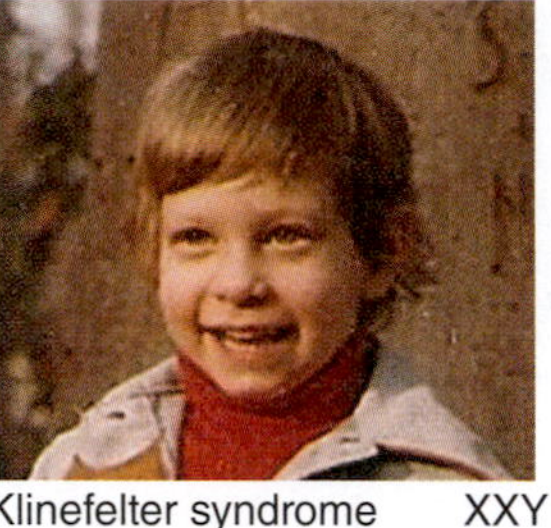

Klinefelter syndrome XXY

Jacob syndrome XYY

When an egg having two X chromosomes is fertilized by an X-bearing sperm, an individual with **poly-X syndrome** results. The body cells have three X chromosomes and therefore 47 chromosomes. Although they tend to have learning disabilities, poly-X females have no apparent physical anomalies, and many are fertile and have children with a normal chromosome count.

When an egg having two X chromosomes is fertilized by a Y-bearing sperm, a male with **Klinefelter syndrome** results. This individual is male in general appearance, but the testes are underdeveloped, and the breasts may be enlarged. The limbs of XXY males tend to be longer than average, muscular development is poor, body hair is sparse, and many XXY males have learning disabilities.

Jacob syndrome occurs in males who are usually taller than average, suffer from persistent acne, and tend to have speech and reading problems. At one time, it was suggested that XYY males were likely to be criminally aggressive, but the incidence of such behavior has been shown to be no greater than that among normal XY males.

Label each karyotype in Figure 12.6 as one of syndromes just discussed. Explain your answers on the lines provided.

Figure 12.6 Sex chromosome anomalies.

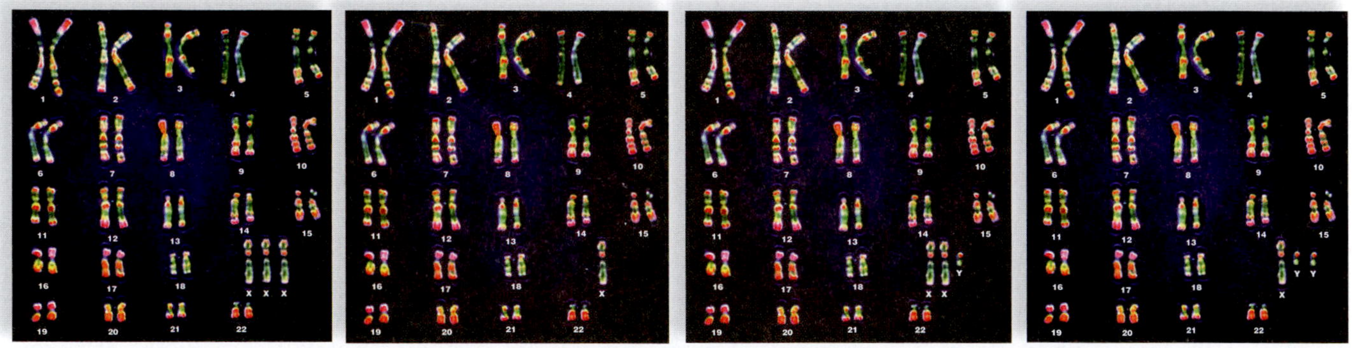

a. ______________________

____

____

b. ______________________

____

____

c. ______________________

____

____

d. ______________________

____

____

Determining the Pedigree

A pedigree shows the inheritance of a genetic disorder within a family and can help determine the inheritance pattern and whether any particular individual has an allele for that disorder. Then a Punnett square can be done to determine the chances of a couple producing an affected child.

The symbols used to indicate normal and affected males and females, reproductive partners, and siblings in a pedigree are shown in Figure 12.7.

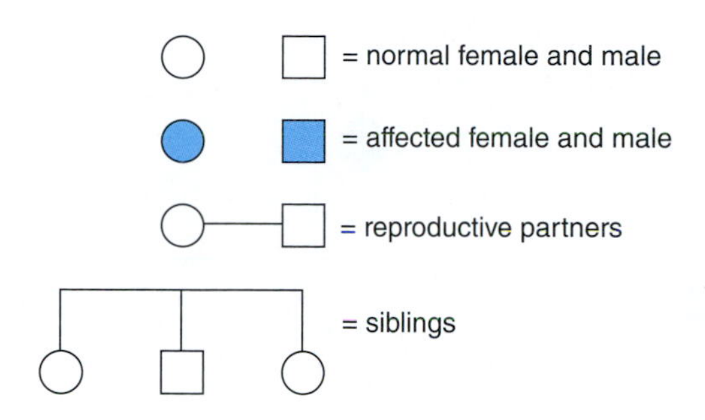

Figure 12.7 Pedigree symbols.

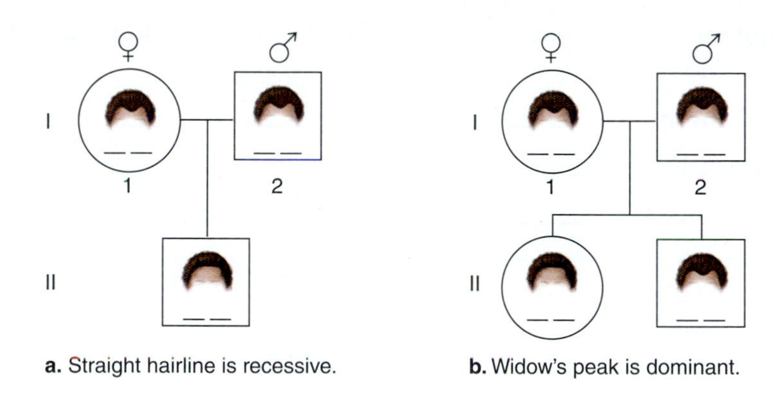

a. Straight hairline is recessive. **b.** Widow's peak is dominant.

Figure 12.8 Autosomal pedigrees.

In a pedigree, Roman numerals indicate the generation, and Arabic numerals indicate particular individuals in that generation. *In Figure 12.8, you are to enter the genotype on the lines provided.* Suppose you wanted to determine the inheritance pattern for straight hairline and you knew which members of a generational family had the trait (Fig. 12.8*a*). The pedigree allows you to determine that straight hairline is autosomal recessive because two parents without this phenotype have a child with the phenotype. This can happen only if the parents are heterozygous and straight hairline is recessive. Similarly, a pedigree allows you to determine that widow's peak is autosomal dominant (Fig. 12.8*b*): A child with this phenotype has at least one parent with the dominant phenotype, but again, heterozygous parents can produce a child without widow's peak. *Give each person in Figure 12.8*a *and* b *a genotype.*

Later you will have an opportunity to see an X-linked recessive pedigree. An X-linked recessive phenotype occurs mainly in males, and it skips a generation because a female who inherits a recessive allele for the condition from her father may have a son with the condition.

Pedigree Analysis

For each of the following pedigrees, decide whether a trait is inherited as an autosomal dominant, autosomal recessive, or X-linked recessive. Then you are asked to decide the genotype of particular individuals in the pedigree.

1. Study the following pedigree:

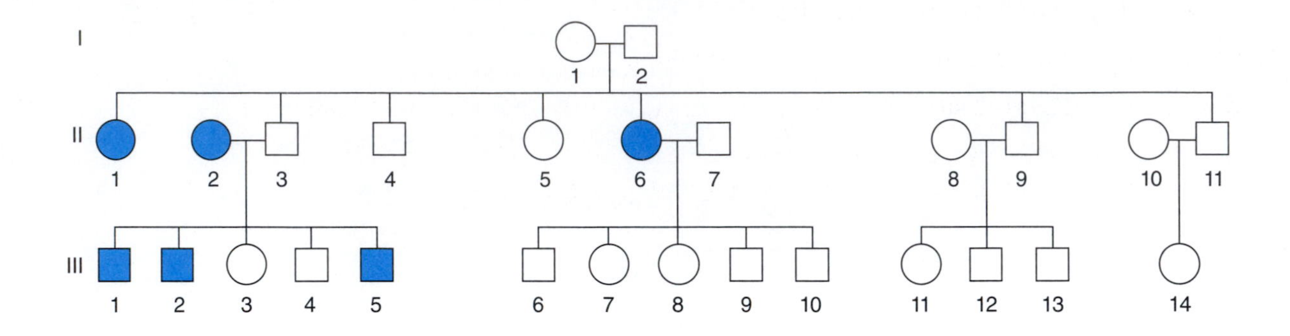

a. Notice that neither of the original parents is affected, but several children are affected. This could happen only if the trait were ______________________________.

b. What is the genotype of the following individuals? Use *A* for the dominant allele and *a* for the recessive allele.

Generation I, individual 1: ______________________________

Generation II, individual 1: ______________________________

Generation III, individual 8: ______________________________

2. Study the following pedigree:

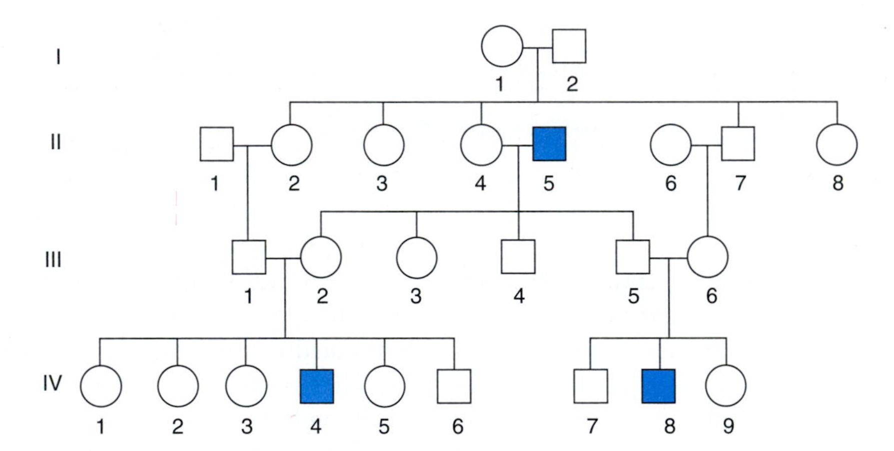

a. Notice that only males are affected. This could happen only if the trait were ______________.

b. What is the genotype of the following individuals?

Generation I, individual 1: ______________________________

Generation II, individual 8: ______________________________

Generation III, individual 1: ______________________________

Construction of a Pedigree

You are a genetic counselor who has been given the following information from which you will construct a pedigree.

1. Your data: Henry has a double row of eyelashes, which is a dominant trait. Both his maternal grandfather and his mother have double eyelashes. Their spouses are normal. Henry is married to Isabella and their first child, Polly, has normal eyelashes. The couple wants to know the chances of any child having a double row of eyelashes.
2. Choose a key for this trait.

   Key: __________ normal eyelashes __________ double row of eyelashes
3. *Construct two pedigrees with symbols only for the underlined persons in step 1.* The pedigrees start with the maternal grandfather and grandmother and end with Polly.

   Pedigree 1 **Pedigree 2**

4. Use the pedigrees you constructed to determine the pattern of inheritance. Pedigree 1: Try out a pattern of autosomal dominant inheritance by assigning appropriate genotypes for an autosomal dominant pattern of inheritance to each person in this pedigree. Pedigree 2: Try out a pattern of X-linked dominant inheritance by assigning appropriate genotypes for this pattern of inheritance to each person in your pedigree. Which pattern is correct? ______________________________
5. Use correct genotypes to show a cross between Henry and Isabella and from your experience with crosses, state the expected phenotypic ratio among the offspring:

Cross	Henry		Isabella	**Phenotypic ratio:**
	__________	×	__________	__________

6. What are the percentage chances that Henry and Isabella will have a child with double eyelashes? ________

   "Change has no memory" and each child has the same chance for double eyelashes. Explain why.

   ____

   ____

Laboratory Review 12

1. If an individual exhibits the dominant trait, do you know the genotype? Why or why not?
____
____

2. Isabella's father does not have freckles, but Mary does. What genotypes could Mary's mother have?
____

3. What are the chances two individuals with an autosomal recessive trait will have a child with this trait?
____

4. Show a cross that would produce a phenotypic ratio of 1:1 among the offspring. _______________
____

5. If the parents are heterozygous for cystic fibrosis, what are the chances of a child having cystic fibrosis?
____
____

6. Tom has blood type AB. Show all possible genotypes for this type blood. _______________
____

7. Mary has blood type A and Don has blood type B; can they be the parents of a child with type O blood? Show why or why not. _______________
____

8. What syndrome is inherited when an egg carrying two X chromosomes is fertilized by a sperm carrying one Y chromosome? _______________
____

9. What is the inheritance pattern in a pedigree if the parents are not affected and a child is affected? Give a genotype for all persons. _______________
____

10. If only males are affected in a pedigree, what is the likely inheritance pattern for the trait? _______________
Draw a three-generation pedigree showing the inheritance of the trait from an affected grandfather to an affected grandson. (No spouses are affected.)

***Concepts of Biology* Website**

Instructors can find lab prep information and answers to all of the laboratory questions in the Laboratory Resource Guide. *Students* can practice their knowledge with quizzes, animations, flashcards, and much more.

www.mhhe.com/maderconcepts3

McGraw-Hill Access Science Website

An online encyclopedia of science and technology that provides information, including videos, that can enhance the laboratory experience.

www.accessscience.com

Human Genetics

LABORATORY

13

DNA Biology and Technology

Learning Outcomes

13.1 DNA Structure and Replication
- Explain how the structure of DNA facilitates replication. 140–41
- Explain how DNA replication is semiconservative. 141–42

13.2 RNA Structure
- Compare DNA and RNA, discussing their similarities as well as what distinguishes one from the other. 143–44

13.3 DNA and Protein Synthesis
- Describe how DNA is able to store the information that specifies a protein. 144
- Compare the events of transcription with those of translation during protein synthesis. 145–47

13.4 Isolation of DNA and Biotechnology
- Describe the procedure for isolating DNA in a test tube. 148
- Describe the process of DNA gel electrophoresis. 149

13.5 Detecting Genetic Disorders
- Understand the relationship between an abnormal DNA base sequence and a genetic disorder. 150
- Suggest two ways to detect a genetic disorder. 151

Introduction

This laboratory pertains to molecular genetics and biotechnology. Molecular genetics is the study of the structure and function of **DNA (deoxyribonucleic acid),** the genetic material. **Biotechnology** is the manipulation of DNA for the benefit of human beings and other organisms.

First we will study the structure of DNA and see how that structure facilitates DNA replication in the nucleus of cells. DNA replicates prior to cell division; following cell division, each daughter cell has a complete copy of the genetic material. DNA replication is also needed to pass genetic material from one generation to the next. You may have an opportunity to use models to see how replication occurs.

Then we will study the structure of **RNA (ribonucleic acid)** and how it differs from that of DNA before examining how DNA, with the help of RNA, specifies protein synthesis. The linear construction of DNA, in which nucleotide follows nucleotide, is paralleled by the linear construction of the primary structure of protein, in which amino acid follows amino acid. Essentially, we will see that the sequence of nucleotides in DNA codes for the sequence of amino acids in a protein. We will also review the role of three types of RNA in protein synthesis. DNA's code is passed to messenger RNA (mRNA), which moves to the ribosomes containing ribosomal RNA (rRNA). Transfer RNA (tRNA) brings the amino acids to the ribosomes, and they become sequenced in the order directed by mRNA.

We now understand that a mutated gene has an altered DNA base sequence, which can lead to a genetic disorder. You will have an opportunity to carry out a laboratory procedure that detects whether an individual is normal, has sickle-cell disease, or is a carrier.

13.1 DNA Structure and Replication

The structure of DNA lends itself to **replication,** the process that makes a copy of a DNA molecule. DNA replication is a necessary part of chromosome duplication, which precedes cell division. It also makes possible the passage of DNA from one generation to the next.

DNA Structure

DNA is a polymer of nucleotide monomers (Fig. 13.1). Each nucleotide is composed of three molecules: deoxyribose (a 5-carbon sugar), a phosphate, and a nitrogen-containing base.

Figure 13.1 Overview of DNA structure.
Diagram of DNA double helix shows that the molecule resembles a twisted ladder. Sugar-phosphate backbones make up the sides of the ladder, and hydrogen-bonded bases make up the rungs of the ladder. Complementary base pairing dictates that A is bonded to T and G is bonded to C and vice versa.

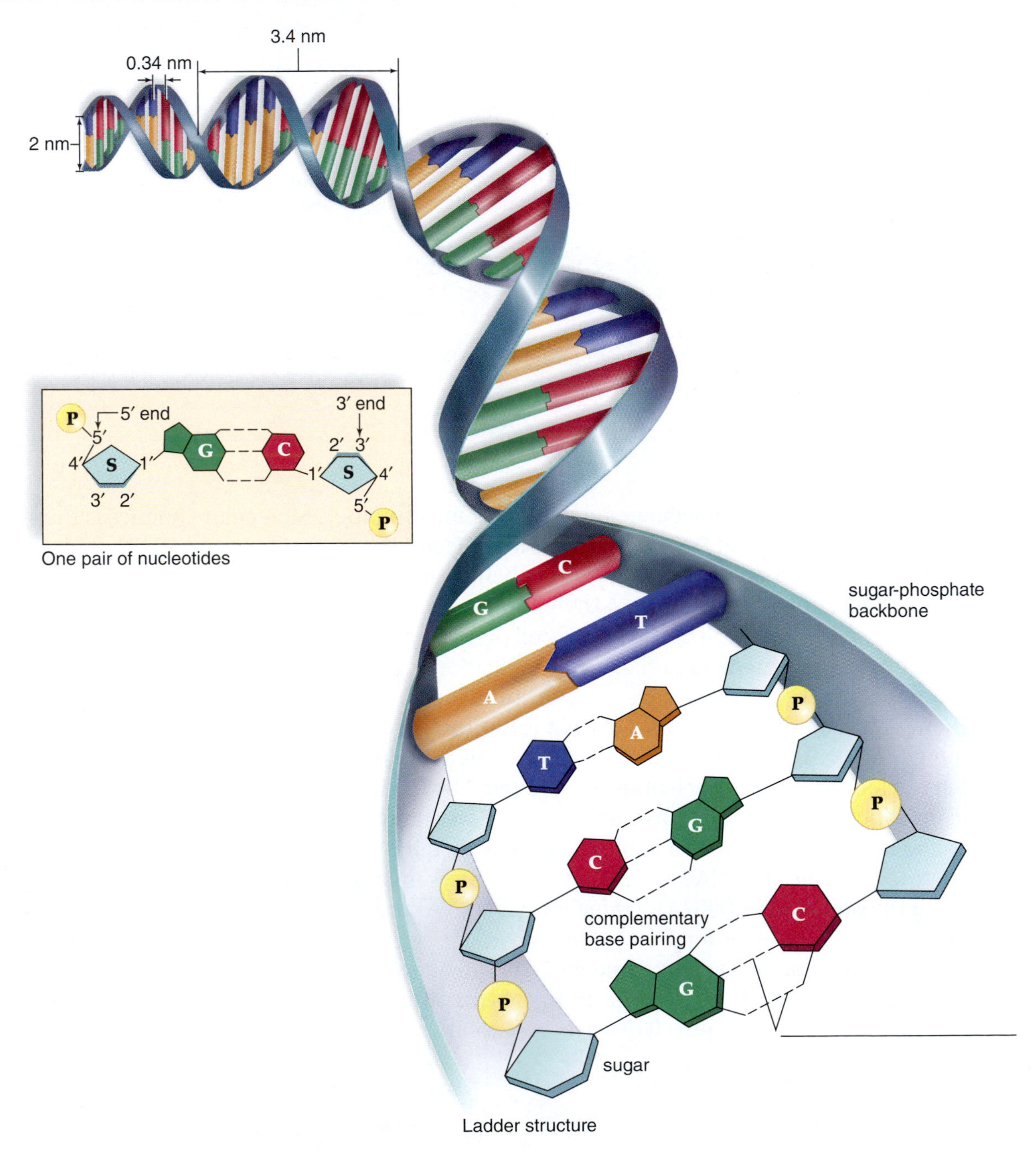

Observation: DNA Structure

1. A boxed nucleotide pair is shown in Figure 13.1. If you are working with a kit, draw a representation of one of your nucleotides here. *Label phosphate, base pair, and deoxyribose in your drawing.*

2. Notice the four types of bases: cytosine (C), thymine (T), adenine (A), and guanine (G). What is the color of each of the four types of bases in Figure 13.1? In your kit? Complete Table 13.1 by writing in the colors of the bases.

Table 13.1 Base Colors

	In Figure 13.1	In Your Kit
Cytosine		
Thymine		
Adenine		
Guanine		

3. Using Figure 13.1 as a guide, join several nucleotides together. Observe the entire DNA molecule. What types of molecules make up the backbone (uprights of ladder) of DNA (Fig. 13.1)? ____________ and ____________________. In the backbone, the phosphate of one nucleotide is bonded to a sugar of the next nucleotide.
4. Using Figure 13.1 as a guide, join the bases together with hydrogen bonds. *Label a hydrogen bond in Figure 13.1.* Dashes are used to represent hydrogen bonds in Figure 13.1 because hydrogen bonds are (strong or weak) ______________.
5. Notice in Figure 13.1 and in your model that the base A is always paired with the base ______________, and the base C is always paired with the base ________________. This is called complementary base pairing.
6. In Figure 13.1, what molecules make up the rungs of the ladder? ______________________________
7. Each half of the DNA molecule is a DNA strand. Why is DNA also called a double helix (Fig. 13.1)?

____

DNA Replication

During replication, the DNA molecule is duplicated so that there are two identical DNA molecules. We will see that complementary base pairing makes replication possible.

Observation: DNA Replication

1. Before replication begins, DNA is unzipped. Using Figure 13.2*a* as a guide, break apart your two DNA strands. What bonds are broken in order to unzip the DNA strands? ___________________________
2. Using Figure 13.2*b* as a guide, attach new complementary nucleotides to each strand using complementary base pairing.
3. Show that you understand complementary base pairing by completing Table 13.2.
4. You now have two DNA molecules (Fig. 13.2*c*). Are your molecules identical?

____

5. Because of complementary base pairing, each new double helix is composed of an __________ strand and a __________ strand. *Write "old" or "new" in 1 to 10, Figure 13.2*a, b, *and* c. *Conservative* means to save something from the past. Why is DNA replication called semiconservative?

____

____

____

Figure 13.2 DNA replication.
Use of the ladder configuration better illustrates how replication takes place. **a.** The parental DNA molecule. **b.** The "old" strands of the parental DNA molecule have separated. New complementary nucleotides available in the cell are pairing with those of each old strand. **c.** Replication is complete.

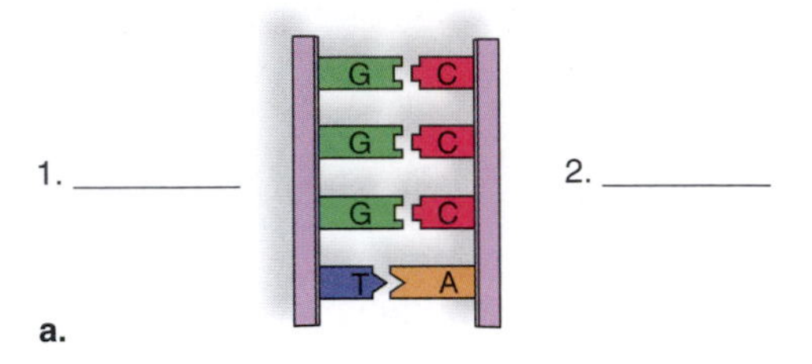

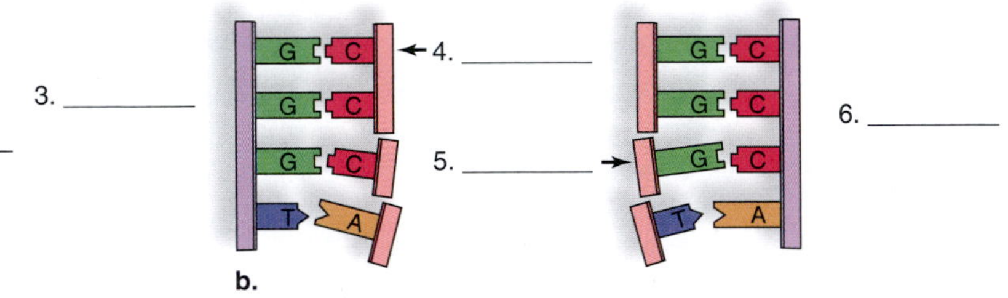

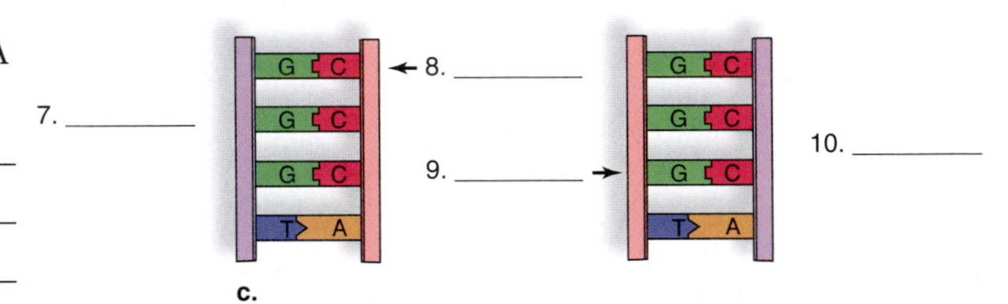

6. Genetic material has to be inherited from cell to cell and organism to organism. Consider that because of DNA replication, a chromosome is composed of two chromatids, and each chromatid is a DNA double helix. The chromatids separate during cell division so that each daughter cell receives a copy of each chromosome. Does replication provide a means for passing DNA from cell to cell and organism to organism? __________ Explain. ___________________________

____

Table 13.2 DNA Replication

Old strand	G	G	G	T	T	C	C	A	T	T	A	A	A	T	T	C	C	A	G	A	A	A	T	C	A	T	A
New strand																											

13.2 RNA Structure

Like DNA, RNA is a polymer of nucleotides (Fig. 13.3). In an RNA nucleotide, the sugar ribose is attached to a phosphate molecule and to a nitrogen-containing base, C, U, A, or G. In RNA, the base uracil replaces thymine as one of the bases. RNA is single stranded, whereas DNA is double stranded.

Figure 13.3 Overview of RNA structure.
RNA is a single strand of nucleotides. *Label the boxed nucleotide as directed in the next Observation.*

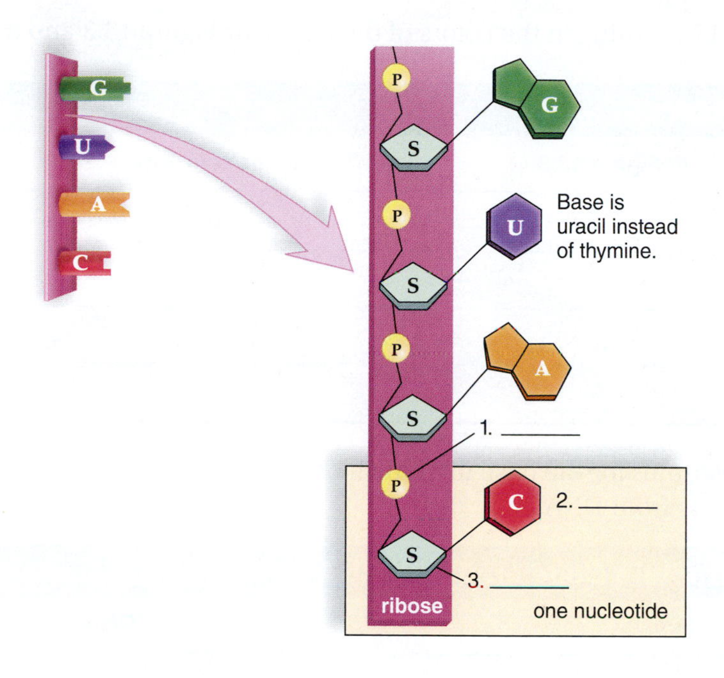

1. Describe the backbone of an RNA molecule. ______
2. Where are the bases located in an RNA molecule? ______
3. Complete Table 13.3 to show the complementary DNA bases for the RNA bases.

Table 13.3 DNA and RNA Bases

RNA bases	C	U	A	G
DNA bases				

Observation: RNA Structure

1. If you are using a kit, draw a nucleotide for the construction of mRNA. *Label the ribose (the sugar in RNA), the phosphate, and the base in your drawing and in 1–3, Figure 13.3.*

2. Complete Table 13.4 by writing in the colors of the bases for Figure 13.3 and for your kit.

Table 13.4 Base Colors

	In Figure 13.3	In Your Kit
Cytosine		
Uracil		
Adenine		
Guanine		

3. The base uracil substitutes for the base thymine in RNA. Complete Table 13.5 to show the several other ways RNA differs from DNA.

Table 13.5 DNA Structure Compared with RNA Structure

	DNA	RNA
Sugar	Deoxyribose	
Bases	Adenine, guanine, thymine, cytosine	
Strands	Double stranded with base pairing	
Helix	Yes	

13.3 DNA and Protein Synthesis

Protein synthesis requires the processes of transcription and translation. During **transcription,** which takes place in the nucleus, an RNA molecule called **messenger RNA (mRNA)** is made complementary to one of the DNA strands. This mRNA leaves the nucleus and goes to the ribosomes in the cytoplasm. Ribosomes are composed of **ribosomal RNA (rRNA)** and proteins in two subunits.

During **translation,** RNA molecules called **transfer RNA (tRNA)** bring amino acids to the ribosome, and they join in the order prescribed by mRNA. This sequence of amino acids was originally specified by DNA. This is the information that DNA, the genetic material, stores.

What is the role of each of these participants in protein synthesis?

DNA ____________________

mRNA ____________________

tRNA ____________________

Transcription

During transcription, complementary RNA is made from a DNA template (Fig. 13.4). A portion of DNA unwinds and unzips at the point of attachment of the enzyme RNA polymerase. A strand of mRNA is produced when complementary nucleotides join in the order dictated by the sequence of bases in DNA. Transcription occurs in the nucleus, and the mRNA passes out of the nucleus to enter the cytoplasm.

Label Figure 13.4. For number 1, note the name of the enzyme that carries out mRNA synthesis. For number 2, note the name of this molecule.

Figure 13.4 Messenger RNA (mRNA). Messenger RNA complementary to a section of DNA forms during transcription.

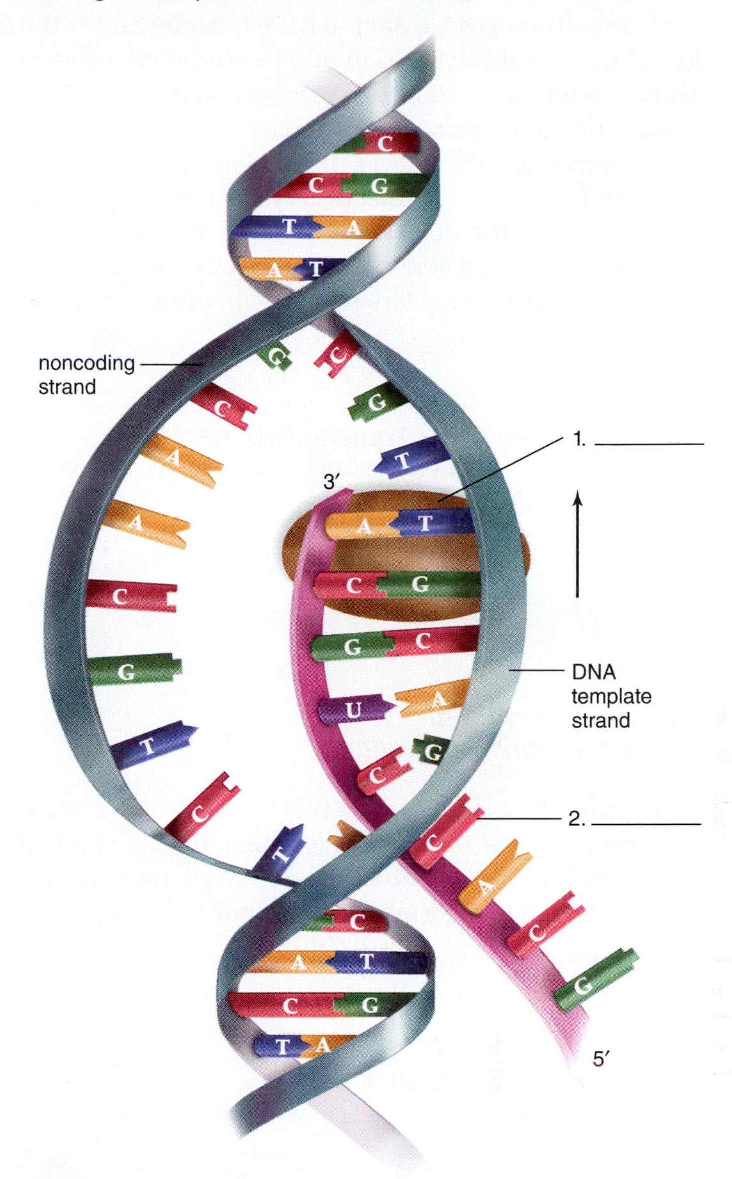

Observation: Transcription

1. If you are using a kit, unzip your DNA model so that only one strand remains. This strand is the **sense strand,** the strand that is transcribed.
2. Using Figure 13.4 as a guide, construct a messenger RNA (mRNA) molecule by first lining up RNA nucleotides complementary to the sense strand of your DNA molecule. Join the RNA nucleotides together to form mRNA.
3. A portion of DNA has the sequence of bases shown in Table 13.6. *Complete Table 13.6 to show the sequence of bases in mRNA.*
4. If you are using a kit, unzip mRNA transcript from the DNA. Locate the end of the strand that will move to

   the ________________ in the cytoplasm.

Table 13.6 Transcription

DNA	T A C A C G A G C A A C T A A C A T
mRNA	

Translation

DNA specifies the sequence of amino acids in a polypeptide because every three bases code for an amino acid. Therefore, DNA is said to have a **triplet code.** The bases in mRNA are complementary to the bases in DNA. Every three bases in mRNA are called a **codon.** One codon of mRNA represents one amino acid. Thus, the sequence of DNA bases serves as the blueprint for the sequence of amino acids assembled to make a protein. The correct sequence of amino acids in a polypeptide is the message that mRNA carries.

Messenger RNA leaves the nucleus and proceeds to the ribosomes, where protein synthesis occurs. Transfer RNA (tRNA) molecules are so named because they transfer amino acids to the ribosomes. Each RNA has a specific tRNA amino acid at one end and a matching **anticodon** at the other end (Fig. 13.5). *Label Figure 13.5, where the amino acid is represented as a colored ball, the tRNA is green, and the anticodon is the sequence of three bases.* (The anticodon is complementary to the mRNA codon.)

Figure 13.5 Transfer RNA (tRNA).
Transfer RNA carries amino acids to the ribosomes.

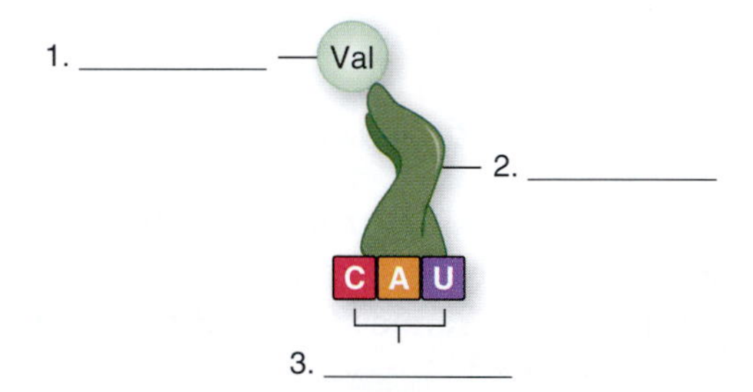

Observation: Translation

1. Figure 13.6 shows seven tRNA–amino acid complexes. Every amino acid has a name; in the figure, only the first three letters of the name are inside the ball. *Using the mRNA sequence given in Table 13.7, number the tRNA–amino acid complexes in the order they will come to the ribosome.*
2. If you are using a kit, arrange your tRNA–amino acid complexes in the order consistent with Table 13.7. *Complete Table 13.7.* Why are the codons and anticodons in groups of three? ____

Figure 13.6 Transfer RNA diversity.
Each type of tRNA carries only one particular amino acid, designated here by the first three letters of its name.

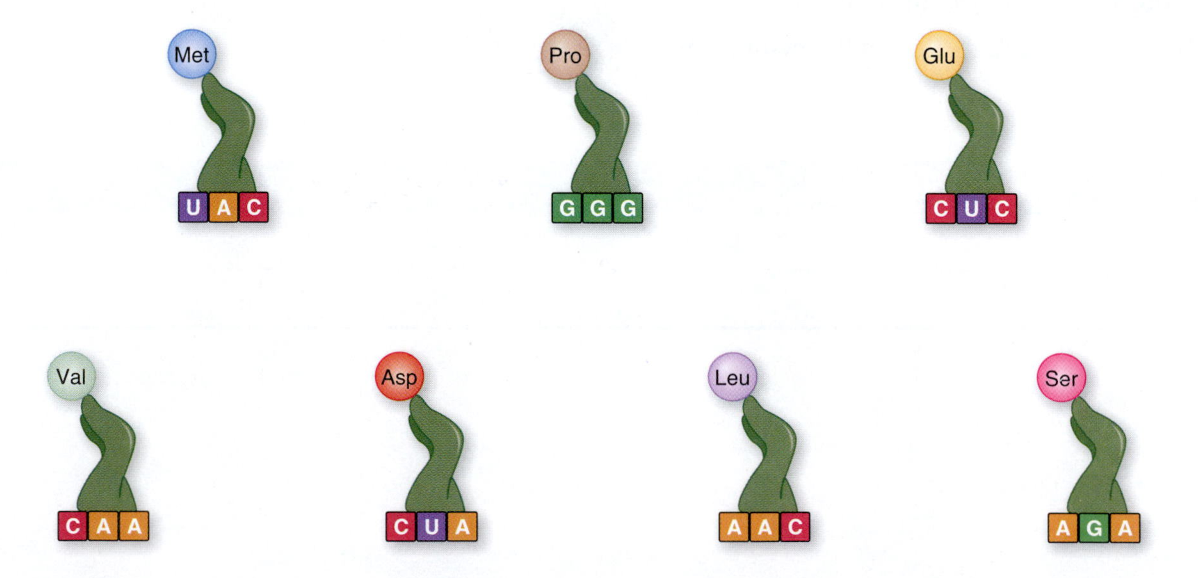

Table 13.7 Translation							
mRNA codons	AUG	CCC	GAG	GUU	GAU	UUG	UCU
tRNA anticodons							
Amino acid*							

*Use three letters only. See Table 13.8 for the full names of these amino acids.

Table 13.8 Names of Amino Acids

Abbreviation	Name
Met	Methionine
Pro	Proline
Asp	Aspartate
Val	Valine
Glu	Glutamate
Leu	Leucine
Ser	Serine

3. Figure 13.7 shows the manner in which the polypeptide grows. A ribosome has room for two tRNA complexes at a time. As the first tRNA leaves, it passes its amino acid or peptide to the next tRNA-amino acid complex. Then the ribosome moves forward, making room for the next tRNA-amino acid. This sequence of events occurs over and over until the entire polypeptide is borne by the last tRNA to come to the ribosome. Then a release factor releases the polypeptide chain from the ribosome. *In Figure 13.7, label the ribosome, the mRNA, and the peptide.*

Figure 13.7 Protein synthesis.
1. A ribosome has a binding site for two tRNA–amino acid complexes. **2.** Before a tRNA leaves, an RNA passes its attached peptide to a newly arrived tRNA–amino acid complex. **3.** The ribosome moves forward, and the next tRNA–amino acid complex arrives.

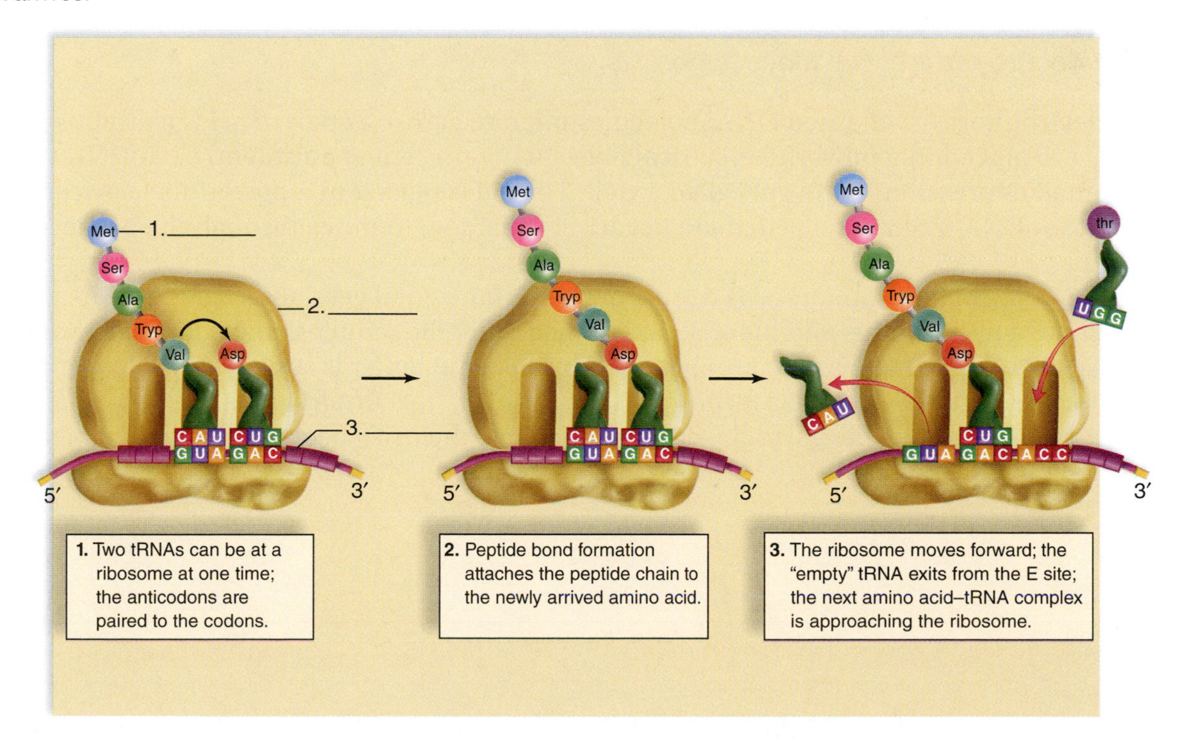

13.4 Isolation of DNA and Biotechnology

In the following experiment, you will isolate DNA from the cells of a fruit or vegetable. It will only be necessary to expose the cells to an agent (dishwasher detergent) that emulsifies membrane in order to "free" the DNA from its enclosures (plasma membrane and nuclear envelope). When transferred to a tube, the presence of NaCl allows DNA to precipitate as a sodium salt. The precipitate forms at the interface between ethanol and the salt solution, and then it floats to the top of the tube where it may be collected.

Experimental Procedure: Isolation of DNA

For experimental procedures, wear safety goggles, gloves, and protective clothing. If any chemical spills on your skin, wash immediately with mild soap and water; flood eyes with water only. Report any spills immediately to your instructor.

1. You will need a slice of fruit or vegetable (i.e., tomato, onion, or apple), a mortar and pestle, and a large clean glass test tube on ice.
2. Crush and grind a slice of fruit or vegetable in a mortar and pestle. Remove the pestle and set aside. Add enough 0.9% NaCl solution to achieve a "soupy" consistency.
3. Add two drops of dishwasher detergent (such as Blue Dawn) to the mixture in the mortar. Swirl until the color of the detergent disappears. Wait 3 to 5 minutes. The solution becomes clear and viscous as the DNA escapes its enclosures.
4. Check the viscosity of the solution in the mortar, and if it is too viscous for you to pipet, add a few more ml of NaCl solution. Use a transfer pipet to move 2 to 3 ml of the DNA from the mortar to the glass tube. Try to pick up clear zones of the solution WITHOUT CELL DEBRIS.
5. Slowly add 5 to 7 ml of ice-cold ethanol down the side of the tube on ice. As you do so, DNA will precipitate *between* the ethanol and the NaCl solutions. Then it will float to the top, where it can be picked up by a fresh transfer pipet. Any "white dust" you see consists of RNA molecules.
6. Use a transfer pipet to place the DNA in a small, clean test tube. Pipet away any extra water/ethanol and air-dry the DNA for a few minutes. Dissolve the DNA in 3 to 4 ml distilled H_2O.
7. Add five drops of phenol red, a pH indicator. The resulting dark pink color confirms the presence of nucleic acid (i.e., the DNA).

Experimental Procedure: Gel Electrophoresis

During gel electrophoresis, charged DNA molecules migrate across a span of gel (gelatinous slab) because they are placed in a powerful electrical field. In the present experiment, each DNA sample is placed in a small depression in the gel called a well. The gel is placed in a powerful electrical field. The electricity causes DNA fragments, which are negatively charged, to move through the gel according to their size.

Almost all DNA gel electrophoresis is carried out using horizontal gel slabs (Fig. 13.8). First, the gel is poured onto a plastic plate, and the wells are formed. After the samples are added to the wells, the gel and the plastic plate are put into an electrophoresis chamber, and buffer is added. The DNA samples begin to migrate after the electrical current is turned on. With staining, the DNA fragments appear as a series of bands spread from one end of the gel to the other according to their size because smaller fragments move faster than larger fragments.

Figure 13.8 Equipment and procedure for gel electrophoresis.

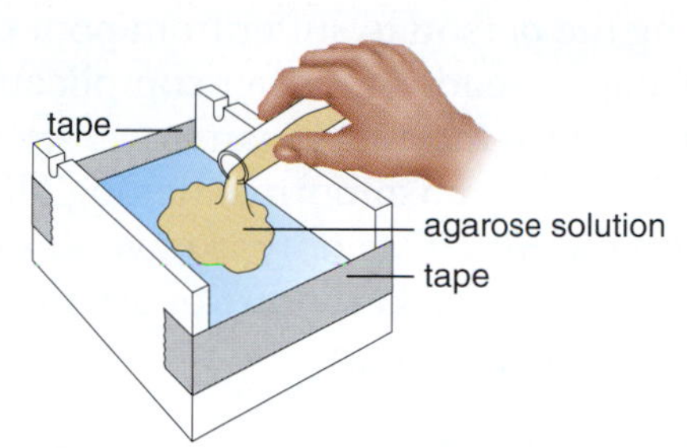

a. Agarose solution poured into casting tray

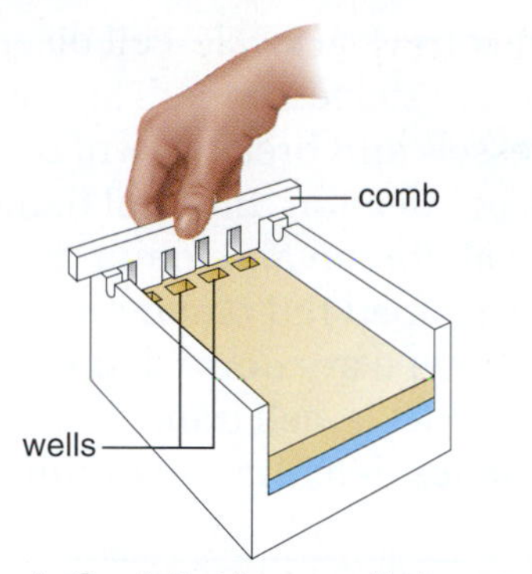

b. Comb that forms wells for samples

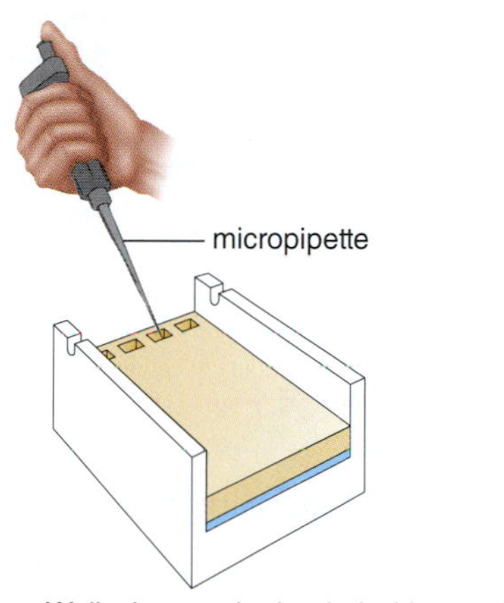

c. Wells that can be loaded with samples

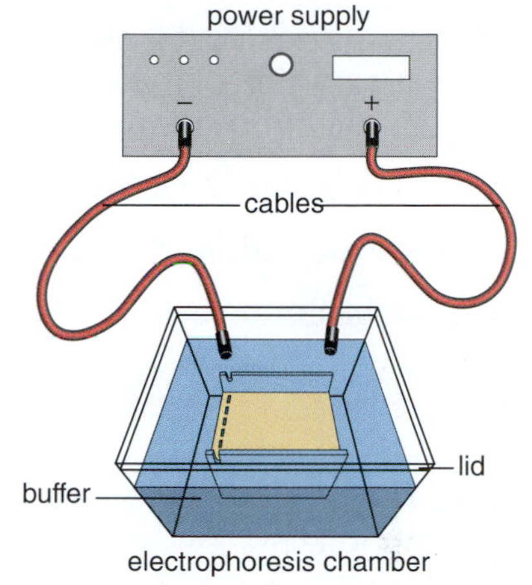

d. Electrophoresis chamber and power supply

Answer the following questions:

1. What is biotechnology? (See page 139.) ______________________________

____

____

2. Speculate how the ability to isolate DNA and run gel electrophoresis of DNA relates to biotechnology.

____

____

____

3. Name a biotechnology product someone you know is now using or taking as a medicine.

____

13.5 Detecting Genetic Disorders

The base sequence of DNA in all the chromosomes is an organism's genome. Now that the Human Genome Project is finished, we know the usual order of all the 3.2 billion nucleotide base pairs in the haploid human genome. Today it is possible to sequence anyone's genome within a relatively short time, and thereby determine what particular base sequence alterations signify that he or she has a disorder or will have one in the future. In this laboratory, you will study the alteration in base sequence that causes a person to have sickle-cell disease.

In persons with sickle-cell disease, the red blood cells aren't biconcave disks like normal red blood cells—they are sickle-shaped. Sickle-shaped cells can't pass along narrow capillary passageways. They clog the vessels and break down, causing the person to suffer from poor circulation, anemia, and poor resistance to infection. Internal hemorrhaging leads to further complications, such as jaundice, episodic pain in the abdomen and joints, and damage to internal organs.

Sickle-shaped red blood cells are caused by an abnormal hemoglobin (Hb^S). Individuals with the Hb^AHb^A genotype are normal; those with the Hb^SHb^S genotype have sickle-cell disease, and those with the Hb^AHb^S have sickle-cell trait. Persons with sickle-cell trait do not usually have sickle-shaped cells unless they experience dehydration or mild oxygen deprivation.

Genetic Sequence for Sickle-Cell Disease

Examine Figure 13.9*a* and *b,* which shows the DNA base sequence, the mRNA codons, and the amino acid sequence for a portion of the gene for Hb^A and the same portion for Hb^S. Today many genetic disorders can be detected by genomic sequencing.

1. In what one base does Hb^A differ from Hb^S? Hb^A ________ Hb^S ________
2. What are the codons that contain this base? Hb^A ________ Hb^S ________
3. What is the amino acid difference? Hb^A ________ Hb^S ________

Figure 13.9 Sickle-cell disease.
a. When red blood cells are normal, the base sequence (in one location) for Hb^A alleles is CTC. **b.** In sickle-cell disease at these locations, it is CAC.

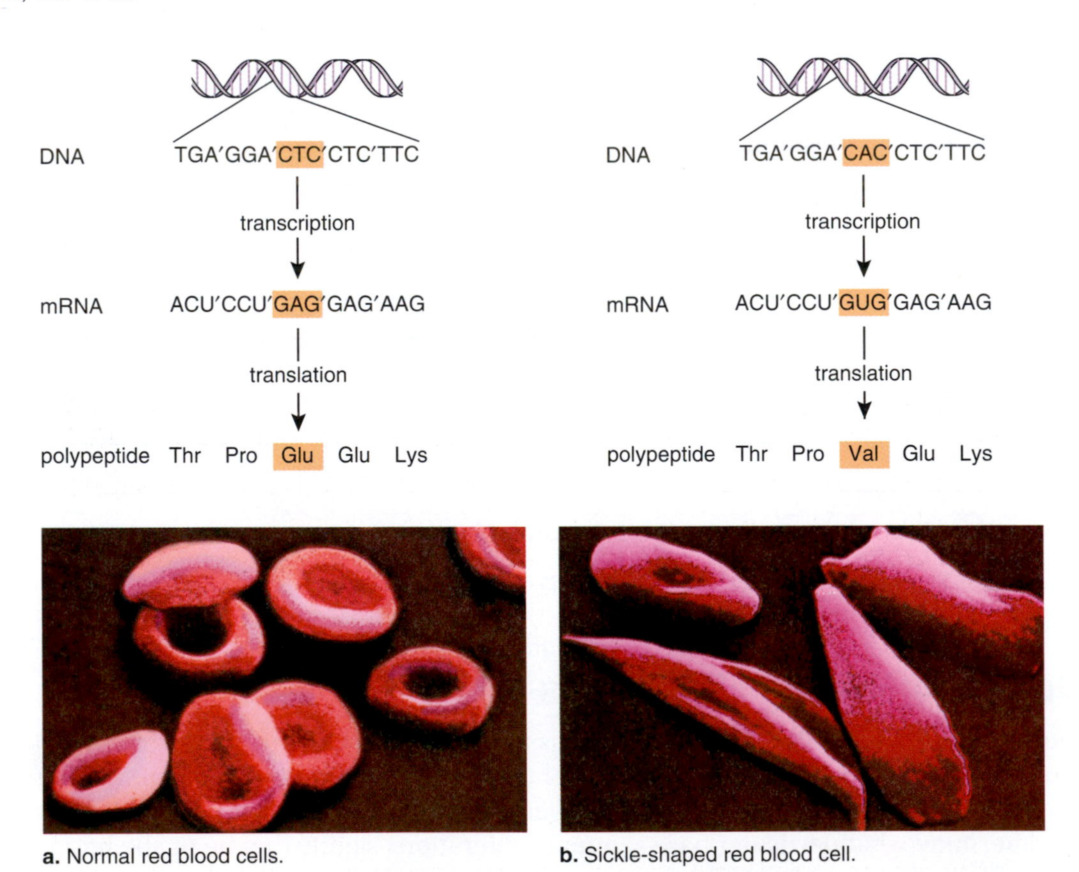

a. Normal red blood cells.

b. Sickle-shaped red blood cell.

This amino acid difference causes the polypeptide chain in sickle-cell hemoglobin to pile up as firm rods that push against the plasma membrane and deform the red blood cell into a sickle shape:

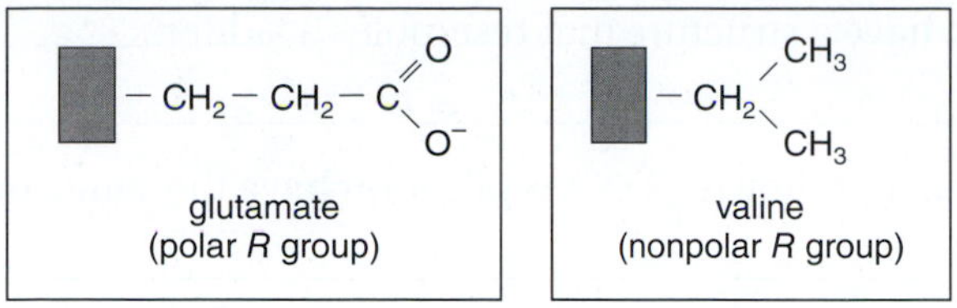

glutamate (polar *R* group)	valine (nonpolar *R* group)
$-CH_2-CH_2-C(=O)O^-$	$-CH_2(CH_3)CH_3$

Detection of Sickle-Cell Disease by Gel Electrophoresis

Three samples of hemoglobin have been subjected to protein gel electrophoresis. Protein gel electrophoresis is carried out in the same manner as DNA gel electrophoresis (see Fig. 13.10) except the gel has a different composition.

1. Sickle-cell hemoglobin (Hb^S) migrates slower toward the positive pole than normal hemoglobin (Hb^A) because the amino acid valine has no polar *R* groups, whereas the amino acid glutamate does have a polar *R* group.
2. In Figure 13.10, which lane contains only Hb^S, signifying that the individual is Hb^SHb^S?

   ____
3. Which lane contains only Hb^A, signifying that the individual is Hb^AHb^A? ________
4. Which lane contains both Hb^S and Hb^A, signifying that the individual is Hb^AHb^S?

   ____

Figure 13.10 Gel electrophoresis of hemoglobins.

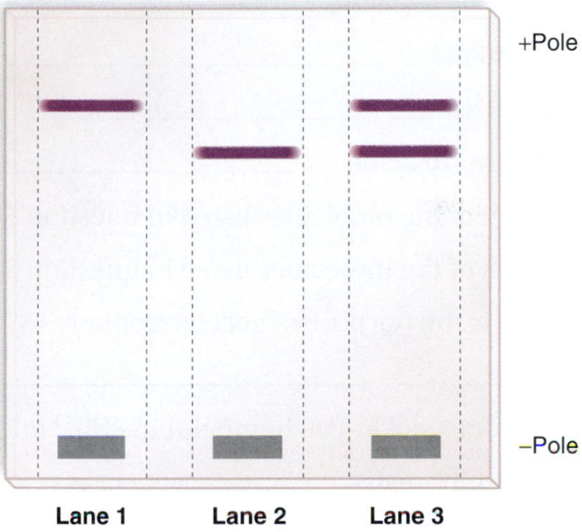

+Pole

–Pole

Lane 1 Lane 2 Lane 3

Detection by Genomic Sequencing

You are a genetic counselor. A young couple seeks your advice because sickle-cell disease occurs among the family members of each. You order DNA base sequencing to be done. The results come back that at one of the loci for normal hemoglobin, each has the abnormal sequence CAC instead of CTC. The other locus is normal. What are the chances that this couple will have a child with sickle-cell disease? ______________

Summary: Detecting Genetic Disorders

- What two methods of detecting sickle-cell disease were described in this section? ______________
- Which method is more direct and probably requires more expensive equipment to do? ______________
- Which method probably preceded the other method as a means to detect sickle-cell disease? ______________

Laboratory Review 13

1. Explain why DNA is said to have a structure that resembles a ladder. ______________________

2. Do the two DNA double helices following DNA replication have the same, or a different, composition? ______________________

3. How is complementary base pairing different when pairing DNA to DNA than when pairing DNA to mRNA? ______________________

4. Explain why the genetic code is called a triplet code. ______________________
5. What role does each of the following molecules play in protein synthesis?
   a. DNA ______________________
   b. mRNA ______________________
   c. tRNA ______________________
   d. Amino acids ______________________
6. Which of the molecules listed in question 5 are involved in transcription? ______________________
7. Which of the molecules listed in question 5 are involved in translation? ______________________
8. What is the purpose of gel electrophoresis? ______________________

9. Why does sickle-cell hemoglobin *(Hb^S)* migrate slower than normal hemoglobin *(Hb^A)* during gel electrophoresis? ______________________
10. Below is a sequence of bases associated with the template DNA strand:
    TAC CCC GAG CTT
    a. Identify the sequence of bases in the mRNA resulting from the transcription of the above DNA sequence.

    b. Identify the sequence of bases in the tRNA anticodon that will bind with the first codon on the mRNA identified above.

Concepts of Biology Website

Instructors can find lab prep information and answers to all of the laboratory questions in the Laboratory Resource Guide. *Students* can practice their knowledge with quizzes, animations, flashcards, and much more.

www.mhhe.com/maderconcepts3

McGraw-Hill Access Science Website

An online encyclopedia of science and technology that provides information, including videos, that can enhance the laboratory experience.

www.accessscience.com

DNA & Biotechnology

LABORATORY

14

Evidence of Evolution

Learning Outcomes

14.1 Evidence from the Fossil Record

- Use the geologic timescale to trace the evolution of life in broad outline. 154–56
- Describe several types of fossils and explain how fossils help establish the sequence in the evolution of life. 156–59
- Explain how scientists use fossils to establish that organisms are related by common descent. 156–59

14.2 Evidence from Comparative Anatomy

- Explain how comparative anatomy provides evidence that humans are related to other groups of vertebrates and also other groups of primates. 160–67
- Compare the human skeleton with the chimpanzee skeleton, and explain the differences on the basis of their different ways of life. 162–65
- Compare hominid skulls and hypothesize a possible evolutionary sequence. 165–66

14.3 Molecular Evidence

- Explain how molecular evidence aids the study of how humans are related to all other groups of organisms on Earth. 167
- Explain how molecular evidence also helps show the degree to which humans are related to other vertebrates and primates. 167–69

Introduction

Evolution is the process by which organisms are related by **common descent:** All organisms can trace their ancestry to the first cells. The process of evolution is amazingly simple: A group of organisms change over time because the members of a group most suited to the natural environment have more offspring than the others in the group. So, for example, among bacteria, those which can withstand an antibiotic leave more offspring, and with time the entire group of bacteria becomes resistant to the antibiotic. Reproduction and therefore evolution have been going on since the first cells appeared on Earth, and through studying (1) the fossil record; (2) comparative anatomy, both anatomical and embryological; and (3) molecular evidence, science is able to show that all organisms are related to one another.

Students digging for fossils.

14.1 Evidence from the Fossil Record

The geologic timescale, which was developed by both geologists and paleontologists, depicts the history of life based on the fossil record (Table 14.1). A **fossil** is any evidence of the existence of an organism in ancient times as opposed to modern times. Paleontologists specialize in removing fossils from the Earth's crust (see page 153). In this section, we will study the geologic timescale and then examine some fossils.

Geologic Timescale

Divisions of the Timescale

Notice that the timescale divides the history of Earth into eras, then periods, and then epochs. The four eras span the greatest amounts of time, and the epochs are the shortest time frames. Notice that only the periods of the Cenozoic era are divided into epochs, meaning that more attention is given to the evolution of primates and flowering plants than to the earlier evolving organisms. List the four eras in the timescale starting with Precambrian time: ______________________________

1. Using the geologic timescale, you can trace the history of life by beginning with Precambrian time at the bottom of the timescale. The timescale indicates that the first cells (the prokaryotes) arose some 3,500 MYA. The prokaryotes evolved before any other group. Why do you read the timescale starting at the bottom? ______________________________
2. The Precambrian time was very long, lasting from the time the Earth first formed until 542 MYA. The fossil record during the Precambrian time is meager, but the fossil record from the Cambrian period onward is rich (for reasons still being determined). This helps explain why the timescale usually does not show any periods until the Cambrian period of the Paleozoic era. You can also use the timescale to check when certain groups evolved and/or flourished.
   Example: During the Ordovician period, the nonvascular plants appear on land, and the first jawless and jawed fishes appear in the seas.
   During the __________ era and the __________ period, the first flowering plants appear. How many million years ago was this? ______________________________
3. On the timescale, note the Carboniferous period. During this period great swamp forests covered the land. These are also called coal-forming forests because, with time, they became the coal we burn today. How do you know that the plants in this forest were not flowering trees as most of our trees are today? ______________________________
   What type of animal was diversifying at this time? ______________________________
4. You should associate the Cenozoic era with the evolution of humans. Among mammals, humans are primates. During what period and epoch did primates appear? ______________________________
   Among primates, humans are hominids. During what period and epoch did hominids appear? ______________________________ The scientific name for humans is *Homo sapiens.*
   What period and epoch is the age of *Homo sapiens?* ______________________________

Dating Within the Timescale

The timescale provides both relative dates and absolute dates. When you say, for example, "Flowering plants evolved during the Jurassic period," you are using relative time, because flowering plants evolved earlier or later than groups in other periods. If you use the dates that are given in millions of years (MYA), you are using absolute time. Absolute dates are usually obtained by measuring the amount of a radioactive isotope in the rocks surrounding the fossils. Why wouldn't you expect to find human fossils and dinosaur fossils together in rocks dated similarly? ______________________________

Table 14.1 The Geologic Timescale: Major Divisions of Geologic Time and Some of the Major Evolutionary Events That Occurred

Era	Period	Epoch	Millions of Years Ago	Plant Life	Animal Life
Cenozoic*		Holocene	0.01–present	Human influence on plant life	Age of *Homo sapiens*
				Significant Mammalian Extinction	
	Quaternary	Pleistocene	1.8–0.01	Herbaceous plants spread and diversify	Presence of Ice Age mammals Modern humans appear
	Tertiary	Pliocene	5.3–1.8	Herbaceous angiosperms flourish	First hominids appear
		Miocene	23–5.3	Grasslands spread as forests contract	Apelike mammals and grazing mammals flourish; insects flourish
		Oligocene	33.9–23	Many modern families of flowering plants evolve	Browsing mammals and monkeylike primates appear
		Eocene	55.8–33.9	Subtropical forests with heavy rainfall thrive	All modern orders of mammals are represented
		Paleocene	65.5–55.8	Flowering plants continue to diversify	Primitive primates, herbivores, carnivores, and insectivores appear
				Mass Extinction: Dinosaurs and Most Reptiles	
Mesozoic	Cretaceous		145.5–65.5	Flowering plants diversify; conifers persist	Placental mammals appear; modern insect groups appear
	Jurassic		199.6–145.5	Flowering plants appear	Dinosaurs flourish; birds appear
				Mass Extinction	
	Triassic		251–199.6	Forests of conifers and cycads dominate	First mammals appear; first dinosaurs appear; corals and molluscs dominate seas
				Mass Extinction	
Paleozoic	Permian		299–251	Gymnosperms diversify	Reptiles diversify; amphibians decline
	Carboniferous		359.2–299	Age of great coal-forming forests: Ferns, club mosses, and horsetails flourish	Amphibians diversify; first reptiles appear; first great radiation of insects
				Mass Extinction	
	Devonian		416–359.2	First seed plants appear Seedless vascular plants diversify	Jawed fishes diversify and dominate the seas; first insects and first amphibians appear
	Silurian		443.7–416	Seedless vascular plants appear	First jawed fishes appear
				Mass Extinction	
	Ordovician		488.3–443.7	Nonvascular land plants appear Marine algae flourish	Invertebrates spread and diversify; jawless fishes (first vertebrates) appear
	Cambrian		542–488.3	First plants appear on land Marine algae flourish	All invertebrate phyla present; first chordates appear
Precambrian Time			630	Oldest soft-bodied invertebrate fossils	
			1,000–700	Protists evolve and diversify	
			2,100	Oldest eukaryotic fossils	
			2,700	O_2 accumulates in atmosphere	
			3,500	Oldest known fossils (prokaryotes)	
			4,570	Earth forms	

*Many authorities divide the Cenozoic era into the Paleogene period (contains the Paleocene, Eocene, and Oligocene epochs) and the Neogene period (contains the Miocene, Pliocene, Pleistocene, and Holocene epochs).

Limitations of the Timescale

Because the timescale tells when various groups evolved and flourished, it might seem that evolution has been a series of events leading only from the first cells to humans. This is not the case; for example, prokaryotes (bacteria and archaea) never declined and are still the most abundant and successful organisms on Earth. Even today, they constitute up to 90% of the total weight of organisms.

Then, too, the timescale lists mass extinctions, but it doesn't tell when specific groups became extinct. **Extinction** is the total disappearance of a species or a higher group; **mass extinction** occurs when a large number of species disappear in a few million years or less. For lack of space, the geologic timescale can't depict in detail what happened to the members of every group mentioned. Figure 14.1 does show how mass extinction affected a few groups of animals. Which of the animals shown in Figure 14.1 suffered the most during the **P-T extinction** (Permian-Triassic extinction)? ______________________________

The **K-T extinction** occurred between the Cretaceous and the Tertiary periods. Which animals shown in Figure 14.1 became extinct during the K-T extinction? ______________________________

Figure 14.1 shows only periods and no eras. *Fill in the eras on the lines provided in the figure.*

Figure 14.1 Mass extinctions.
Five significant mass extinctions and their effects on the abundance of certain forms of marine and terrestrial life. The width of the horizontal bars indicates the varying number of each life-form considered.

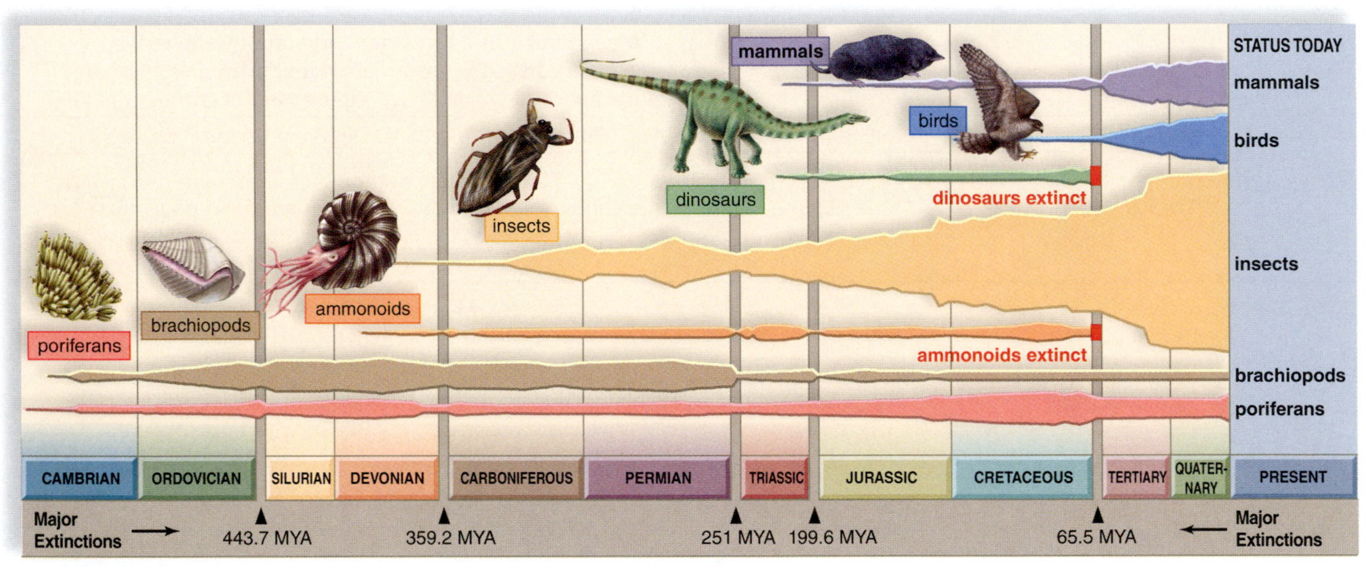

Fossils

The fossil record depends heavily on anatomical data to show the evolutionary changes that have occurred as one group gives rise to another group. Why would that be? ______________________________

____

Invertebrate Fossils

The **invertebrates** are animals without a backbone (see Fig. 14.2). Even the chordates, which is the phylum that contains the **vertebrates** (animals with a backbone), contain a couple of insignificant groups of invertebrates.

In order to familiarize yourself with possible invertebrate fossils, a few of the most common fossilized invertebrate groups are depicted in Figure 14.2.

Figure 14.2 Invertebrate fossils.

Fossilized ammonite, a mollusc

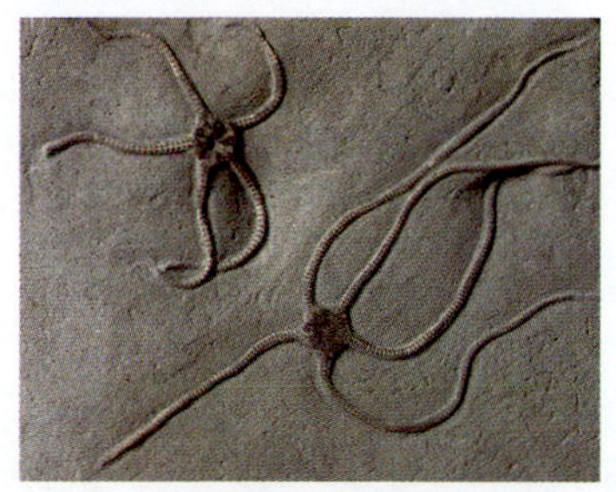

Fossilized brittle stars, echinoderms

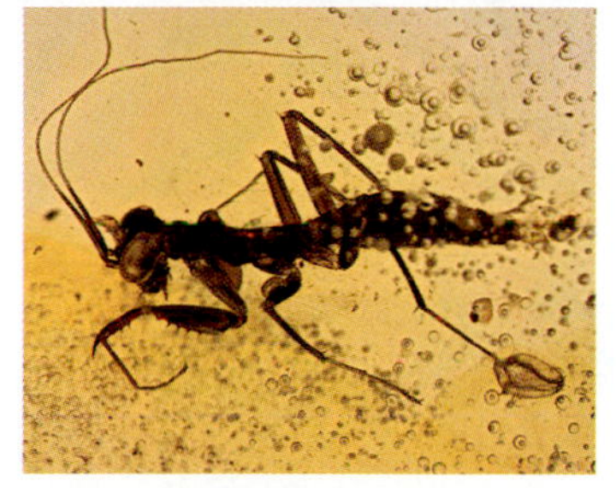

Fossilized mantis in amber

Fossilized pseudoscorpion, an arthropod

Fossilized snails, molluscs

Fossilized trilobites, arthropods

Observation: Invertebrate Fossils

1. Obtain a box of selected fossils. If the fossils are embedded in rocks, examine the rock until you have found the fossil. Fossils are embedded in rocks because the sediment that originally surrounded them hardened over time. Most fossils consist of hard parts such as shells, bones, or teeth because these parts are not consumed or decomposed over time. One possible reason the Cambrian might be rich in fossils is that organisms now had ______________________________ whereas before they did not.
2. The kit you are using, or your instructor, will identify which of the fossils are invertebrate animals. List the names of these fossils in Table 14.2 and give a description of the fossil. Use the Geologic Timescale (Table 14.1) to tell the era and period they were most abundant; list the fossils from the latest (top) to earliest (bottom).

Table 14.2 Invertebrate Fossils

Type of Fossil	Era, Period	Description

Observation: Vertebrate Fossils

The various groups of vertebrates are shown in Figure 14.3. Today it is generally agreed that birds are reptiles rather than being a separate group. That means that the major groups of vertebrates are (1) various types of fishes (jawless, jawed, cartilaginous and bony fishes), (2) amphibians such as frogs and salamanders, (3) reptiles such as lizards and crocodiles, and (4) mammals. There are many types of mammals from whales to mice to humans. The fossil record can rely on skeletal differences, such as limb structure, to tell if an animal is a mammal.

In order to familiarize yourself with possible vertebrate fossils, a few of the most common fossilized vertebrate groups are depicted in Figure 14.3. Which of the fossils available to you are vertebrates? ______________

____

Use Table 14.1 to associate each fossil with the particular era and period when this type animal was most abundant. Fill in Table 14.3 according to sequence of the time frames from the latest *(top)* to earliest *(bottom).*

Fossilized bird, a reptile

Fossilized bony fish

Fossilized deerlike mammal

Fossilized frog, an amphibian

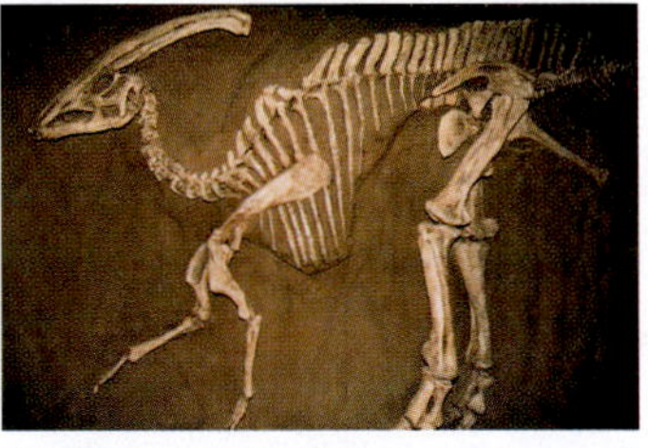

Fossilized duckbill dinosaur, a reptile

Fossilized snake, a reptile

Figure 14.3 Vertebrate fossils.

Table 14.3 Vertebrate Fossils

Type of Fossil	Era, Period	Description

Observation: Plant Fossils

See Figure 14.4, which shows the evolution of plants including the bryophytes, ferns and their allies, gymnosperms, and angiosperms. The fossil record for plants is not as good as that for invertebrates and vertebrates because plants have no hard parts that are easily fossilized. In order to familiarize yourself with possible plant fossils, a few of the most common fossilized plant groups are depicted in Figure 14.4.

Fossilized ferns

Fossilized flower

Fossilized maple leaf

Fossilized poplar leaf

Fossilized sassafras leaf

Fossilized early seed plant leaves

Figure 14.4 Plant fossils.

Plants that have no hard parts become fossils when their impressions are filled in by minerals. Use Table 14.1 to associate each fossilized plant in your kit with a particular era and period. Assume trees are flowering plants and associate them with the era and period when flowering plants were most abundant. Fill in Table 14.4 according in sequence from the latest *(top)* to earliest *(bottom)*.

Table 14.4 Plant Fossils

Type of Fossil	Era, Period	Description of Fossil

Summary of Evidence from the Fossil Record

In this section, you studied the geologic timescale and various fossils represented in the record. The geologic timescale gives powerful evidence of evolution because

1. Fossils are ______________________________.
2. Younger fossils and not older fossils are more like ______________________________.
3. In short, the fossil record shows that ______________________________.

14.2 Evidence from Comparative Anatomy

In the study of evolutionary relationships, parts of organisms are said to be **homologous** if they exhibit similar basic structures and embryonic origins (Fig. 14.5). If parts of organisms are similar in function only, they are said to be **analogous.** Only homologous structures indicate an evolutionary relationship and are used to classify organisms.

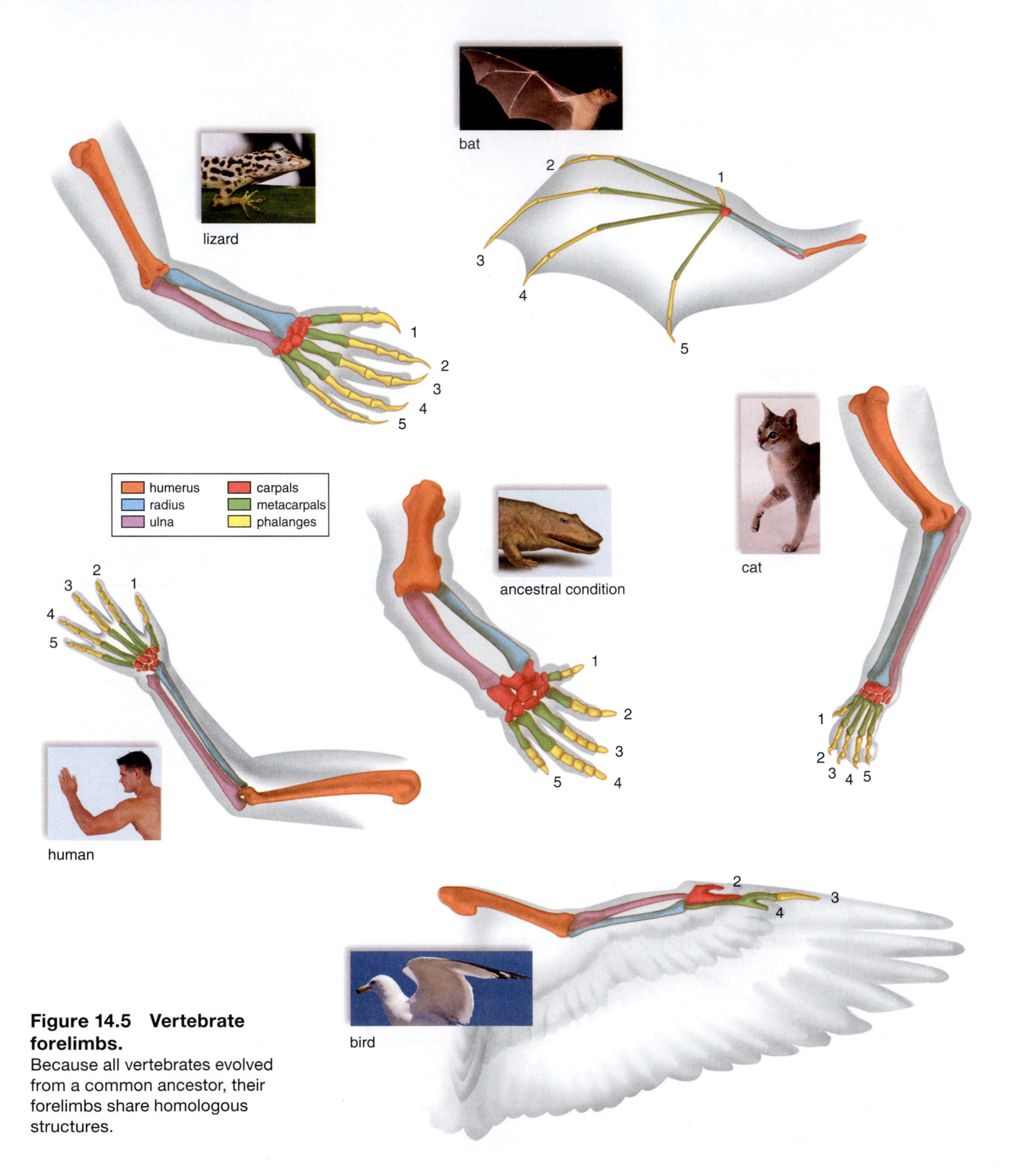

Figure 14.5 Vertebrate forelimbs.
Because all vertebrates evolved from a common ancestor, their forelimbs share homologous structures.

Comparison of Adult Vertebrate Forelimbs

The limbs of vertebrates are homologous structures (Fig. 14.5). The similarity of homologous structures is explained by descent from a common ancestor.

Observation: Vertebrate Forelimbs

1. Find the forelimb bones of the ancestral vertebrate in Figure 14.5. The basic components are the humerus (h), ulna (u), radius (r), carpals (c), metacarpals (m), and phalanges (p) in the five digits.
2. *Label the corresponding forelimb bones of the lizard, the bird, the bat, the cat, and the human.*
3. Fill in Table 14.5 to indicate which bones in each specimen appear to most resemble the ancestral condition and which most differ from the ancestral condition.
4. Adaptation to a way of life can explain the modifications that have occurred. Relate the change in bone structure to mode of locomotion in two examples.

   Example 1: ______________________________

   Example 2: ______________________________

Table 14.5 Comparison of Vertebrate Forelimbs

Animal	Bones That Resemble Common Ancestor	Bones That Differ from Common Ancestor
Lizard		
Bird		
Bat		
Cat		
Human		

Conclusion: Vertebrate Forelimbs

- Vertebrates are descended from a ________________, but they are adapted to ________________.

Comparison of Vertebrate Embryos

The anatomy shared by vertebrates extends to their embryological development. During early developmental stages, all animal embryos resemble each other closely but as development proceeds the different types of vertebrates take on their own shape and form. In Figure 14.6, the reptile and bird embryo resemble each other more than either resembles a fish. What does that tell you about their evolutionary relationship? ______________________________

____

In the following observation, you will see that as embryos all vertebrates have a postanal tail, a dorsal spinal cord, pharyngeal pouches, and various organs. In aquatic animals, pharyngeal pouches become functional gills (Fig. 14.6). In humans, the first pair of pouches becomes the cavity of the middle ear and auditory tube, the second pair becomes the tonsils, and the third and fourth pairs become the thymus and parathyroid glands.

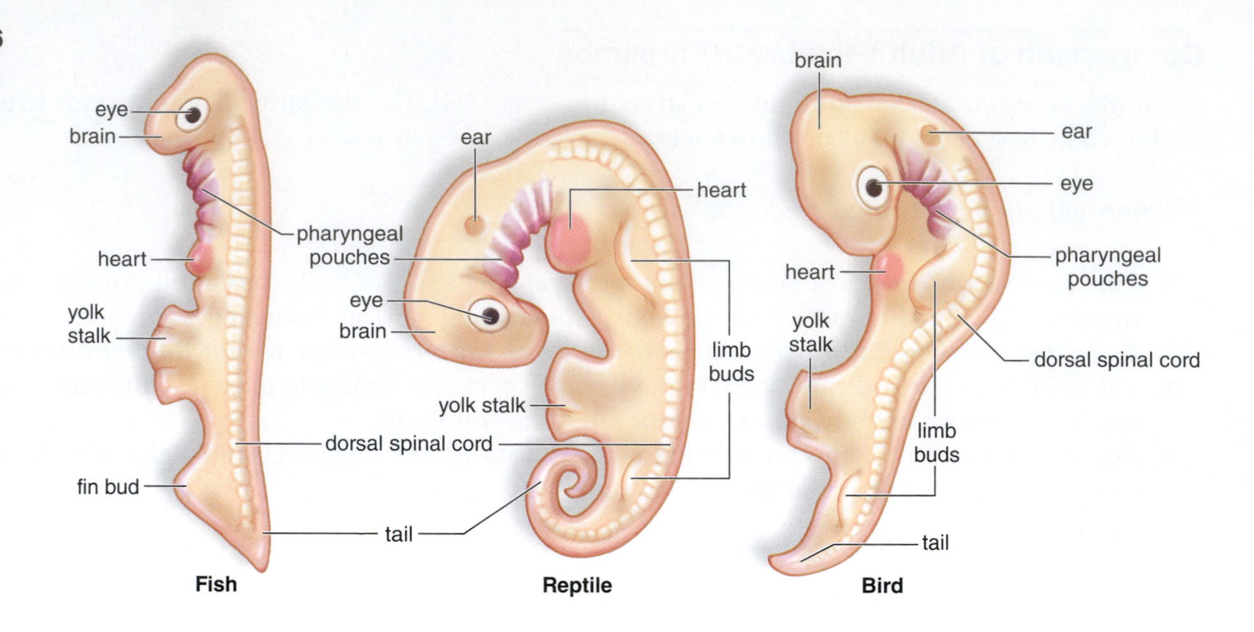

Figure 14.6 Vertebrate embryos.

Observation: Vertebrate Embryos

1. Obtain prepared slides of vertebrate embryos at comparable stages of development. Observe each of the embryos using a stereomicroscope.
2. List five similarities of the embryos:

   a. ___

   b. ___

   c. ___

   d. ___

   e. ___

Conclusion: Vertebrate Embryos

Vertebrate embryos resemble one another because ___

___.

Comparison of Chimpanzee and Human Skeletons

Chimpanzees and humans are closely related, as is apparent from an examination of their skeletons. However, they are adapted to different ways of life. Chimpanzees are adapted to living in trees and are herbivores—they eat mainly plants. Humans are adapted to walking on the ground and are omnivores—they eat both plants and meat.

Observation: Chimpanzee and Human Skeletons

Posture

Chimpanzees are arboreal and climb in trees. While on the ground, they tend to knuckle-walk, with their hands bent. Humans are terrestrial and walk erect. In Table 14.6, compare:

1. **Head and torso:** Where are the head and trunk with relation to the hips and legs—thrust forward over the hips and legs or balanced over the hips and legs (see Fig. 14.7)?

Table 14.6 Comparison of Chimpanzee and Human Postures		
Skeletal Part	**Chimpanzee**	**Human**
1. Head and torso		
2. Spine		
3. Pelvis		
4. Femur		
5. Knee joint		
6. Foot: Opposable toe		
Arch		

2. **Spine:** Which animal has a long and curved lumbar region, and which has a short and stiff lumbar region? How does this contribute to an erect posture in humans? ________________

____

3. **Pelvis:** Chimpanzees sway when they walk because lifting one leg throws them off balance. Which animal has a narrow and long pelvis, and which has a broad and short pelvis? Record your observations in Table 14.6.
4. **Femur:** In humans, the femur better supports the trunk. In which animal is the femur angled between articulations with the pelvic girdle and the knee? In which animal is the femur straight with no angle? Record your observations in Table 14.6.
5. **Knee joint:** In humans, the knee joint is modified to support the body's weight. In which animal is the femur larger at the bottom and the tibia larger at the top? Record your observations in Table 14.6.
6. **Foot:** In humans, the foot is adapted for walking long distances and running with less chance of injury. In which animal is the big toe opposable? ________ How does an opposable toe assist chimpanzees?

____

Figure 14.7 Human and chimpanzee skeletons.

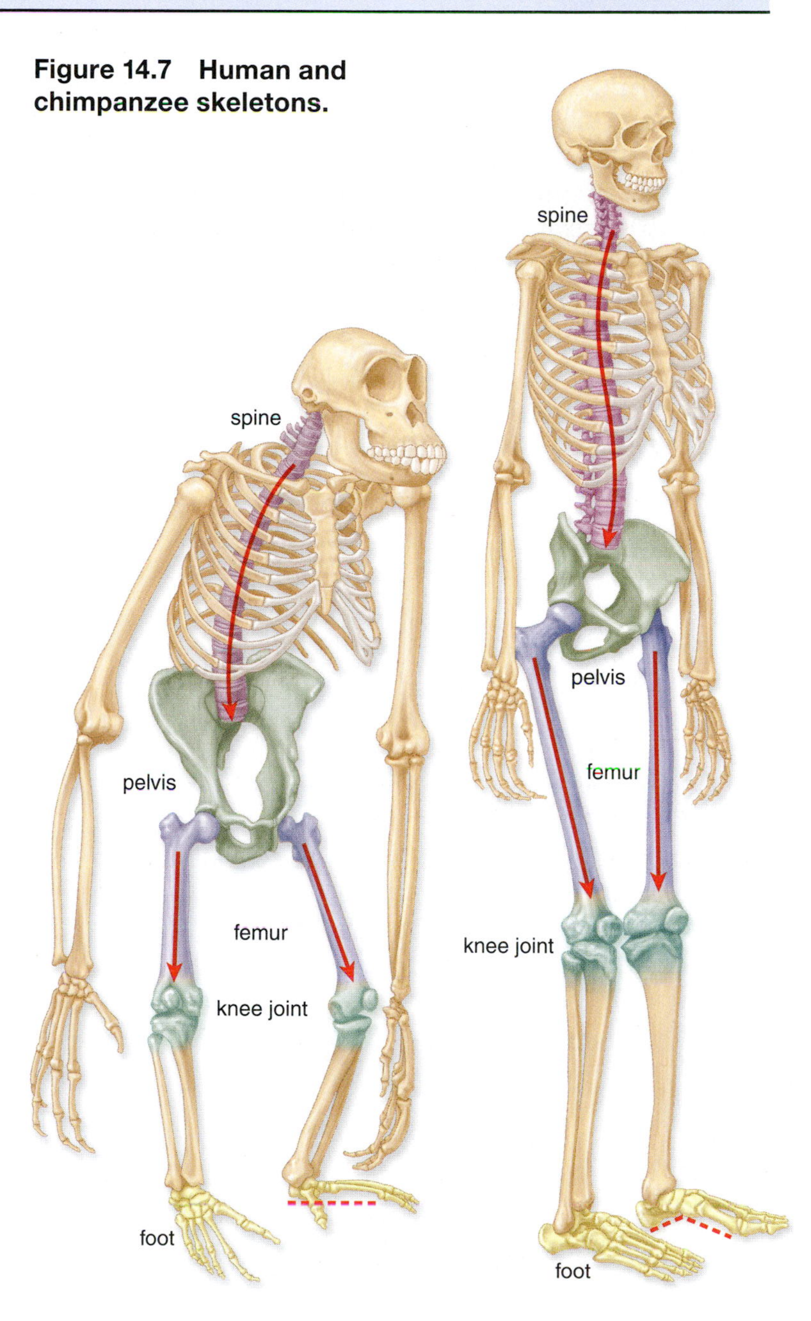

Which foot has an arch? ____________ How does an arch assist humans? ____________

____

Record your observations in Table 14.6.

7. How does the difference in the position of the foramen magnum, a large opening in the base of the skull for the spinal cord, correlate with the posture and stance of the two organisms?

____

Skull Features

Humans are omnivorous. A diet rich in meat does not require strong grinding teeth or well-developed facial muscles. Chimpanzees are herbivores, and a vegetarian diet requires strong teeth and strong facial muscles that attach to bony projections. Compare the skulls of the chimpanzee and the human in Figure 14.8 and answer the following questions:

1. **Supraorbital ridge:** For which skull is the supraorbital ridge (the region of frontal bone just above the eye socket) thicker? Record your observations in Table 14.7.
2. **Sagittal crest:** Which skull has a sagittal crest, a projection for muscle attachments that runs along the top of the skull? Record your observation in Table 14.7.
3. **Frontal bone:** Compare the slope of the frontal bones of the chimpanzee and human skulls. How are they different? Record your observations in Table 14.7.
4. **Teeth:** Examine the teeth of the adult chimpanzee and adult human skulls. Are the incisors (two front teeth) vertical or angled? Do the canines overlap the other teeth? Are the molars larger or moderate in size? Record your observations in Table 14.7
5. **Chin:** What is the position of the mouth and chin in relation to the profile for each skull? Record your observations in Table 14.7.

Figure 14.8 Chimpanzee and human skulls.

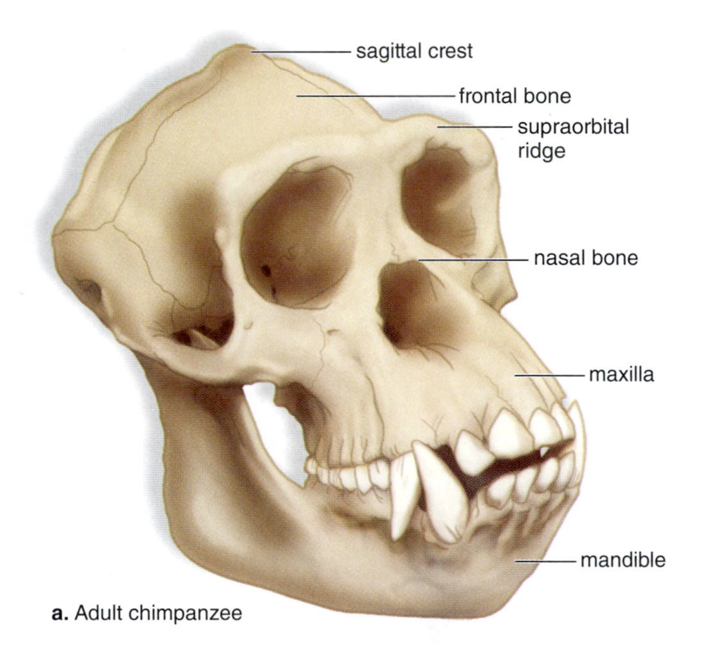

a. Adult chimpanzee

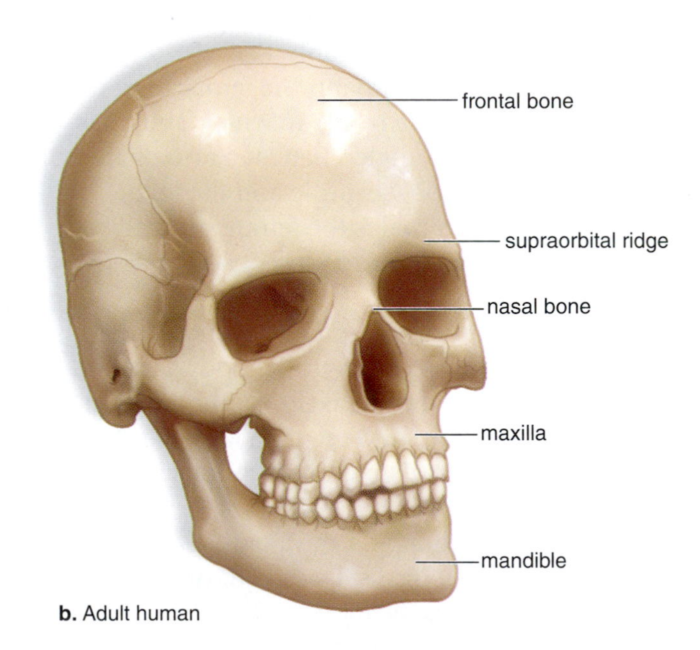

b. Adult human

Table 14.7 Skull Features of Chimpanzees and Humans		
Feature	**Chimpanzee**	**Human**
1. Supraorbital ridge		
2. Sagittal crest		
3. Slope of frontal bone		
4. Teeth		
5. Chin		

Conclusions: Chimpanzee and Human Skeletons

- Do your observations show that the skeletal differences between chimpanzees and humans can be related to posture? ________ Explain. ________________________________
- Do your observations show that diet can be related to the skull features of chimpanzees and humans? ________ Explain. ________________________________

Comparison of Hominid Skulls

The designation *hominid* includes humans and primates that are humanlike. Paleontologists have uncovered several fossils dated from 7.5 MYA (millions of years ago) to 30,000 years BP (before present), when humans called Cro-Magnons arose that are virtually identical to modern humans *(Homo sapiens)* (Fig. 14.9).

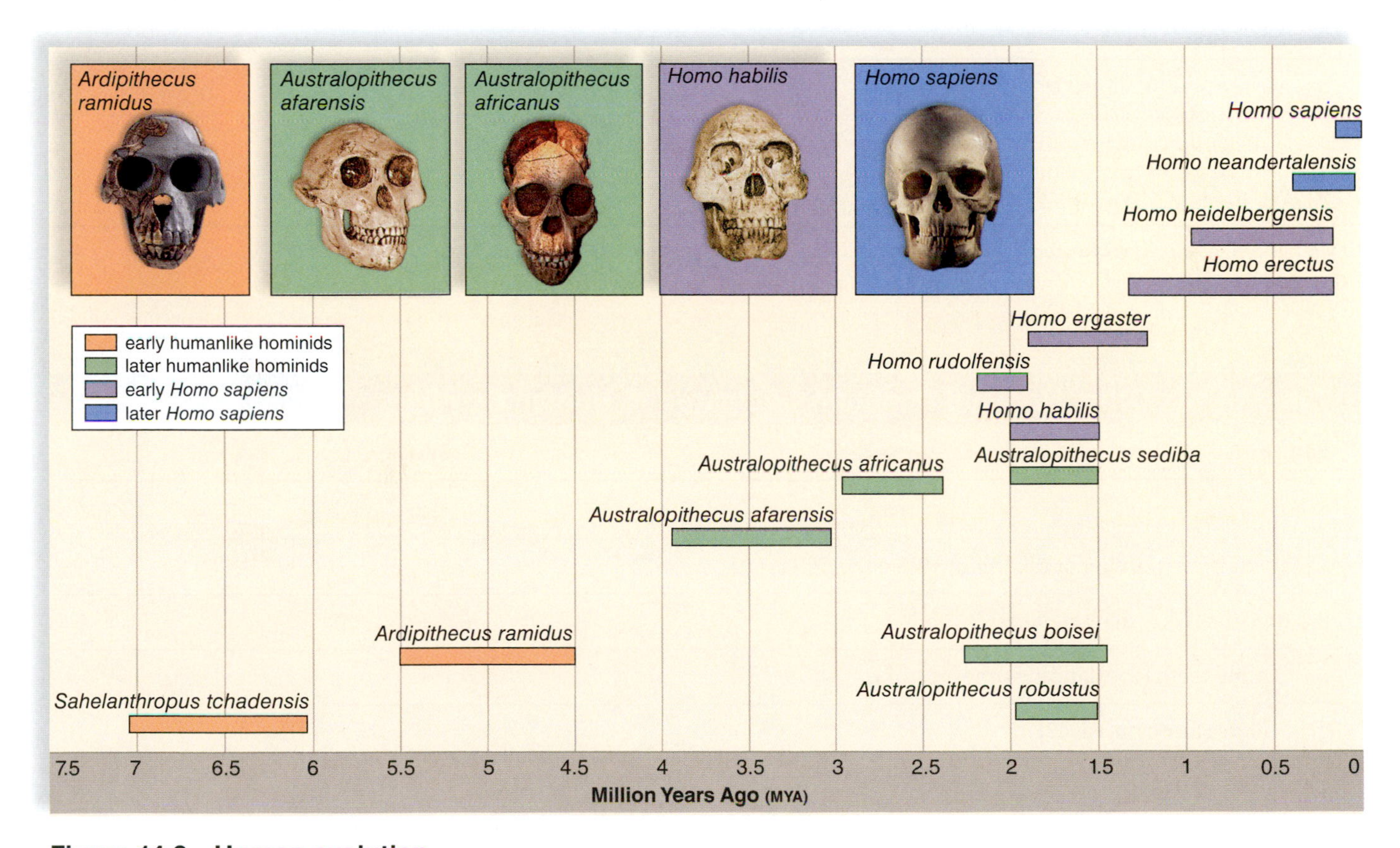

Figure 14.9 Human evolution.

Observation of Hominid Skulls

Several of the skulls noted in Figure 14.9 are on display. Use Tables 14.8, 14.9, and 14.10 to record data pertaining to the cranium (or braincase), the face, and the teeth. Compare the early skulls on display with a modern human skull. For example, is the forehead like or more flat than the human skull? List at least three skulls you examined for each feature.

Table 14.8 Other Hominid Craniums Compared to Human Cranium

Feature	Skulls		
	1.	2.	3.
a. Frontal bone (like or more flat?)			
b. Supraorbital ridge (divided or continuous?)			
c. Sagittal crest (present?)			
d. Mastoid process (flat or projecting?)			

Table 14.9 Other Hominid Faces Compared to Human Face

Feature	Skulls		
	1.	2.	3.
a. Nasal bones (raised or flat?)			
b. Nasal opening (larger?)			
c. Chin (projecting forward?)*			
d. Width of face (wider?)**			

*If your instructor directs you to, measure from the edge of the foramen magnum to between the incisors.

**If your instructor directs you to, measure the width of the face from mid-zygomatic arch to the other mid-arch.

Table 14.10 Other Hominid Dentition Compared to Human Dentition

Feature	Skulls		
	1.	2.	3.
a. Teeth rows (parallel or diverging from each other?)			
b. Incisors (vertical or angled?)			
c. Canine teeth (overlapping other teeth?)			
d. Molars (more massive?)			

Conclusions: Hominid Skulls

- Do your data appear to be consistent with the evolutionary sequence of the hominids in Figure 14.9? Explain. ________________
- Report here any data you collected that would indicate a particular hominid was closer in time to humans than indicated in Figure 14.9. ________________
____
- Report here any data you collected that would indicate a particular hominid was more distant in time from humans than indicated in Figure 14.9. ________________
____

14.3 Molecular Evidence

Molecular data substantiate the comparative and developmental data that biologists have accumulated over the years. The activity of *Hox* genes differ and this can account, for example, for why vertebrates have a dorsally placed nerve cord while it is ventrally placed in invertebrates. Sequencing DNA data show which organisms are closely related. Molecular data among primates is of extreme interest because it can help determine which of the primates we are most closely related to. Chromosomal and genetic data allow us to conclude that we are more closely related to chimpanzees than to other types of apes.

In this section, we note that scientists can compare the amino acid sequence in proteins to determine the degree to which any two groups of organisms are related. The sequence of amino acids in cytochrome *c,* a carrier of electrons in the electron transport chain found in mitochondria, has been determined per a variety of organisms. On the basis of the number of amino acid *differences* reported in Figure 14.10, it is concluded that the evolutionary relationship between humans and these organisms decreases in the order stated: monkeys, pigs, ducks, turtles, fishes, moths, and yeast. This conclusion agrees with the sequence of dates these organisms are found in the fossil record. Why can comparing amino acid data lead to the same conclusions as comparing DNA data?

____
____

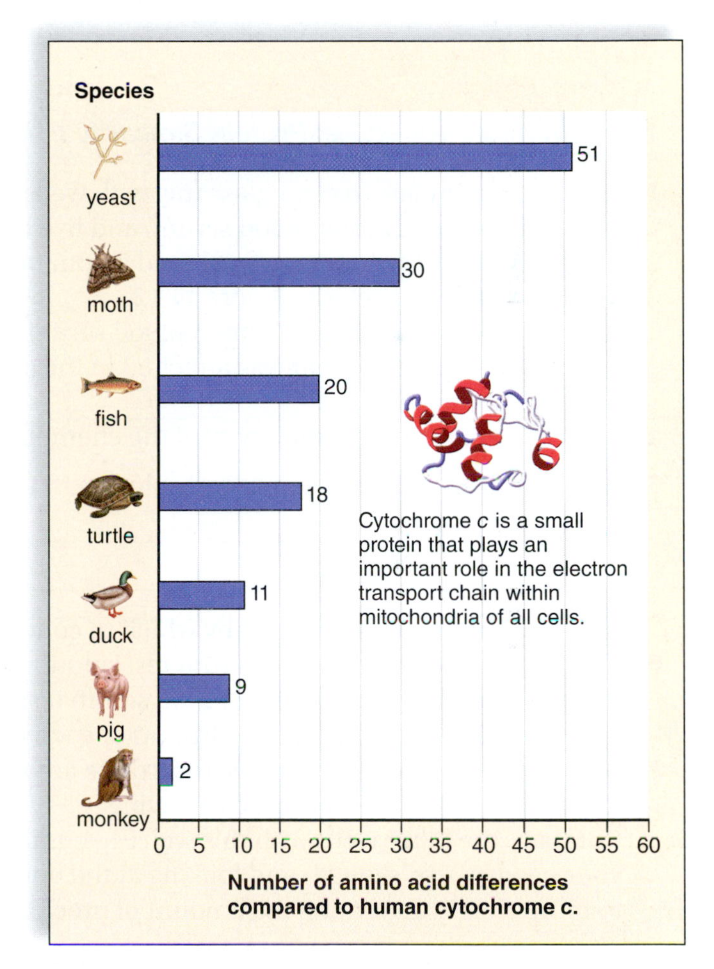

Figure 14.10 Amino acid differences in cytochrome *c*.
The few differences found in cytochrome *c* between monkeys and humans show that of these organisms, humans are most closely related to monkeys.

Protein Similarity Evidence

The immune system makes **antibodies** (proteins) that react with foreign proteins, termed **antigens.** Antigen-antibody reactions are specific. An antibody will react only with a particular antigen. In today's procedure, it is assumed rabbit antibodies to human antigens are in rabbit serum (Fig. 14.11). When these antibodies are allowed to react against the antigens of other animals, the stronger the antibody-antigen reaction (determined by the amount of precipitate), the more closely related the animal is to humans.

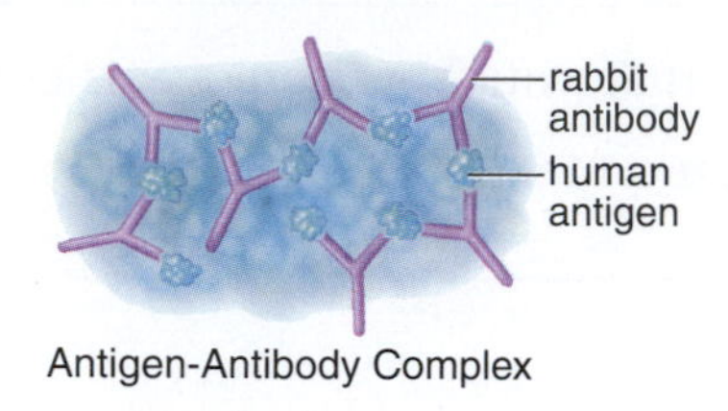

Figure 14.11 Antigen-antibody reaction.
When antibodies react to antigens, a precipitate appears.

Experimental Procedure: Protein Similarity Evidence

1. Obtain a chemplate (a clear glass tray with wells), one bottle of synthetic human blood serum, one bottle of synthetic rabbit blood serum, and five bottles (I to V) of blood serum test solution.
2. Put two drops of synthetic rabbit blood serum in each of the six wells in the chemplate. *Label the wells 1 to 6.* See yellow circles in Figure 14.12.
3. Add two drops of synthetic human blood serum to each well. See red circles in Figure 14.12. Stir with the plastic stirring rod that was attached to the chemplate. The rabbit serum now contains antibodies against human antigens.
4. Rinse the stirrer. (The large cavity of the chemplate may be filled with water to facilitate rinsing.)
5. Add four drops of blood serum test solution III (contains human antigens) to well 6. Describe what you see. ____

   ____

   This well will serve as the basis by which to compare all the other samples of test blood serum.
6. Now add four drops of blood serum test solution I to well 1. Stir and observe. Rinse the stirrer. Do the same for each of the remaining blood serum test solutions (II to V)—adding II to well 2, III to well 3, and so on. Be sure to rinse the stirrer after each use.
7. At the end of 10 and 20 minutes, record the amount of precipitate in each of the six wells in Figure 14.12. Well 6 is recorded as having ++++ amount of precipitate after both 10 and 20 minutes. Compare the other wells with this well (+ = trace amount; 0 = none). Holding the plate slightly above your head at arm's length and looking at the underside toward an overhead light source will allow you to more clearly determine the amount of precipitate.

Figure 14.12 Protein similarity.
The greater the amount of precipitate, the more closely related an animal is to humans.

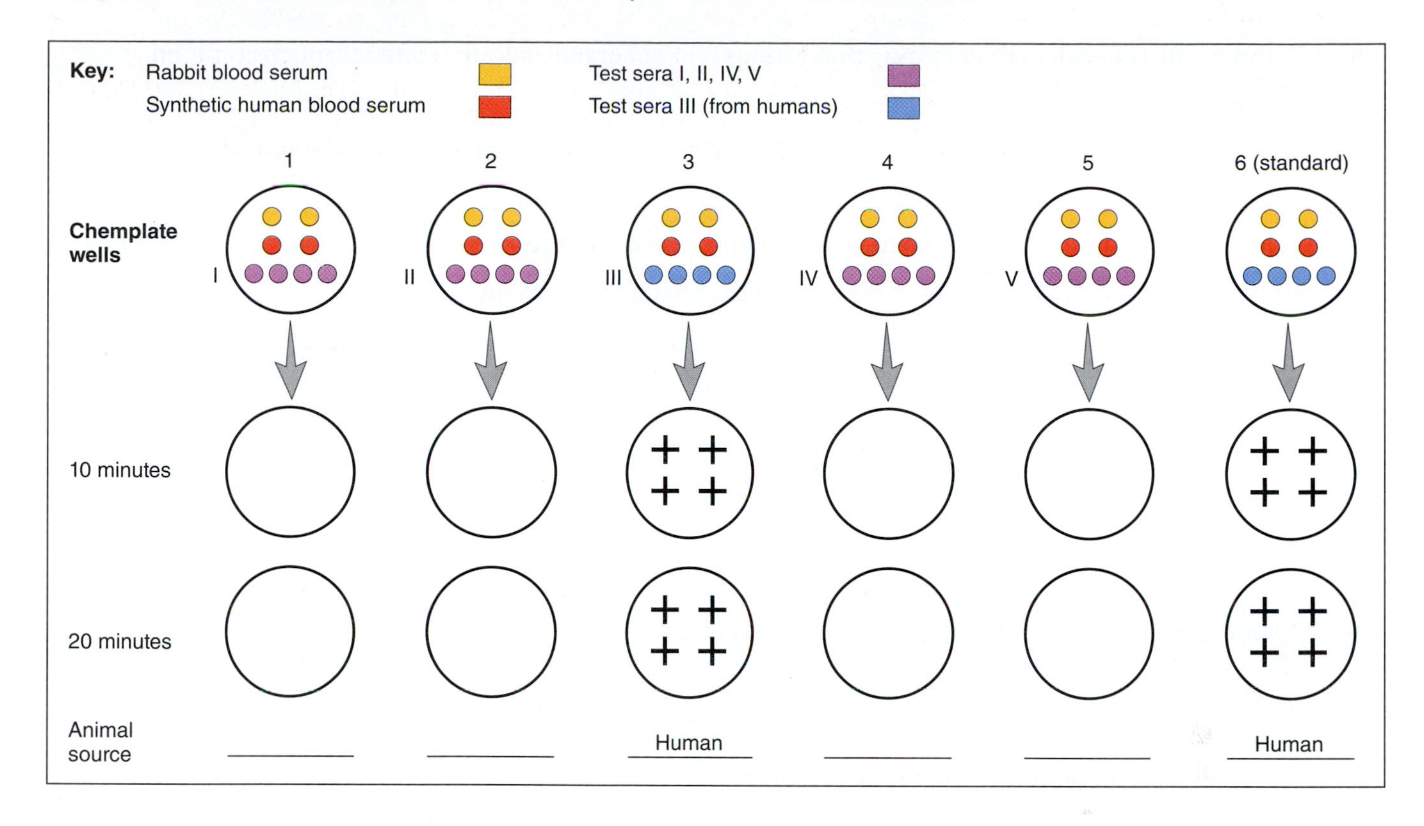

Conclusions: Protein Similarity Evidence

- The last row in Figure 14.12 tells you that the test serum in well 3 is from a human. How do your test results confirm this? ______
- Aside from humans, the test sera (supposedly) came from a pig, a monkey, an orangutan, and a chimpanzee. Which is most closely related to humans—the pig or the chimpanzee? ______
- Judging by the amount of precipitate, complete the last row in Figure 14.12 by indicating which serum you believe came from which animal. On what do you base your conclusions? ______

  ______
- In Section 14.2, a comparison of bones in vertebrate forelimbs showed that vertebrates share a common ancestor. Molecular evidence shows us that of the vertebrates studied, ______ and ______ are most closely related. In evolution, closely related organisms share a recent common ancestor. Humans share a more recent common ancestor with chimpanzees than they do with pigs.

Laboratory Review 14

1. List three types of evidence suggesting that various types of organisms are related through common descent. ______
2. Why would you *not* expect a fossil buried millions of years ago to look exactly like a modern-day organism? ______
3. A horseshoe crab has changed little in approximately 200 million years of existence. Would you expect to find that the environment of the horseshoe crab has changed minimally? Explain. ______
4. If a characteristic is found in bacteria, fungi, pine trees, snakes, and humans, when did it most likely evolve? ______ Why? ______
5. What are homologous structures, and what do they show about relatedness? ______
6. Why do humans and chicks develop similarly to reptiles? ______
7. What do DNA mutations have to do with amino acid changes in a protein? ______
8. How can the antigen-antibody reaction help determine the degree of relatedness between species? ______
9. Using plus (+) symbols, show the amount of reaction you would expect when an antibody against human serum is tested against sera from a pig, monkey, and chimpanzee. ______
10. Define the following terms:
    a. Fossil ______
    b. Common descent ______
    c. Adaptation ______
    d. Molecular evidence ______

Concepts of Biology Website

Instructors can find lab prep information and answers to all of the laboratory questions in the Laboratory Resource Guide. *Students* can practice their knowledge with quizzes, animations, flashcards, and much more.

www.mhhe.com/maderconcepts3

McGraw-Hill Access Science Website

An online encyclopedia of science and technology that provides information, including videos, that can enhance the laboratory experience.

www.accessscience.com

LABORATORY

15

Evolution of Bacteria, Protists, and Fungi

Learning Outcomes

15.1 Bacteria

- Relate the structure of a bacterium to its ability to cause disease. 172–73
- Recognize colonies of bacteria on agar plates. 174–75
- Identify the three commonly recognized shapes of bacteria. 175
- Recognize *Gloeocapsa, Oscillatoria,* and *Anabaena* as cyanobacteria. 175–77

15.2 Protists

- Identify and give examples of green, brown, red, and golden-brown algae (diatoms). 177–81
- Describe the structure and importance of diatoms and dinoflagellates. 181–82
- Distinguish protozoans on the basis of locomotion. 182–85

15.3 Fungi

- Describe the general characteristics of fungi. 186
- Compare sexual reproduction of black bread mold with that of a mushroom. 187–89
- Give evidence that fungi are agents of human disease. 189

Introduction

The history of life began with the evolution of the prokaryotic cell. Although the prokaryotic cell contains genetic material, it is not located in a nucleus, and the cell also lacks any other type of membranous organelle. At one time prokaryotes were believed to be a unified group, but based on molecular data, they are now divided into two major groups—**domain Bacteria** and **domain Archaea.** The eukaryotic cell is more closely related to the archaea than the bacteria. Eukaryotes in **domain Eukarya** have a membrane-bounded nucleus and membranous organelles. Prokaryotes resemble each other structurally but are metabolically diverse. Eukaryotes, on the other hand, are structurally diverse and exist as protists, fungi, plants, and animals (Fig. 15.1).

Figure 15.1 The living world.
Prokaryotes are represented in this illustration by the bacteria. The protists, fungi, plants, and animals are all eukaryotes.

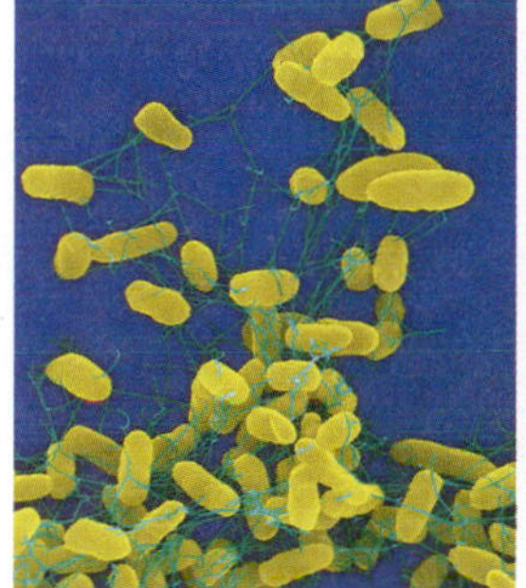

Bacteria 2,750×

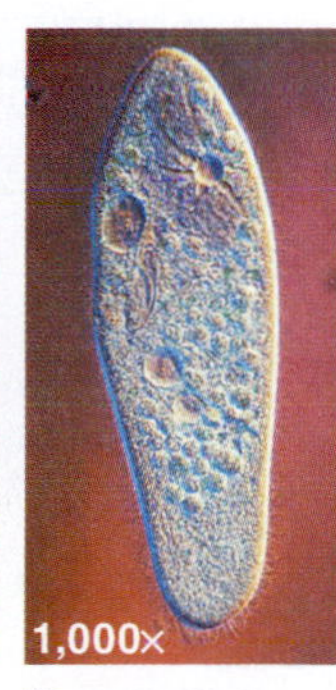

Paramecium, a protist

Morel, a fungus

Sunflower, a plant

Snow goose, an animal

15.1 Bacteria

In this laboratory, you will first relate the general structure of a bacterium to its ability to cause disease. The specific shape, growth habit, and staining characteristics of bacteria are often used to identify them. Therefore, you will observe a variety of bacteria using the microscope. Aside from their medical importance, bacteria are essential in ecosystems because, along with fungi, they are decomposers that break down dead organic remains, and thereby return inorganic nutrients to plants.

Pathogenic Bacteria

Pathogenic bacteria are infectious agents that cause disease. Infectious bacteria are able to invade and multiply within a host. Some also produce a toxin. Antibiotic therapy is often an effective treatment against a bacterial infection.

We will explore how it is possible to relate the structure of a bacterium to its ability to be invasive and avoid destruction by the immune system. We will also consider what morphophysiological attributes allow bacteria to be resistant to antibiotics and to pass the necessary genes on to other bacteria.

Observation: Structure of a Bacterium

1. Study the generalized structure of a bacterium in Figure 15.2 and, if available, examine a model or view a video of a bacterium.
2. Identify the following:
   - a. **Capsule:** A gel-like coating outside the cell wall. Capsules often allow bacteria to stick to surfaces such as teeth. They also prevent phagocytic white blood cells from taking them up and destroying them.
   - b. **Pili:** Hairlike bristles that allow adhesion to surfaces. This can be how a bacterium clings to and gains access to the body prior to an infection.
   - c. **Conjugation pilus:** An elongated, hollow appendage used to transfer DNA to other cells. Genes that allow bacteria to be resistant to antibiotics can be passed in this manner.
   - d. **Flagellum:** A rotating filament that pushes the cell forward.
   - e. **Cell wall:** A structure that provides support and shapes the cell. The cell wall contains a substance called **peptidoglycan.** Antibiotics that prevent the formation of a cell wall are most effective against Gram-positive rather than Gram-negative bacteria. Gram-positive bacteria have a thick cell wall that stains purple with the Gram stain, and Gram-negative bacteria have a thin cell wall that stains red with the Gram stain.

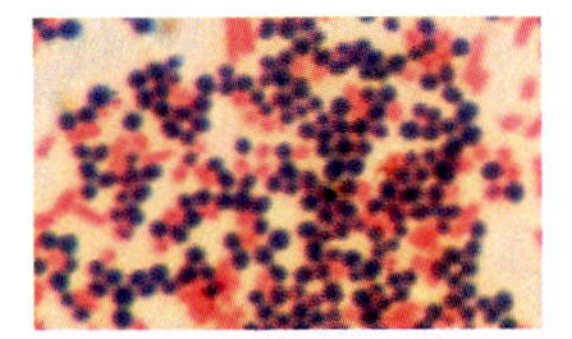
Gram stain results

   - f. **Plasma membrane:** A sheet that surrounds the cytoplasm and regulates entrance and exit of molecules. Resistance to antibiotics can be due to plasma membrane alterations that do not allow the drug to bind to the membrane or cross the membrane, or to a plasma membrane that increases the elimination of the drug from the bacteria.
   - g. **Ribosomes:** Site of protein synthesis. Some bacteria possess antibiotic-inactivating enzymes that make them resistant to antibiotics.
   - h. **Nucleoid:** The location of the bacterial chromosome.

Figure 15.2 Generalized structure of a bacterium.

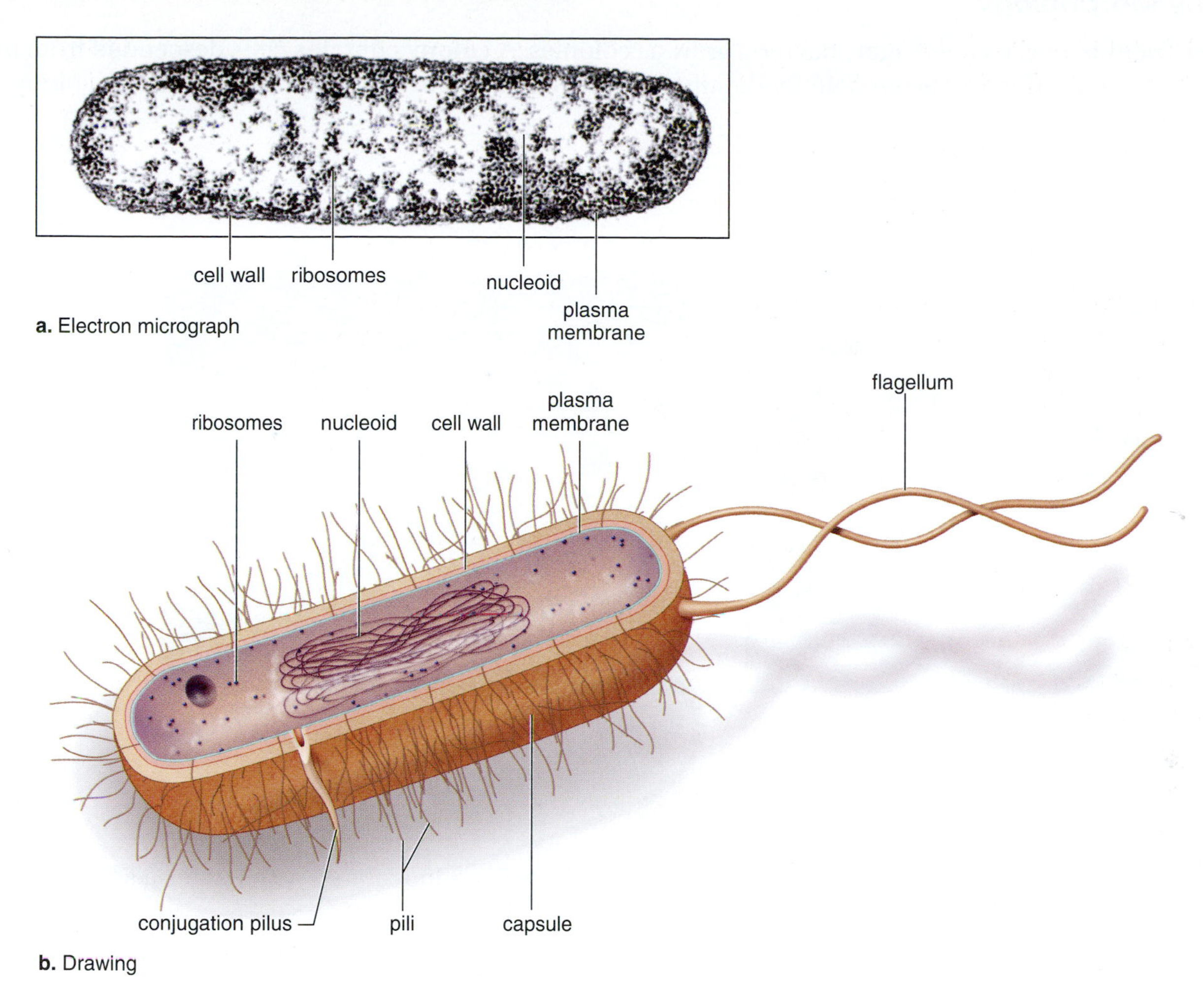

Also, some bacteria contain plasmids, small rings of DNA that replicate independently of the chromosomes and can be passed to other bacteria. Genes that allow bacteria to be resistant to antibiotics are often located in a plasmid.

Conclusions: Structure of a Bacterium

- Which portions of a bacterial cell aid the ability of a bacterium to cause infections?

  ____

- Which portions of a bacterial cell aid the ability of a bacterium to be resistant to antibiotics?

  ____

Colony Morphology

On a nutrient material called **agar,** bacteria grow as colonies. A **colony** contains cells descended from one original cell. Sometimes, it is possible to identify the type of bacterium by the appearance of the colony (Fig. 15.3).

Figure 15.3 Colony morphology.
Colonies of bacteria on agar plates.

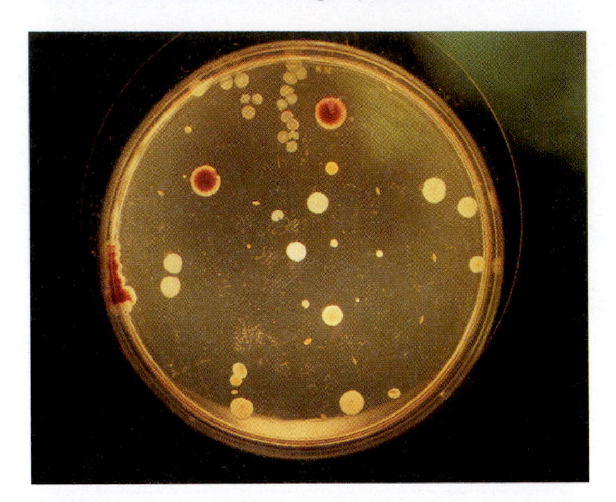

a.

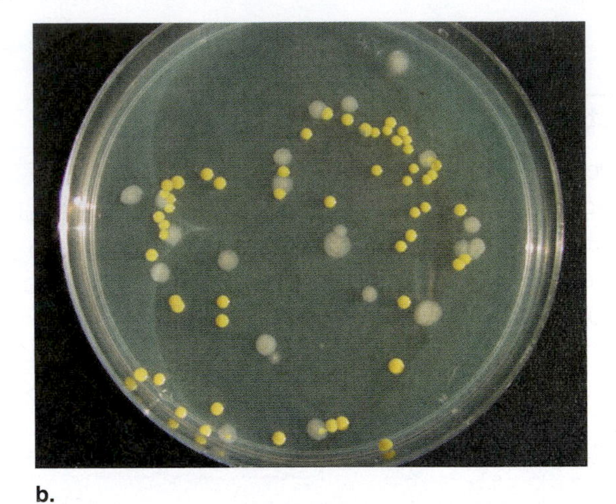

b.

Observation: Colony Morphology

1. Do not remove the covers and view agar plates that have been inoculated with bacteria and then incubated. Notice the "colonies" of bacteria growing on the plates.
2. Compare the colonies' color, surface, and margin, and note your observations in Table 15.1. It is not necessary to identify the type of bacteria.

Table 15.1 Agar Plates

Plate Number	Description of Colonies

Experimental Procedure: Colony Morphology

1. If available, obtain a sterile agar plate, and inoculate the plate with your thumbprint, or use a swab and inoculate the plate with material from around your teeth or inside your nose. Put your name on the plate, and place it where directed by your instructor. Remember to view the plate next laboratory period and describe what you see. ______________________________

   ____

2. If available, obtain a sterile agar plate, and expose it briefly (at most for 10 minutes) anywhere you choose, such as in the library, your room, or your car. Describe the appearance of your plate after exposure. ______________________________

____

Conclusions

- What have you discovered from this exercise? ______________________________

____

Shape of Bacterial Cell

Most bacteria are found in three basic shapes: **spirillum** (spiral or helical), **bacillus** (rod), and **coccus** (round or spherical) (Fig. 15.4). Bacilli may form long filaments, and cocci may form clusters or chains. Some bacteria form endospores. An **endospore** contains a copy of the genetic material encased by heavy protective spore coats. Spores survive unfavorable conditions and germinate to form vegetative cells when conditions improve.

Observation: Shape of Bacterial Cell

1. View the microscope slides of bacteria on display. What magnification is required to view bacteria?

____

2. Using Figure 15.4 as a guide, identify the three different shapes of bacteria. ______________________________
3. Do any of the slides on display show bacterial cells with endospores? ______________________________

What is an endospore, and why does it have survival value? ______________________________

____

Figure 15.4 Diversity of bacteria.
a. Spirillum, a spiral-shaped bacterium. **b.** Bacilli, rod-shaped bacteria. **c.** Cocci, round bacteria.

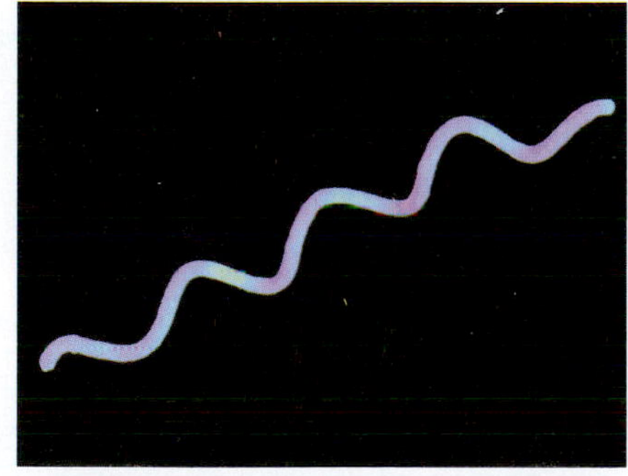

a. Spirillum: *Spirillum volutans* SEM 3,520×

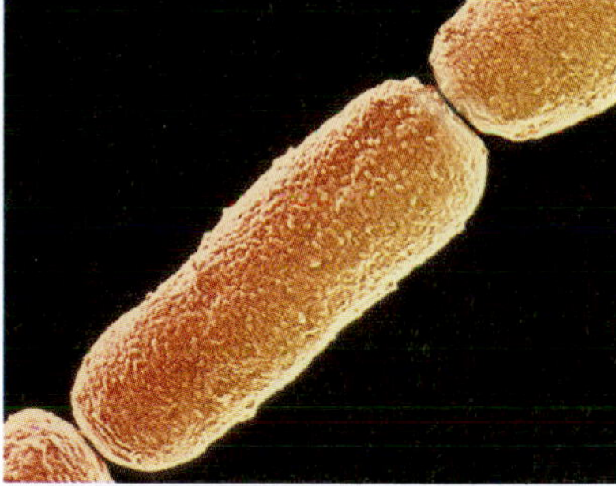

b. Bacilli: *Bacillus anthracis* SEM 35,000×

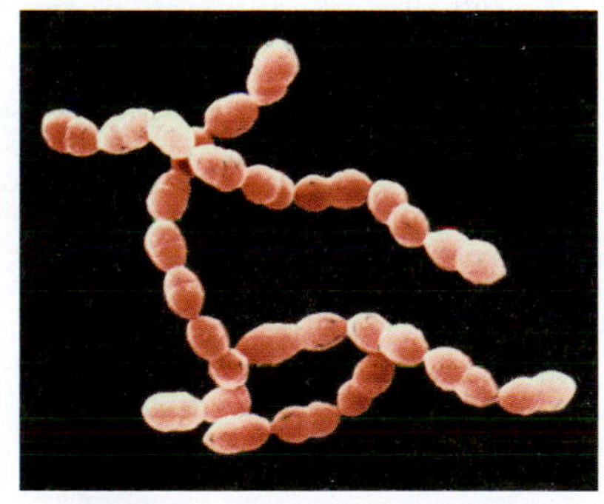

c. Cocci: *Streptococcus thermophilus* SEM 6,250×

Cyanobacteria

Cyanobacteria were formerly called blue-green algae because their general growth habit and appearance through a compound light microscope are similar to green algae. Electron microscopic study of cyanobacteria, however, revealed that they are structurally similar to other bacteria, particularly other photosynthetic bacteria. Although cyanobacteria do not have chloroplasts, they do have thylakoid membranes, where photosynthesis occurs.

Observation: Cyanobacteria

Gloeocapsa

1. Prepare a wet mount of a *Gloeocapsa* culture, if available, or examine a prepared slide, using high power (45×) or oil immersion (if available). The single cells adhere together because each is surrounded by a sticky, gelatinous sheath (Fig. 15.5).
2. What is the estimated size of a single cell? ______________________________

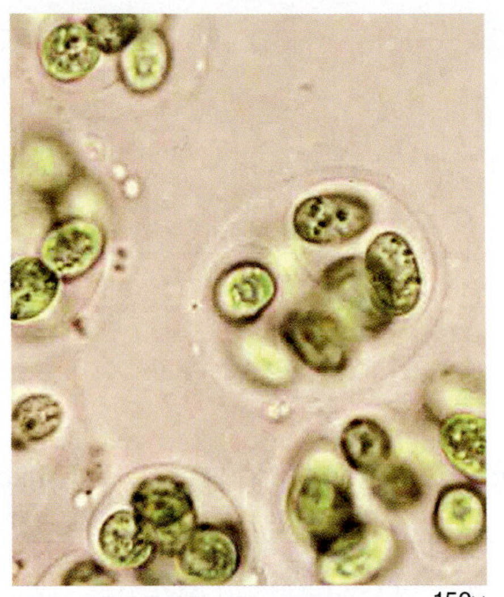

150×

a. Micrograph at low magnification.

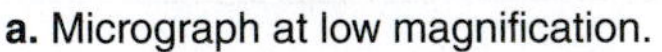

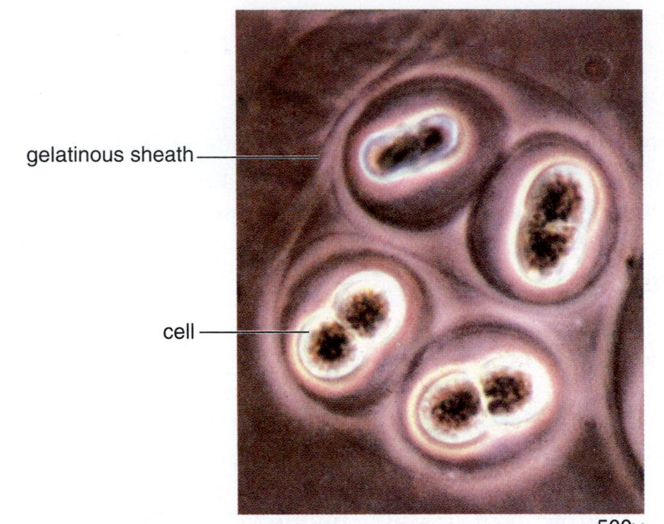

500×

b. Micrograph at high magnification.

Figure 15.5 ***Gloeocapsa.***

Oscillatoria

1. Prepare a wet mount of an *Oscillatoria* culture, if available, or examine a prepared slide, using high power (45×) or oil immersion (if available). This is a filamentous cyanobacterium with individual cells that resemble a stack of pennies (Fig. 15.6).
2. *Oscillatoria* takes its name from the characteristic oscillations that you may be able to see if your sample is alive. If you have a living culture, are oscillations visible? ______________________________

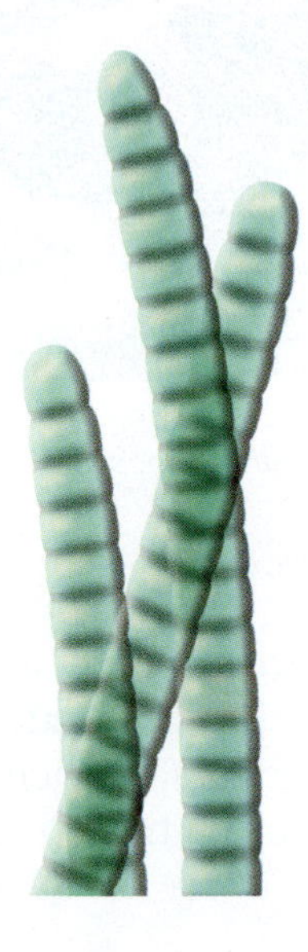

250×

Figure 15.6 ***Oscillatoria.***

Anabaena

1. Prepare a wet mount of an *Anabaena* culture, if available, or examine a prepared slide, using high power (45×) or oil immersion (if available). This is also a filamentous cyanobacterium, although its individual cells are barrel-shaped (Fig. 15.7).
2. Note the thin nature of this strand. If you have a living culture, what is its color? ______________________

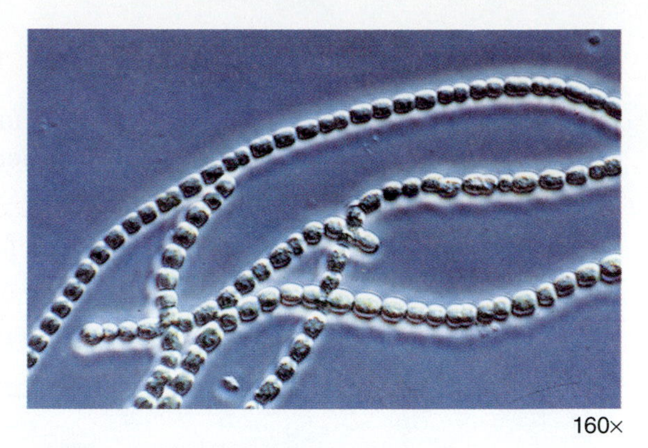

Figure 15.7 ***Anabaena.***

15.2 Protists

Protists were the first eukaryotes to evolve. Their diversity and complexity make it difficult to categorize them. However, the diagram in Figure 15.8 may be helpful.

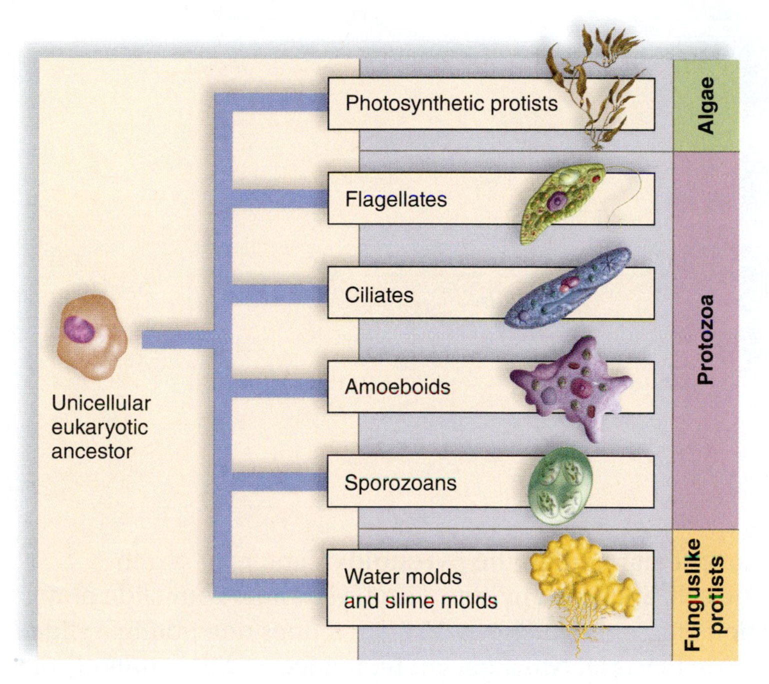

Figure 15.8 Major groups of protists. Because the precise evolutionary relationships between these groups are not yet known, they are grouped here by major shared characteristics.

Algae

Algae is a term that is used for aquatic organisms that photosynthesize in the same manner as land plants. All photosynthetic protists contain green chlorophyll, but they also may contain other pigments that mask the chlorophyll color, and this accounts for their common names—the green algae, red algae, brown algae, and golden-brown algae.

Observation: Green Algae

If available, view a video showing the many forms of green algae. Notice that green algae can be single cells, filaments, colonies, or multicellular sheets. You will examine a filamentous form *(Spirogyra)* and a colonial form *(Volvox)*. A **colony** is a loose association of cells.

1. *Spirogyra* is a filamentous alga, lives in fresh water, and often is seen as a green scum on the surface of ponds and lakes. The most prominent feature of the cells is the spiral, ribbonlike chloroplast (Fig. 15.9). How do you think *Spirogyra* got its name? ______________________________

____

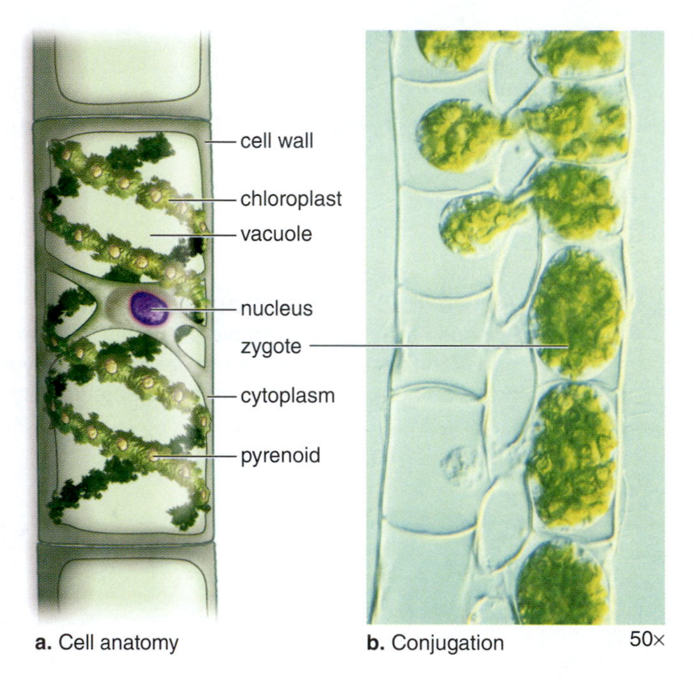

Figure 15.9 ***Spirogyra.***
a. *Spirogyra* is a filamentous green alga, in which each cell has a ribbonlike chloroplast. **b.** During conjugation, the cell contents of one filament enter the cells of another filament. Zygote formation follows.

Spirogyra's chloroplast contains a number of circular bodies, the **pyrenoids,** centers of starch polymerization. The nucleus is in the center of the cell, anchored by cytoplasmic strands. Your slide may show **conjugation,** a sexual means of reproduction illustrated in Figure 15.9*b*. If it does not, obtain a slide that does show this process. Conjugation tubes form between two adjacent filaments, and the contents of one set of cells enter the other set. As the nuclei fuse, a zygote is formed. The zygote overwinters, and in the spring, meiosis and, subsequently, germination occur. The resulting adult protist is therefore haploid.

Make a wet mount of live *Spirogyra,* or observe a prepared slide.

Volvox is a green algal colony. It is motile (capable of locomotion) because the thousands of cells that make up the colony have flagella. These cells are connected by delicate cytoplasmic extensions (Fig. 15.10).

Volvox is capable of both asexual and sexual reproduction. Certain cells of the adult colony can divide to produce **daughter colonies** (Fig. 15.10) that reside for a time within the parental colony. A daughter colony escapes the parental colony by releasing an enzyme that dissolves away a portion of the matrix of the parental colony. During sexual reproduction, some colonies of *Volvox* have cells that produce sperm, and others have cells that produce eggs. The resulting zygote undergoes meiosis and the adult *Volvox* is haploid.

Figure 15.10 ***Volvox.***
Volvox is a colonial green alga. The adult *Volvox* colony often contains daughter colonies, asexually produced by special cells.

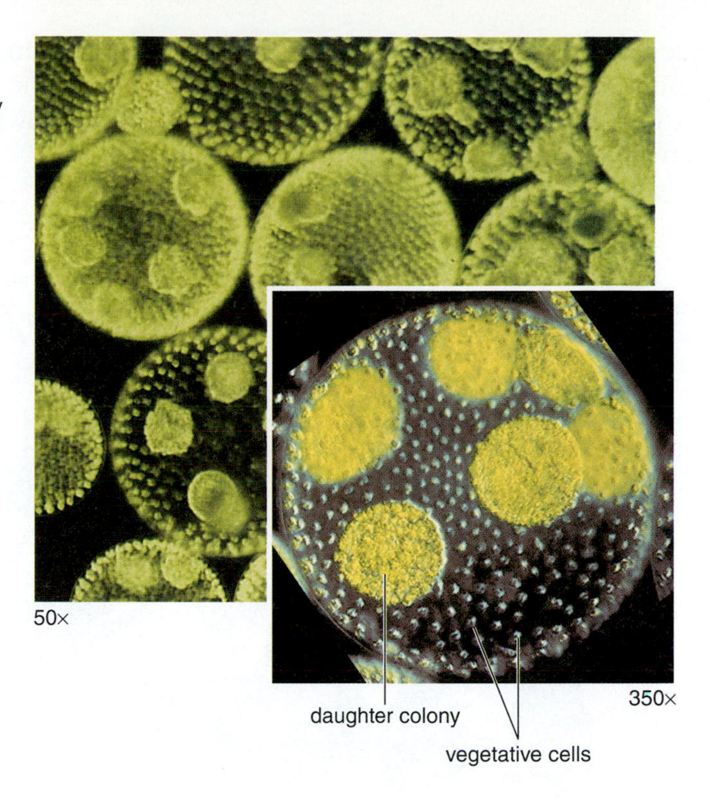

Using a depression slide, make a wet mount of live *Volvox,* or study a prepared slide.

In Table 15.2, list the genus names of each of the green algae specimens available, and give a brief description.

Table 15.2 Green Algae Diversity

Specimen	Description

Observation: Brown Algae

Brown algae are commonly called *seaweed,* along with the multicellular green and red algae. Brown algae contain brown pigments that mask chlorophyll's green color. These algae are large and have specialized parts.

Fucus is called rockweed because it is seen attached to rocks at the seashore when the tide is out (Fig. 15.11). If available, view a preserved specimen. Note the dichotomously branched body plan, so called because the **stipe** repeatedly divides into two branches (Fig. 15.11). Note also the **holdfast** by which the alga anchors itself to the rock; the **air vesicles,** or bladders, that help hold the thallus erect in the water; and the **receptacles,** or swollen tips. The receptacles are covered by small raised areas, each with a hole in the center. These areas are cavities in which the sex organs are located, with the gametes escaping to the outside through the holes. *Fucus* is unique among algae in that as an adult it is diploid (2n) and always reproduces sexually.

If available, study preserved specimens of other brown algae (Fig. 15.11). *Laminaria* algae are called **kelps.**

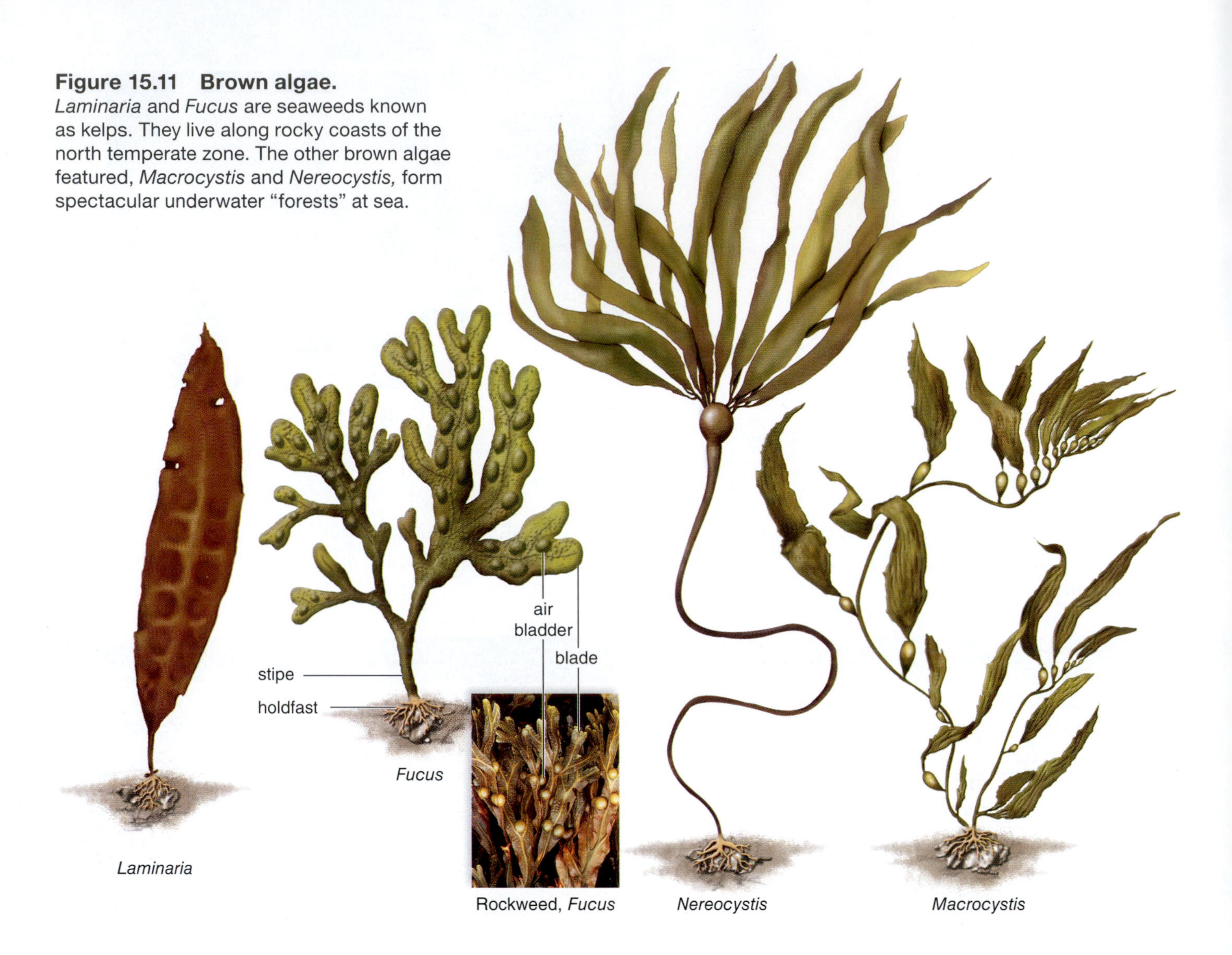

Figure 15.11 Brown algae. *Laminaria* and *Fucus* are seaweeds known as kelps. They live along rocky coasts of the north temperate zone. The other brown algae featured, *Macrocystis* and *Nereocystis,* form spectacular underwater "forests" at sea.

Observation: Red Algae

Like most brown algae, the red algae (Fig. 15.12) are multicellular, but they occur chiefly in warmer seawater, growing both in shallow waters and as deep as light penetrates. Some forms of red algae are filamentous, but more often, they are complexly branched with a feathery, flat, and expanded or ribbonlike appearance. Coralline algae are red algae that have cell walls impregnated with calcium carbonate ($CaCO_3$).

In Table 15.3, list the genus names of each of the brown and red algae specimens available, and give a brief description.

Figure 15.12 Red algae.
Generally, red algae are smaller and more delicate than brown algae. **a.** *Sebdenia* has a pronounced filamentous structure. **b.** Calcium carbonate is deposited in the walls of the red alga, *Corallina.*

a.

b.

Table 15.3 Brown and Red Algae

Specimen	Genus	Description
1		
2		
3		
4		

Observation: Diatoms

Diatoms (golden-brown algae) have a yellow-brown pigment that, in addition to chlorophyll, gives them their color.

The cell wall of **diatoms** is in two sections, with the larger one fitting over the smaller as a lid fits over a box. Since the cell wall is impregnated with silica, diatoms are said to "live in glass houses." The glass cell walls of diatoms do not decompose, so they accumulate in thick layers subsequently mined as diatomaceous earth and used in filters and as a natural insecticide. Diatoms, being photosynthetic and extremely abundant, are important food sources for the small heterotrophs (organisms that must acquire food from external sources) in both marine and freshwater environments.

Make a wet mount of live diatoms, or view a prepared slide (Fig. 15.13). Describe what you see:

350×

Figure 15.13 Diatoms.
Diatoms, photosynthetic protists of the oceans.

____

____

Observation: Dinoflagellates

Dinoflagellates are photosynthetic, but they have two flagella; one is free, but the other is located in a transverse groove that encircles the animal. The beating of these flagella causes the organism to spin like a top. The cell wall, when present, is frequently divided into closely joined polygonal plates of cellulose. At times there are so many of these organisms in the ocean that they cause a condition called "red tide." The toxins given off in these red tides cause widespread fish kills and can cause paralysis in humans who eat shellfishes that have fed on the dinoflagellates.

Make a wet mount of live dinoflagellates or view a prepared slide (Fig. 15.14). Describe what you see.

___

___

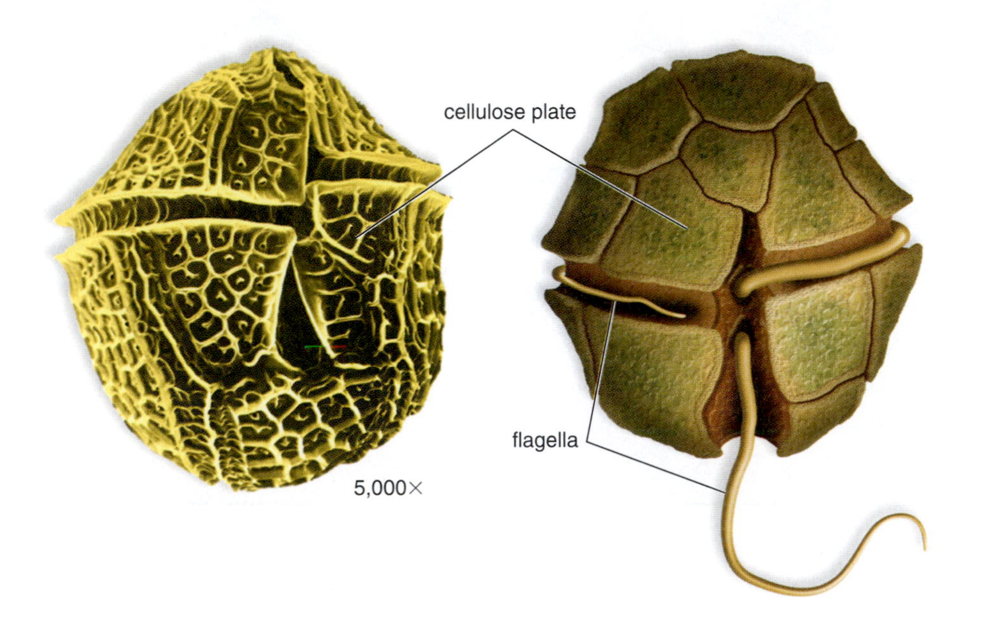

Figure 15.14 Dinoflagellates. Dinoflagellates such as *Gonyaulax* have cellulose plates.

Protozoans

The term *protozoan* refers to unicellular eukaryotes and is often restricted to heterotrophic organisms that ingest food by forming **food vacuoles.** Other vacuoles, such as **contractile vacuoles** that rid the cell of excess water, are also typical. Usually protozoans have some form of locomotion; some, such as amoebas, use **pseudopods** for locomotion and feeding; some, such as paramecia, move by **cilia;** and some, such as trypanosomes, use **flagella** (Fig. 15.15).

Plasmodium vivax, a common cause of malaria, is an apicomplexan, a protozoan that contains a special organelle called an apicoplast. The apicoplast assists the parasite in penetrating host cells. This type of protozoan is also often called a **sporozoan** because it goes through an asexual phase in which it exists as particulate spores (Fig. 15.15*d*). During its asexual phase, *Plasmodium vivax* lives inside red blood cells and the chills and fever of malaria occur when the red blood cells burst to release the spores. Sporozoans have no obvious means of locomotion as do the other types of protozoans illustrated in Figure 15.15. How do sporozoans differ from other types of protozoans? ___

___

___

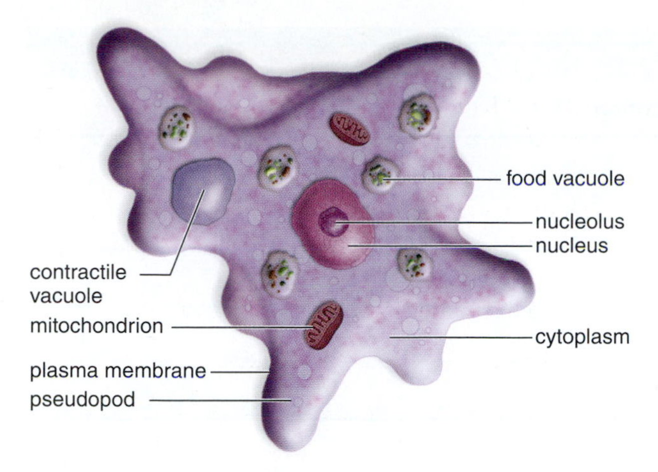

a. *Amoeba* moves by pseudopods.

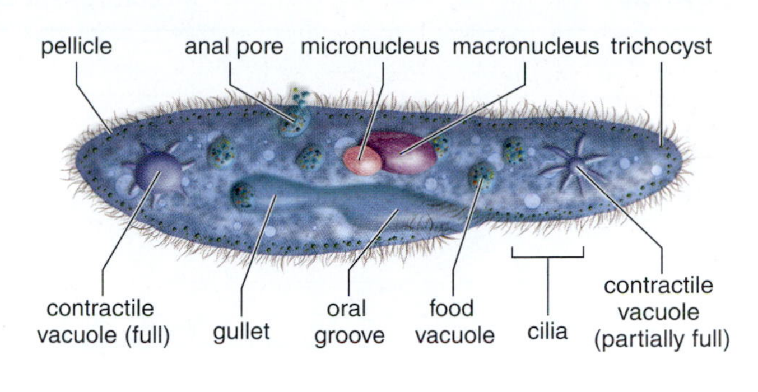

b. *Paramecium* moves by cilia.

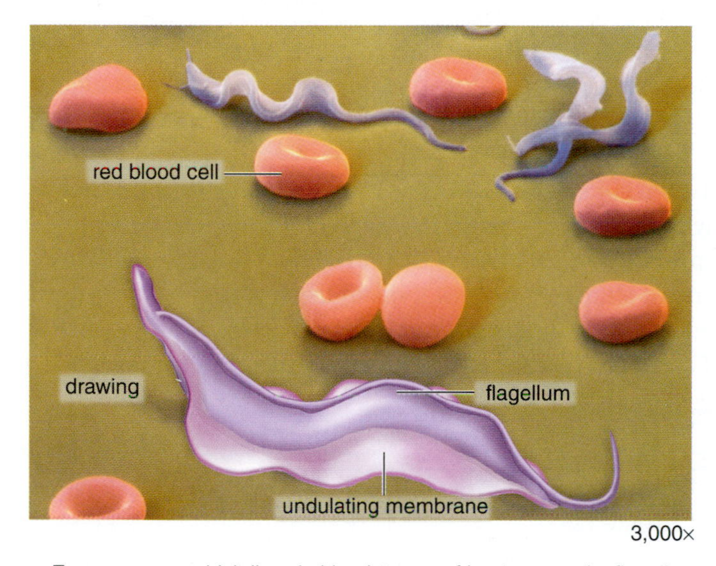

c. *Trypanosoma*, which lives in bloodstream of host, moves by flagella.

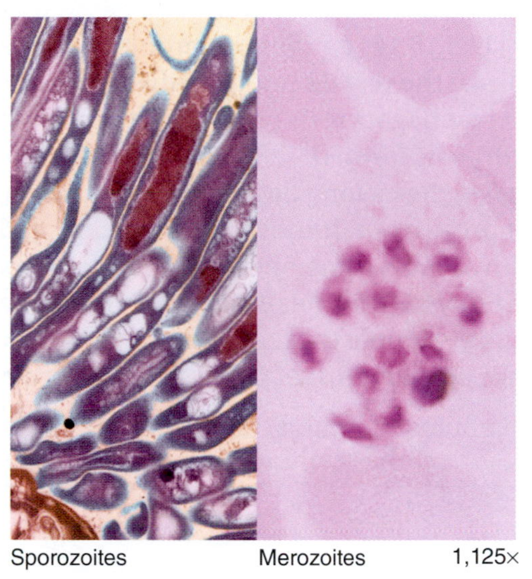

d. *Plasmodium* exists as nonmotile spores.

Figure 15.15 Protozoan diversity.
Protozoans are motile by the means illustrated, except for sporozoans such as *Plasmodium.* **a.** *Amoeba.* **b.** *Paramecium.* **c.** *Trypanosoma* in host. **d.** *Plasmodium* exists in several stages; merozoites invade red blood cells and divide to produce more spores.

Observation: Protozoans

Individual Protozoans

You may already have had the opportunity to observe a protozoan such as *Paramecium* or *Euglena* in Laboratory 2. However, your instructor may want you to observe these organisms again. *Euglena* have flagella but many also have chloroplasts (Fig. 15.16), and therefore they do not match our definition of a protozoan. Watch a video if available, and note the various forms of protozoans. Prepare wet mounts or examine prepared slides of protozoans as directed by your instructor. Complete Table 15.4, listing the structures for locomotion in the types of protozoans you have observed.

Table 15.4 Heterotrophic Protists

Name	Structures for Locomotion	Observations

Observation: Euglena

Euglena (Fig. 15.16) typifies the problem of classifying protists. One third of all *Euglena* genera have chloroplasts; the rest do not. This discrepancy can be explained: The chloroplasts are probably green algae taken up by phagocytosis (engulfing them). A pyrenoid is a region of the chloroplast where a special type of carbohydrate is formed.

Euglena has a long flagellum that projects out of a vaselike indentation and a much shorter one that does not project out. It moves very quickly, and you will be advised to add Protoslo to your wet mount to slow it down. Like some of the protozoans discussed previously, *Euglena* is bounded by a flexible pellicle made of protein. This means it can also assume all sorts of shapes. *Euglena* lives in fresh water and contains a contractile vacuole that collects water and then contracts, ridding the body of excess water.

Make a wet mount of *Euglena* by using a drop of a *Euglena* culture and adding a drop of Protoslo (methyl cellulose solution) onto the slide to slow it down. Describe what you see. ______________________

____

____

Figure 15.16 ***Euglena.***
Euglena is a unicellular, flagellated protist.

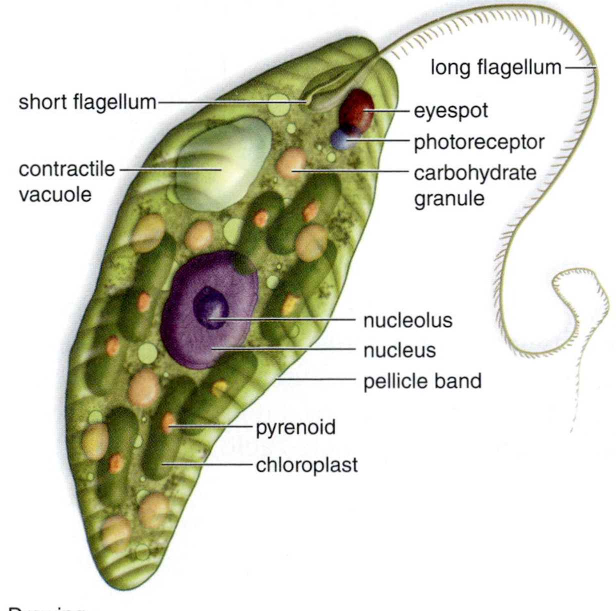

a. Drawing

b. Photomicrograph

Observation: Pond Water

1. Make a wet mount of pond water by taking a drop from the bottom of a container of pond water.
2. Scan the slide for organisms: Start at the upper left-hand corner, and move the slide forward and back as you work across the slide from left to right.
3. Experiment by using all available objective lenses, by focusing up and down with the fine-adjustment knob, and by adjusting the light so that it is not too bright.
4. Identify the organisms you see by consulting Figure 15.17, and use any pictorial guides provided by your instructor.

Figure 15.17 Microorganisms found in pond water. Drawings are not actual sizes of the organisms.

15.3 Fungi

Fungi (kingdom Fungi) (Fig. 15.18) are **saprotrophs** that release their digestive enzymes into the environment. Both fungi and bacteria are often referred to as "organisms of decay" because as they break down dead organic matter they release inorganic nutrients for plants. A fungal body, called a **mycelium,** is composed of many strands, called **hyphae** (Fig. 15.19). Sometimes, the nuclei within a hypha are separated by walls called septa.

Fungi produce windblown **spores** (small, haploid bodies with a protective covering) when they reproduce sexually or asexually.

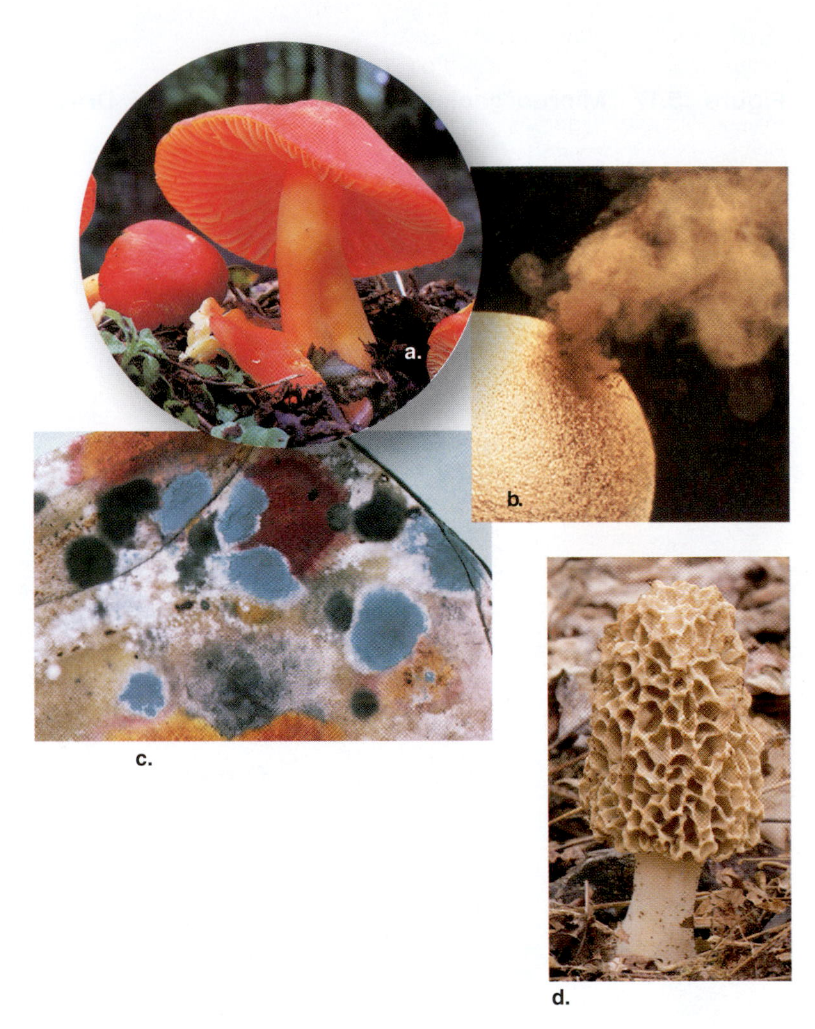

Figure 15.18 Diversity of fungi.
a. Scarlet hood, an inedible mushroom. **b.** Spores exploding from a puffball. **c.** Common bread mold. **d.** Morel, an edible fungus.

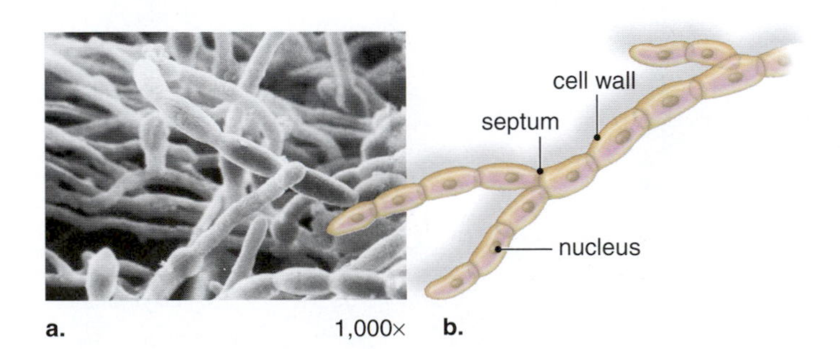

Figure 15.19 Body of a fungus.
a. The body of a fungus is called a mycelium. **b.** A mycelium contains many individual chains of cells, and each chain is called a hypha.

Black Bread Mold

In keeping with its name, black bread mold grows on bread and any other type of bakery goods. Notice in Figure 15.20 the sporangia at the tips of aerial hyphae that produce spores in both the asexual and sexual life cycles. A **zygospore** is diploid (2n); otherwise, all structures in the asexual and sexual life cycles of bread mold are haploid (n).

Figure 15.20 Black bread mold. The mycelium of this mold (1) uses sporangia to produce windblown spores. (2) During sexual reproducing, the ends of plus and minus hyphae fuse as (3) fertilization occurs. (4) The resulting zygospore undergoes meiosis to produce spores that germinate on the bread to produce a new mycelium (5).

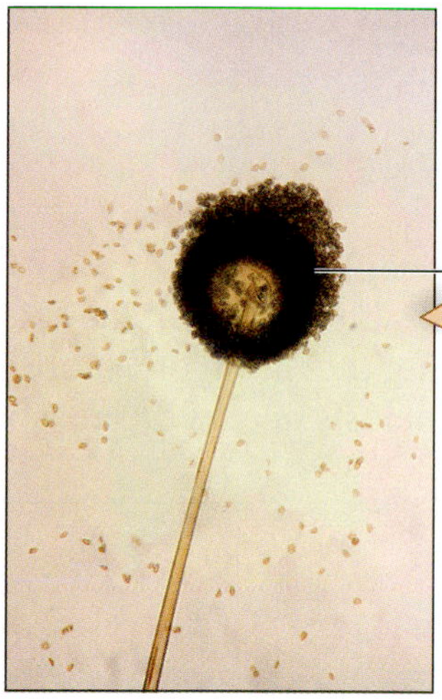

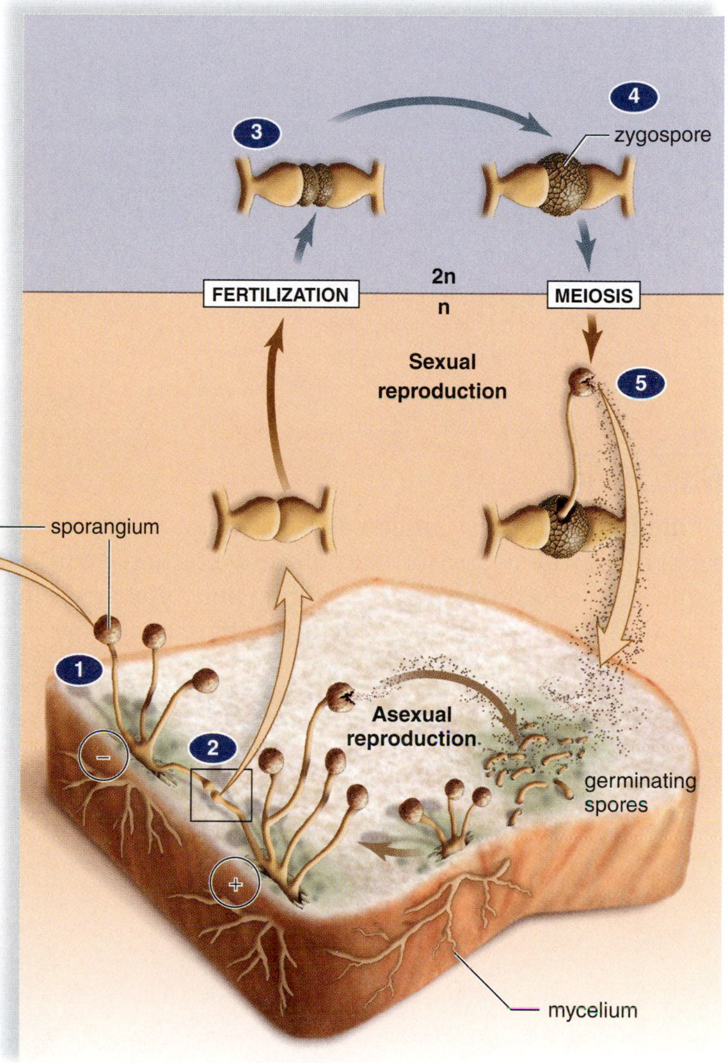

Observation: Black Bread Mold

1. If available, examine bread that has become moldy. Do you recognize black bread mold on the bread?
2. Obtain a petri dish that contains living black bread mold. Observe with a stereomicroscope. *Label the mycelium and a sporangium in Figure 15.21*a.
3. View a prepared slide of *Rhizopus,* using both a stereomicroscope and the low-power setting of a light microscope. The absence of cross walls in the hyphae is an identifying feature of zygospore fungi. *Label the mycelium and zygospore in Figure 15.21*b.

Figure 15.21 Microscope slides of black bread mold.
a. Asexual life-cycle structures. **b.** Sexual life-cycle structures.

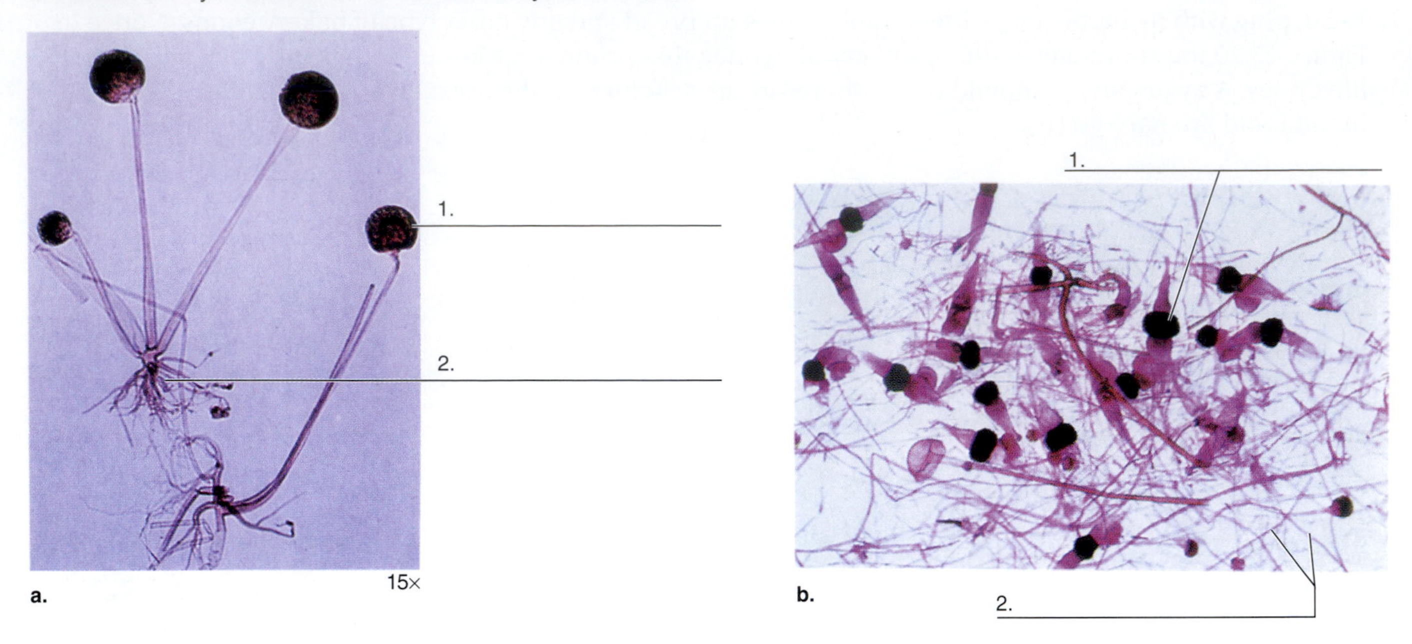

Club Fungi

Club fungi are just as familiar as black bread mold to most laypeople because they include the mushrooms. A gill mushroom consists of a stalk and a terminal cap with gills on the underside (Fig. 15.22). The stalk and cap, called a **basidiocarp,** is a fruiting body that arises following the union of + and − hyphae. The gills bear basidia, club-shaped structures where nuclei fuse, and meiosis occurs during spore production. The spores are called **basidiospores.**

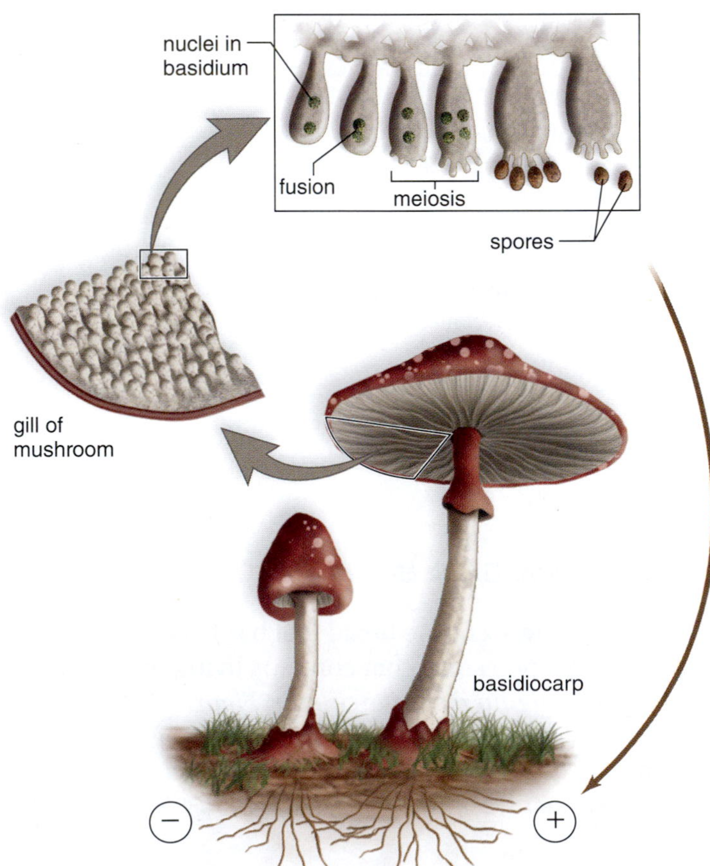

Figure 15.22 Sexual reproduction produces mushrooms.
Fusion of + and – hyphae tips results in hyphae that form the mushroom (a fruiting body). The nuclei fuse in clublike structures attached to the gills of a mushroom, and meiosis produces spores.

Observation: Mushroom

1. Obtain an edible mushroom—for example, *Agaricus*—and identify as many of the following structures as possible:
   a. **Stalk:** The upright portion that supports the cap.
   b. **Annulus:** A membrane surrounding the stalk where the immature (button-shaped) mushroom was attached.
   c. **Cap:** The umbrella-shaped basidiocarp of the mushroom.
   d. **Gills:** On the underside of the cap, radiating lamellae on which the basidia are located.
   e. **Basidia:** On the gills, club-shaped structures where basidiospores are produced.
   f. **Basidiospores:** Spores produced by basidia.
2. View a prepared slide of a cross section of *Coprinus.* Using all three microscope objectives, look for the gills, basidia, and basidiospores.
3. Can you see individual hyphae in the gills? ______________________________
4. Are the basidiospores inside or outside of the basidia? ______________________________
5. What type of nuclear division does the zygote undergo to produce the basidiospores? ______________
____
6. Can you suggest a reason for some of the basidia having fewer than four basidiospores? ___________
____
7. What happens to the basidiospores after they are released? ______________________________
____

Fungi and Human Diseases

Fungi cause a number of human diseases. Oral thrush is a yeast infection of the mouth common in newborns and AIDS patients (Fig. 15.23*a*). Ringworm is a group of related diseases caused by the fungus *Tinea.* The fungal colony grows outward, forming a ring of inflammation (Fig. 15.23*b*). Athlete's foot is a form of *Tinea* that affects the foot, mainly causing itching and peeling of the skin between the toes (Fig. 15.23*c*).

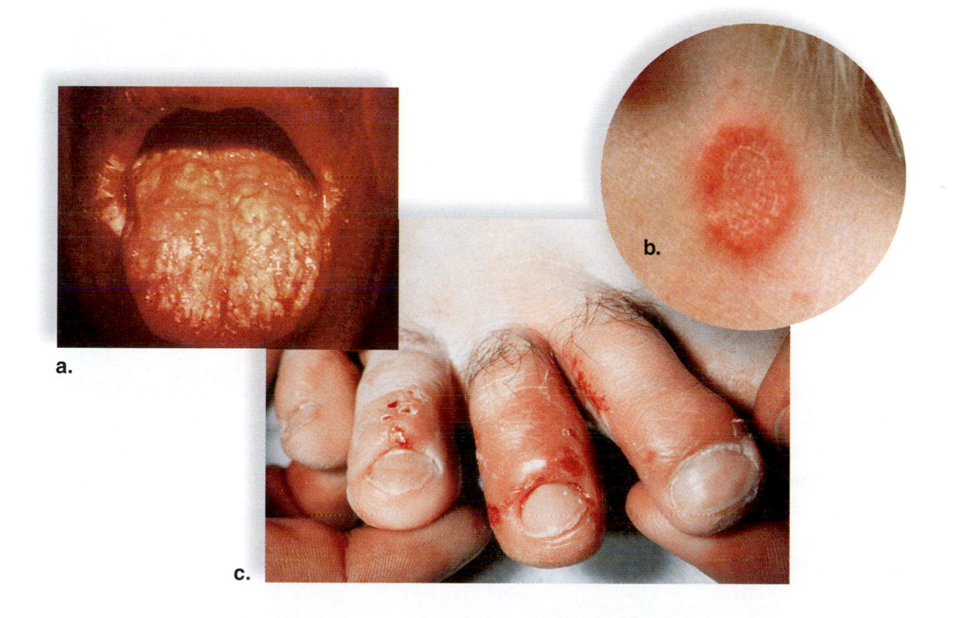

Figure 15.23 Human fungal diseases.
a. Thrush, or oral candidiasis, is characterized by the formation of white patches on the tongue.
b. Ringworm and **c.** athlete's foot are caused by *Tinea* spp.

Laboratory Review 15

1. What role do bacteria and fungi play in ecosystems? ______________________________

2. What type of semisolid medium is used to grow bacteria? ______________________________
3. What is the scientific name for spherical bacteria? ______________________________
4. It is sometimes said that diatoms live in what kind of "houses"? ______________________________
5. What type of nutrition do algae have? ______________________________
6. Name a colonial alga studied today. ______________________________
7. Bacteria have what substance in their cell walls? ______________________________

8. What color are Gram-negative bacteria following Gram staining? ______________________________
9. Once called the blue-green algae, cyanobacteria are now classified as what? ______________________________
10. What do you call the projection that allows amoeboids to move and feed? ______________________________
11. In what type environment are you likely to find brown algae, such as *Fucus*? ______________________________

12. The stalk and cap of a mushroom are scientifically termed what? ______________________________
13. Describe the saprotrophic nutrition of fungi. ______________________________

14. What do fungi produce during both sexual and asexual reproduction? ______________________________
15. Why aren't all the organisms studied today in the domain Eukarya? ______________________________

16. In general, how does sexual reproduction differ from asexual reproduction among fungi? ______________________________

LABORATORY

16

Plant Evolution

Learning Outcomes

16.1 The Evolution and Diversity of Land Plants

- Understand the relationship of charophytes to land plants and the main events in the evolution of land plants as they become adapted to the land environment. 192–93
- Describe the land plant life cycle and the concept of the dominant generation. 193
- Contrast the role of meiosis in the land plant and animal life cycles. 193

16.2 Seedless Plants

- Use the moss, including its life cycle, to exemplify the adaptations of nonvascular plants to life on land. 194–95
- Use lycophytes to exemplify the first plants to have vascular tissue and true roots, stems, and leaves. 196
- Use the fern, including its life cycle, to exemplify the adaptations of seedless vascular plants to life on land. 197–98

16.3 Seed Plants

- Name and describe four types of gymnosperms. 199
- Use the pine, including its life cycle, to exemplify the adaptations of seed plants to life on land. 200–202
- Describe the many ways in which flowering plants, including their life cycle, are adapted to the land environment. 202–05

Introduction

Among aquatic green algae, the **charophytes** are most closely related to the plants that now live on land. The charophytes have several features that would have promoted the evolution of land plants. Plants are multicellular photosynthetic eukaryotes whose evolution is marked by increasing adaptations to a land existence. The great success of flowering plants can be attributed to their adaptations to varied land environments. The number and kinds of flowering plants is much greater than that of all the other groups of plants (Fig. 16.1).

Adaptation to a land environment includes the ability to prevent excessive loss of water into the atmosphere; to obtain and transport water and nutrients to all parts of the plant; to support a large body against the pull of gravity; and to reproduce without dependence on external water.

With regard to humans, consider that skin protects us from drying out, blood transports water and nutrients about the body, the skeleton supports us, males have a penis for delivering flagellated sperm to the female, and the embryo and fetus are protected from drying out within the uterus of the female.

Figure 16.1 The flowering plants.
The flowering plants provide us with much of our food including grains, potatoes, and beans. Here, we see the flower and the fruit of a watermelon plant.

16.1 The Evolution and Diversity of Land Plants

Figure 16.2 shows the evolution of land plants. What evolutionary events led to adaptation of plants to a land existence? ____

____

Figure 16.2 Evolutionary relationships among the plants.

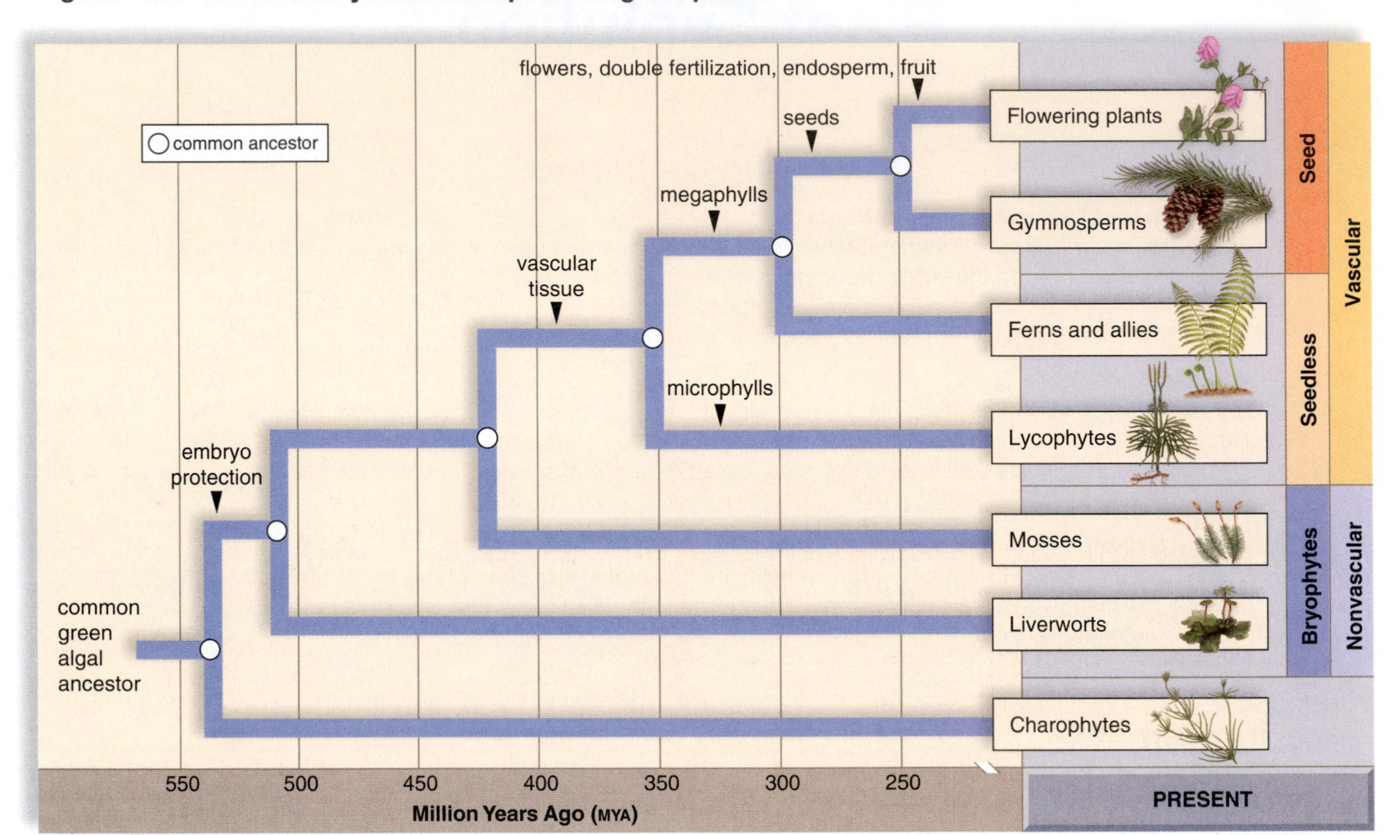

Algal Ancestor of Land Plants

In the evolutionary tree, the charophytes are green algae that share a common ancestor with land plants. This common ancestor may have resembled a Charales, such as *Chara,* which can live in warm, shallow ponds that occasionally dry up. Adaptation to such periodic desiccation may have facilitated the ability of certain members of the common ancestor population to invade land, and with time, become the first land plants.

Observation: Chara

Examine a living *Chara* (Fig. 16.3). How does it superficially resemble a land plant? ____

____

Chara is a filamentous green alga that consists of a primary branch and many side branches (see Fig. 16.2). Each branch has a series of very long cells. Note where one cell ends and the other begins. Measure the length of one cell. _______________ Gently pick up and handle *Chara* while you are examining it. What does it feel like? _______________ Its cell walls are covered with calcium carbonate deposit.

Figure 16.3 ***Chara.***
Chara is an example of a stonewort, the type of green alga believed to be most closely related to the land plants.

Conclusions: Chara

- What characteristics cause *Chara* to resemble land plants? ______________________________

____

- Why are *Chara* called stoneworts? ______________________________

Alternation of Generations

Land plants have a two-generation life cycle called **alternation of generations,** in which one generation is the **dominant generation**; it is longer lasting and the one we recognize as the plant.

1. The **sporophyte** (diploid) **generation** produces haploid spores by meiosis. Spores develop into a haploid generation, the gametophyte.
2. The **gametophyte** (haploid) **generation** produces **gametes** (eggs and sperm) by mitosis. The gametes then unite to form a diploid zygote.

Figure 16.4 contrasts the plant life cycle (alternation of generations) with the animal life cycle **(diploid).**

1. In the plant life cycle, meiosis occurs during the production of ______________________________.
2. In the human life cycle, meiosis occurs during the production of ______________________________.
3. In the plant life cycle, the generation that produces gametes is (n or 2n) ______________________________.
4. In the human life cycle, the individual that produces gametes is (n or 2n) ______________________________.

Figure 16.4 Plant and animal life cycles.

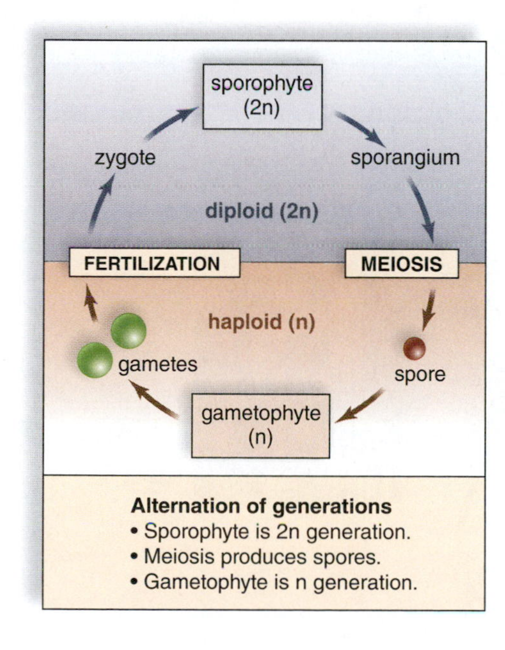

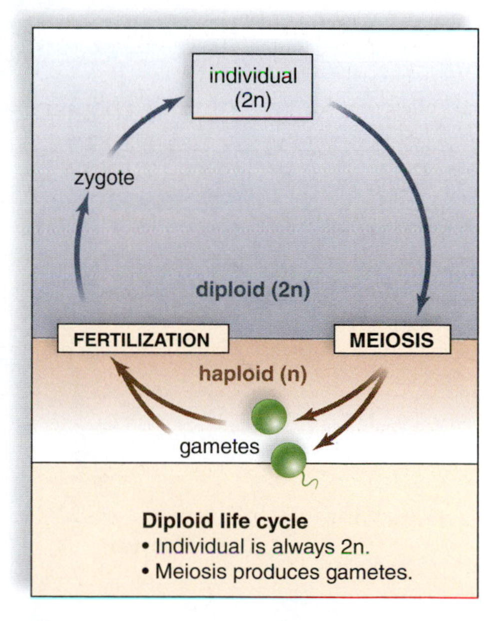

16.2 Seedless Plants

Mosses and ferns are both seedless plants. Mosses and their relatives, called the bryophytes, are low-lying plants, called the nonvascular plants because they lack vascular tissue. Ferns, characterized by large leaves, do have vascular tissue, but even so, share other characteristics with the bryophytes. For example, they are both seedless plants.

Mosses

The bryophytes (mosses and liverworts) were the first plants to live on land. The gametophyte is dominant in these nonvascular plants. The gametophyte produces eggs within archegonia and swimming sperm in

Figure 16.5 Life cycle of the moss.
In mosses, the gametophyte consists of leafy shoots that produce flagellated sperm within antheridia and eggs within archegonia. Following fertilization, the sporophyte, consisting of a stalk and capsule (a sporangium), is dependent on the gametophyte.

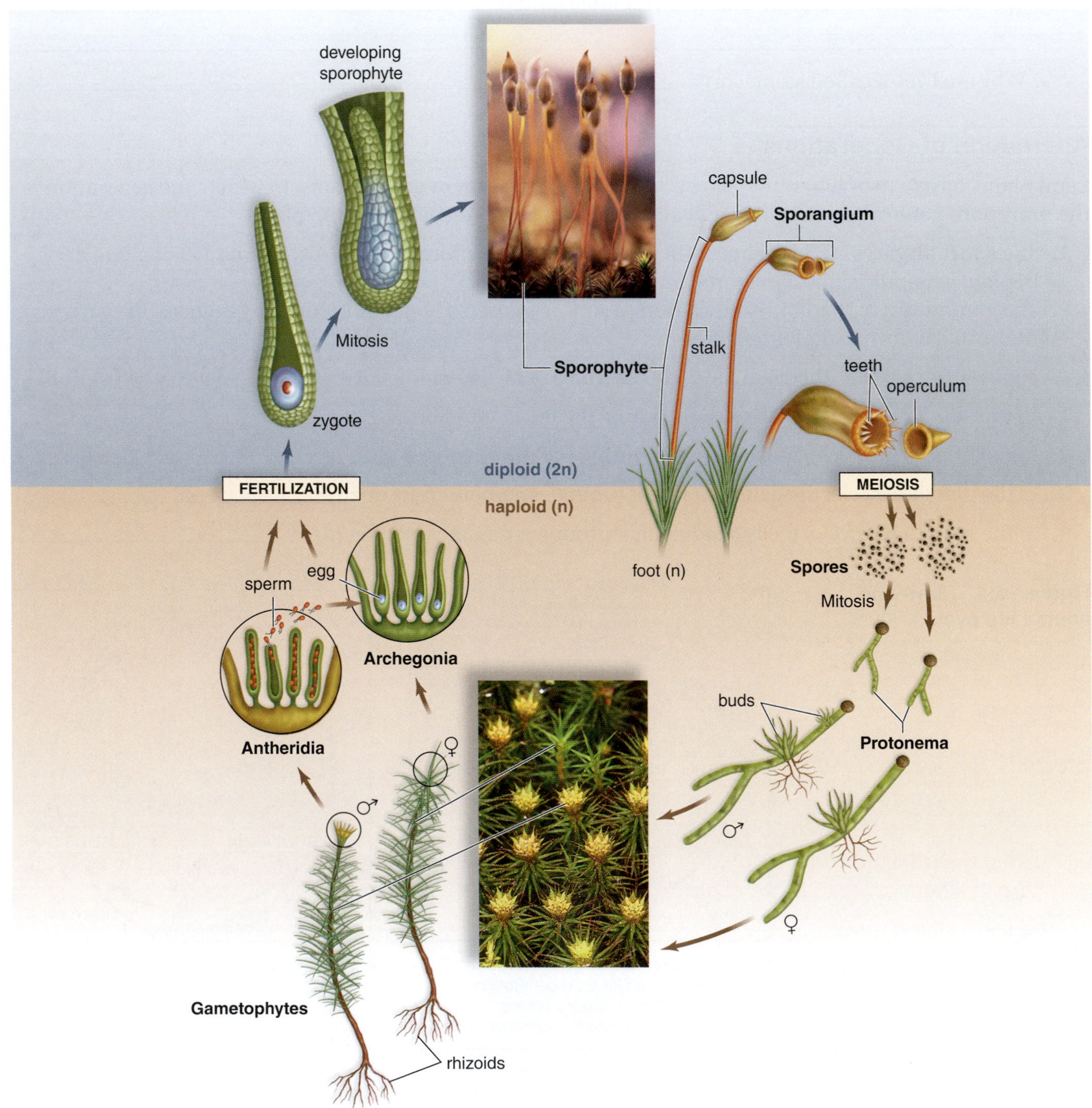

antheridia (Fig. 16.5). The bryophytes are dependent on external moisture to ensure fertilization because the sperm must swim to the egg. However, the zygote developing within the archegonia is protected from drying out. The bryophytes have another adaptation to life on land in that the spores, produced by the dependent sporophyte, are windblown. The spores disperse the gametophyte.

Life Cycle of Mosses

Study the life cycle of the moss (Fig. 16.5) and find the gametophyte. Examine a living gametophyte or a plastomount of this generation. Describe its appearance. ____________________

____________________ Considering that this is the generation we refer to as the "moss," what generation is dominant in mosses? ____________________

Describe the sporophyte. ____________________

Observation: Moss Life Cycle

1. Study the prepared slide of the female shoot head (top of female shoot) and the male shoot head (top of the male shoot). Find an **archegonium** in a female shoot head and locate an egg in at least one of these. Find an **antheridium** in a male shoot head. An antheridium is filled with many flagellated sperm. When sperm produced by the antheridia swim in a film of water to the eggs in the archegonia, zygotes result. A zygote develops into a new sporophyte.
2. Examine the plastomount of a shoot with a sporophyte attached. The sporophyte is dependent on the female shoot. Why female? ____________________
3. Examine a slide of a longitudinal section through the sporophyte of the moss. Identify the stalk and sporangium. What is being produced in the sporangium? ____________ By what process? ____________________

The Life Cycle of a Moss

1. Which generation is haploid? ____________ Which is diploid? ____________

   Which generation is dominant in mosses? ____________ Which generation is dependent? ____________________
2. Is there any evidence of vascular tissue in the moss sporophyte? ____________________
3. When spores germinate, what generation begins to develop? ____________________
4. Why is it proper to say that, in the moss, spores disperse the plant? ____________________
5. By what means are spores disseminated? ____________________

Adaptation of Mosses to the Land Environment

Which of these is an indication that mosses are well adapted to life on land? Write "yes" if the feature is an adaptation to land and "no" if the feature is not an adaptation to living on land.

Lack of vascular tissue. __________ Flagellated sperm. __________

Body covered by a cuticle that prevents drying out. __________ Spores are windblown. __________

Egg and embryo protected by female shoot. __________

Observation: Lycophytes

Lycophytes (phylum Lycophyta) are commonly called **club mosses.** Lycophytes are representative of the first vascular plants. They have an aerial stem and a horizontal root (rhizome with attached rhizoids), both of which have vascular tissue. The leaves are called **microphylls** because they are small and have only one strand of vascular tissue.

Ground Pines

1. Examine a living or preserved specimen of *Lycopodium* (Fig. 16.6).
2. Note the shape and the size of the microphylls and the branches of the stems.
3. Note the terminal clusters of leaves, called **strobili,** that are club-shaped and bear sporangia.
4. *Label strobili, leaves, stem, and rhizoids in Figure 16.6.*
5. Examine a prepared slide of a *Lycopodium* that shows the sporangia with spores inside. The spore develops into a tiny microscopic gametophyte that remains in the soil.

Figure 16.6 ***Lycopodium.***
In the club moss *Lycopodium*, green photosynthetic stems are covered by scalelike leaves, and spore-bearing leaves are clustered in strobili.

Ferns

All the other plants to be studied in this laboratory are vascular plants. Ferns are seedless vascular plants in which the dominant sporophyte possesses vascular tissue. The windblown spores develop into an independent gametophyte that is water dependent because it lacks vascular tissue and also because it produces flagellated sperm. The sperm must swim from the antheridia to the archegonia, where the eggs are produced. The zygote, protected from drying out within an archegonium, develops directly into the sporophyte.

Observation: Fern Life Cycle

1. Study the life cycle of the fern (Fig. 16.7) and find the sporophyte and gametophyte. This large, complexly divided megaphyll (large leaf) is known as a **frond.** Fronds arise from an underground stem. Examine a preserved specimen of a frond and on the underside, notice brownish **sori** (sing., **sorus**),

Figure 16.7 Fern life cycle.
The frond is the dominant sporophyte generation that produces windblown spores. A spore gives rise to an independent gametophyte. The prothallus produces gametes. When the gametes fuse, a new sporophyte begins to develop.

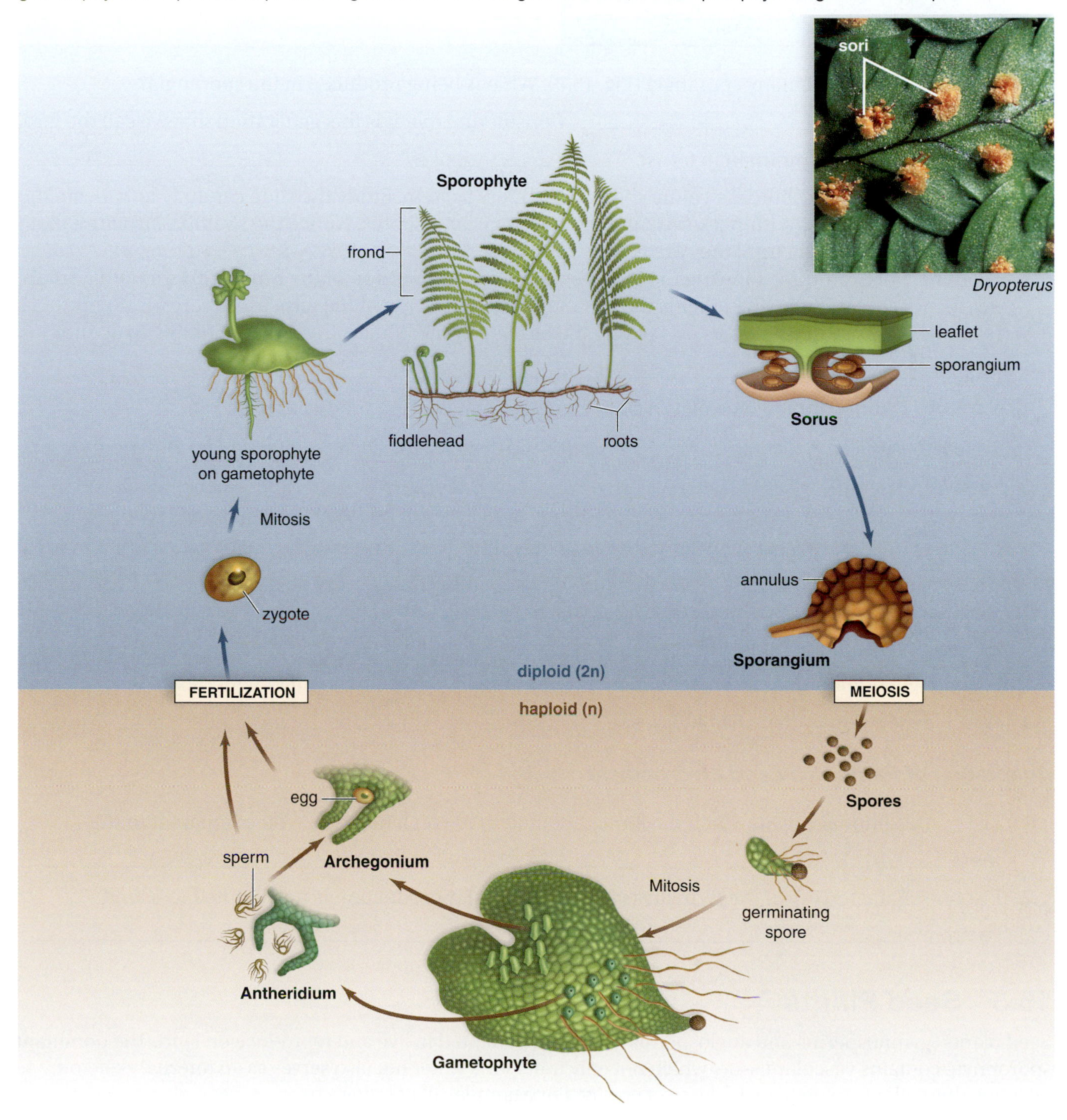

Figure 16.8 Underside of frond leaflets showing sori.

each one a cluster of many sporangia (Fig. 16.8). What is being produced in the sporangia? ______________________ Considering that it is this generation that we call the fern, what generation is dominant in ferns? ______________________

2. Examine a prepared slide of a young sporangium cross section. Study the slide carefully and locate the fern leaf and a sorus. Within a sorus, find the sporangia and spores. Notice the shelflike structure that protects the sporangia until they are mature.
3. Examine a plastomount showing the fern life cycle. Notice a portion of the frond with sori and a small heart-shaped structure. The latter is the gametophyte generation of the fern. Most persons do not realize that this structure exists as a part of the fern life cycle. What is the function of this structure? ______________________
4. Examine a whole mount slide of a fern archegonia. What is being produced inside an archegonium? ______________ If you focus up and down very carefully on an archegonium, you may be able to see an egg inside.
5. Examine a whole mount slide of fern antheridia. What is being produced inside the antheridia? ______________ When sperm produced by the antheridia swim to the archegonia in a film of water, what results? ______________ The latter develop into what generation? ______________

The Life Cycle of a Fern

1. Is either generation in the fern dependent for any length of time on the other generation? ______________
2. Which generation is dispersed in ferns? ______________ How? ______________

Adaptation of Ferns to a Land Environment

1. List one additional way in which the fern is adapted to life on land, when you compare it to the moss. ______________________
2. List one characteristic of the fern illustrating that sexual reproduction is not adapted to a land environment. ______________________

16.3 Seed Plants

Seed plants (gymnosperms and angiosperms) are further adapted to live and reproduce on land. The dominant sporophyte contains vascular tissue, which not only transports water but also serves as an internal skeleton, allowing these plants to oppose the force of gravity. For example, all of today's trees are seed plants.

Gymnosperms

The gymnosperms are usually evergreen trees in which the sporangia are found on **cones.** The four groups of living gymnosperms are **cycads, ginkgoes, gnetophytes,** and **conifers** (Fig. 16.9). All of these plants have ovules and subsequently develop seeds that are exposed on the surface of cone scales or analogous structures. (Because the seeds are not enclosed by fruit, gymnosperms are said to have "naked seeds.") Early gymnosperms were present in the swamp forests of the Carboniferous period, and they became dominant during the Triassic period. Today, living gymnosperms are classified into 780 species, the most plentiful being the conifers.

Conifers (phylum Coniferophyta) consist of about 575 species of trees, many of them evergreens such as pines, spruces, firs, cedars, hemlocks, redwoods, cypresses, yews, and junipers. Vast areas of northern temperate regions are covered in evergreen coniferous forests. The tough, needlelike leaves of pines conserve water because they have a thick cuticle and recessed stomata (openings for gas exchange).

Figure 16.9 Gymnosperm diversity.

Cycad, *Encephalartos humlis*
Female plant with large seed cones

Ginkgo, *Ginkgo biloba*
Female maidenhair tree with seeds

Gnetophyte, *Ephedra*
Branched shrub with scalelike leaves

Conifer, *Picea*
Spruce tree with pollen cones and seed cones

Life Cycle of Pine Trees

The life cycle of a pine tree (Fig. 16.10) illustrates that seed plants are no longer dependent on external water to ensure fertilization. In seed plants, the dominant sporophyte produces two types of spores, termed the microspore and the megaspore. Microspore mother cells produce **microspores** by meiosis, and each develops into a male gametophyte generation, the pollen grain. During **pollination,** pollen grains are dispersed in the vicinity of female gametophytes.

A megaspore mother cell within an **ovule** produces **megaspores** by meiosis. Only one of these develops into a female gametophyte that produces an egg. The 2n zygote (a sporophyte) is still within the ovule, which becomes a seed. The ability of seeds to withstand harsh conditions until the environment is again favorable for growth largely accounts for the dominance of seed plants today.

Study the life cycle of the pine tree and describe the sporophyte. ______________________________

____

Figure 16.10 Life cycle of pine.

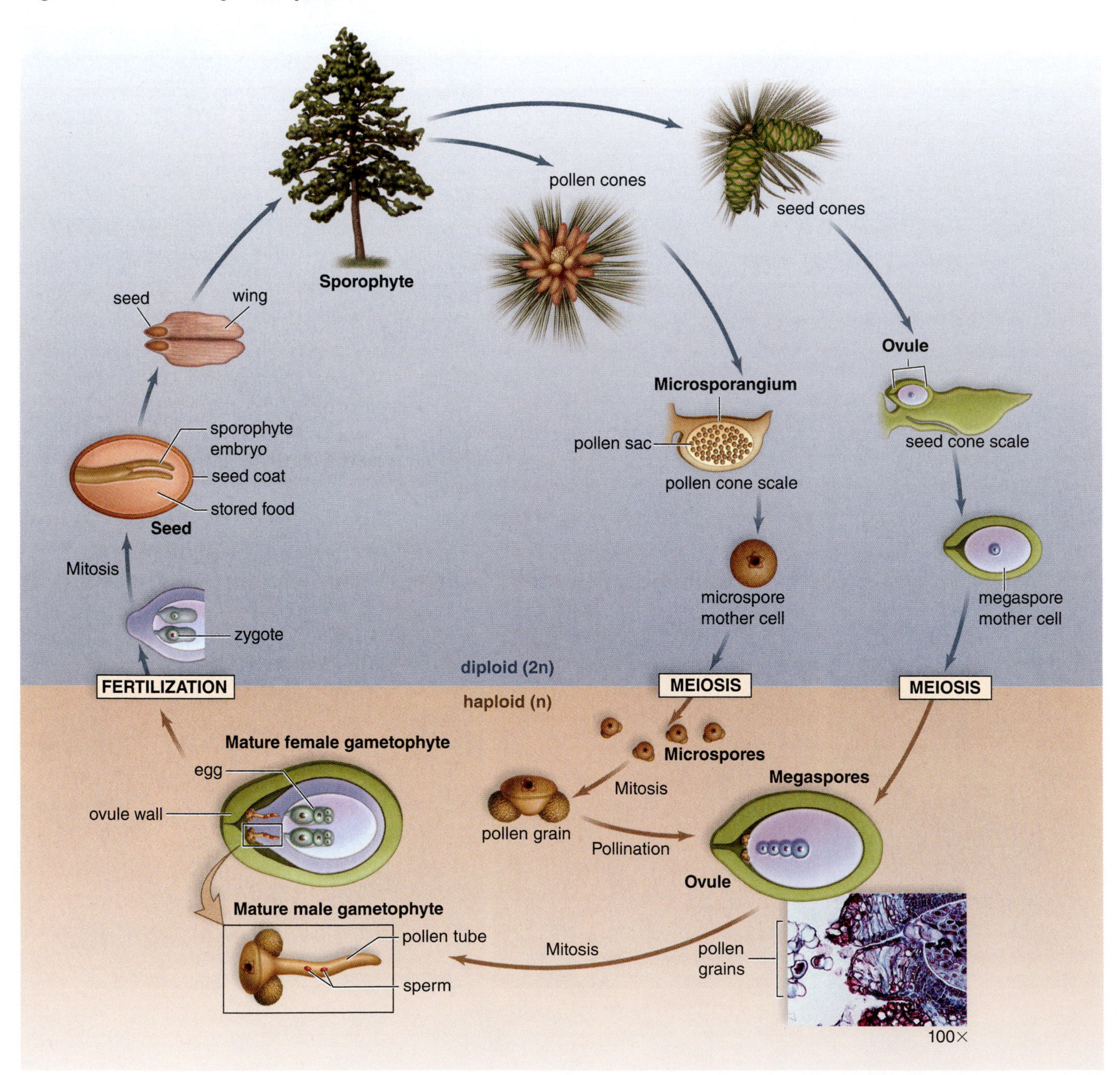

Observation: Pine Cones

1. Observe the pollen and seed cones on display (Fig. 16.11). Compare their relative size and structure. Remove a single scale (sporophyll) from the male cone and from the seed cone, which has been heated so that the cone opens. Observe with a stereomicroscope. Note the two pollen sacs located on the lower surface of each scale from the pollen cone. Note also the two ovules that may have developed into seeds on the upper surface of each scale from the seed cone.

Figure 16.11 Pine cones.
a. The scales of pollen cones bear pollen sacs, where microspores become pollen grains. **b.** The scales of seed cones bear ovules that develop into winged seeds.

2. Examine the prepared slide of a longitudinal section through a mature pollen cone and label Figure 16.12. On the lower surface of each scale are the pollen sacs in which the microspore mother cells produce microspores. Each microspore develops into a male gametophyte, a pollen grain. Under high power, focus on a pollen grain and note the external wings and interior cells. One of these will divide to produce two cells, one of which is the sperm nucleus and the other of which forms the pollen tube, through which a sperm travels to the egg.

Figure 16.12 Pine pollen cone.
Pollen cones bear **a.** pollen sacs in which microspores develop into pollen grains. **b.** Enlargement of pollen grains.

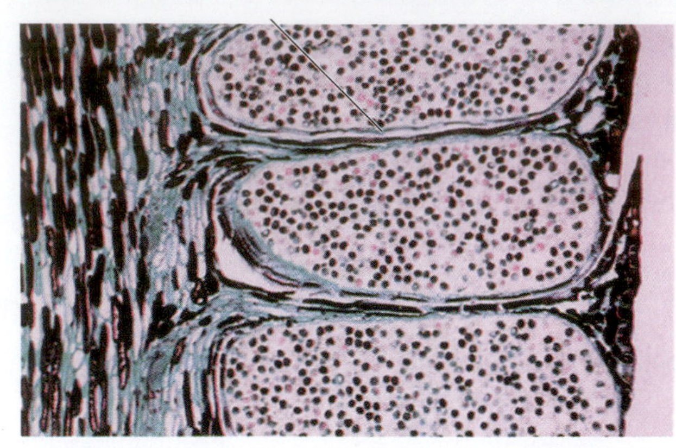

a. Longitudinal section through pine pollen cone, showing pollen sacs.

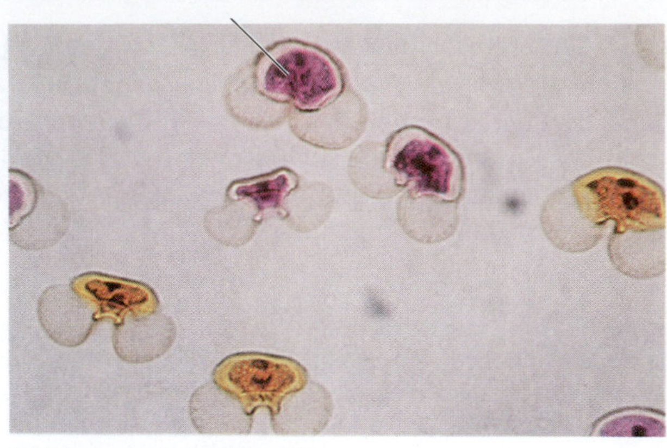

b. Enlargement of pollen grains.

3. Examine the prepared slide of a longitudinal section through a seed cone and label Figure 16.13. In some ovules, you will see a megaspore mother cell surrounded by nutritive cells. You may also be able to find pollen grains just outside or within the integuments of an ovule. At about the time the megaspore mother cell is undergoing meiosis, the scales swell and open to allow the wind-dispersed pollen grains to enter. The megaspore undergoes a series of mitotic divisions and develops into the female gametophyte that contains two archegonia, each of which encloses a single large egg.

What generation is now within the ovule? ______________________________

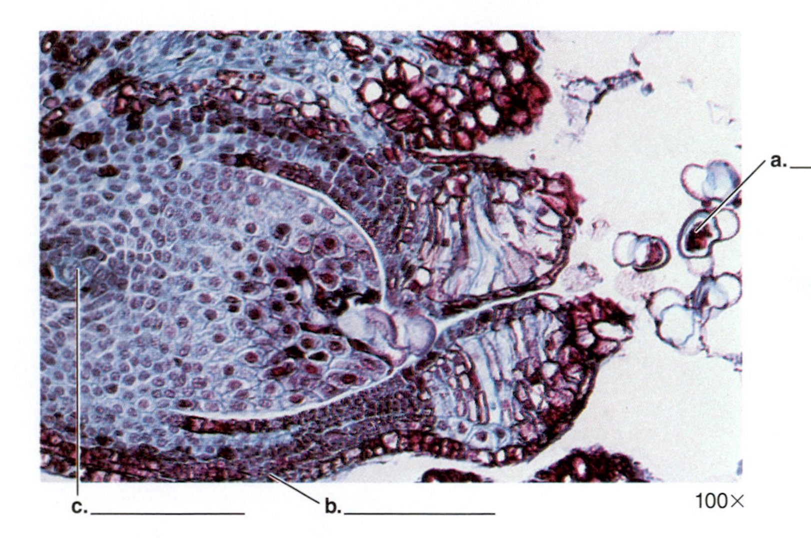

Figure 16.13 Seed cone.
Seed cones bear ovules, shown here in longitudinal section. Note pollen grains near the entrance.

4. Following fertilization, a seed contains the embryonic plant and nutrient material within a seed coat. If available, examine pine seeds. The seeds of gymnosperms are windblown. In seed plants, seeds disperse the sporophyte. If you wish, dissect a seed and, with the help of a hand lens, attempt to find the embryo.

Angiosperms

The angiosperms are the flowering plants. The life cycle of a flowering plant (Fig. 16.14) is like that of the pine tree except for these innovations:

- The often brightly colored flower contains the pollen sacs and ovules. Pollen may be windblown or carried by animals (e.g., insects).

- Flowering plants practice **double fertilization.** A mature pollen grain contains two sperm; one fertilizes the egg, and the other joins with the two polar nuclei to form **endosperm** (3n), which serves as food for the developing embryo.
- Flowering plants have seeds enclosed within fruits. Fruits protect the seeds and aid in seed dispersal. Sometimes, animals eat the fruits, and after the digestion process, the seeds are deposited far away from the parent plant. The term **angiosperm** means "covered seeds." The seeds of angiosperms are found in fruits, which develop from parts of the flower.

  Which generation of a flowering plant bears flowers and fruits? ______________________________

  ____

Figure 16.14 Flowering plant life cycle.
The parts of the flower involved in reproduction are the anthers of stamens and the ovules in the ovary of a carpel.

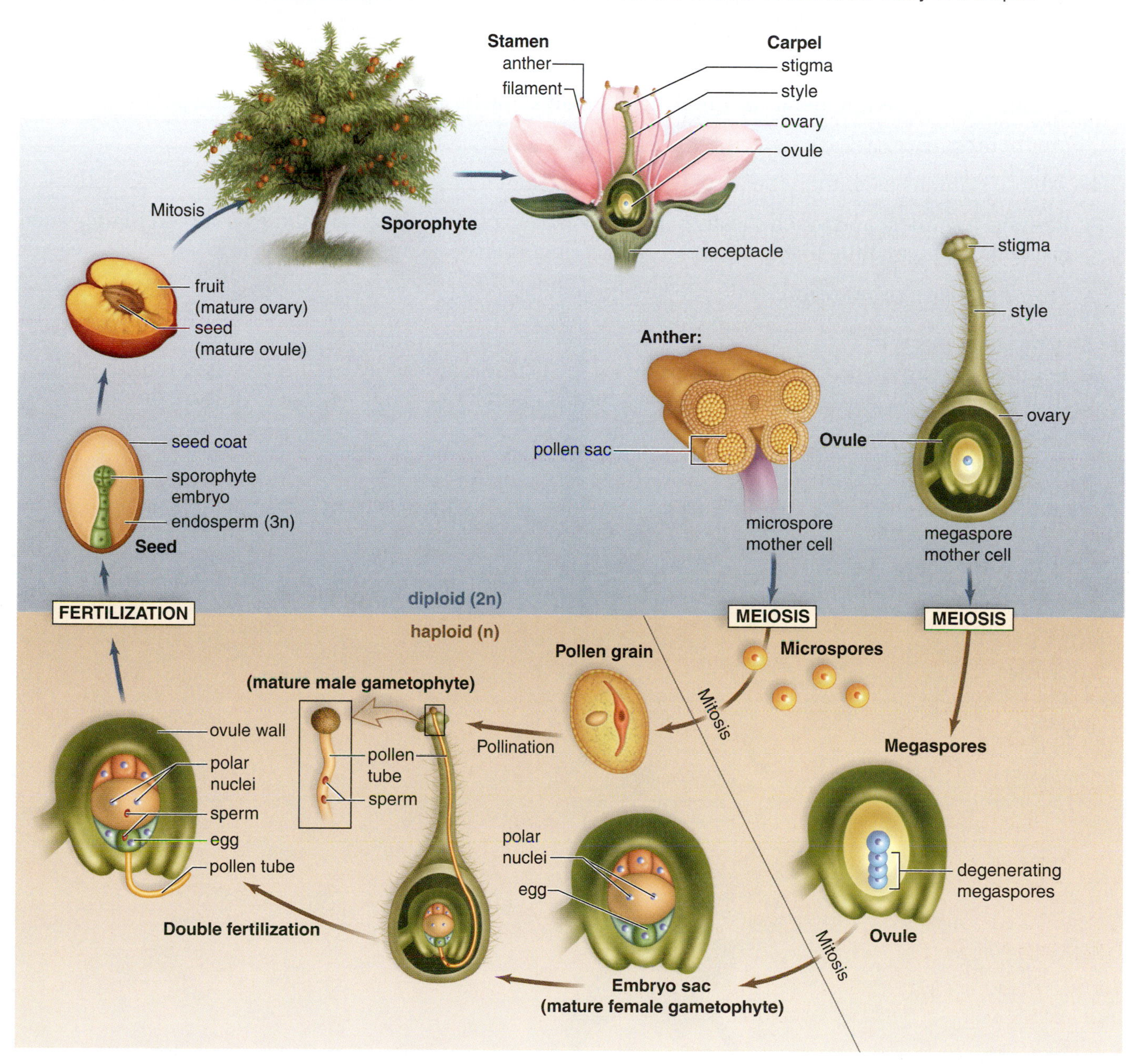

Observation: A Flower

1. With the help of Figure 16.15, identify the following structures on a model of a flower:
   a. **Receptacle:** The portion of a stalk to which the flower parts are attached.
   b. **Sepals:** An outermost whorl of modified leaves, collectively termed the calyx. Sepals are green in most flowers. They protect a bud before it opens.
   c. **Petals:** Usually colored leaves that collectively comprise the corolla.
   d. **Stamen:** A swollen terminal **anther** and the slender supporting **filament**. The anther contains two pollen sacs, where microspores develop into microgametophytes (pollen grains).
   e. **Carpel:** A modified sporophyll consisting of a swollen basal ovary; a long, slender **style** (stalk); and a terminal **stigma** (sticky knob).
   f. **Ovary:** The enlarged part of the carpal that develops into a fruit.
   g. **Ovule:** The structure within the ovary where a megaspore develops into a female gametophyte (embryo sac). The ovule becomes a seed.
2. Carefully inspect a fresh flower. What is the common name of your flower? ______________________
3. Remove the sepals and petals by breaking them off at the base. How many sepals and petals are there? ______________________
4. Are the stamens taller than the carpel? ______________________
5. Remove a stamen, and touch the anther to a drop of water on a slide. If nothing comes off in the water, crush the anther a little to release some of its contents. Place a coverslip on the drop, and observe with low- and high-power magnification. What are you observing? ______________________

   ____
6. Remove the carpel by cutting it free just below the base. Make a series of thin cross sections through the ovary. The ovary is hollow, and you can see nearly spherical bodies inside. What are these bodies? ______________________

Figure 16.15 Generalized flower.
A flower has four main kinds of parts: sepals, petals, stamens, and a carpel. A stamen has an anther and filament. A carpel has a stigma, style, and ovary. An ovary contains ovules.

7. As per Figure 8.5, is your flower a monocot or a eudicot? ________________

8. Remove a stamen and touch the anther to a drop of water on a slide. If nothing comes off in the water, crush the anther a little to squeeze out some of its contents. Place a coverslip on the drop and observe with low and high powers of the microscope. The somewhat spherical cells with thick walls are pollen grains. How is pollen dispersed? ________________

   Flowering plants provide nectar to insects, whose mouthparts are adapted to acquiring the nectar from this particular species of plant. This is a mutualistic relationship. What does a pollinator do for the plant? ________________

9. Remove the carpel by cutting it free just below the base. Make a series of thin cross sections through the ovary. The ovary is hollow and in this cavity there are small, nearly spherical bodies much larger than pollen grains. These are ovules. Remove an ovule that may be still attached by a stalk to the ovary wall. Place the ovule on a drop of water on a slide; cover with a cover glass; press firmly on the top of the cover glass with a clean eraser so as to smear the ovule into a thin mass. Observe with the microscope. Do you see cells within the ovule? ________________

10. At maturity, an angiosperm seed contains the embryo and possibly some nutrient material in the form of endosperm covered by a seed coat. A fruit is a ripened ovary, together with any accessory flower parts that may be associated with the ovary. If available, examine a pea pod and an apple. Find the remnants of the flower parts still attached to the fruit. Open the pea pod and slice the apple. Which portion of each should be associated with the ovules? ________________ Which with the ovary? ________________ *Label flower remnants, fruit, and seed in the following diagram.* Can you think of a biological advantage to producing fruits? ____________ To producing a fleshy fruit? ____________

    ____

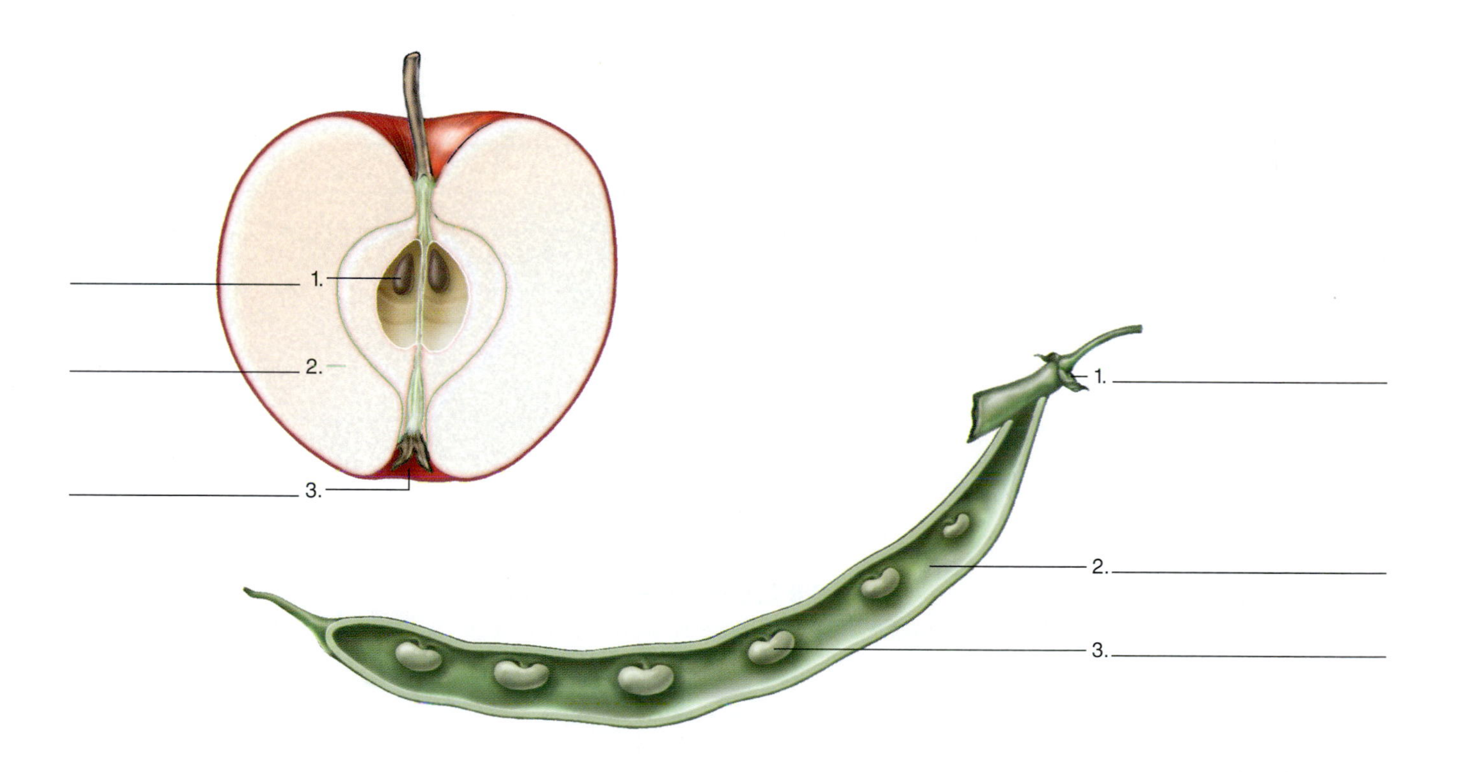

Laboratory Review 16

1. Name one way that each type plant shows an adaptation to the land environment not seen before:
   a. Moss ______
   b. Lycophyte ______
   c. Fern ______
   d. Pine tree ______
   e. Flowering plant ______
2. The gametophyte is not protected in which of the plants listed in question 1?
   ______
3. In what way do flowering plants protect the female gametophyte? ______
   ______
   The male gametophyte? ______
4. In the flowering plant life cycle
   a. What becomes of the ovule? ______
   b. What becomes of the ovary? ______
5. Windblown spores are an adaptation to the land environment in which three types of plants studied in this laboratory? ______
6. Why is it beneficial to have the sporophyte dominant in plants adapted to the land environment?
   ______
7. Which generation is in a seed? ______
8. Why are flagellated sperm a drawback in plants that live on land? ______
   ______
9. When does meiosis occur in the life cycle of plants? ______
   ______
10. What type of cell division produces the gametophyte generation and the gametes in plants? ______
    ______

LABORATORY

17

Early Invertebrate Evolution

Learning Outcomes

Introduction

- Distinguish between invertebrate and vertebrate animals. 207

17.1 Evolution of Animals

- Discuss the evolution of animals in terms of symmetry, number of tissue layers, and pattern of development. 208–09

17.2 Sponges

- Identify a sponge and describe the anatomy of a sponge. 209–10
- Identify various sponges according to the type of spicule. 210
- Show that the anatomy and behavior of a sponge aid its survival and ability to reproduce. 211

17.3 Cnidarians

- Describe the typical life cycle of a cnidarian. 212
- Show that the anatomy and behavior of a hydra aid its survival and ability to reproduce. 213–14
- Realize that cnidarians are diverse; some exist only as a polyp and some exist only as a medusa. 215

17.4 Flatworms

- Describe the anatomy and behavior of a free-living flatworm and relate its behavior to the animal's way of life. 216–18
- Contrast the complexity of a hydra with that of a planarian. 218
- Describe the life cycle of a tapeworm and contrast its anatomy with that of a planarian. 219–20

17.5 Roundworms

- List the anatomical features of all roundworms (e.g., *Ascaris*) and tell how a roundworm reproduces. 220
- Explain the diversity of roundworms on the basis of adaptation to various environments. 221–22

17.6 Rotifers

- Identify rotifers and describe their anatomy and behavior. 222–24

Introduction

In our survey of the animal kingdom, we will see that animals are very diverse in structure. Even so, all animals are multicellular and **heterotrophic,** which means their food consists of organic molecules made by other organisms. Corresponding to their need to acquire food, animals have some means of locomotion by use of muscle fibers. Animals are always diploid, and during sexual reproduction, the embryo undergoes specific developmental stages.

While we tend to think of animals in terms of **vertebrates** (e.g., dogs, fishes, squirrels), which have a backbone, most animal species are those that lack a backbone, commonly known as **invertebrates.** In this laboratory, we will examine those invertebrates that lack a true body cavity, called a **coelom.** A survey of the rest of the animal kingdom follows in Laboratory 18.

Planning Ahead To see hydra and planarians feed, have students observe the animals at the start of lab, add food, and then check frequently until food engulfment occurs.

17.1 Evolution of Animals

Today, molecular data are used in addition to comparative anatomy to trace the evolutionary history of animals. These data tell us that all animals share a common ancestor. This common ancestor was most likely a choanoflagellate, a protist consisting of a colony of flagellated cells. All but one of the phyla depicted in the tree (Fig. 17.1) consists of only invertebrates—the chordates contain a few invertebrates and also the vertebrates.

Figure 17.1 Evolutionary tree of animals.

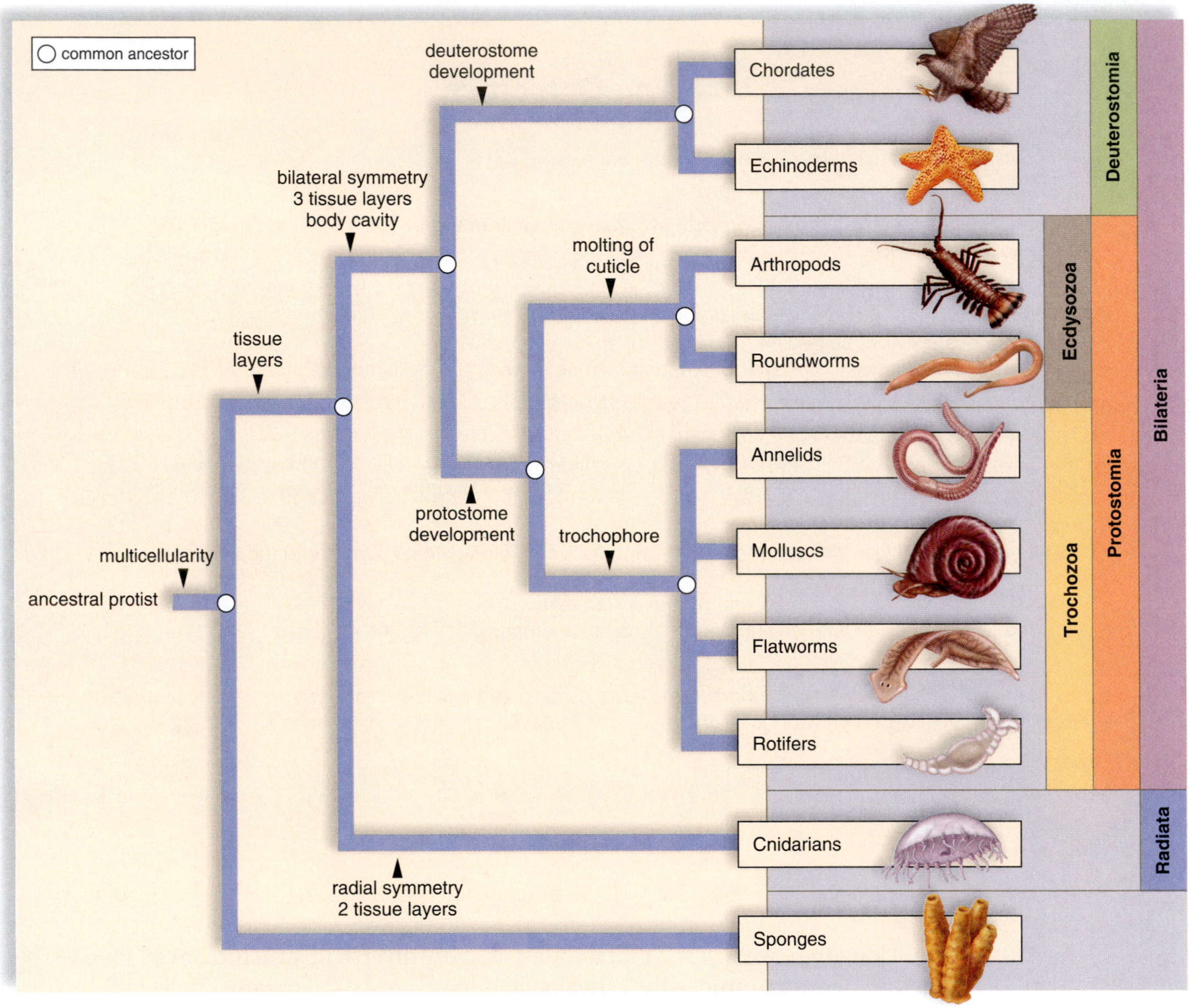

Certain anatomical features of animals are used in the tree. The first feature of interest is formation of tissue layers. Sponges have no true tissue layers. Which phyla in the tree have only two tissue layers? ________________ The other phyla have three tissue layers.

Another feature of interest is **symmetry. Asymmetry** means the animal has no particular symmetry. Which phyla have **radial symmetry,** in which, as in a wheel, two identical halves are obtained no matter how the animal is longitudinally sliced? ________________ The other phyla have **bilateral symmetry,** which means the adult animal has a definite right half and left half.

Finally, complex animals are either **protostomes** (first opening during development is the mouth) or **deuterostomes** (second opening during development is the mouth [the first one is the anus]). Which pattern of development do the flatworms, rotifers, and roundworms—animals included in this laboratory—have? ________________

17.2 Sponges

Sponges (phylum Porifera) live in water, mostly marine, attached to rocks, shells, and other solid objects. An individual sponge is typically shaped like a tube, cup, or barrel. Sponges grow singly or in colonies whose overall appearances vary widely. A single sponge can become a colony by asexual budding.

Anatomy of Sponges

Sponges consist of loosely organized cells and have no well-defined tissues. They are asymmetrical or radially symmetrical and **sessile** (immotile). They can reproduce asexually by budding or fragmentation, but they also reproduce sexually by producing eggs and sperm.

Sponges have a few types of specialized cells. Most notably they have flagellated **collar cells.** The movement of their flagella keep water moving through the pores into the central cavity and out the osculum of a sponge (Fig. 17.2). Collar cells also take in suspended food particles from the water and digest them for the benefit of all the other cells in a sponge.

Observation: Anatomy of Sponges

Preserved Sponge

1. Examine a preserved sponge (Fig. 17.2*a*). Note the main excurrent opening **(osculum)** and the multiple incurrent pores. Water is constantly flowing in through the pores and out the osculum. *Label the arrows in Figure 17.2*a *to indicate the flow of water.* Use the labels *water out* and *water in* through pores.
2. Examine a sponge specimen cut in half. Note the central cavity and the sponge wall. The wall is convoluted in some sponges, and the pores line small canals. Does this particular sponge have pore-lined canals? ________________
3. You may be able to see **spicules,** fine projections over the body and especially encircling the osculum. Does this sponge have spicules? ________________

Figure 17.2 Sponge anatomy.
a. Movement of water through pores into the central cavity and out the osculum is noted. **b.** Collar cells line the central cavity of a sponge and the movement of their flagella keeps the water moving through the sponge. **c.** Draw an enlargement of spicules here.

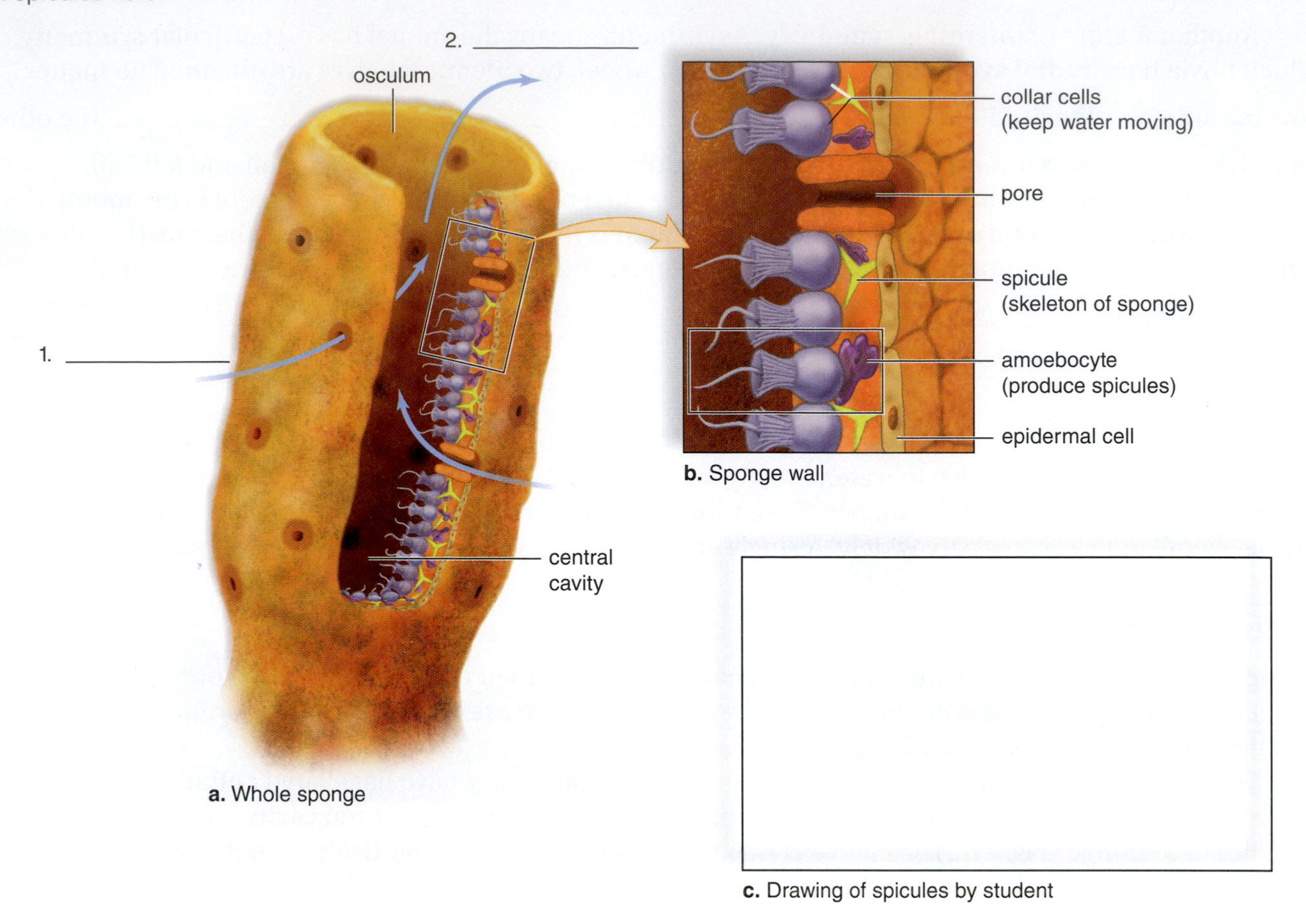

a. Whole sponge

b. Sponge wall

c. Drawing of spicules by student

Prepared Slides

1. Examine a prepared slide of *Grantia*.

   a. Find the collar cells that line the interior (Fig. 17.2*b*). A sponge is a **sessile filter feeder.** Collar cells phagocytize (engulf) tiny bits of food that come through the pores along with the water flowing through the sponge. They then digest the food in food vacuoles. Explain the expression *sessile filter feeder.* ____

   ____

   b. Do you see any spicules? Do they project from the wall of a sponge? ____________

   c. Depending on the sponge, spicules are made of either calcium carbonate, silica (glass), or protein. Calcium carbonate and silica produce hard, sharp spicules. Name two possible advantages of spicules to a sponge. ____

   ____

2. Examine a prepared slide of sponge spicules. What do you see? ____________

   *Draw a sketch of four spicules, each having a different appearance, in the space provided in Figure 17.2*c.

Diversity of Sponges

Sponges are very diverse and come in many shapes and sizes. Some sponges live in fresh water although most live in the sea and are a prominent part of coral reefs, areas of abundant sea life discussed in the next section. Zoologists have described over 5,000 species of sponges, which are grouped according to the type of spicule (Fig. 17.3).

a. Calcareous sponge, *Clathrina canariensis*

b. Bath sponge, *Xestospongia testudinaria*

c. Glass sponge, *Euplectella aspergillum*

Figure 17.3 Diversity of sponges.
a. Calcareous (chalk) sponges have spicules of calcium carbonate. **b.** Bath sponges have a skeleton of spongin. **c.** Glass sponges have glassy spicules.

Conclusions: Anatomy of Sponges

- The anatomy and behavior of a sponge aid its survival and its ability to reproduce. How does a sponge
  - **a.** Protect itself from predators? ______________________________
  - **b.** Acquire and digest food? ______________________________
  - **c.** Reproduce asexually and sexually? ______________________________

17.3 Cnidarians

Cnidaria (phylum Cnidaria) are tubular or bell-shaped animals that live in shallow coastal waters, except for the oceanic jellyfishes. Two basic body forms are seen among cnidaria. The mouth of a **polyp** is directed upward, while the mouth of a jellyfish, or **medusa,** is directed downward. At one time, both body forms may have been a part of the life cycle of all cnidaria. When both are present, the sessile polyp stage produces medusae, and this motile stage produces egg and sperm (Fig. 17.4). Today in some cnidaria, one stage is dominant and the other is reduced; in other species, one form is absent altogether. Regardless, all cnidaria are radially symmetrical. How can radial symmetry be a benefit to an animal? ______________________________

____

How can a life cycle that involves two forms, called **polymorphism,** be of benefit to an animal, especially if one stage is sessile (stationary)? ______________________________

____

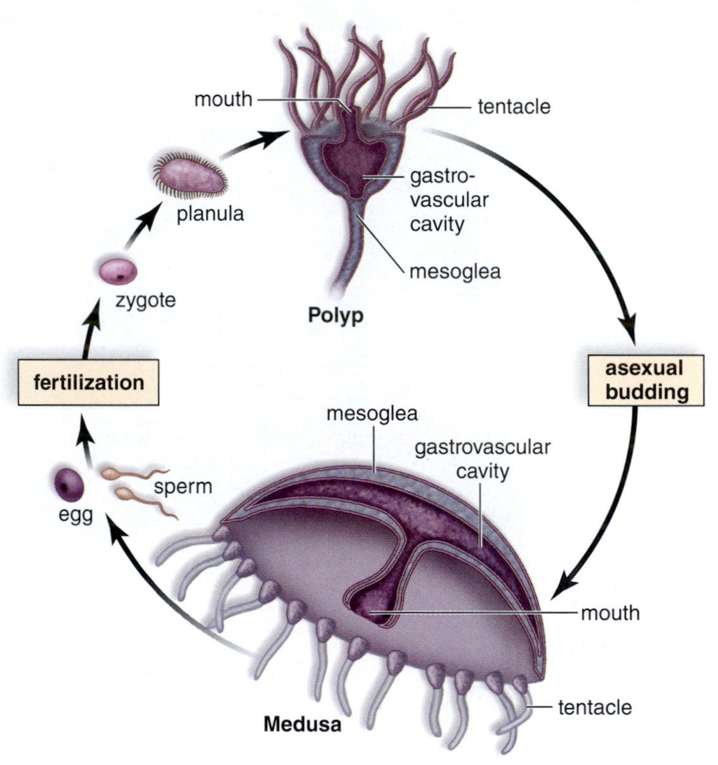

Figure 17.4 The life cycle of a cnidarian. Some cnidarians have both a polyp stage and a medusa stage; in others, one stage may be dominant or absent altogether.

Anatomy of *Hydra*

Figure 17.5 shows the anatomy of a hydra, which will be studied as a typical cnidarian. Hydras exist only as sessile polyps; there is no alternate stage. Note the **tentacles** that surround the **mouth,** the large **gastrovascular cavity,** and the basal disk. A gastrovascular cavity has a single opening that is used as both an entrance for food and an exit for wastes.

Figure 17.5 Anatomy of *Hydra.*
Hydra typifies the anatomy of a cnidarian.

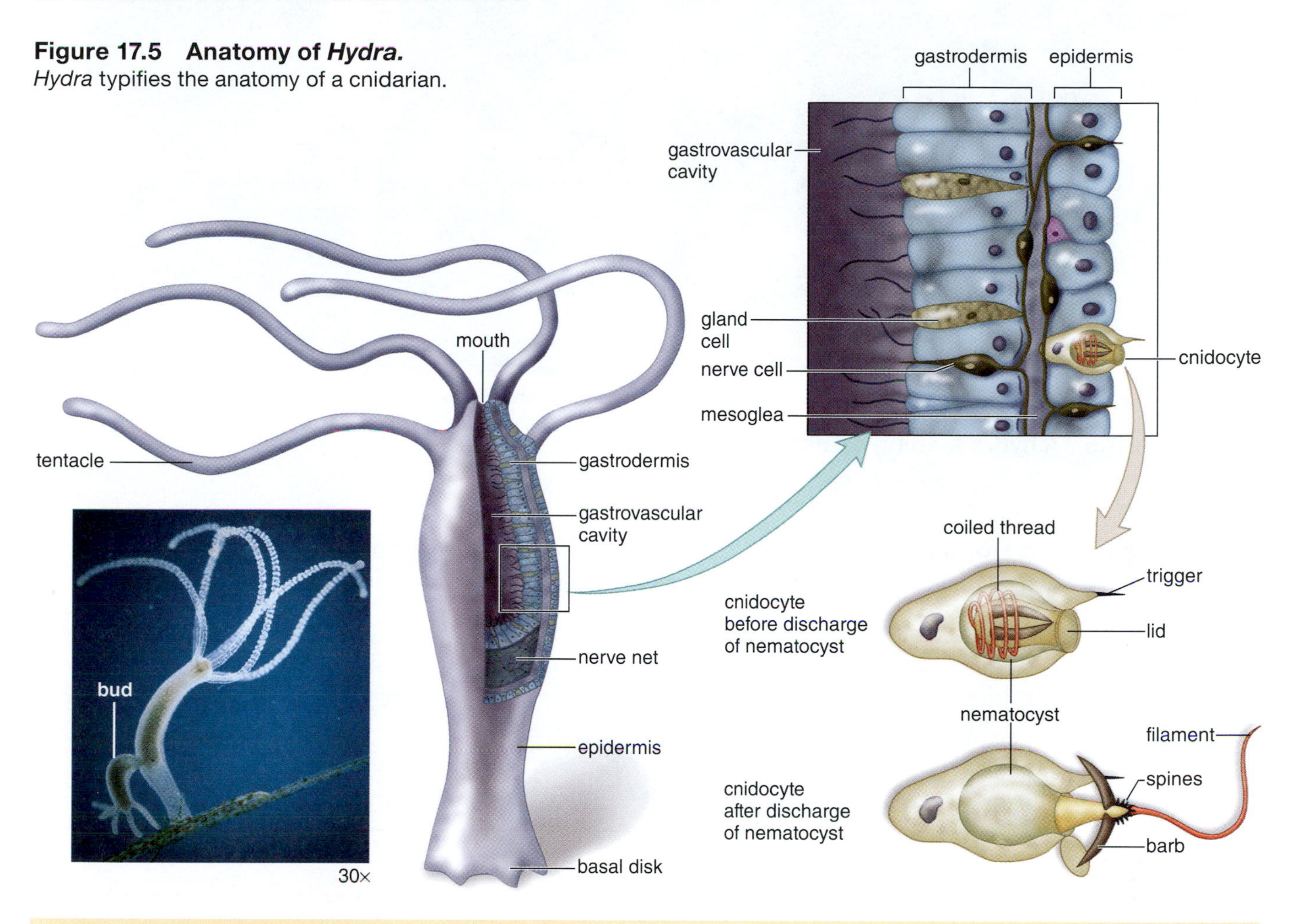

Observation: Anatomy of Hydra

Preserved Hydra

1. With the aid of a hand lens, examine preserved specimens of *Hydra.* Hydras typically reproduce asexually by budding. Do you see any evidence of buds that are developing directly into small hydras? __________ The body wall can also produce ovaries and testes that produce eggs and sperm. The testes are generally located near the attachment of the tentacles; the ovaries appear farther down on the trunk, toward the basal disk.

Prepared Slide of Hydra

1. Examine prepared slides of cross and longitudinal sections of *Hydra.* With the help of Figure 17.5, note the epidermis, the mesoglea (a gelatinous material between the two tissue layers), and the gastrodermis, which lines the gastrovascular cavity. Switch to high power. Do you find any cells? __________ Describe them. ____

____

Daphnia is caught by tentacles.

Daphnia is shown inside gastrovascular cavity.

Figure 17.6 A hydra feeding on daphnia *(Daphnia)*.

Living Hydra

1. Observe a living *Hydra* in a small petri dish for a few minutes. Most often a hydra is attached to a hard surface by its basal disk. A hydra can move, however, by turning somersaults:

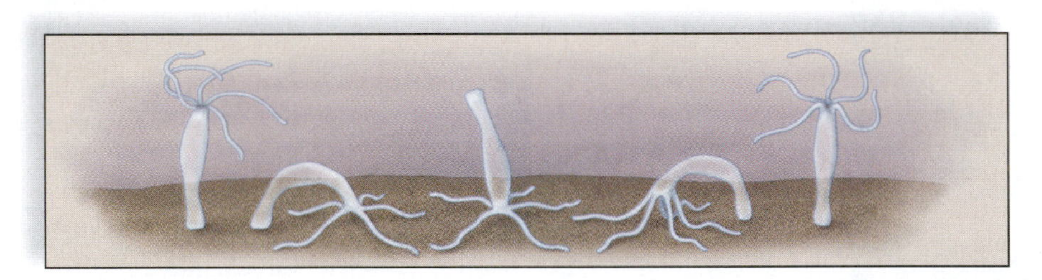

What is the current behavior of your hydra? ______________________________

____

After a few minutes, tap the edge of the petri dish. What is the reaction of your hydra? __________

____

2. When hydras feed, tentacles capture food, which is stuffed into the gastrovascular cavity (Fig. 17.6). If food is available, watch a hydra feed.
3. Mount a living *Hydra* on a depression glass slide with a coverslip and examine a tentacle. Unique to cnidarians are specialized stinging cells, called **cnidocytes**, which give the phylum its name. Each cnidocyte has a fluid-filled capsule called a **nematocyst** (see Fig. 17.5, far right), which contains a long, spirally coiled hollow thread. The threads trap and/or sting prey. Note the cnidocytes as swellings on the tentacles. Add a drop of vinegar (5% acetic acid) and note what happens to the cnidocytes. Did your hydra discard any nematocysts? __________ Describe. ______________

____

Of what benefit is it to *Hydra* to have cnidocysts? ______________________________

____

Conclusions: Anatomy of Hydra

- The anatomy and behavior of a hydra aid its survival and its ability to reproduce. How does a hydra
  - **a.** Acquire and digest food? ______________________________

    ____
  - **b.** Protect itself from predators? ______________________________

    ____
  - **c.** Reproduce asexually and sexually? ______________________________

    ____

Diversity of Cnidarians

Cnidarians consist of a large number of mainly marine animals (Fig. 17.7). Sea anemones, sometimes called the flowers of the sea, are solitary polyps often found in coral reefs, areas of biological abundance in shallow tropical seas. Stony corals have a calcium carbonate skeleton that contributes greatly to the building of coral reefs. Portuguese man-of-war is a colony of modified polyps and medusae. Jellyfishes are a part of the zooplankton, suspended animals that serve as food for larger animals in the ocean.

Figure 17.7 Cnidarian diversity.

a. Sea anemone, *Corynactis*

b. Cup coral, *Tubastrea*

c. Portuguese man-of-war, *Physalia*

d. Jellyfish, *Aurelia*

17.4 Flatworms

Flatworms (phylum Platyhelminthes) are bilaterally symmetrical animals that can be either free-living or parasitic. Free-living flatworms, called planarians, are more complex than the cnidarians. In addition to the germ layers ectoderm and endoderm, mesoderm is also present. Flatworms are usually **hermaphroditic**, which means they possess both male and female sex organs (Fig. 17.8*c*). Why is it advantageous for an animal to be hermaphroditic? ____

____

Planarians practice cross-fertilization when the penis of one is inserted into the genital pore of the other. The fertilized eggs are enclosed in a cocoon and hatch as tiny worms in two or three weeks.

Planarians

Planarians such as *Dugesia* live in lakes, ponds, and streams, where they feed on small, living or dead organisms. In planarians, the three germ layers give rise to various organs aside from the reproductive organs (Fig. 17.8). The three-part **gastrovascular cavity** ramifies throughout the body; the excretory organs called **flame cells** (because their cilia reminded early investigators of a flickering flame of a candle) collect fluids from inside the body and send via a tube to an excretory pore; and the nervous system contains a brain and lateral nerve cords connected by transverse nerves. Therefore it is called **ladder-like**. *Complete the labels in both 17.8*b *and* d. *Label excretory canal, brain, and nerve cord.* Why would you expect an animal that lives in fresh water to have a well-developed excretory system? ____________________________

____

Figure 17.8 Planarian anatomy.
a. When a planarian extends the pharynx, food is sucked up into a gastrovascular cavity that branches throughout the body. **b.** The excretory system has flame cells. **c.** The reproductive system has both male (blue) and female (pink) organs. **d.** The nervous system looks like a ladder.

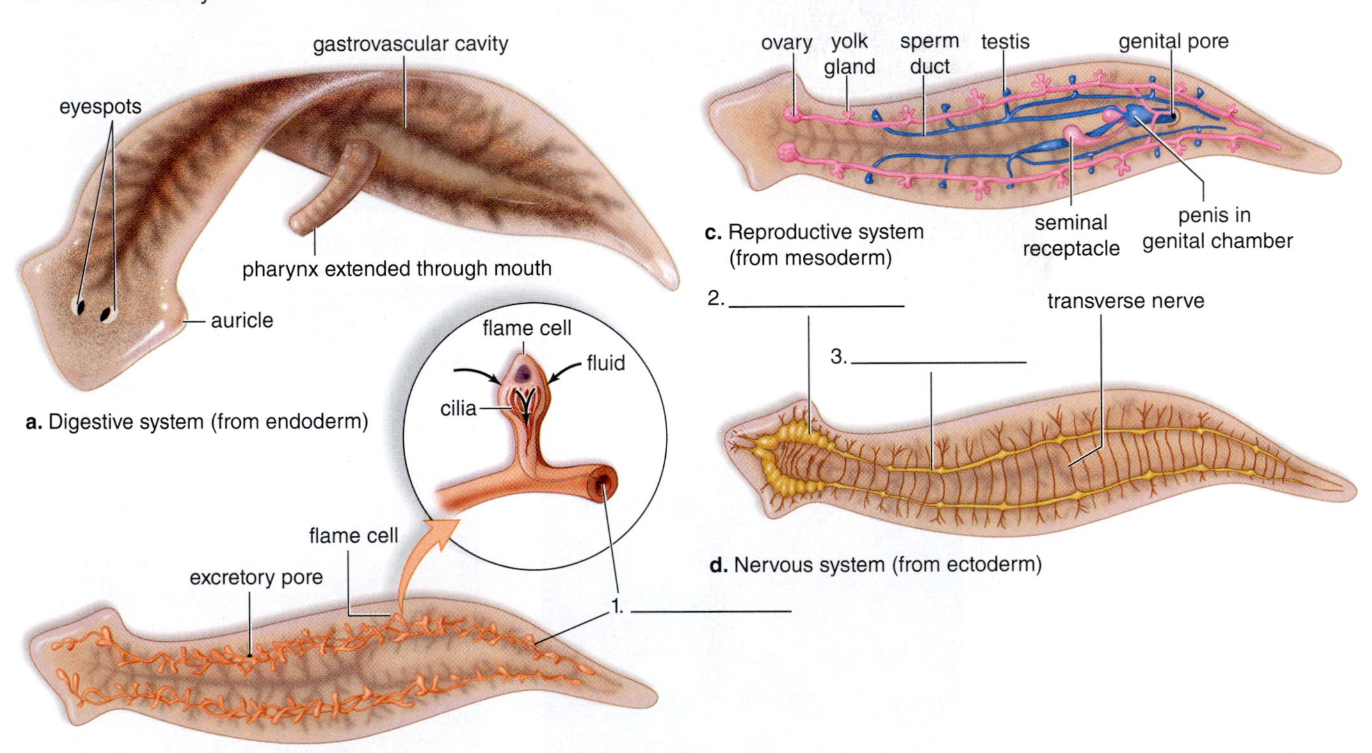

Observation: Planarians

Prepared Slides

1. Examine a whole mount of a planarian that shows the branching gastrovascular cavity (Fig. 17.8*a*). What is the advantage of a gastrovascular cavity that ramifies through the body? ______________________

____

2. Examine a cross section of a planarian under the microscope. Can you locate the structures shown in Figure 17.9? Does a planarian have a body cavity for its internal organs? ______________________ Explain. ______________________

____

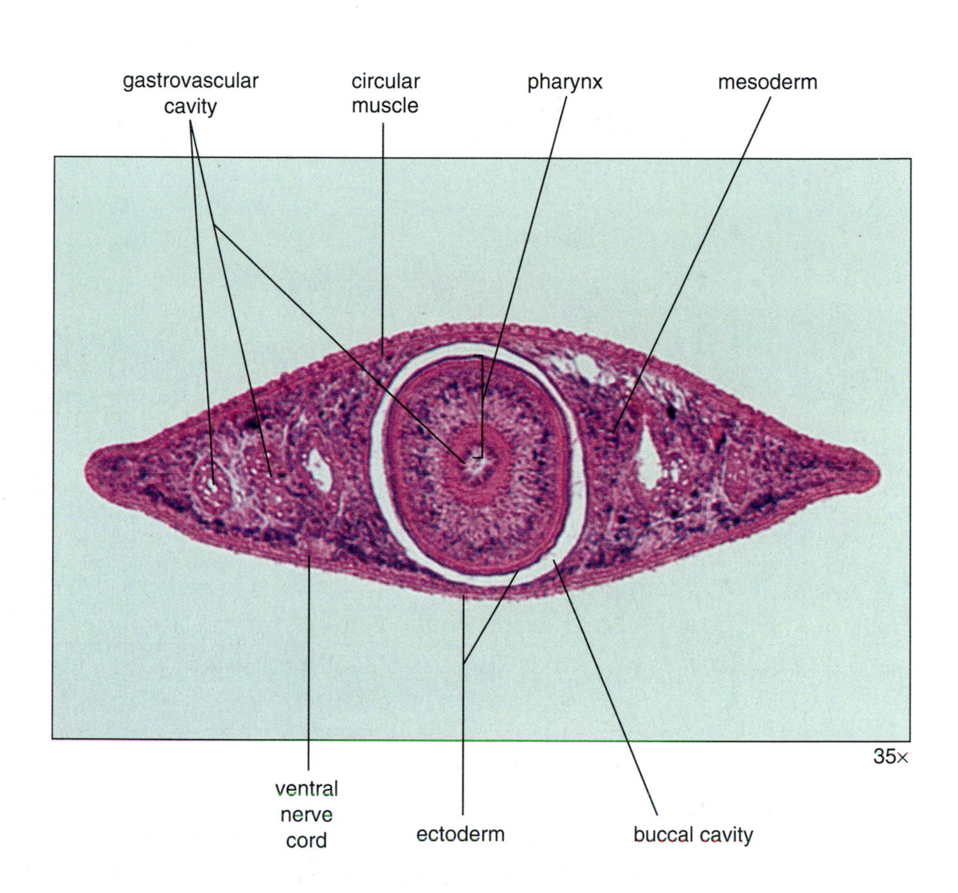

Figure 17.9 Planarian micrograph. Cross section of a planarian at the pharynx.

Living Planarian

1. Examine the behavior of a living planarian (Fig. 17.10) in a petri dish. Describe the behavior of the animal. ______________________

____

2. Does the animal move in a definite direction? __________ Why is it advantageous for a predator such as a planarian to have bilateral symmetry and a definite head region? ______________________

____

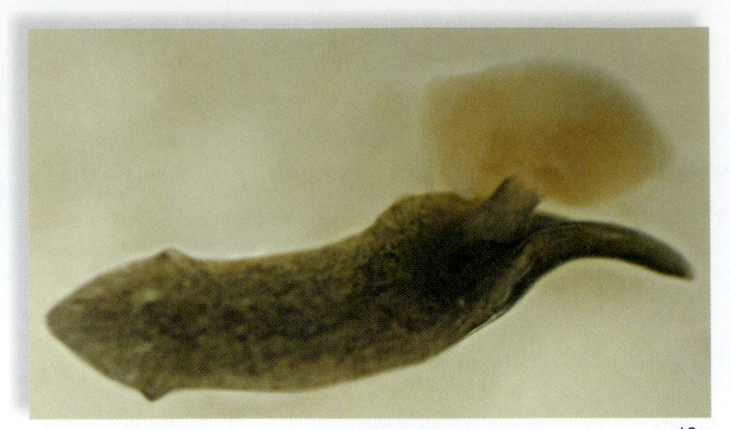

Planarian feeding 10×

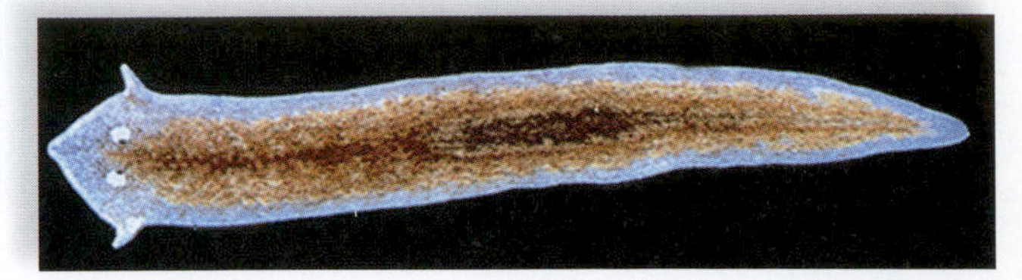

Gastrovascular cavity

Figure 17.10 Micrograph of planarian feeding and the gastrovascular cavity.

3. Gently touch the animal with a probe. What three types of cells must be present for flatworms to be able to respond to stimuli and move about? ______________________________

____

The auricles on the side of the head are sense organs. Flatworms have well-developed muscles and a nervous system consisting of a brain and nerves.

4. If a strong light is available, shine it on the animal. What part of the animal would be able to detect light? __________ How does the animal respond to the light? ______________________________

____

5. Offer the worm some food, such as a small piece of liver, and describe its manner of eating.

____

Roll the animal away from its food and note the pharynx extending from the body (Fig. 17.10).

6. Transfer your worm to a concave depression slide and cover with a coverslip. Examine with a microscope and note the cilia on the ventral surface. Numerous gland cells secrete a mucous material that assists movement. Describe the mode of locomotion. ______________________________

____

Conclusions: Planarians

- Planarians, with three germ layers, are more complex than cnidarians. Contrast a hydra with a planarian by stating in Table 17.1 significant organ differences between them.
- Planarians have no respiratory or circulatory system. As with cnidarians, each individual __________ takes care of its own needs for these two life functions.

Table 17.1 Contrasts Between a Hydra and a Planarian

	Digestive System	Excretory System	Nervous Organization
Hydra			
Planarian			

Tapeworms

Tapeworms are parasitic flatworms known as cestodes. They live in the intestines of vertebrate animals, including humans (Fig. 17.11). The worms consist of a **scolex** (head), usually with suckers and hooks, and **proglottids** (segments of the body). Ripe proglottids detach and pass out with the host's feces, scattering fertilized eggs on the ground. If pigs or cattle happen to ingest these, larvae called bladder worms develop and eventually become encysted in muscle, which humans may then eat in poorly cooked or raw meat. A bladder worm that escapes from a cyst develops into a mature tapeworm attached to the intestinal wall.

1. How do humans get infected with the pig tapeworm? ______________________________

2. What is the function of a tapeworm's hooks and suckers? ______________________________

3. Why would you expect a tapeworm to have a reduced digestive system? ______________________________

4. Proglottids mature into "bags of eggs." Given the life cycle of the tapeworm, why might a tapeworm have an expanded reproductive system compared to a planarian? ______________________________

Figure 17.11 Life cycle of the tapeworm *Taenia*.
The pig host is the means by which the worm is dispersed to the human host.

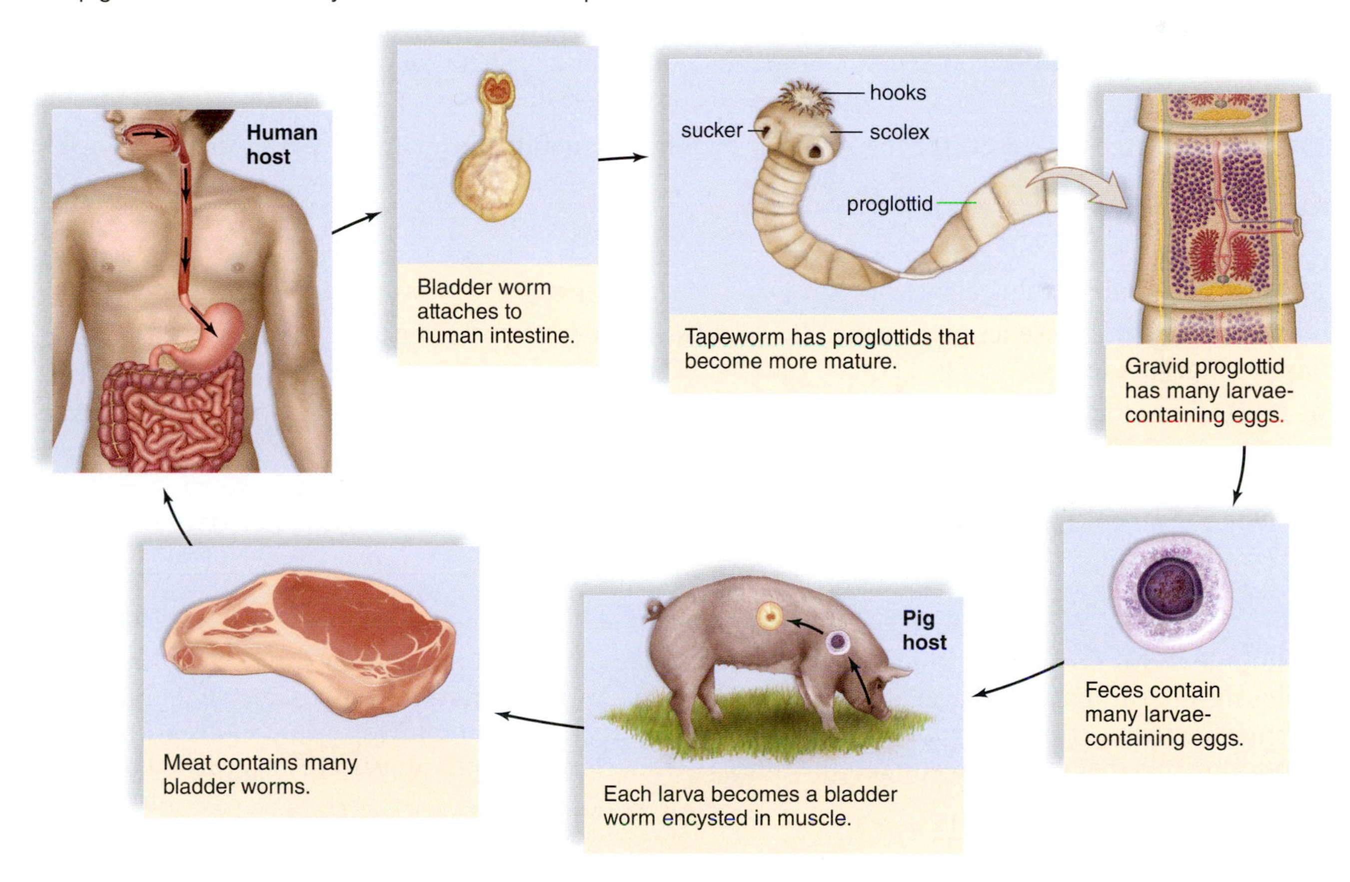

Observation: Tapeworm Anatomy

1. Examine a preserved specimen and/or slide of *Taenia pisiformis*, a tapeworm.
2. With the help of Figure 17.12, identify the scolex, with hooks and suckers, and the proglottids.

Figure 17.12 Anatomy of *Taenia*.
The adult worm is modified for its parasitic way of life. It consists of a scolex and many proglottids, which become bags of eggs.

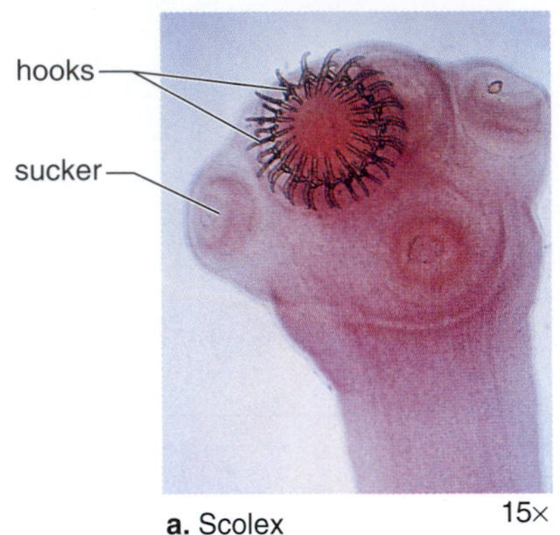

a. Scolex 15×

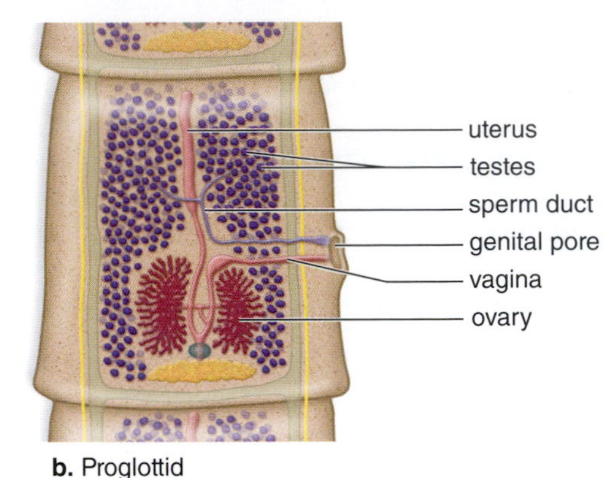

b. Proglottid

17.5 Roundworms

Like planarians, **roundworms** (phylum Nematoda) have three germ layers, bilateral symmetry, and various organs, including a well-developed nervous system. Both planarians and roundworms are nonsegmented; the body has no repeating units. In addition, roundworms have the following features:

1. **Complete digestive tract:** The digestive tract has both a mouth and an anus. What is the advantage of this? ____

   ____

2. **Pseudocoelom:** A body cavity, which allows space for the organs, is incompletely lined with mesoderm.

How would the presence of these two features lead to complexity? For example, how would a complete digestive system lead to a greater number of specialized organs such as both a small intestine that assists digestion and a large intestine that assists elimination? ____

____

How would a spacious body cavity promote the evolution of a greater number of diverse internal organs such as a pancreas and a liver? ____

____

Roundworms are found in all aquatic habitats and in damp soil. Some even survive in hot springs, deserts, and cider vinegar. They parasitize (take nourishment from) both plants and animals. They are significant crop pests and also cause disease in humans. Both pinworms and hookworms are roundworms that cause intestinal difficulties; trichinosis and elephantiasis are also caused by roundworms.

Ascaris

Ascaris, a large, primarily tropical intestinal parasite, is often studied as an example of this phylum.

Observation: Ascaris

Examine preserved specimens of *Ascaris,* both male and female (Fig. 17.13). In roundworms, the sexes are separate. The male is smaller and has a curved posterior end. Be sure to examine specimens of each sex.

Figure 17.13 Roundworm anatomy.
a. Photograph of male *Ascaris.* **b.** Male reproductive system. **c.** Photograph of female *Ascaris.* **d.** Female reproductive system.

a. Male *Ascaris*

c. Female *Ascaris*

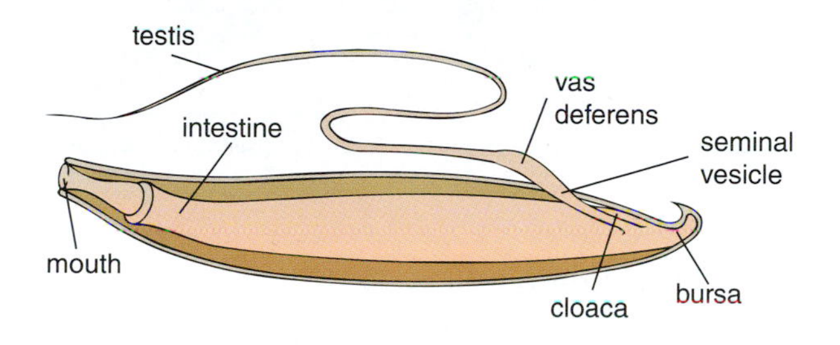

b. Male reproductive system

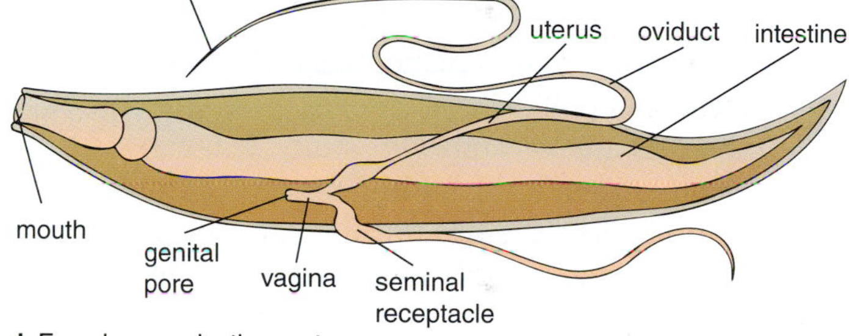

d. Female reproductive system

Trichinella

Trichinella is a parasitic roundworm that causes the disease **trichinosis.** When pigs or humans eat raw or undercooked pork infected with *Trichinella* cysts, juvenile worms are released in the digestive tract where they penetrate the wall of the small intestine and mature sexually. After male and female worms mate, females produce juvenile worms that migrate and form cysts in various muscles (Fig. 17.14). A human with trichinosis has muscular aches and pains that can lead to death if the respiratory muscles fail.

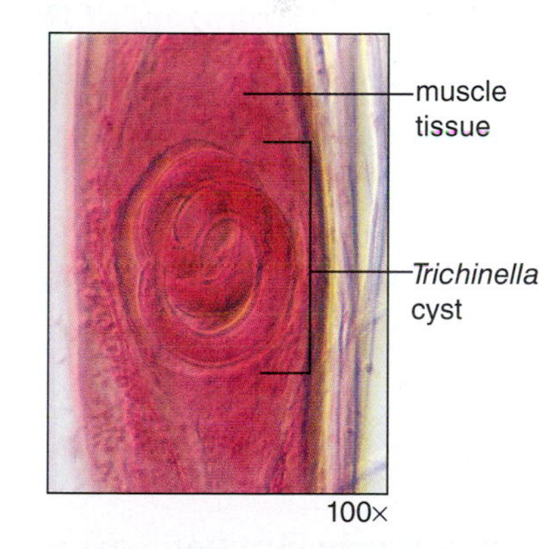

Figure 17.14 Larva of the roundworm *Trichinella* embedded in a muscle.
A larva coils in a spiral and is surrounded by a sheath derived from a muscle fiber.

Observation: Trichinella

1. Examine preserved, infected muscle or a slide of infected muscle, and locate the *Trichinella* cysts, which contain the juvenile worms.
2. How can trichinosis be prevented in humans? ______________________________

____
3. How can pig farmers help to stamp out trichinosis so that humans are not threatened by the disease?

____

Filarial Worm

A roundworm called a **filarial worm** infects lymphatic vessels and blocks the flow of lymph. The condition is called **elephantiasis** because when a leg is affected, it becomes massively swollen.

Vinegar Eels

Vinegar eels are tiny, free-living nematodes that can live in unpasteurized vinegar.

1. Examine live vinegar eels, and observe their active, whiplike swimming movements. This thrashing motion may be a result of nematodes having longitudinal muscles only; they lack circular muscles.
2. Select a few larger vinegar eels for further study, and place them in a small drop of vinegar on a clean microscope slide. If the eels are too active for study, you can slow them by briefly warming them or by adding methyl cellulose.
3. Try to observe the tubular digestive tract, which begins with the mouth and ends with the anus. Also, you may be able to see some of the reproductive organs, particularly in a large female vinegar eel.

Conclusion: Anatomy of Roundworms

- Nematodes are extremely plentiful, in terms of both their variety and their overall number. From your knowledge of adaptive radiation, explain why there might be so many different types of nematodes. ______________________________

____

____

17.6 Rotifers

Rotifers (phylum Rotifera) are common and abundant freshwater animals. They are important constituents of the plankton of lakes, ponds, and streams, and are a significant food source for many species of fish and other animals.

Like roundworms, rotifers have a pseudocoelom and a complete digestive tract with a mouth and anus. The corona is a crown of cilia around the mouth. To some, the movement of the cilia resembles that of a rapid wheel; and in Latin, *rotifer* means "wheel-bearer." The cilia draw water into the mouth and from there food is ground up by trophi (jaws) before entering the stomach from which nondigested remains pass through the cloaca and anus (Fig. 17.15).

Observation: Rotifers

Living Specimen

1. Use a pipet to obtain a living rotifer specimen near a clump of vegetation. Place the liquid and rotifer on a concave depression slide. Do not add a coverslip. Study the animal's behavior and appearance.
2. Describe the rotifer's behavior. ____

____
3. Observe the elongated, cylindrical body that can be divided into three general regions: the head, the trunk, and a posterior foot. The rotifers have no true segmentation, but the cuticle covering the body can be divided into a number of superficial segments.
4. Observe the telescoping of the segments when the animal retracts its head.
5. Note also the large, ciliated **corona** at the anterior end. Most rotifers have a conspicuous corona that serves both for locomotion and for feeding. It creates a current that brings smaller microorganisms (e.g., algae, protozoans, bacteria) close enough to be swallowed.
6. With the help of Figure 17.15, try to identify some of the rotifer's internal organs. List those you observe here. ____

____

Figure 17.15 Live *Philodina*, a common rotifer.
a. Micrograph. **b.** Line art.

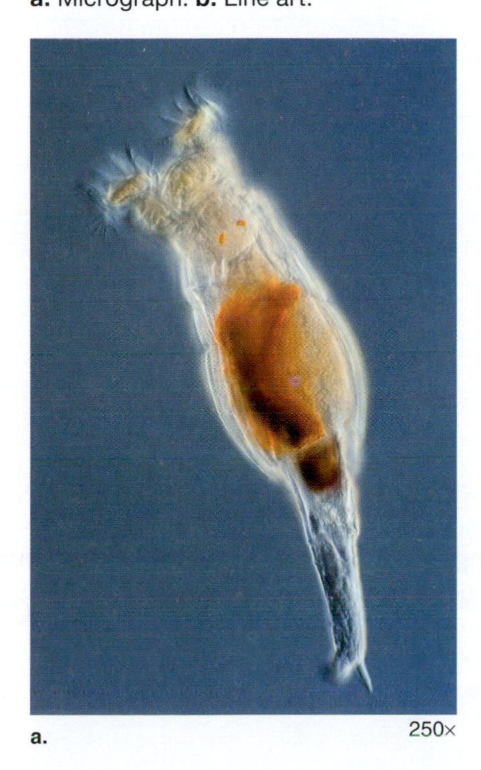

a.

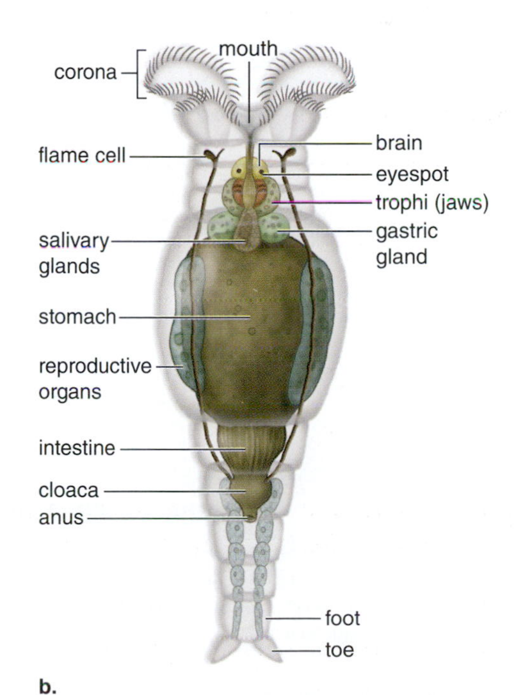

b.

Laboratory Review 17

1. Distinguish invertebrate from vertebrate. ______________________________

2. A hydra has radial symmetry, but a planarian has bilateral symmetry. Explain the difference. ______
____

3. The animals studied today have some sort of defense. Contrast the defense of a sponge with that of a hydra.
____
____

4. If a cnidarian has both a polyp stage and a medusa stage, what is the life-cycle function of the medusa stage? ______________________________

5. Why would you expect an animal with bilateral symmetry to be more active than one with radial symmetry?
____
____

6. Name two contrasting anatomical features (other than symmetry) that distinguish a planarian from a hydra.

Planarian	**Hydra**

7. Name a group of animals that is usually hermaphroditic and a group that is usually dioecious (sexes are separate). ______________________________
____

8. How does the process of acquiring food in planarians differ from the process in tapeworms? ______
____

9. Why would you expect a free-living roundworm to have a nervous system? ______________________________
____

10. Name two types of parasitic roundworms and one type of roundworm that is free-living. ______
____

LABORATORY

18

Later Invertebrate and Vertebrate Evolution

Learning Outcomes

18.1 Molluscs
- Describe the general characteristics of molluscs and the specific features of selected groups. 226
- Identify and/or locate external and internal structures of a clam. 227–29
- Compare clam anatomy to squid anatomy, and tell how each is adapted to its way of life. 229–30

18.2 Annelids
- Describe the general characteristics of annelids and the specific features of the three major groups. 231
- Identify and/or locate external structures of an earthworm and also internal structures in a cross-section slide. 232–33

18.3 Arthropods
- Describe the general characteristics of arthropods. 234
- Contrast the anatomy of the crayfish and the grasshopper, indicating how each is adapted to its way of life. 236–37
- Describe metamorphosis, and compare complete with incomplete metamorphosis. 238–39

18.4 Echinoderms
- Describe the general characteristics of echinoderms. 239
- Identify and locate external and internal structures of a sea star. 240–41

18.5 Chordates
- State the four characteristics that all chordates have in common. 242
- Discuss the evolutionary tree of chordates in terms of seven innovations that appear as evolution occurs. 242
- Categorize vertebrates into six major groups. 243

Introduction

Among animals, the **acoelomates** have no body cavity, the **pseudocoelomates** have a body cavity incompletely lined with mesoderm, and the **coelomates** have a body cavity completely lined with mesoderm. (**Mesoderm** is an embryonic tissue that performs many functions, including lining the coelom.) A coelom offers many advantages. (1) The digestive system and body wall can move independently. (2) Internal organs can become more complex. (3) Coelomic fluid can assist respiration, circulation, and excretion and serve as a hydrostatic skeleton.

All the phyla considered in today's laboratory are coelomates. The molluscs, annelids, and arthropods are protostomes, animals in which the first *(protos)* embryonic opening becomes the mouth *(stoma),* while the echinoderms and vertebrates are deuterostomes. In deuterostomes, the first opening becomes the anus, and the second *(deutero)* opening becomes the mouth.

18.1 Molluscs

Most **molluscs** (phylum Mollusca) are marine, but there are also some freshwater and terrestrial molluscs (Fig. 18.1). Among molluscs, the grazing marine herbivores, known as **chitons**, have a body flattened dorsoventrally covered by a shell consisting of eight plates (Fig. 18.1*a*). The **bivalves** contain marine and freshwater sessile filter feeders, such as clams and scallops, with a body enclosed by a shell consisting of two valves (Fig. 18.1*b*). The **gastropods** contain marine, freshwater, and terrestrial species. In snails, the shell, if present, is coiled (Fig. 18.1*c*). The **cephalopods** contain marine active predators, such as squids and nautiluses. Tentacles are about the head (Fig. 18.1*d*).

All molluscs have a three-part body consisting of (1) a muscular **foot** specialized for various means of locomotion; (2) **visceral mass** that includes the internal organs; and (3) a **mantle,** a thin tissue that encloses the visceral mass and may secrete a shell. **Cephalization** is the development of a head region. *On the lines provided in Figure 18.1, write "cephalization" or "no cephalization" as appropriate for this mollusc.*

a. Chitons, *Tonicella*

____

Figure 18.1 Molluscan diversity.
a. You can see the exoskeleton of this chiton but not its dorsally flattened foot. **b.** A scallop doesn't have a foot but it does have strong adductor muscles to close the shell. In this specimen, the edge of the mantel bears tentacles and many blue eyes. **c.** A gastropod, such as a snail, is named for the location of its large foot beneath the visceral mass. **d.** In a cephalopod, such as this nautilus, a funnel (its foot) opens in the area of the tentacles and allows it to move by jet propulsion.

b. Scallop, *Argopecten,* is a bivalve.

____

c. Snail, *Helix,* is a gastropod.

____

d. Nautilus, *Nautilus,* is a cephalopod.

____

Anatomy of a Clam

Clams are bivalved because they have right and left shells secreted by the mantle. Clams have no head, and they burrow in sand by extending a **muscular foot** between the valves. Clams are **filter feeders** and feed on debris that enters the mantle cavity. In the visceral mass, the blood leaves the heart and enters sinuses (cavities) by way of anterior and posterior aortas. There are many different types of clams. The one examined here is the freshwater clam *Venus.*

Observation: Anatomy of a Clam

External Anatomy

1. Examine the external shell (Fig. 18.2) of a preserved clam *(Venus)*. The shell is an **exoskeleton.**
2. Find the posterior and anterior ends. The more pointed end of the **valves** (the halves of the shell) is the posterior end.
3. Determine the clam's dorsal and ventral regions. The valves are hinged together dorsally.
4. What is the function of a heavy shell? ______________________________

Figure 18.2 External view of the clam shell.

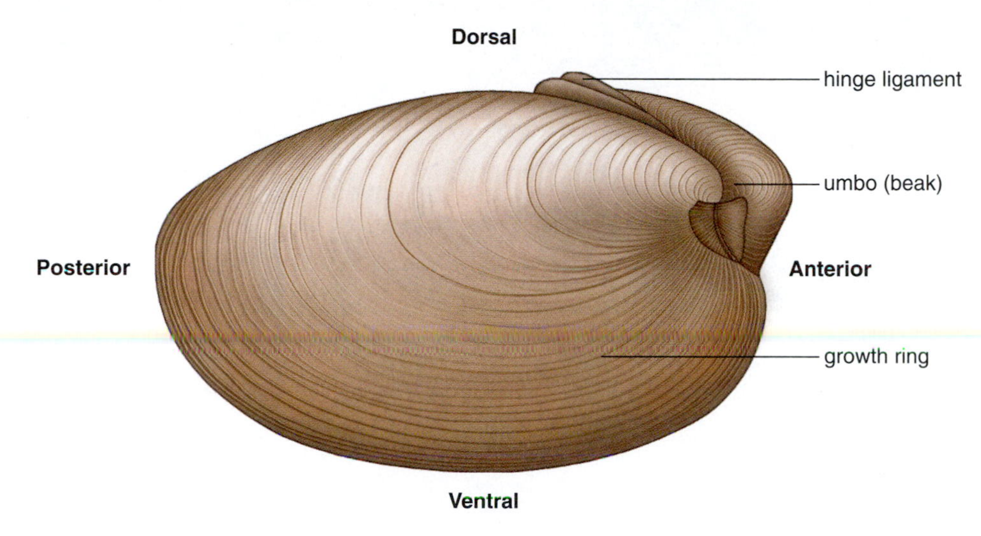

Internal Anatomy

1. Place the clam in the dissecting pan, with the **hinge ligament** and **umbo** (blunt dorsal protrusion) down. Carefully separate the **mantle** from the right valve by inserting a scalpel into the slight opening of the valves. What is a mantle? ______________________________

____

2. Insert the scalpel between the mantle and the valve you just loosened.
3. The **adductor muscles** hold the valves together. Cut the adductor muscles at the anterior and posterior ends by pressing the scalpel toward the dissecting pan. After these muscles are cut, the valve can be carefully lifted away. What is the advantage of powerful adductor muscles? ____________

____

4. Examine the inside of the valve you removed. Note the concentric lines of growth on the outside, the hinge teeth that interlock with the other valve, the adductor muscle scars, and the mantle line. The inner layer of the shell is mother-of-pearl.
5. Examine the rest of the clam (Fig. 18.3) attached to the other valve. Notice the adductor muscles and the mantle, which lie over the visceral mass and foot.
6. Bring the two halves of the mantle together. Explain the term *mantle cavity*. ____________

____

7. Identify the **incurrent** (more ventral) and **excurrent siphons** at the posterior end (Fig. 18.3). Explain how water enters and exits the mantle cavity. ______________________________

____

Figure 18.3 Anatomy of a bivalve.
The mantle has been removed to reveal the internal organs. **a.** Drawing. **b.** Dissected specimen.

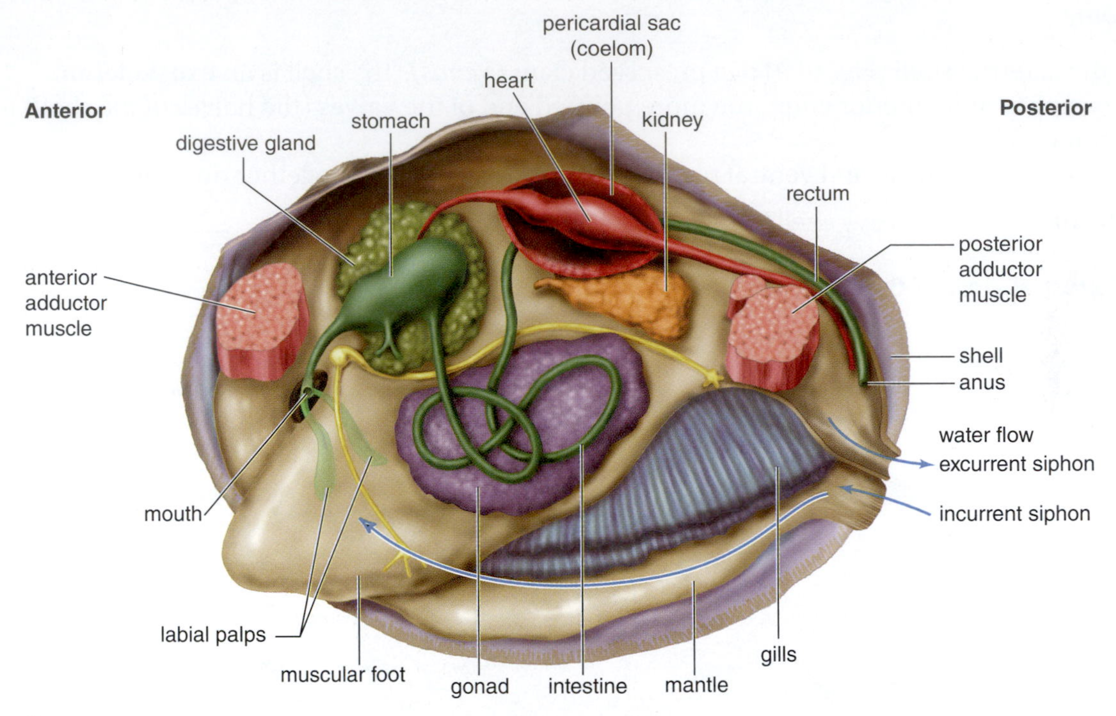

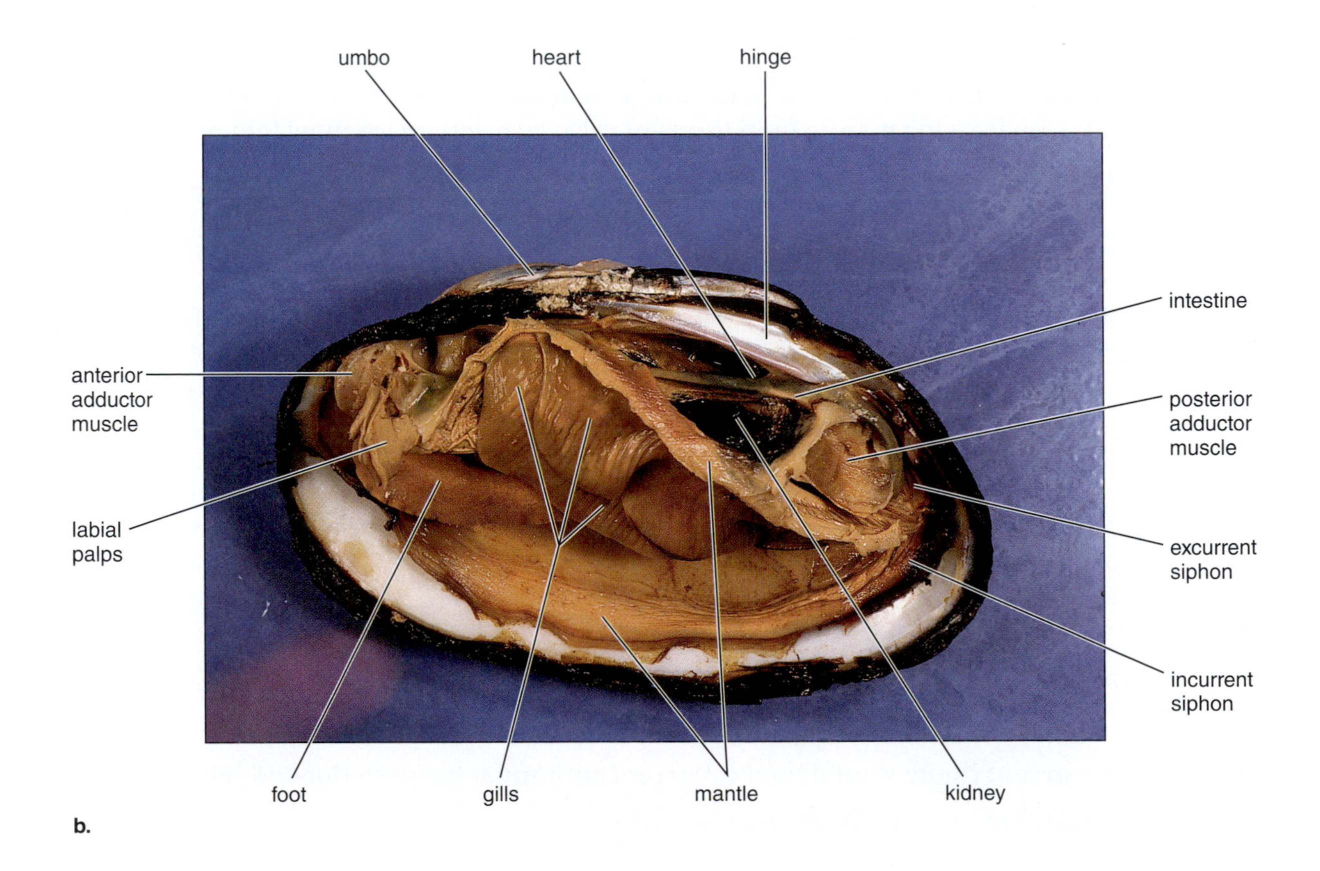

8. Cut away the free-hanging portion of the mantle to expose the **gills.** Does the clam have a respiratory organ? ______

   If so, what type of respiratory organ? ______
9. A mucous layer on the gills entraps food particles brought into the mantle cavity, and the cilia on the gills convey these food particles to the mouth. Why is the clam called a filter feeder?

   ______

   ______
10. The nervous system is composed of three pairs of ganglia (located anteriorly, posteriorly, and in the foot), all connected by nerves. The clam does not have a brain. A ganglion contains a limited number of neurons, whereas a brain is a large collection of neurons in a definite head region.
11. Identify the **foot,** a tough, muscular organ for locomotion, and the **visceral mass,** which lies above the foot and is soft and plump. The visceral mass contains the digestive and reproductive organs.
12. Identify the **labial palps** that channel food into the open mouth.
13. Identify the **anus,** which discharges into the excurrent siphon.
14. Find the **intestine** by its dark contents. Trace the intestine forward until it passes into a sac, the clam's only evidence of a coelom.
15. Locate the **pericardial sac (pericardium)** that contains the heart. The intestine passes through the heart. The heart pumps blood into the aortas, which deliver it to blood sinuses in the tissues.

    A clam has an **open circulatory system.** Explain. ______

    ______
16. Cut the visceral mass and the foot into exact left and right halves, and examine the cut surfaces. Identify the digestive glands, greenish-brown; the stomach, embedded in the digestive glands; and the intestine, which winds about in the visceral mass. Reproductive organs are also present.

Anatomy of a Squid

Squids are cephalopods because they have a well-defined head; the foot became the funnel surrounded by two arms and the many tentacles about the head. The head contains a brain and bears sense organs. The squid moves quickly by jet propulsion of water, which enters the mantle cavity by way of a space that encircles the head. When the cavity is closed off, water exits by means of the funnel. Then the squid moves rapidly in the opposite direction.

The squid seizes fish with its tentacles; the mouth has a pair of powerful, beaklike jaws and a **radula,** a beltlike organ containing rows of teeth. The squid has a **closed circulatory system** composed of vessels and three hearts, one of which pumps blood to all the internal organs, while the other two pump blood to the gills located in the mantle cavity.

Observation: Anatomy of a Squid

1. Examine a preserved squid.
2. Refer to Figure 18.4 for help in identifying the mouth (defined by beaklike jaws and containing a radula) and the tentacles and arms, which encircle the mouth.
3. Locate the head with its sense organs, notably the large, well-developed eye.
4. Find the funnel, where water exits from the mantle cavity, causing the squid to move backward.
5. If a dissected squid is available, note the heart, gills, and blood vessels.
6. Compare clam anatomy to squid anatomy by completing Table 18.1.

Figure 18.4 Anatomy of a squid.
The squid is an active predator and lacks the external shell of a clam. It captures fish with its tentacles and bites off pieces with its jaws. A strong contraction of the mantle forces water out the funnel, resulting in "jet propulsion."

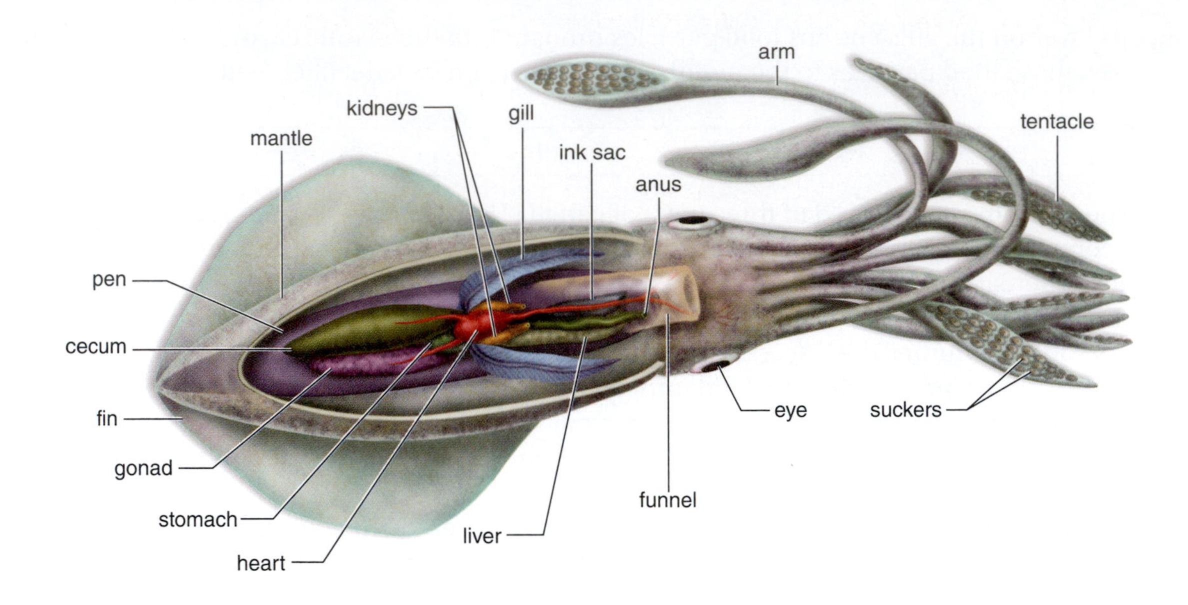

Table 18.1 Comparison of Clam to Squid

	Clam	Squid
Feeding mode		
Skeleton		
Circulation		
Cephalization		
Locomotion		
Nervous system	Three separate ganglia	

Conclusions: Comparison of Clam to Squid

- Explain how both clams and squids are adapted to their way of life.

____

____

18.2 Annelids

Annelids (phylum Annelida) are the segmented worms, so called because the body is divided into a number of segments and has a ringed appearance (Fig. 18.5). The circular and longitudinal muscles work against the fluid-filled coelom to produce changes in width and length. Therefore, annelids are said to have a **hydrostatic skeleton.**

Among annelids, **polychaetes** have many slender bristles called **setae.** The polychaetes, almost all marine, are plentiful from the intertidal zone to the ocean depths. They are quite diverse, ranging from jawed forms that are carnivorous to fanworms that live in tubes and extend feathery filaments when filter feeding. Earthworms are called **oligochaetes** because they have few setae. Earthworms, which have a worldwide distribution in almost any soil, occur in large number, and reach a length of as much as 3 meters. **Leeches,** annelids without setae, include the medicinal leech, which has been used in the practice of bloodletting for centuries. Most people simply called them bloodsuckers.

*Show that the annelids are the segmented worms by labeling a segment in Figure 18.5*a, b, d. In which group would you expect the animals to be predators based on the type of head region? ________________

Figure 18.5 Annelid diversity.
The annelids include **a.** the earthworm, an oligochaete; **b.** the marine clam worm, and **c.** the giant fanworm, which are polychaetes; and **d.** the leech, well known for being a bloodsucker.

a. Earthworms, *Lumbricus*, mating

head
(sense organs
and jaws)

b. Clam worm, *Nereis*

c. Christmas tree worm, *Spirobranchus*

d. Leech, *Hirudo*

Anatomy of an Earthworm

Earthworms are segmented in that the body has a series of ringlike segments. Earthworms have no head, and burrow in the soil by alternately expanding and contracting segments along the length of the body.

Earthworms are scavengers that feed on decaying organic matter in the soil. They have a well-developed coelomic cavity, providing room for a well-developed digestive tract.

Observation: External Anatomy of an Earthworm

1. Examine a live or preserved specimen of an earthworm. Locate the small projection that sticks out over the mouth. Has cephalization occurred? __________ Explain. ______________________
____

2. Count the total number of segments, beginning at the anterior end. The sperm duct openings are on segment 15 (somite XV) (Fig. 18.6). The enlarged section around a short length of the body is the **clitellum.** The clitellum secretes mucus that holds the worms together during mating. It also functions as a cocoon, in which fertilized eggs hatch and young worms develop. The anus is located on the worm's terminal segment.

3. Lightly pass your fingers over the earthworm's ventral and lateral sides. Do you feel the setae? ________
____

   Earthworms insert these slender bristles into the soil. Setae, along with circular and longitudinal muscles, enable the worm to locomote. Explain the action. ______________________
____

Figure 18.6 External anatomy of an earthworm.

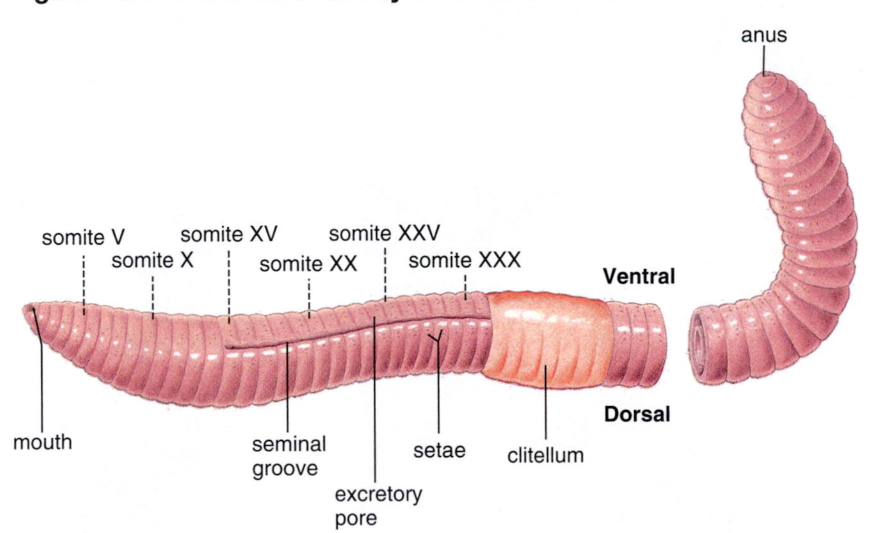

Observation: Internal Anatomy of an Earthworm

1. Obtain a prepared slide of a cross section of an earthworm (Fig. 18.7). Examine the slide under the dissecting microscope and under the light microscope.
2. Identify the following structures.
   a. **Body wall:** A thick outer circle of tissue, consisting of the **cuticle** and the **epidermis.**
   b. **Coelom:** A cavity for internal organs as also occurs in vertebrates.
   c. **Intestine:** An inner circle with a suspended fold.
   d. **Typhlosole:** A fold that increases the intestine's surface area.
   e. **Ventral nerve cord:** A white, threadlike structure.
   f. **Nephridia:** Excretory tubules in every segment are good evidence of segmentation.
   g. **Blood vessels:** Part of a closed circulatory system as also seen in vertebrates. Vertebrates have a dorsal nerve cord.
3. How does the typhlosole help in nutrient absorption? ______________________________

____

Figure 18.7 Cross section of an earthworm.
Cross-section micrograph as it would appear under the microscope.

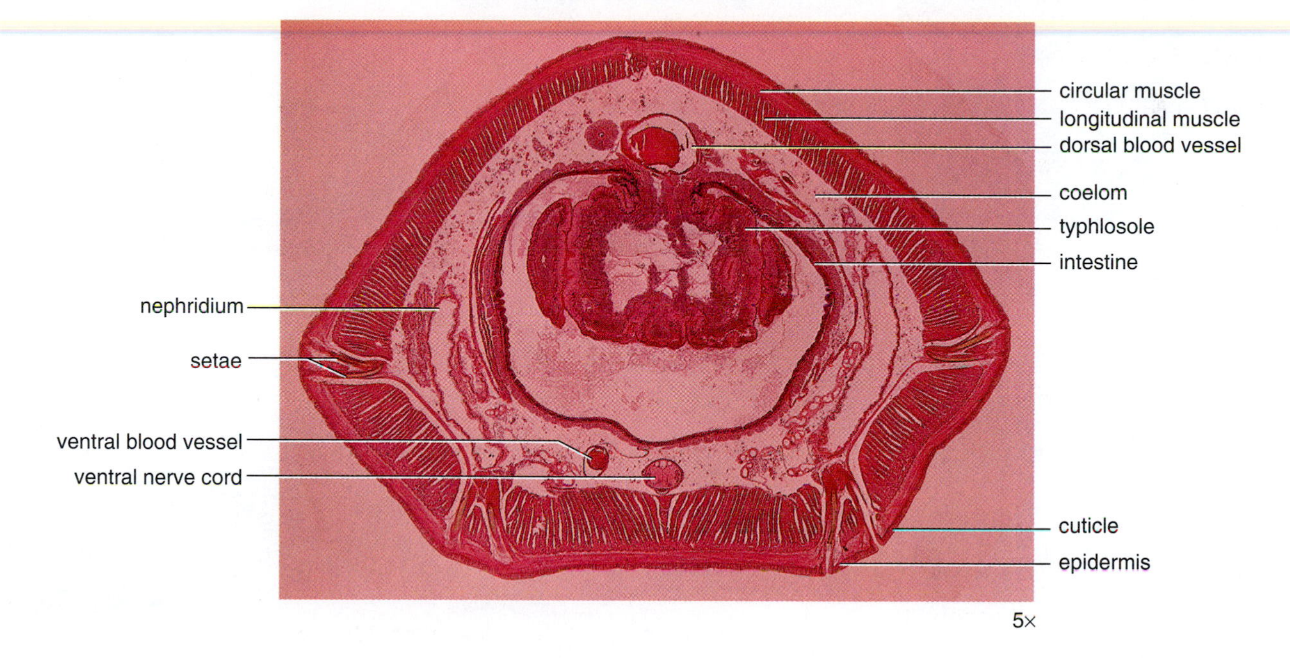

Virtual Lab Earthworm Dissection
A virtual lab called Earthworm Dissection is available on the *Concepts of Biology* website **www.mhhe.com/maderconcepts3**. Follow the directions given to click and drag labels to external and internal illustrations of earthworm anatomy.

18.3 Arthropods

Arthropods (phylum Arthropoda) have paired, jointed appendages and a hard exoskeleton that contains chitin. The chitinous exoskeleton consists of hardened plates separated by thin, membranous areas that allow movement of the body segments and appendages.

Figure 18.8*a* features insects and relatives. **Insects** with three pairs of legs, with or without wings, and three distinct body regions comprise 95% of all arthropods. **Millipedes** have two pairs of legs per segment, while **centipedes** have one pair of legs per segment. Figure 18.8*b* features spiders and relatives. Spiders and scorpions have four pairs of legs, no antennae, and a cephalothorax (head and thorax are fused). The horseshoe crab is a living fossil. It has remained unchanged for thousands of years. The **crustaceans** (Fig. 18.8*c*), which include crabs, shrimp, and lobsters, have three to five pairs of legs, and two pairs of antennae. Barnacles are unusual, in that their legs are used to gather food.

For each animal in Figure 18.8, circle the obvious types of appendages.

Figure 18.8 Arthropod diversity.
a. Insects, millipedes, and centipedes are related. **b.** Spiders, scorpions, and horseshoe crabs are related. **c.** Crabs, shrimp, and barnacles, among others, are crustaceans.

a. Insects and relatives

Honeybee, *Apis mellifera*

Millipede, *Ophyiulus pilosus*

Centipede, *Scolopendra* sp.

b. Spider and relatives

Spider, *Argiope rafaria*

Scorpion, *Hadrurus hirsutus*

Horseshoe crab, *Limulus polyphemus*

c. Crustaceans

Crab, *Cancer productus*

Shrimp, *Stenopus* sp.

Barnacles, *Lepas anatifera*

Anatomy of a Crayfish

Crayfish belong to the group of arthropods called crustaceans. Crayfish are adapted to an aquatic existence. They are known to be scavengers, but they also prey on other invertebrates. The mouth is surrounded by appendages modified for feeding, and there is a well-developed digestive tract. In contrast to vertebrates, there is a ventral nerve cord.

Observation: Anatomy of a Crayfish

External Anatomy

1. In a preserved crayfish, identify the chitinous **exoskeleton.** With the help of Figure 18.9, identify the head, thorax, and abdomen. Together, the head and thorax are called the **cephalothorax;** the cephalothorax is covered by the **carapace.** Has specialization of segments occurred? ____________

   Explain. ____
2. Find the **antennae,** which project from the head. At the base of each antenna, locate a small, raised nipple containing an opening for the **green glands,** the organs of excretion. Crayfish excrete a liquid nitrogenous waste.
3. Locate the **compound eyes,** composed of many individual units for sight. Do crayfish demonstrate cephalization? __________ Explain. ______________________________
4. Identify the six pairs of appendages around the mouth for handling food.
5. Find the five pairs of walking legs attached to the cephalothorax. The most anterior pair is modified as pincerlike claws.
6. Locate the five pairs of **swimmerets** on the abdomen. In males, the anterior two pairs are stiffened and folded forward. They are claspers that aid in the transfer of sperm during mating.
7. In the female, identify the **seminal receptacles,** a swelling located between the bases of the third and fourth pairs of walking legs. Sperm from the male are deposited in the seminal receptacles. In the male, identify the opening of the sperm duct located at the base of the fifth walking leg. Find the last abdominal segment, which bears a pair of broad, fan-shaped **uropods** that, together with a terminal extension of the body, form a tail. Has specialization of appendages occurred? __________ Explain. ______________________________

Figure 18.9 External anatomy of a crayfish.

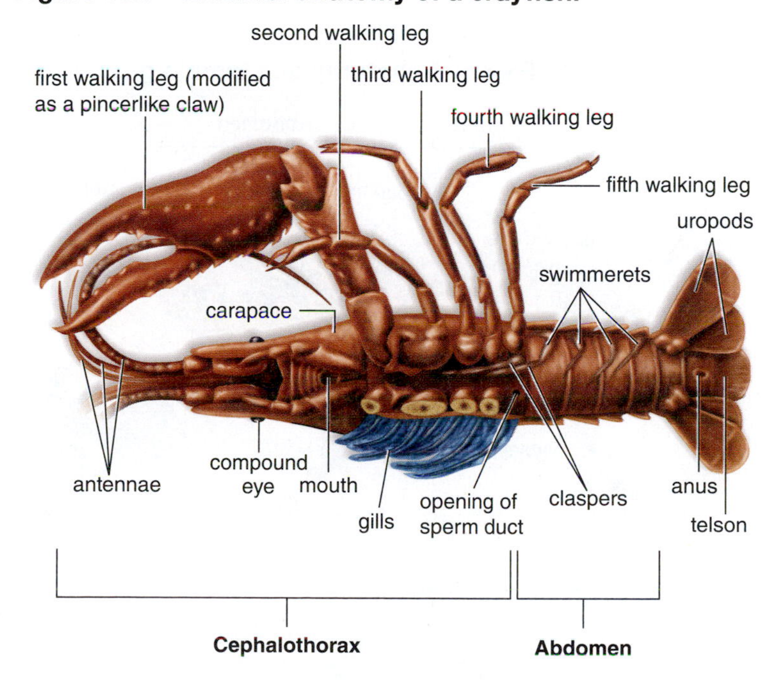

Anatomy of a Grasshopper

The grasshopper is an **insect.** All insects have a head, a thorax, and an abdomen. Their appendages always include (1) three pairs of jointed legs and usually (2) two antennae as sensory organs. Grasshoppers are adapted to live on land. Wings and jumping legs are suitable for locomotion on land; **Malpighian tubules** save water by secreting a solid nitrogenous waste; the **tracheae** are tiny tubules that deliver air directly to the muscles; and the male has a penis with attached claspers to deliver sperm to the seminal receptacles of a female so they do not dry out.

Observation: Anatomy of a Grasshopper

External Anatomy

1. Obtain a preserved grasshopper *(Romalea),* and study its external anatomy with the help of Figure 18.10. Identify the head, thorax, and abdomen.
2. Use a hand lens or dissecting microscope to examine the grasshopper's special sense organs of the **head.** Identify the **antennae** (a pair of long, jointed feelers), the **compound eyes,** and the three dotlike **simple eyes.** The labial palps, labeled in Figure 18.10, have sense organs for tasting food.
3. Note the sturdy **mouthparts,** which are used for chewing plant material. A grasshopper's mouthparts are quite different from those of a piercing and sucking insect.
4. Locate the leathery **forewings** and the inner, membraneous **hindwings** attached to the **thorax**. Which pair of legs is used for jumping? ____________ How many segments does each leg have? ______
5. Is locomotion in the grasshopper adapted to land? ____________________
   Explain. ____________________
   ____
6. In the **abdomen**, identify the **tympana** (sing., **tympanum**), one on each side of the first abdominal segment (Fig. 18.10). The grasshopper detects sound vibrations with these membranes.
7. Locate the **spiracles,** along the sides of the abdominal segments. These openings allow air to enter the tracheae, which constitute the respiratory system.
8. Find the **ovipositors** (Figs. 18.10 and 18.11*a*), four curved and pointed processes projecting from the abdomen of the female. These are used to dig a hole in which eggs are laid. The male has a **penis** with **claspers** used during copulation (Fig. 18.11*b*).

Figure 18.10 External anatomy of a female grasshopper.

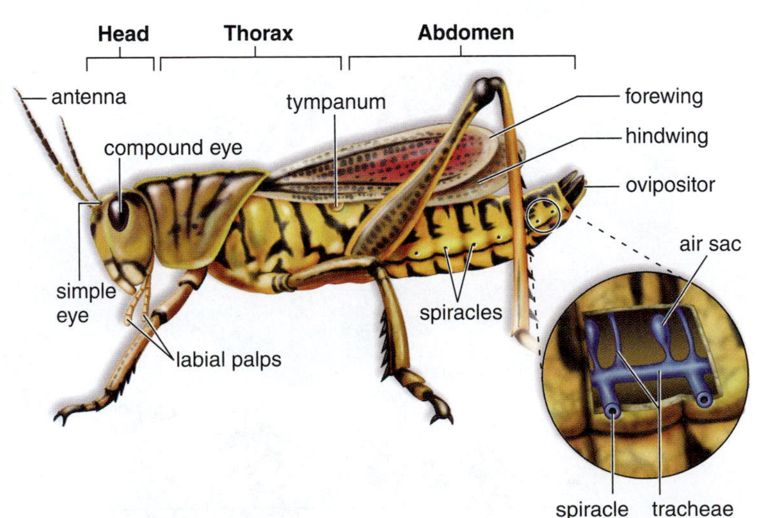

Figure 18.11 Grasshopper genitalia.
a. Females have an ovipositor. **b.** Males have claspers at the distal end of the penis.

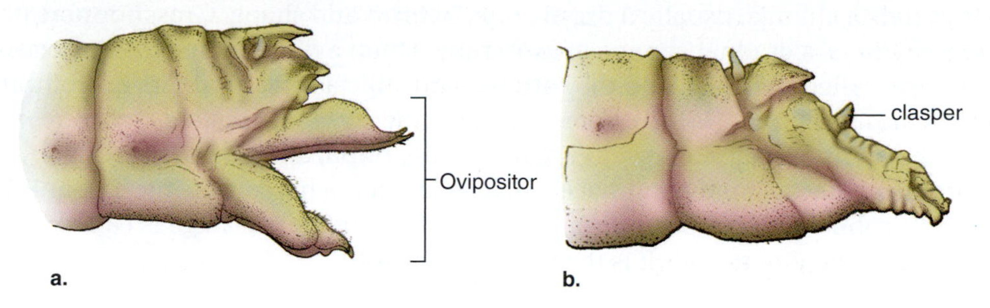

Internal Anatomy

Observe a longitudinal section of a grasshopper if available on demonstration. Try to locate the structures shown in Figure 18.12. What is the function of Malpighian tubules? ______________________________

Figure 18.12 Internal anatomy of a female grasshopper.

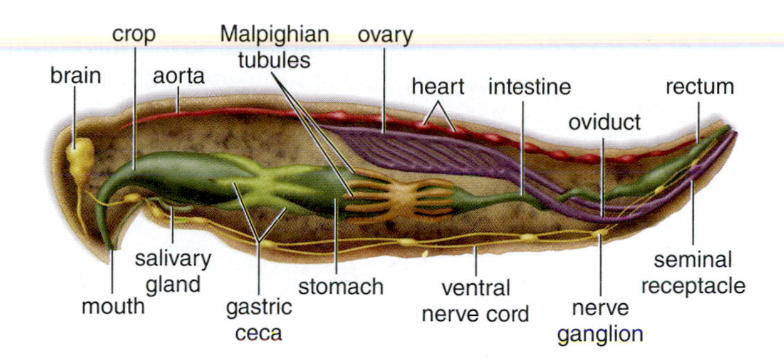

Comparison of Crayfish to Grasshopper

Compare the adaptations of a crayfish to those of a grasshopper by completing Table 18.2. Put a star beside each item that indicates an adaptation to life in the water (crayfish) and to life on land (grasshopper). Check with your instructor to see if you identified the maximum number of adaptations.

Table 18.2 Comparison of Crayfish to Grasshopper

	Crayfish	Grasshopper
Locomotion		
Respiration		
Sense organs		
Nervous system		
External reproductive features Male Female		

Grasshopper Metamorphosis

Metamorphosis means a change, usually a drastic one, in form and shape. Grasshoppers undergo *incomplete metamorphosis,* a gradual change in form rather than a drastic change. The immature stages of the grasshopper are called **nymphs,** and they are recognizable as grasshoppers even though they differ somewhat in shape and form (Fig. 18.13*a*). Some insects undergo what is called *complete metamorphosis,* in which case they have three stages of development: **larvae, pupa,** and **adult** (Fig. 18.13*b*). Metamorphosis occurs during the pupa stage when the animal is enclosed within a hard covering. The animals best known for complete metamorphosis are the butterfly and the moth, whose larval stage is called a caterpillar and whose pupa stage is the cocoon; the adult is the butterfly or moth.

Observation: Grasshopper Metamorphosis

1. Use Figure 18.13 to add the grasshopper and the moth to Table 18.3.
2. Examine any specific life cycle displays or plastomounts that illustrate complete and incomplete metamorphosis and add these examples to Table 18.3.

Figure 18.13 Metamorphosis.
a. During incomplete metamorphosis of a grasshopper, a series of **nymphs** leads to a full-grown grasshopper. **b.** During complete metamorphosis of a moth, a series of larvae lead to pupation. The **adult** hatches out of the pupa.

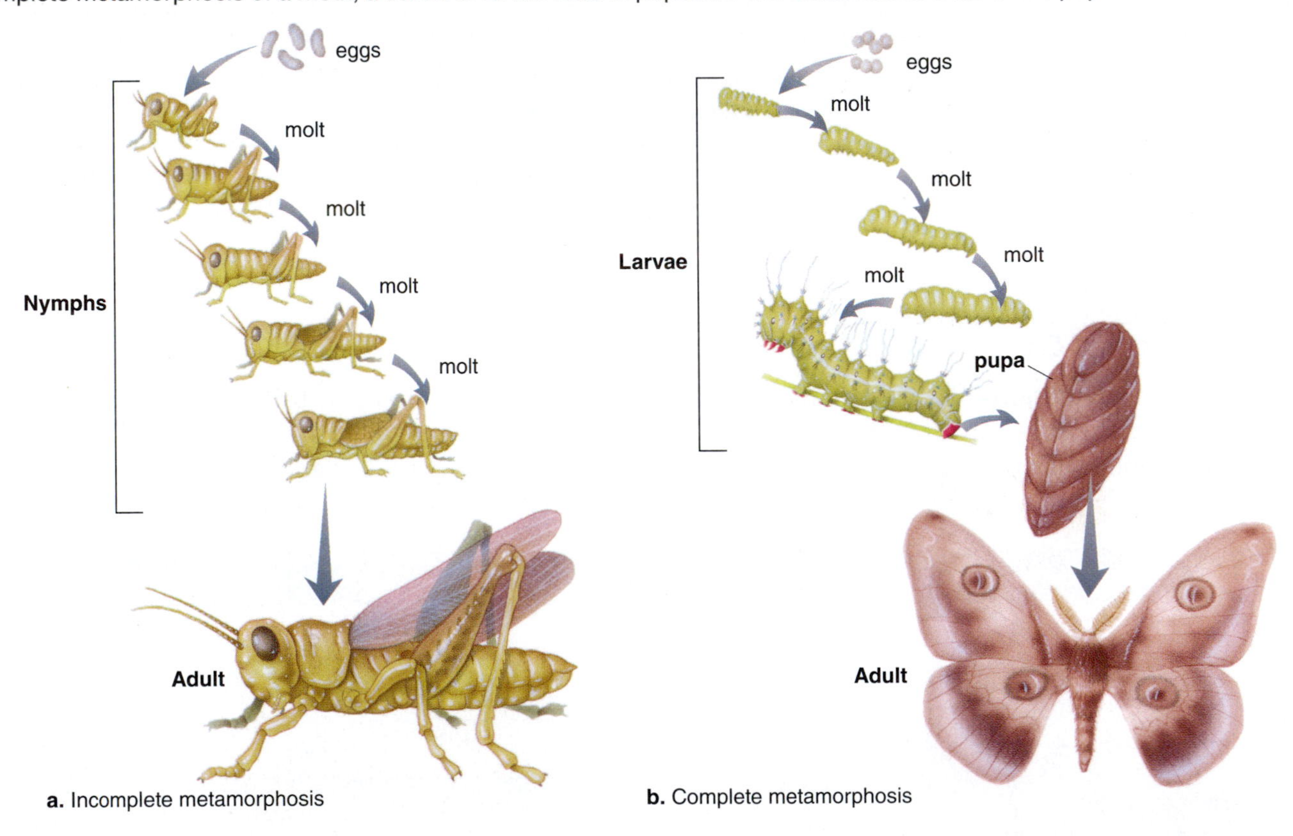

Table 18.3 Insect Metamorphosis	
Specimen	**Complete or Incomplete Metamorphosis**

18.4 Echinoderms

The echinoderms (phylum Echinodermata) are the only invertebrate group that shares deuterostome development with the vertebrates (see Fig. 17.1). Unlike the vertebrates, all **echinoderms** are marine, and they dwell on the seabed, either attached to it, like sea lilies, or creeping slowly over it. The name *echinoderm* means "spiny-skinned," and most members of the group have defensive spines on the outside of their bodies. The spines arise from an **endoskeleton** composed of calcium carbonate plates. The endoskeleton supports the body wall and is covered by living tissue that may be soft (as in sea cucumbers) or hard (as in sea urchins).

Especially note that (1) adult echinoderms are radially symmetrical; with generally five points of symmetry arranged around the axis of the mouth, and (2) the echinoderms' most unique feature is their **water vascular system.** In those echinoderms in which the arms make contact with the substratum, the **tube feet** associated with the water vascular system are used for locomotion. In other echinoderms, the tube feet are used for gas exchange and food gathering.

Echinoderms belong to one of five groups: sea lilies and feather stars; sea stars; brittle stars; sea urchins and sand dollars; and sea cucumbers (Fig. 18.14). *Where appropriate in Figure 18.14, write "ORS" for obvious radial symmetry or "RSNO" for radial symmetry not obvious on the lines provided.*

Figure 18.14 Echinoderm diversity.

a. Bennett's feather star, *Oxycomanthus bennetti* ____________

b. Sea star, *Pentaceraster cumingi* ____________

c. Brittle star, *Ophiopholis aculeata* ____________

d. Sea urchin, *Stronglocentrotus pranciscanus* ______

e. Sand dollar, *Dendraster excentricus* ____________

f. Sea cucumber, *Pseudocolochirus* sp. ____________

Anatomy of a Sea Star

Sea stars (starfish) usually have five arms that radiate from a central disk. The mouth is normally oriented downward, and when sea stars feed on clams, they use the suction of their tube feet to force the shells open a crack. Then they evert a portion of the stomach, which releases digestive juices into the mantle cavity. Partially digested tissues are taken up, and digestion continues in the stomach and in the digestive glands found in the arms.

Observation: Anatomy of a Sea Star

External Anatomy

1. Place a preserved sea star in a dissecting pan so that the aboral side is uppermost.
2. With the help of Figure 18.15, identify the **central disk** and five arms. What type of symmetry does a sea star have? ________________
3. With the oral side uppermost, find the mouth, located in the center and protected by spines. Why is this side of the sea star called the oral side? ________________________________

Figure 18.15 Anatomy of a sea star.
a. Diagram, and **b.** image of dissected sea star. Both show the aboral side. **c.** Image of cut arm. **d.** Canals and tube feet of water vascular system. Both seen from the aboral side.

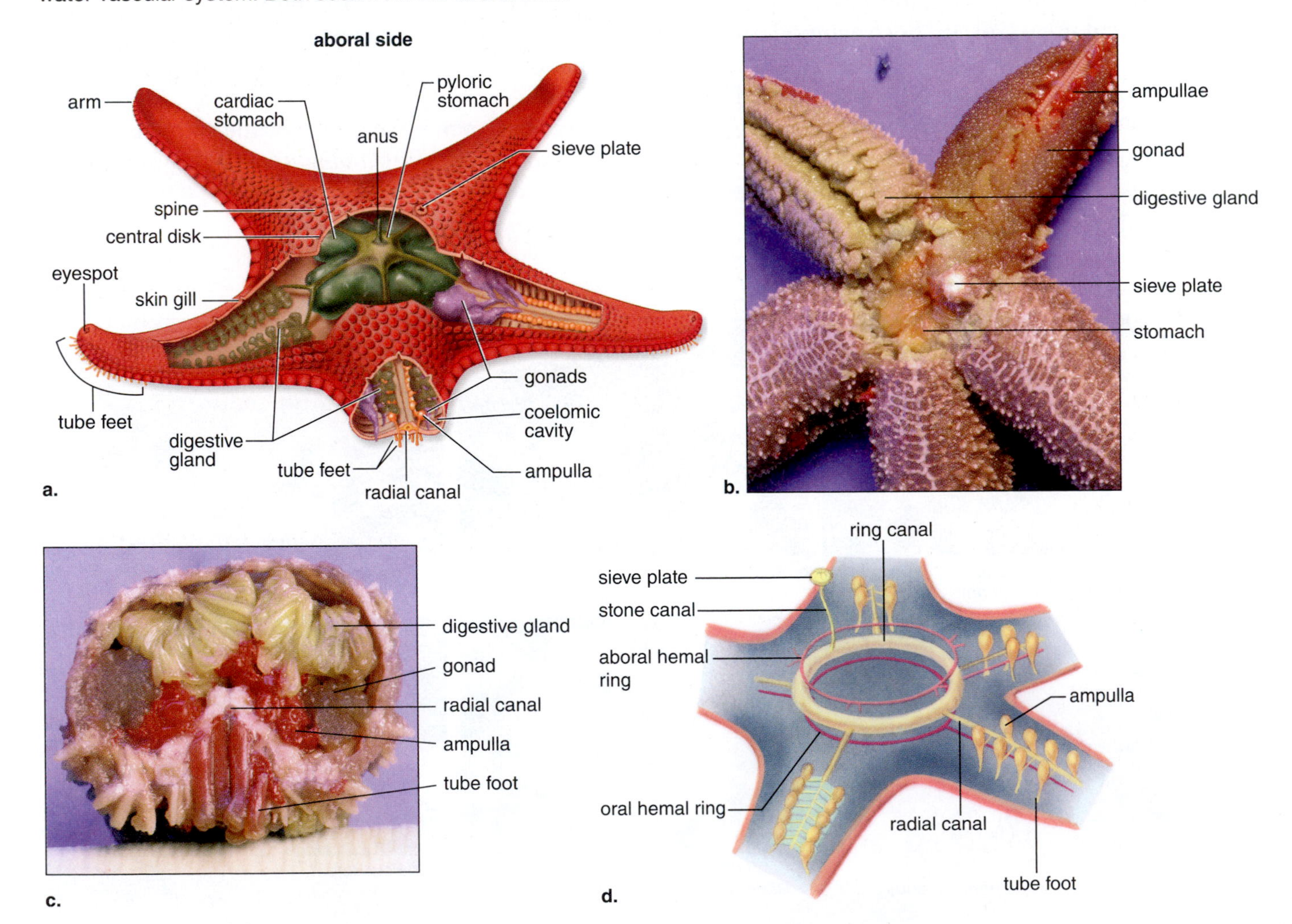

4. Locate the groove that runs along the middle of each arm and the **tube feet** (suctionlike disks) in rows on either side of the groove. Pluck away the tube feet from one area.

   How many rows of feet are there? __________

5. Turn the sea star over to its aboral side. Locate the anal opening (Fig. 18.15). Why is this side of the sea star called the aboral side? ____

   ____

6. Lightly run your fingers over the spines extending from calcium carbonate plates that lie buried in the body wall beneath the surface. The plates form an endoskeleton of the animal.
7. Identify the **sieve plate (madreporite),** a brownish, circular spot between two arms where water enters the water vascular system.

Internal Anatomy

Refer to Figure 18.14*a* and *b* as you dissect the sea star following these instructions:

1. Place the sea star so that the aboral side is uppermost. Cut the tip of one of the arms and, with scissors, carefully cut through the body wall along each side of this arm.
2. Carefully lift up the upper body wall. Separate any internal organs that may be adhering so that all internal organs are left intact.
3. Cut off the body wall near the central disk, but leave the sieve plate (madreporite) in place. Remove the body wall of the central disk, being careful not to injure the internal organs.
4. Identify the digestive system. The mouth leads into a short **esophagus,** which is connected to the saclike **cardiac stomach.** When a sea star eats, the cardiac stomach sticks out through the sea star's mouth and starts digesting the contents of a clam or oyster. Above the cardiac stomach is the **pyloric stomach,** where digestion continues. The pyloric stomach leads to a short intestine. Each arm contains one pair of **digestive glands.** To which stomach do the digestive glands attach? __________
5. Cut off a portion of an arm, and examine the cut edge (Fig. 18.14*c*). Identify the digestive glands and remove them.
6. Identify the **gonads** extending into the arm. What is the function of gonads? __________________

   ____

   It is not possible to distinguish male sea stars from females by this observation.
7. Remove both stomachs.

In the **water vascular system** (Fig. 18.15*d*), you have already located the sieve plate and tube feet. Now identify:

1. **Stone canal:** Takes water from the sieve plate to the ring canal.
2. **Ring canal:** Surrounds the mouth and takes water to the radial canals.
3. **Radial canals:** Send water into the ampullae, muscular sac attached to the tube feet. When the ampullae contract, water enters the tube feet.

What is the function of the water vascular system? ______________________________

____

18.5 Chordates

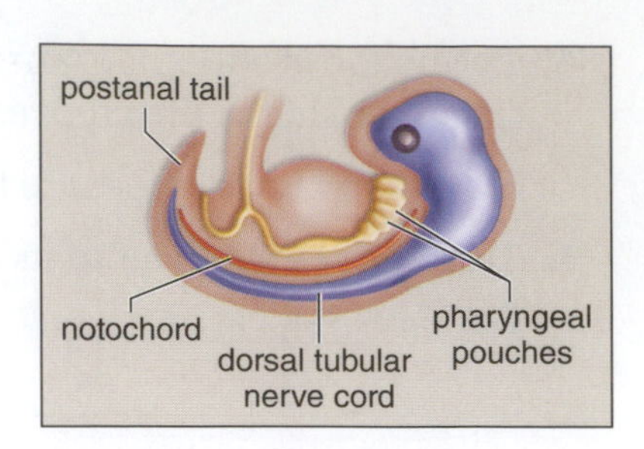

All **chordates** have (1) a dorsal tubular nerve cord; (2) a dorsal supporting rod, called a notochord, at some time in their life history; (3) a postanal tail (e.g., tailbone or coccyx); and, in chordates that breathe by means of gills, (4) pharyngeal pouches that become gill slits. In terrestrial chordates, these pouches are modified for other purposes.

Evolutionary Tree

The evolutionary tree of chordates (Fig. 18.16) shows that some chordates, notably the **tunicates** and **lancelets,** are not vertebrates. These chordates retain a supporting rod called the **notochord** and are called the **invertebrate chordates.** Explain the term invertebrate chordates. ______________________________

The other animal groups in Figure 18.17 are **vertebrates** in which the notochord has been replaced by the vertebral column. Vertebrates are segmented as evidenced by their vertebral column. Fishes include three groups: The **jawless fishes** were the first to evolve, followed by the **cartilaginous fishes** and then the **bony fishes.** The bony fishes include the **ray-finned fishes** (the largest group of vertebrates) and the **lobe-finned fishes.** The first lobe-finned fishes had a bony skeleton, fleshy appendages, and a lung.

Figure 18.16 Evolutionary tree of the chordates.
Evolution of chordates is marked by these seven derived innovations.

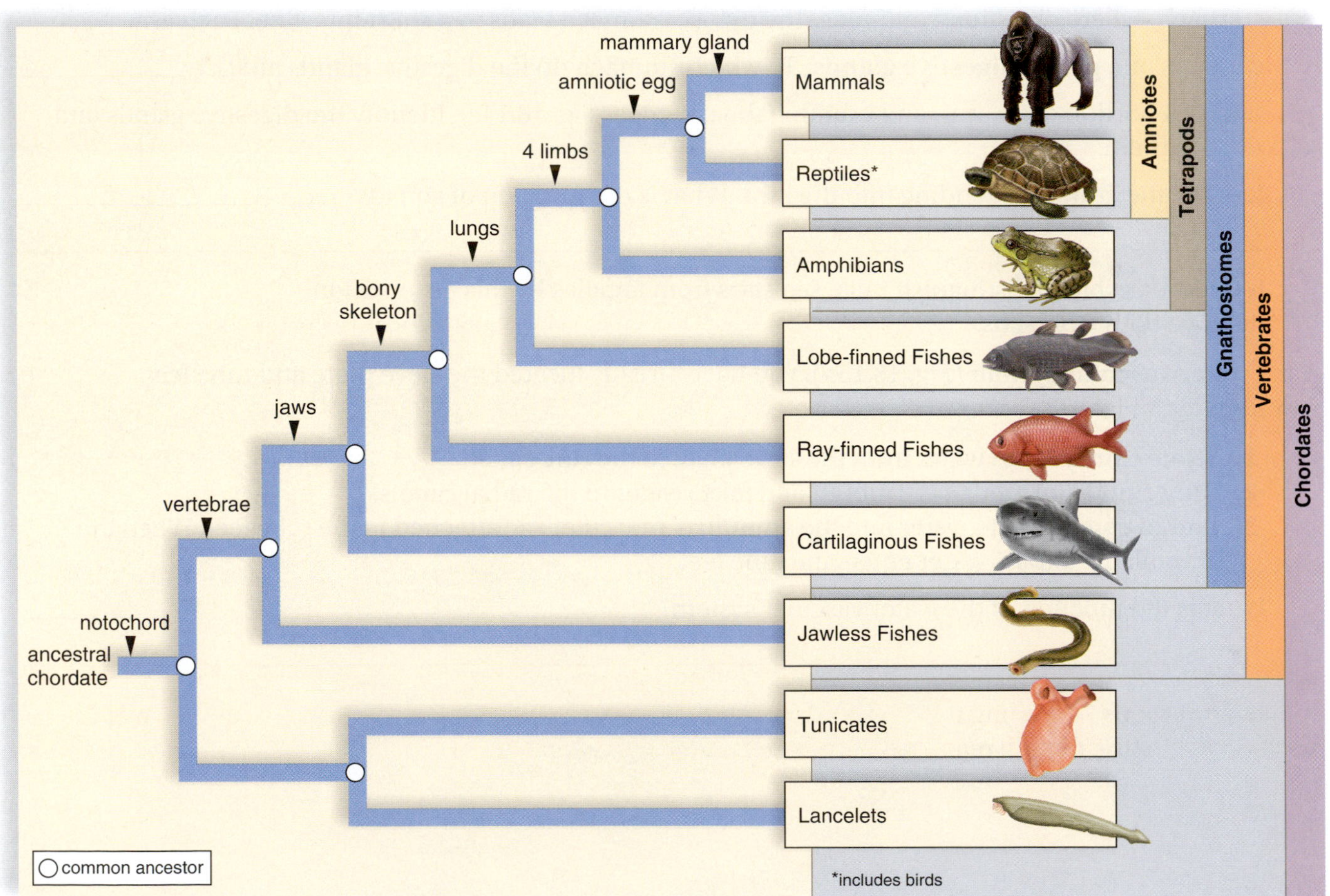

Figure 18.17 Vertebrate groups.

Blue shark

Cartilaginous fishes
Lack operculum and swim bladder; tail fin usually asymmetrical (sharks, skates, and rays)

Blueback butterflyfish

Bony fishes
Operculum; swim bladder or lungs; tail fin usually symmetrical: lung-fishes, lobe-finned fishes, and ray-finned fishes (herring, salmon, sturgeon, eels, and sea horse)

Northern leopard frog

Amphibians
Tetrapods with nonamniotic egg; nonscaly skin; some show metamorphosis; three-chambered heart (salamanders, frogs, and toads)

Pearl River redbelly turtle

Reptiles
Tetrapods with amniotic egg; scaly skin (snakes, lizards, turtles, and tortoises)

Scissor-tailed flycatcher

Birds
Now grouped with reptiles; tetrapods with feathers; bipedal with wings; double circulation (sparrows, penguins, and ostriches)

Gray fox

Mammals
Tetrapods with hair, mammary glands; double circulation; teeth differentiated: monotremes (spiny anteater and duckbill platypus), marsupials (opossum and kangaroo), and placental mammals (whales, rodents, dogs, cats, elephants, horses, bats, and humans)

These lobe-finned fishes lived in shallow pools and gave rise to the amphibians. What three innovations called out in Figure 18.16 evolved among fishes? ____

____

The terrestrial vertebrates are all **tetrapods** because they have four limbs. The limbs of tetrapods are ________________ appendages just like those of arthropods. Amphibians still return to the water to reproduce, but **reptiles** are fully adapted to life on land because among other features they produce an **amniotic egg.** The amniotic egg is so named because the embryo is surrounded by an amniotic membrane that encloses amniotic fluid. Therefore, amniotes develop in an aquatic environment of their own making. Do all animals develop in a water environment? __________ Explain your answer. ____________________

____

In most **mammals**, including humans, the fertilized egg develops inside the female, where the unborn receives nutrients from the maternal bloodstream at the placenta.

Laboratory Review 18

1. Name two types of organisms (not dissected) that belong to each of these phyla:

a. Mollusca ______________________

b. Annelida ______________________

c. Arthropoda ______________________

d. Echinodermata ______________________

e. Chordata ______________________

2. Name the type of foot and location of the foot in a clam and squid. ______________________

3. Associate the type of foot in a clam and squid with their way of life. ______________________

4. Both externally and internally, earthworms show evidence of segmentation. How? ______________________

5. Name one obvious adaptation of a crayfish and one adaptation of a grasshopper to their environment.

____

6. How do insects assist the transport of oxygen to flight muscles? ______________________

7. How do you know that the grasshopper adaptation you provided in question 5 is an adaptation to life on land? ______________________

8. What system is unique to echinoderms and what is its function? ______________________

9. What innovation during the course of evolution can you properly associate with these classes of vertebrates?

a. Ray-finned fishes ______________________

b. Amphibians ______________________

c. Reptiles ______________________

10. What type of skeleton do both arthropods and vertebrates share? ______________________

LABORATORY

19

Vertebrate Body Tissues

Learning Outcomes

19.1 Epithelial Tissue
- Identify slides and models or diagrams of various types of epithelium. 248–50
- Tell where a particular type of epithelium is located in the vertebrate body, and state a function. 248–50

19.2 Connective Tissue
- Identify slides and models or diagrams of various types of connective tissue. 251–54
- Tell where a particular connective tissue is located in the vertebrate body, and state a function. 251–54

19.3 Muscular Tissue
- Identify slides and models or diagrams of three types of muscular tissue. 255–56
- Tell where each type of muscular tissue is located in the vertebrate body, and state a function. 255–56

19.4 Nervous Tissue
- Identify a slide and model or diagram of a neuron. 257
- Tell where nervous tissue is located in the vertebrate body, and state a function. 257

19.5 Organ Level of Organization
- Identify a slide of the intestinal wall and the tissue layers in the wall, and state a function for each tissue. 258
- Identify a slide of skin and the two regions of skin, and state a function for each region of skin. 259

Introduction

All organisms are composed of cells. Groups of cells that have the same structural characteristics and perform the same functions are called tissues. Figure 19.1 shows the four categories of tissues in the vertebrate body. An organ is composed of different types of tissues, and various organs form organ systems.

The micrographs of tissues in this laboratory were obtained by viewing prepared slides with a light microscope. Preparation required the following sequential steps:

1. **Fixation:** The tissue is immersed in a preservative solution to maintain the tissue's existing structure.
2. **Embedding:** Water is removed with alcohol, and the tissue is impregnated with paraffin wax.
3. **Sectioning:** The tissue is cut into extremely thin slices by an instrument called a microtome. Longitudinal sections (l.s.) run the length of the tissue; cross sections (c.s.) run across the tissue.
4. **Staining:** The tissue is immersed in dyes that stain different structures; therefore, when viewing a slide, you are not observing the true color.

Figure 19.1 The major tissues in the human body.
The many kinds of tissues in the human body are grouped into four types: epithelial tissue, muscular tissue, nervous tissue, and connective tissue.

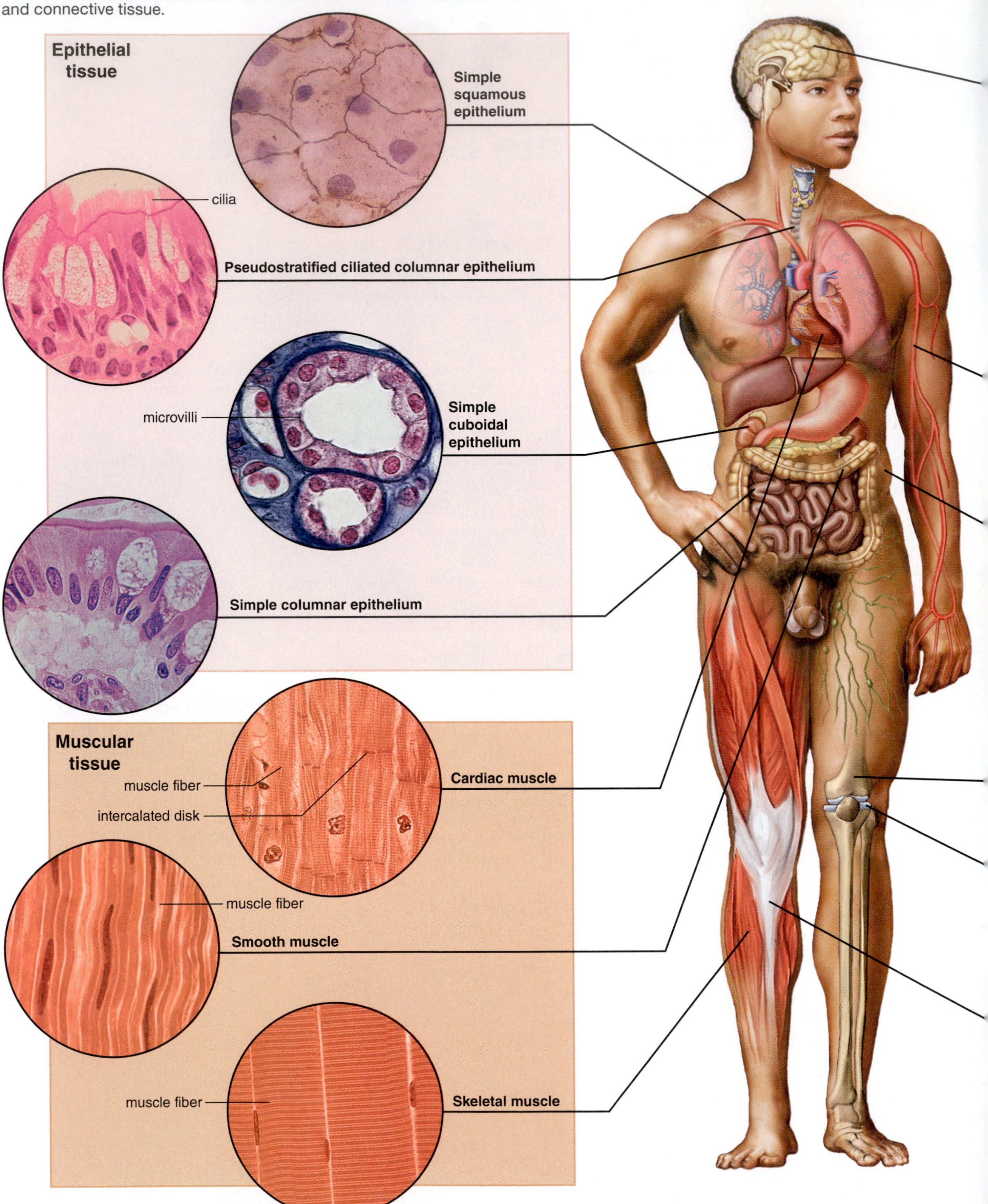

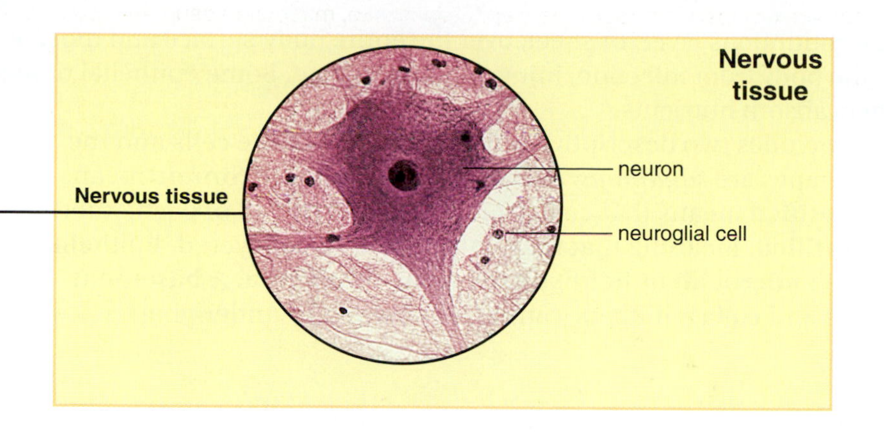
Nervous tissue
Nervous tissue
neuron
neuroglial cell

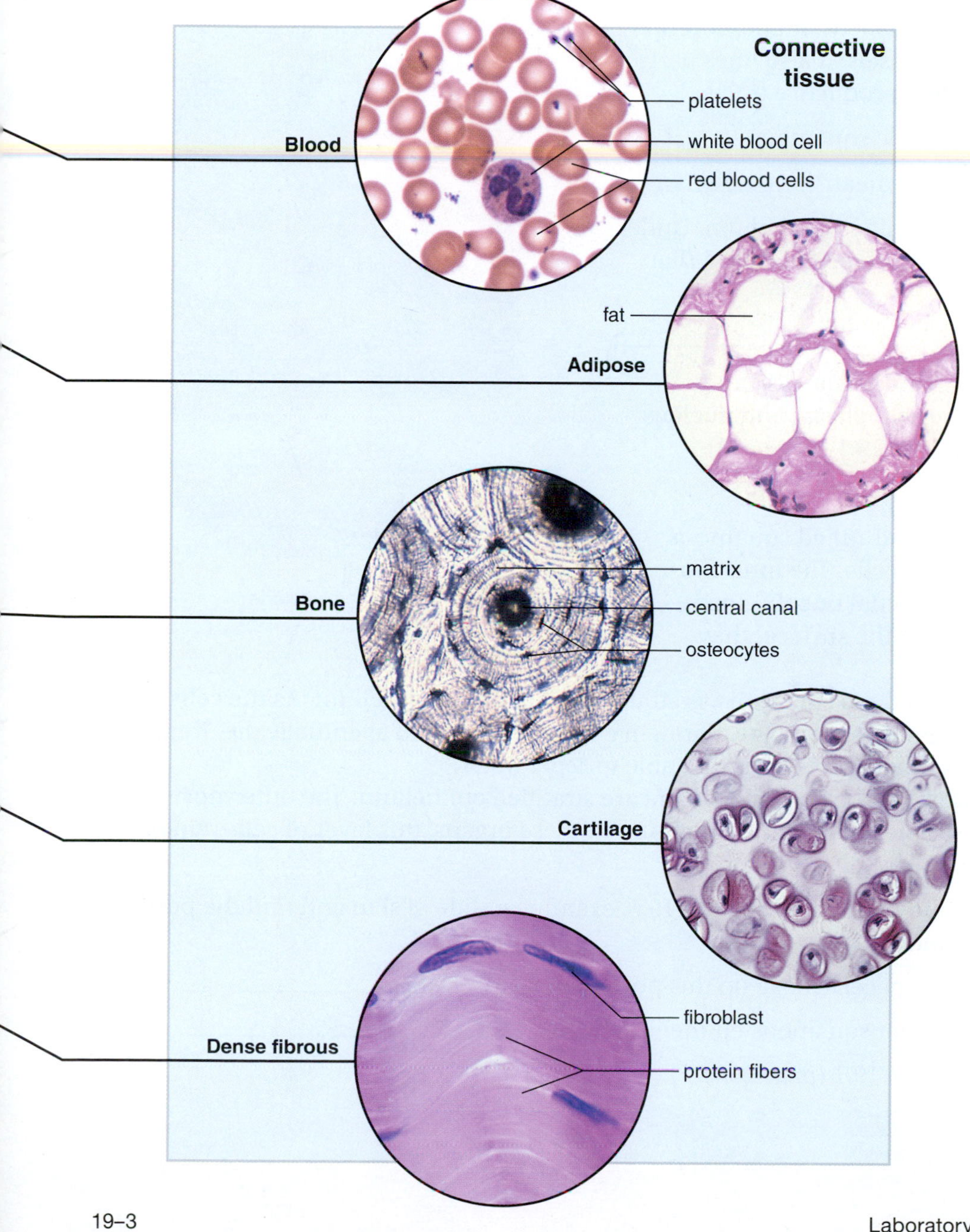
Connective tissue
Blood
platelets
white blood cell
red blood cells
fat
Adipose
Bone
matrix
central canal
osteocytes
Cartilage
Dense fibrous
fibroblast
protein fibers

19.1 Epithelial Tissue

Epithelial tissue (epithelium) forms a continuous layer, or sheet, over the entire body surface and most of the body's inner cavities. It protects the body from infection, injury, and drying out. Some epithelial tissues produce and release secretions. Others absorb nutrients.

The name of an epithelial tissue includes two descriptive terms: the shape of the cells and the number of layers. The three possible shapes are *squamous, cuboidal,* and *columnar.* **Simple** means that there is only one layer of cells; **stratified** means that cell layers are placed on top of each other. Some epithelial tissues are **pseudostratified,** meaning that they only appear to be layered. Epithelium may also have cellular extensions called **microvilli** or hairlike extensions called **cilia.** A **basement membrane** consisting of glycoproteins and collagen fibers joins an epithelium to underlying connective tissue.

Observation: Simple and Stratified Squamous Epithelium

Simple Squamous Epithelium

Simple squamous epithelium is a single layer of thin, flat, many-sided cells, each with a central nucleus. It lines internal cavities, the heart, and all the blood vessels. It also lines parts of the urinary, respiratory, and male reproductive tracts.

1. Study a model or diagram of simple squamous epithelium. What does squamous mean? ______________
2. Examine a prepared slide of squamous epithelium. Under low power, note the close packing of the flat cells. What shapes are the cells?

   ____
3. Under high power, examine an individual cell, and identify the plasma membrane, cytoplasm, and nucleus.
4. Add a sketch of this tissue to Table 19.1 (page 250).

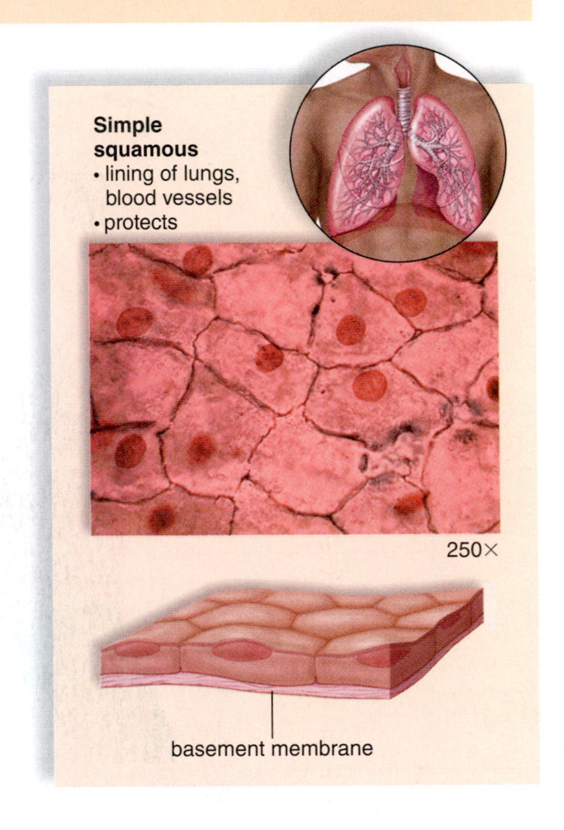

Stratified Squamous Epithelium

As would be expected from its name, stratified squamous epithelium consists of many layers of cells. The innermost layer produces cells that are first cuboidal or columnar in shape, but as the cells push toward the surface, they become flattened.

The outer region of the skin, called the epidermis, is stratified squamous epithelium. As the cells move toward the surface, they flatten, begin to accumulate a protein called **keratin,** and eventually die. Keratin makes the outer layer of epidermis tough, protective, and able to repel water.

The linings of the mouth, throat, anal canal, and vagina are stratified epithelium. The outermost layer of cells surrounding the cavity is simple squamous epithelium. In these organs, this layer of cells remains soft, moist, and alive.

1. Either now or when you are studying skin in Section 19.5, examine a slide of skin and find the portion of the slide that is stratified squamous epithelium.
2. Approximately how many layers of cells make up this portion of skin? ______________
3. Which layers of cells best represent squamous epithelium? ______________
4. Add a sketch of this tissue to Table 19.1 (page 250).

Observation: Simple Cuboidal Epithelium

Simple cuboidal epithelium is a single layer of cube-shaped cells, each with a central nucleus. It is found in tubules of the kidney and in the ducts of many glands, where it has a protective function. It also occurs in the secretory portions of some glands—that is, where the tissue produces and releases secretions.

1. Study a model or diagram of simple cuboidal epithelium.
2. Examine a prepared slide of simple cuboidal epithelium. Move the slide until you locate cube-shaped cells that line a lumen (cavity). Are these cells ciliated?

   ____

3. When cuboidal epithelium has an absorptive function, **microvilli,** tiny projections of the plasma membrane, cause the cells to have a fuzzy appearance called a **brush-border.**
4. Add a sketch of this tissue to Table 19.1 (page 250).

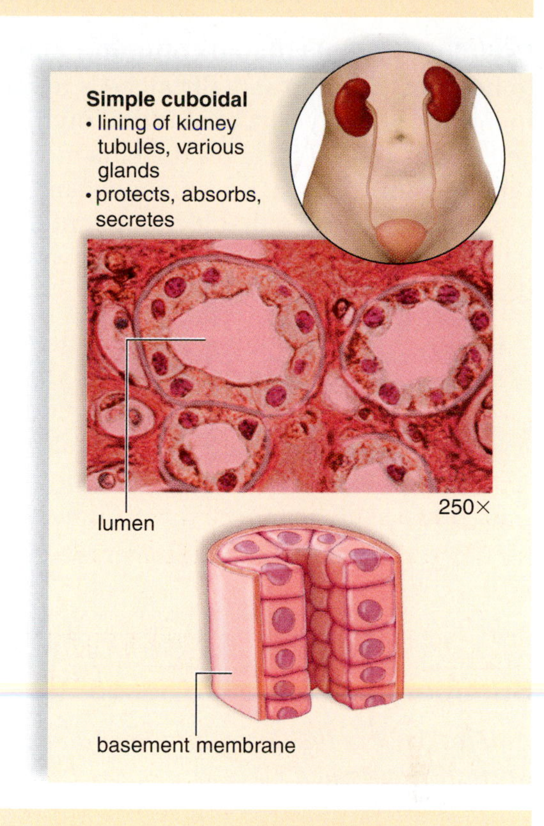

Observation: Simple Columnar Epithelium

Simple columnar epithelium is a single layer of tall, cylindrical cells, each with a nucleus near the base. This tissue, which lines the digestive tract from the stomach to the anus, protects, secretes, and allows absorption of nutrients.

1. Study a model or diagram of simple columnar epithelium.
2. Examine a prepared slide of simple columnar epithelium. If this is epithelium from the small intestine, note the brush border of the cells. Find tall and narrow cells that line a lumen. Under high power, focus on an individual cell. Identify the plasma membrane, the cytoplasm, and the nucleus. Epithelial tissues are attached to underlying tissues by a basement membrane composed of extracellular material containing protein fibers.
3. The tissue you are observing contains mucus-secreting cells. Search among the columnar cells until you find a **goblet cell,** so named because of its goblet-shaped, clear interior. This region contains mucus, which may be stained a light blue. In the living animal, the mucus is discharged into the gut cavity and protects the lining from digestive enzymes.
4. Add a sketch of this tissue to Table 19.1 (page 250).

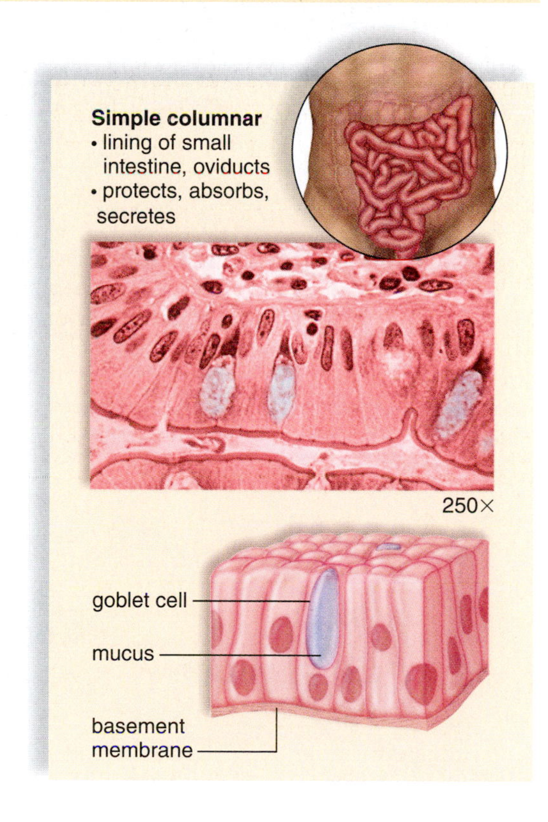

Observation: Pseudostratified Ciliated Columnar Epithelium

Pseudostratified ciliated columnar epithelium appears to be layered, while actually all cells touch the basement membrane. Many cilia are located on the free end of each cell. In males and females, the cilia move sex cells along tubes. In the trachea, the cilia move mucus and debris up toward the throat so that it cannot enter the lungs. Smoking destroys these cilia, but they will grow back if smoking is discontinued.

1. Study a model or diagram of pseudostratified ciliated columnar epithelium.
2. Examine a prepared slide of pseudostratified ciliated columnar epithelium. Concentrate on the part of the slide that resembles the model. Identify the cilia.
3. Add a sketch of this tissue to Table 19.1.

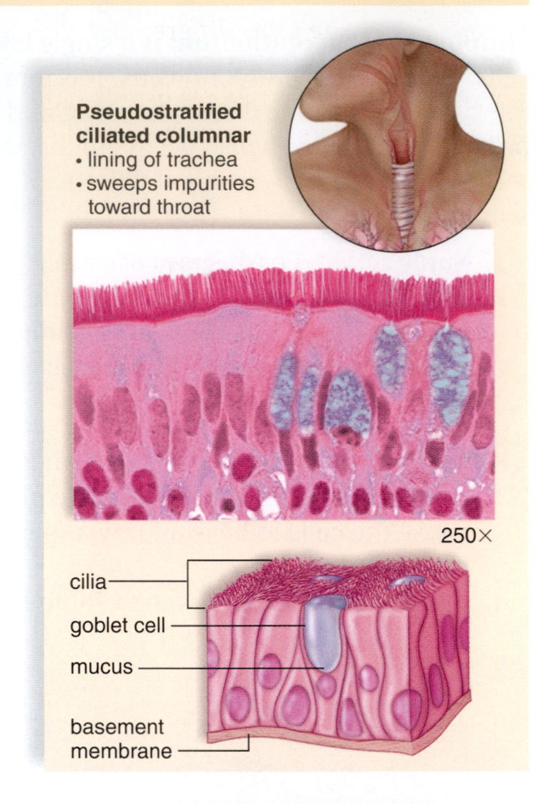

Summary of Epithelial Tissue

Add a sketch in the first column of Table 19.1 under each type of epithelial tissue.

Table 19.1 Epithelial Tissue

Sketch	Structure	Function	Location
Simple squamous	Flat, pancake-shaped	Filtration, diffusion, osmosis	Walls of capillaries, lining of blood vessels, air sacs of lungs, lining of internal cavities
Stratified squamous	Innermost layers are cuboidal or columnar; outermost layers are flattened	Protection, repel water	Skin, linings of mouth, throat, anal canal, vagina
Simple cuboidal	Cube-shaped	Protection, secretion, absorption	Surface of ovaries, linings of ducts and glands, lining of kidney tubules
Simple columnar	Columnlike—tall, cylindrical nucleus at base	Protection, secretion, absorption	Lining of uterus, tubes of digestive tract
Pseudostratified ciliated columnar	Looks layered but is not; ciliated	Protection, secretion, movement of mucus and sex cells	Linings of respiratory passages

19.2 Connective Tissue

Connective tissue joins different parts of the body together. There are four general classes of connective tissue: connective tissue proper, bone, cartilage, and blood. All types of connective tissue consist of cells surrounded by a matrix that usually contains fibers. Elastic fibers are composed of a protein called elastin. Collagenous fibers contain the protein collagen.

Observation: Connective Tissue

There are several different types of connective tissue. We will study loose fibrous connective tissue, dense fibrous connective tissue, adipose tissue, bone, cartilage, and blood. **Loose fibrous connective tissue** (sometimes called areolar tissue) supports epithelium and also many internal organs, such as muscles, blood vessels, and nerves. Its presence allows organs to expand. **Dense fibrous connective tissue** (sometimes called white fibrous tissue) contains many collagenous fibers packed together, as in tendons, which connect muscles to bones, and in ligaments, which connect bones to other bones at joints.

1. Examine a slide of loose fibrous connective tissue, and compare it to the figure below *(left)*. What is the function of loose fibrous connective tissue? ___________________________
2. Examine a slide of dense fibrous connective tissue, and compare it to the figure below *(right)*. What two kinds of structures in the body contain dense fibrous connective tissue? ___________________________

____
3. Add sketches of these tissues to Table 19.2 (page 254).

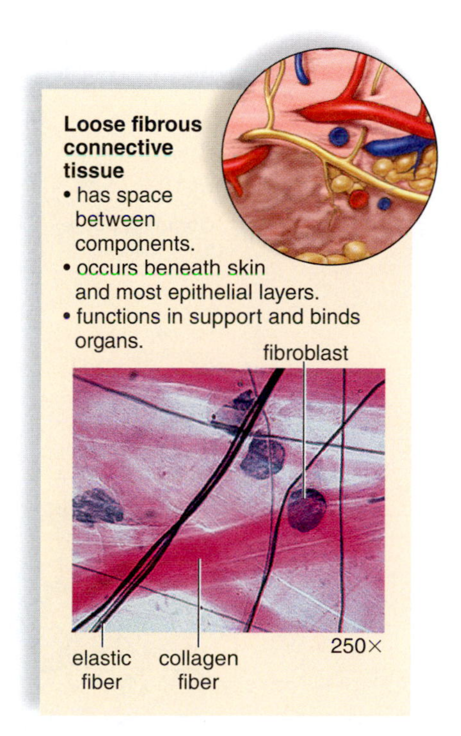

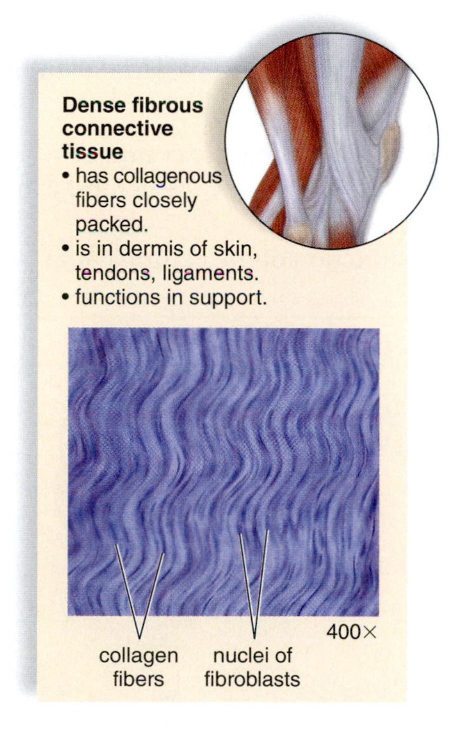

Observation: Adipose Tissue

In **adipose tissue,** the cells have a large, central, fat-filled vacuole that causes the nucleus and cytoplasm to be at the perimeter of the cell. Adipose tissue occurs beneath the skin, where it insulates the body, and around internal organs, such as the kidneys and heart. It cushions and helps protect these organs.

1. Examine a prepared slide of adipose tissue. Why is the nucleus pushed to one side? ______________________________
____
2. State a location for adipose tissue in the body.
____
____
What are two functions of adipose tissue at this location?
____
____
3. Add a sketch of this tissue to Table 19.2 (page 254).

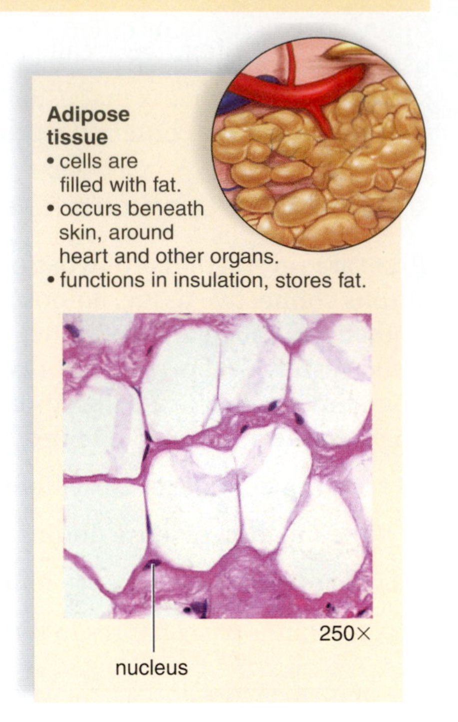

Observation: Compact Bone

Compact bone is found in the bones that make up the skeleton. It consists of **osteons** (Haversian system) with a **central canal,** and concentric rings of spaces called **lacunae,** connected by tiny crevices called **canaliculi.** The central canal contains a nerve and blood vessels, which service bone. The lacunae contain bone cells called **osteocytes,** whose processes extend into the canaliculi. Separating the lacunae is a matrix that is hard because it contains minerals, notably calcium salts. The matrix also contains collagenous fibers.

1. Study a model or diagram of compact bone. Then look at a prepared slide and identify the central canal, lacunae, and canaliculi.
2. What is the function of the central canal and canaliculi?
____
____
3. Add a sketch of this tissue to Table 19.2 (page 254).

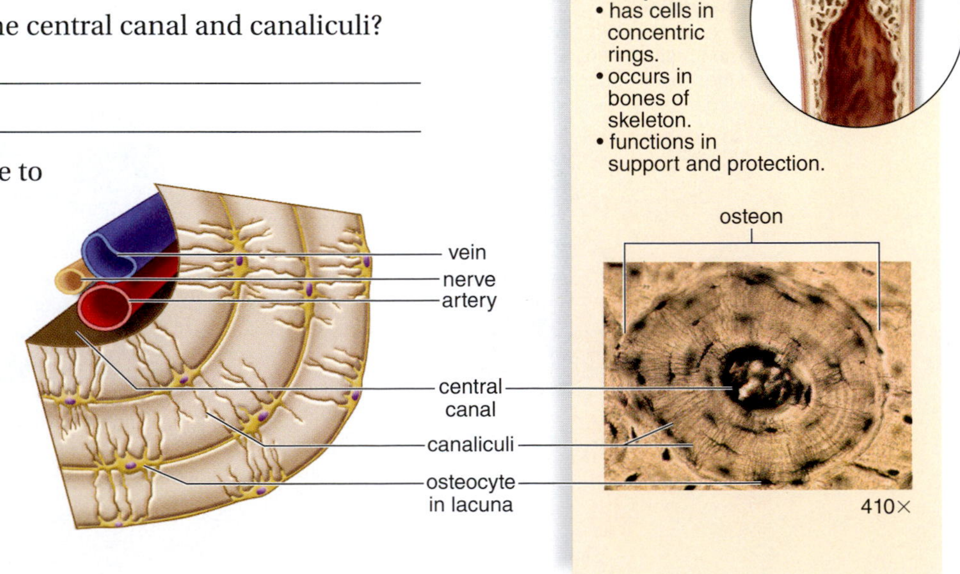

Observation: Hyaline Cartilage

In **hyaline cartilage,** cells called **chondrocytes** are found in twos or threes in lacunae. The lacunae are separated by a flexible matrix containing weak collagenous fibers.

1. Study the diagram and photomicrograph of hyaline cartilage in the figure at the right. Then study a prepared slide of hyaline cartilage, and identify the matrix, lacunae, and chondrocytes.
2. Compare compact bone and hyaline cartilage. Which of these types of connective tissue is more organized? ______________

   Why? ______________________________
3. Which of these two types of connective tissue lends more support to body parts? __________ Why do you say so? ______________

   ____
4. Add a sketch of this tissue to Table 19.2 (page 254).

Hyaline cartilage
- has cells in lacunae.
- occurs in nose and walls of respiratory passages; at ends of bones, including ribs.
- functions in support and protection.

chondrocyte within lacunae

matrix

250×

Observation: Blood

Blood is a connective tissue in which the matrix is an intercellular fluid called **plasma. Red blood cells** (erythrocytes) have a biconcave appearance and lack a nucleus. These cells carry oxygen combined with the respiratory pigment hemoglobin. **White blood cells** (leukocytes) have a nucleus and are typically larger than the more numerous red blood cells. These cells fight infection. Platelets (thrombocytes) are tiny cell fragments which are involved in blood clotting.

1. Study a prepared slide of human blood. With the help of Figure 19.2, identify the red blood cells and the white blood cells, which appear faint because of the stain.
2. Try to identify a neutrophil, which has a multilobed nucleus, and a lymphocyte, which is the smallest of the white blood cells, with a spherical or slightly indented nucleus.
3. Add a sketch of this tissue to Table 19.2 (page 254).

Figure 19.2 Blood cells.
Red blood cells are more numerous than white blood cells. White blood cells can be separated into five distinct types.

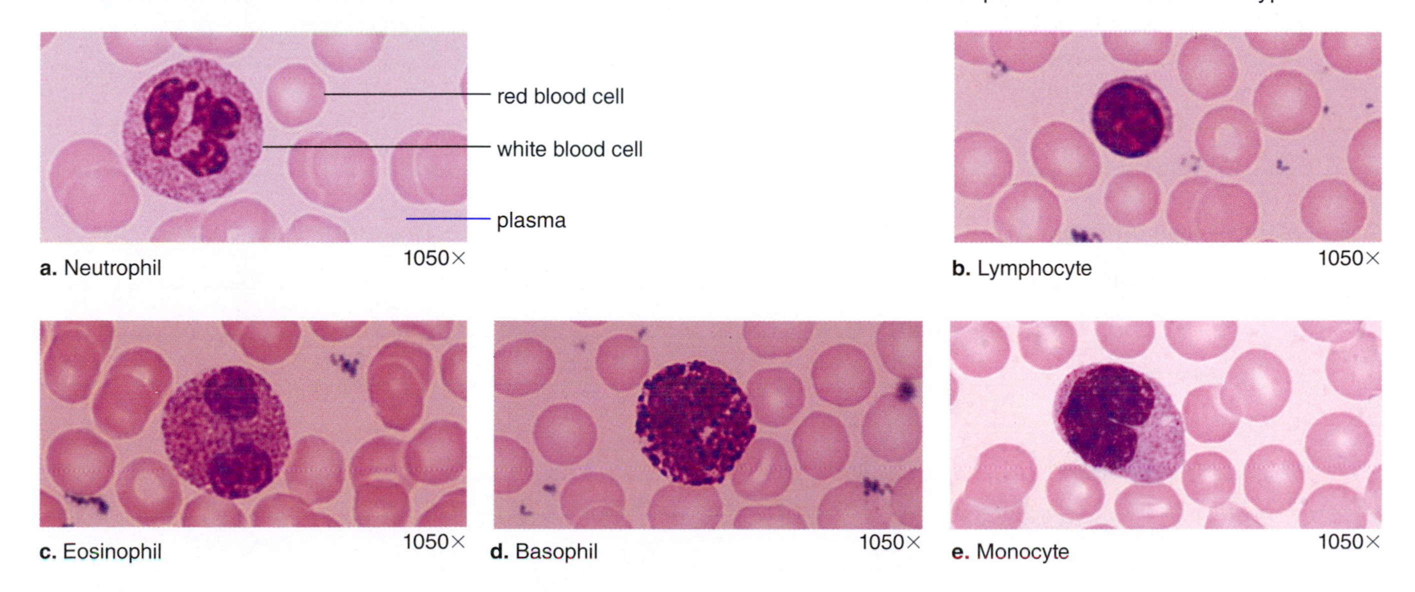

a. Neutrophil **b.** Lymphocyte **c.** Eosinophil **d.** Basophil **e.** Monocyte

Summary of Connective Tissue

1. *Add a sketch in the first column of Table 19.2 under each type of connective tissue.*

Table 19.2 Connective Tissue

Sketch	Structure	Function	Location
Loose fibrous	Fibers are widely separated	Binds organs together	Between the muscles; beneath the skin; beneath most epithelial layers
Dense fibrous	Fibers are closely packed	Binds organs together, binds muscle to bone, binds bone to bone	Tendons, ligaments
Adipose tissue	Large cell with fat-filled vacuole; nucleus pushed to one side	Insulation, fat storage, cushioning, and protection	Beneath the skin; around the kidney and heart; in the breast
Compact bone	Concentric circles	Support, protection	Bones of skeleton
Hyaline cartilage	Cells in lacunae	Support, protection	Nose, ends of bones, rings in walls of respiratory passages; between ribs and sternum
Blood	Red and white cells floating in plasma	RBCs carry oxygen and hemoglobin for respiration; WBCs fight infection	Blood vessels

2. Working with others in a group, decide how the structure of each connective tissue suits its function.

a. Loose fibrous connective tissue ______________________________

b. Dense fibrous connective tissue ______________________________

c. Adipose tissue ______________________________

d. Compact bone ______________________________

e. Hyaline cartilage ______________________________

f. Blood ______________________________

19.3 Muscular Tissue

Muscular (contractile) tissue is composed of cells called muscle fibers. Muscular tissue has the ability to contract, and contraction usually results in movement. The body contains skeletal, cardiac, and smooth muscle.

Observation: Skeletal Muscle

Skeletal muscle occurs in the muscles attached to the bones of the skeleton. The contraction of skeletal muscle is said to be **voluntary** because it is under conscious control. Skeletal muscle is striated; it contains light and dark bands. The striations are caused by the arrangement of contractile filaments (actin and myosin filaments) in muscle fibers. Each fiber contains many nuclei, all peripherally located.

1. Study a model or diagram of skeletal muscle, and note that striations are present. You should see several muscle fibers, each marked with striations.
2. Examine a prepared slide of skeletal muscle. Using high power, locate the striations. Bringing the slide in and out of focus may also help.
3. Add a sketch of this tissue to Table 19.3 (page 256).

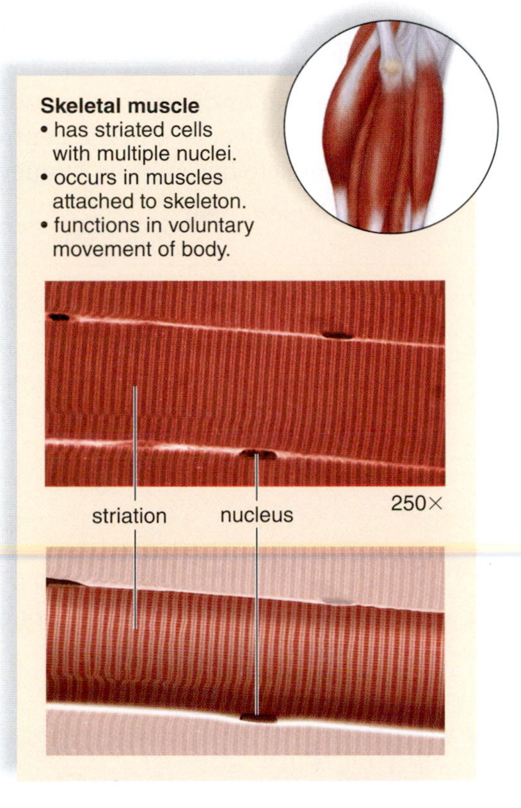

Observation: Cardiac Muscle

Cardiac muscle is found only in the heart. It is called **involuntary** because its contraction does not require conscious effort. Cardiac muscle is striated in the same way as skeletal muscle. However, the fibers are branched and bound together at **intercalated disks,** where their folded plasma membranes touch. This arrangement aids communication between fibers.

1. Study a model or diagram of cardiac muscle, and note that striations are present.
2. Examine a prepared slide of cardiac muscle. Using high power, find an **intercalated disk**. What is the function of cardiac muscle? ______________________________

   ____

   ____

3. Add a sketch of this tissue to Table 19.3 (page 256).

Cardiac muscle
• has branching, striated cells, each with a single nucleus.
• occurs in the wall of the heart.
• functions in the pumping of blood.
• is involuntary.

intercalated disk
nucleus
250×

Observation: Smooth Muscle

Smooth muscle is sometimes called **visceral muscle** because it makes up the walls of the internal organs, such as the intestines and the blood vessels. Smooth muscle is involuntary because its contraction does not require conscious effort.

1. Study a model or diagram of smooth muscle, and note the shape of the cells and the centrally placed nucleus. Smooth muscle has spindle-shaped cells. What does *spindle-shaped* mean? ______________________________
____
____
2. Examine a prepared slide of smooth muscle. Distinguishing the boundaries between the different cells may require you to take the slide in and out of focus.
3. Add a sketch of this tissue to Table 19.3.

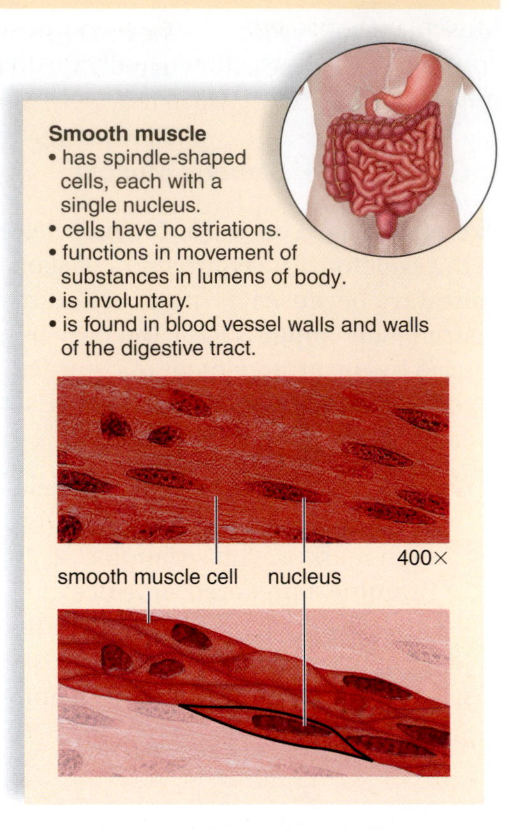

Summary of Muscular Tissue

1. *Add a sketch in the first column of Table 19.3 under each type of muscular tissue.*

Table 19.3 Muscular Tissue

Sketch	Striations (Yes or No)	Branching (Yes or No)	Conscious Control (Yes or No)
Skeletal			
Cardiac			
Smooth			

2. How does it benefit an animal that skeletal muscle is voluntary, while cardiac and smooth muscle are involuntary? ______________________________
____

19.4 Nervous Tissue

Nervous tissue is found in the brain, spinal cord, and nerves. Nervous tissue receives and integrates incoming stimuli before conducting nerve impulses, which control the glands and muscles of the body. Nervous tissue is composed of two types of cells: **neurons** that transmit messages and **neuroglia** that support and nourish the neurons. Motor neurons, which take messages from the spinal cord to the muscles, are often used to exemplify typical neurons. Motor neurons have several **dendrites,** processes that take signals to a **cell body,** where the nucleus is located, and an **axon** that takes nerve impulses away from the cell body.

Observation: Nervous Tissue

1. Study a model or diagram of a neuron, and identify the dendrites, cell body, nucleus, and axon (Fig. 19.3*a*). Long axons are called nerve fibers.
2. *In Figure 19.3*b, *label the dendrites, cell body, nucleus, and axon. Also label neuroglia.*
3. Explain the appearance and function of the parts of a motor neuron:
   a. Dendrites ____
   ____
   b. Cell body ____
   ____
   c. Axon ____
   ____

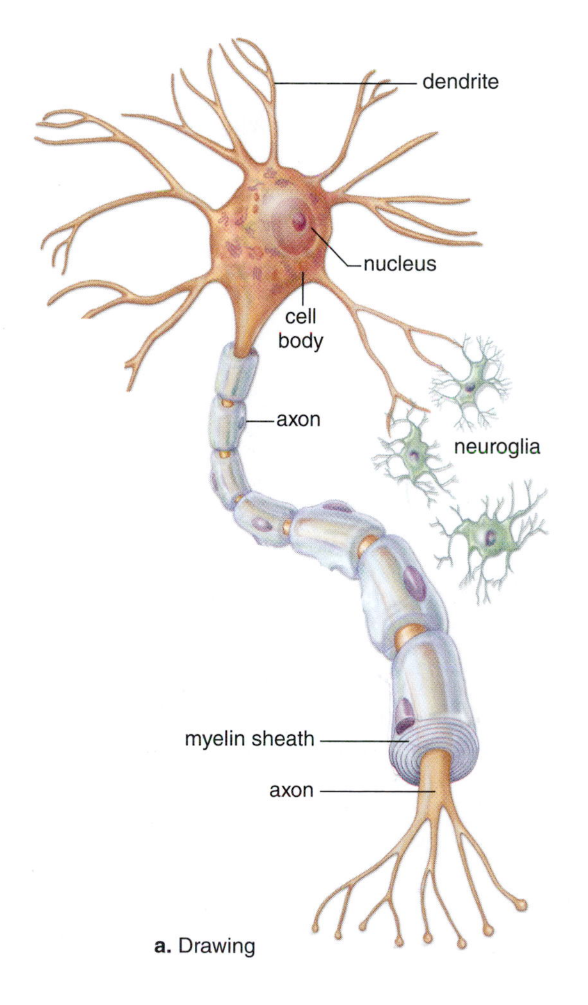

a. Drawing

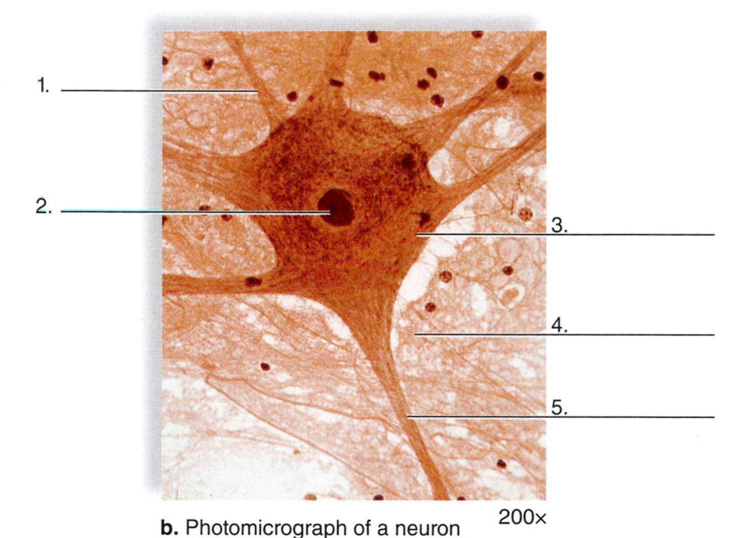

b. Photomicrograph of a neuron

Figure 19.3 Motor neuron anatomy.

19.5 Organ Level of Organization

Organs are structures composed of two or more types of tissue that work together to perform particular functions. You may tend to think that a particular organ contains only one type of tissue. For example, muscular tissue is usually associated with muscles and nervous tissue with the brain. However, muscles and the brain also contain other types of tissue—for example, loose connective tissue and blood. Here we will study the compositions of two organs—the intestine and the skin.

Intestine

The **intestine,** a part of the digestive system, processes food and absorbs nutrient molecules.

Observation: Intestinal Wall

Study a slide of a cross section of intestinal wall. With the help of Figure 19.4, identify the following layers:

1. **Mucosa** (mucous membrane layer): This layer, which lines the central lumen (cavity), is made up of columnar epithelium overlying a layer of connective tissue. The epithelium has a brush border and is glandular—that is, it secretes mucus from goblet cells and digestive enzymes from the rest of the epithelium. The membrane is arranged in deep folds (fingerlike projections) called **villi,** which increase the small intestine's absorptive surface.
2. **Submucosa** (submucosal layer): This connective tissue layer contains nerve fibers, blood vessels, and lymphatic vessels. The products of digestion are absorbed into these blood and lymphatic vessels.
3. **Muscularis** (smooth muscle layer): Circular muscular tissue and then longitudinal muscular tissue are found in this layer. Rhythmic contraction of these muscles causes **peristalsis,** a wavelike motion that moves food along the intestine.
4. **Serosa** (serous membrane layer): In this layer, a thin sheet of connective tissue underlies a thin, outermost sheet of squamous epithelium. This membrane is part of the **peritoneum,** which lines the entire abdominal cavity.

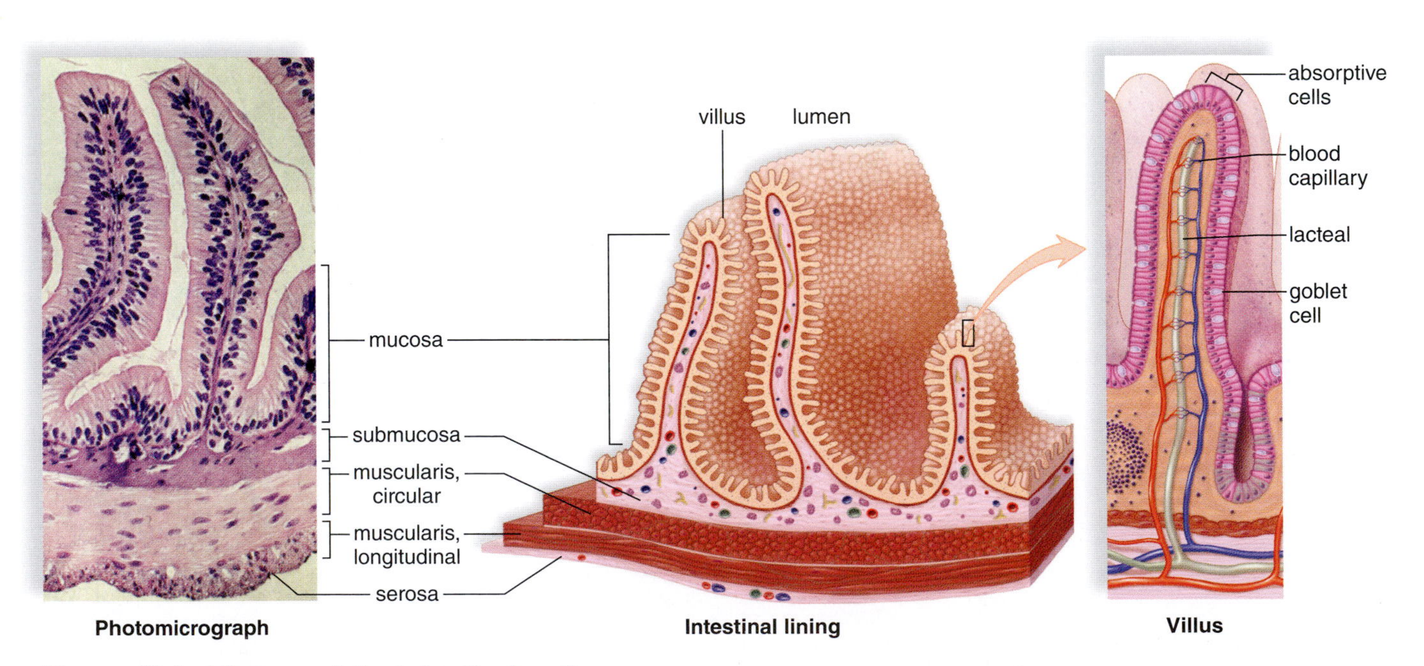

Figure 19.4 Tissues of the intestinal wall.

Skin

The skin covers the entire exterior of the human body. Skin functions include protection, water retention, sensory reception, body temperature regulation, and vitamin D synthesis.

Observation: Skin

Study a model or diagram and also a prepared slide of the skin. With the help of Figure 19.5, identify the two skin regions and the subcutaneous layer.

1. **Epidermis:** This region is composed of stratified squamous epithelial cells. The outer cells of the epidermis are nonliving and create a waterproof covering that prevents excessive water loss. These cells are always being replaced because an inner layer of the epidermis is composed of living cells that constantly produce new cells.
2. **Dermis:** This region is a connective tissue containing blood vessels, nerves, sense organs, and the expanded portions of oil (sebaceous) and sweat glands and hair follicles.

   List the structures you can identify on your slide:

   ____

   ____

3. **Subcutaneous layer:** This is a layer of loose connective tissue and adipose tissue that lies beneath the skin proper and serves to insulate and protect inner body parts. This layer is not part of the skin.

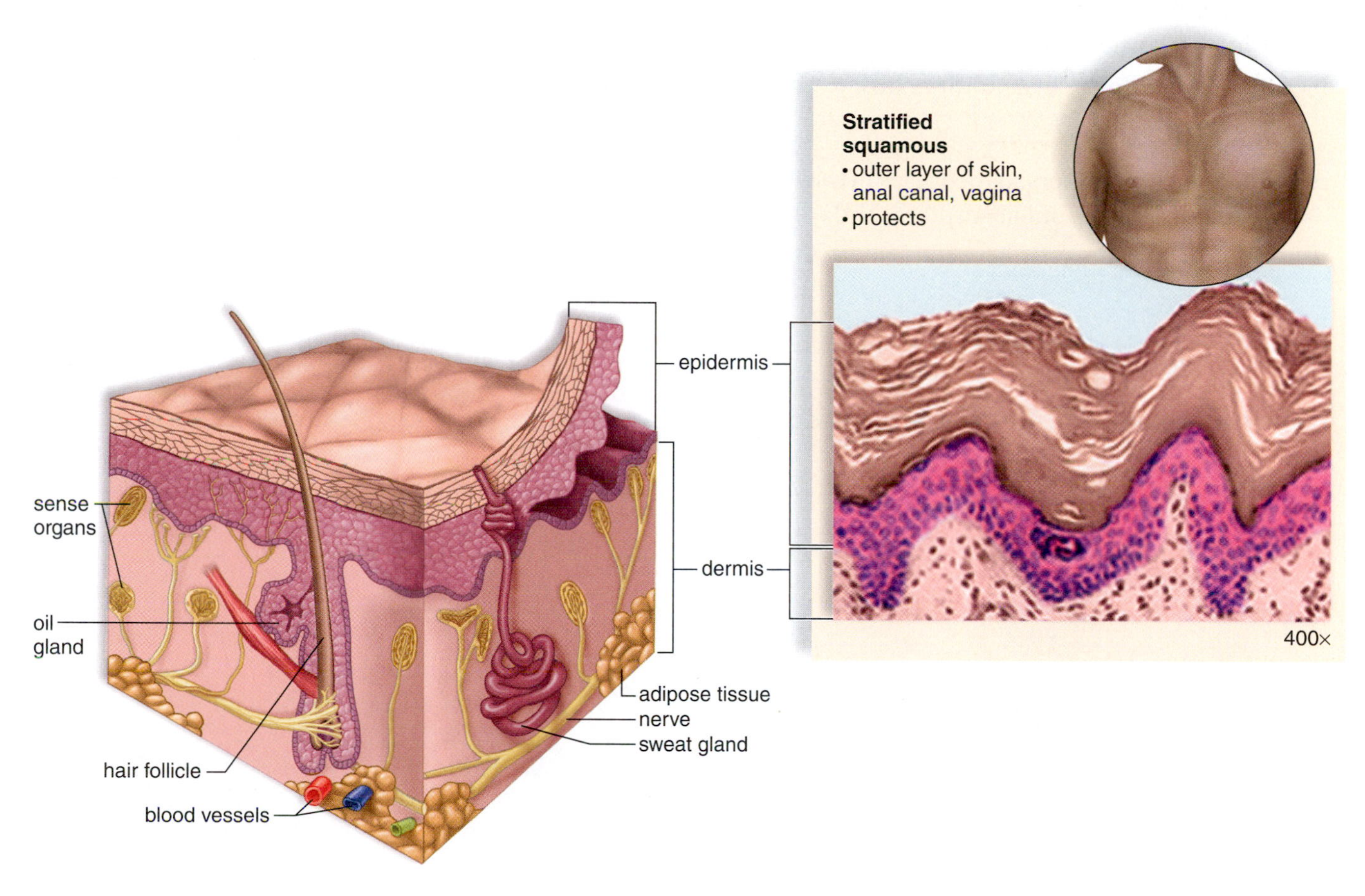

Figure 19.5 Human skin.
Human skin contains two regions, the epidermis and the dermis.

Laboratory Review 19

1. Show that structure suits functions by describing how the structure of squamous epithelium is appropriate to its function. ______

______

2. The cuboidal epithelium lining kidney tubules reabsorbs nutrients into the blood. How does its structure assist this function? ______

______

3. The epithelium lining the trachea sweeps debris up into the throat. What kind of epithelium might this be?

______

4. Defend classifying blood as a connective tissue. ______

______

5. The type of connective tissue that holds bones together has what appearance? ______

______

6. Why is it beneficial for smooth muscle to be involuntary? ______

______

7. The arrangement of contractile filaments in skeletal muscle gives it what kind of appearance? ______

______

8. The very long axon of a motor neuron serves what function? ______

______

9. Match these numbers to the layers of the intestine: (1) permits rhythmical movement, (2) supports blood vessels, (3) carries out absorption, and (4) provides protective covering.

______ ______

______ ______

10. What makes skin impenetrable? ______ How is this an adaptation to a land environment? ______

______

LABORATORY

20

Basic Mammalian Anatomy I

Learning Outcomes

Introduction

In this laboratory, you will dissect a fetal pig. Alternately your instructor may choose to have you observe a pig that has already been dissected. Both pigs and humans are mammals; therefore, you will be studying mammalian anatomy. The period of pregnancy, or gestation, in pigs is approximately 17 weeks (compared with an average of 40 weeks in humans). The piglets used in class will usually be within 1 to 2 weeks of birth.

The pigs may have a slash in the right neck region, indicating the site of blood drainage. A red latex solution may have been injected into the **arterial system,** and a blue latex solution may have been injected into the **venous system** of the pigs. If so, when a vessel appears red, it is an artery, and when a vessel appears blue, it is a vein.

As a result of this laboratory, you should gain an appreciation of which organs work together. For example, the liver and the pancreas aid the digestion of fat in the small intestine.

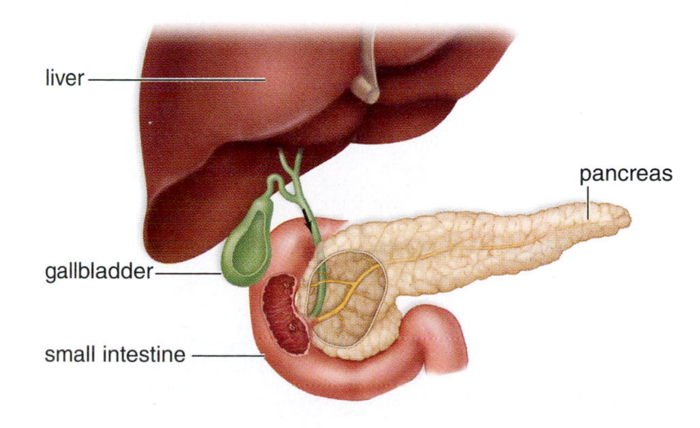

20.1 External Anatomy

Mammals are characterized by the presence of mammary glands and hair. Mammals also occur in two distinct sexes, males and females, often distinguishable by their external **genitals,** the reproductive organs.

Both pigs and humans are placental mammals, which means that development occurs within the uterus of the mother. An **umbilical cord** stretches externally between the fetal animal and the **placenta,** where carbon dioxide and organic wastes are exchanged for oxygen and organic nutrients.

Pigs and humans are tetrapods—that is, they have four limbs. Pigs walk on all four of their limbs; in fact, they walk on their toes, and their toenails have evolved into hooves. In contrast, humans walk only on the feet of their legs.

Observation: External Anatomy

Body Regions and Limbs

> ⚠ **Latex gloves** Wear protective safety goggles, latex gloves, and protective clothing when handling preserved animal organs. Exercise caution when using sharp instruments during this experiment. Wash hands thoroughly upon completion of this experiment.

1. Place a pig in a dissecting pan, and observe the following body regions: the rather large head; the short, thick neck; the cylindrical trunk with two pairs of appendages (forelimbs and hindlimbs); and the short tail (Fig. 20.1*a*). The tail is an extension of the vertebral column.
2. Examine the four limbs, and feel for the joints of the digits, wrist, elbow, shoulder, hip, knee, and ankle.
3. Determine which parts of the forelimb correspond to your arm, elbow, forearm, wrist, and hand.
4. Do the same for the hindlimb, comparing it with your thigh and leg.
5. The pig walks on its toenails, which would be like a ballet dancer on "tiptoe." Notice how your heel touches the ground when you walk. Where is the heel of the pig? ____________________

Umbilical Cord

1. Locate the umbilical cord arising from the ventral (toward the belly) portion of the abdomen.
2. Note the cut ends of the umbilical blood vessels. If they are not easily seen, cut the umbilical cord near the end and observe this new surface.
3. What is the function of the umbilical cord? ____________________
____

Nipples and Hair

1. Locate the small **nipples,** the external openings of the **mammary glands.** The nipples are *not* an indication of sex, since both males and females possess them. How many nipples does a pig have? ______
When is it advantageous for a pig to have so many nipples? ____________________
____
2. Can you find hair on the pig? __________ Where? ____________________
____

Directional Terms for Dissecting Fetal Pig

Anterior: toward the head end
Posterior: toward the hind end
Ventral: toward the belly
Dorsal: toward the back

Figure 20.1 External anatomy of the fetal pig.
a. Body regions and limbs. **b, c.** The sexes can be distinguished by the external genitals.

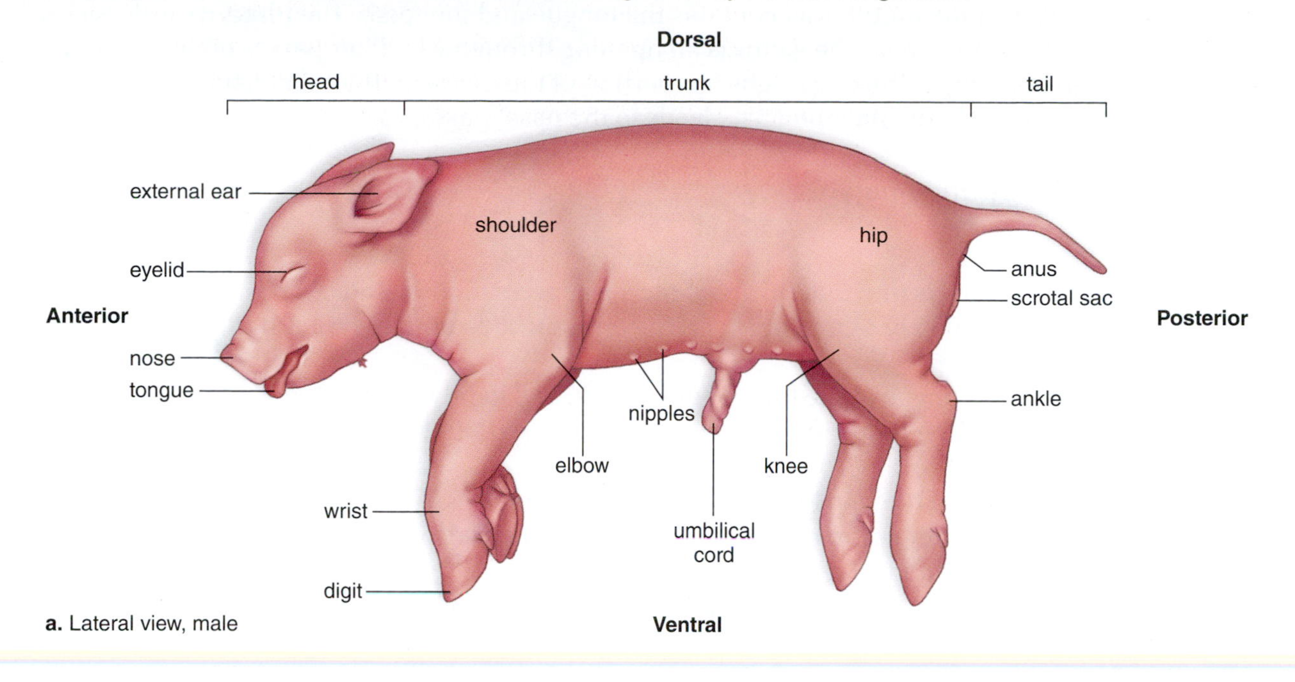

a. Lateral view, male

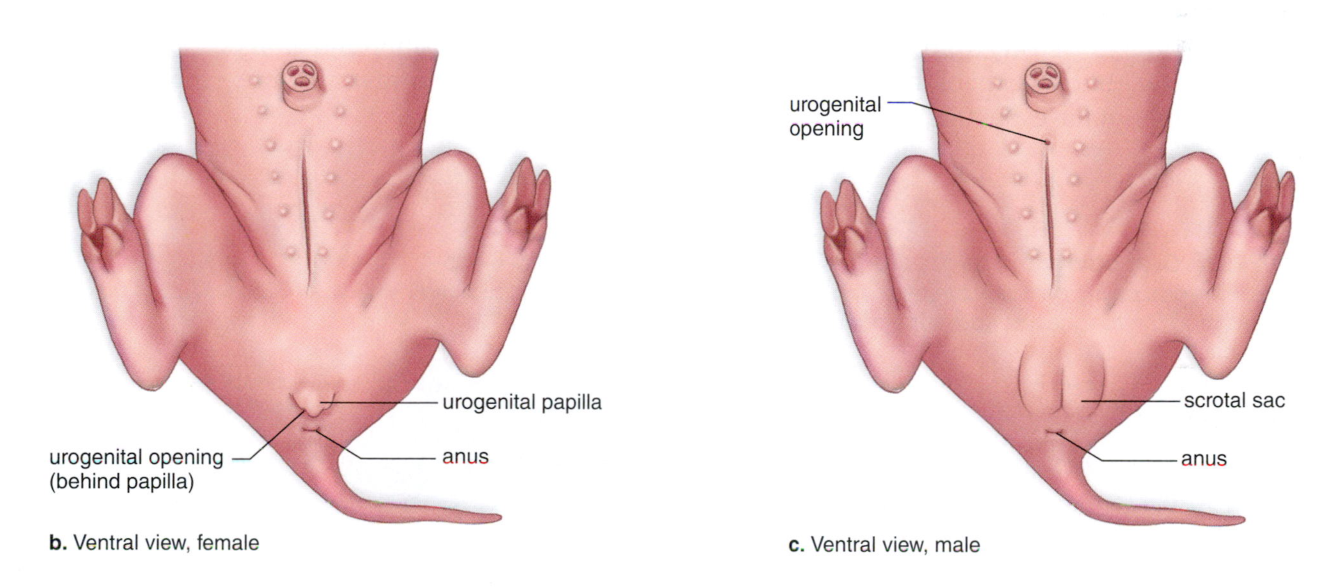

b. Ventral view, female

c. Ventral view, male

Anus and External Genitals

1. Locate the **anus** under the tail. Name the organ system that ends in the opening called the anus. ____________
2. In females, locate the **urogenital opening,** just anterior to the anus, and a small, fleshy **urogenital papilla** projecting from the urogenital opening (Fig. 20.1*b*).
3. In males, locate the urogenital opening just posterior to the umbilical cord. The duct leading to it runs forward from between the legs in a long, thick tube, the **penis,** which can be felt under the skin. In males, the urinary system and the genital system are always joined (Fig. 20.1*c*).
4. You are responsible for identifying pigs of both sexes. What sex is the pig you are examining? ____________

   Be sure to look at a pig of the opposite sex that another group of students is dissecting.

20.2 Oral Cavity and Pharynx

The **oral cavity** is the space in the mouth that contains the tongue and the teeth. The **pharynx** is dorsal to the oral cavity and has three openings: The glottis is an opening through which air passes on its way to the **trachea** (the windpipe) and lungs. The esophagus is a portion of the digestive tract that leads through the neck and thorax to the stomach. The **nasopharynx** leads to the nasal passages.

Observation: Oral Cavity and Pharynx

Oral Cavity

1. Insert a sturdy pair of scissors into one corner of the specimen's mouth, and cut posteriorly (toward the hind end) for approximately 4 cm. Repeat on the opposite side until the mouth is open as in Figure 20.2.
2. Place your thumb on the tongue at the front of the mouth, and gently push downward on the lower jaw. This will tear some of the tissue in the angles of the jaws so that the mouth will remain partly open (Fig. 20.2).
3. Note small, underdeveloped teeth in both the upper and lower jaws. Care should be taken because teeth can be very sharp. Other embryonic, nonerupted teeth may also be found within the gums. The teeth are used to chew food.
4. Examine the tongue, which is partly attached to the lower jaw region but extends posteriorly and is attached to a bony structure at the back of the oral cavity (Fig. 20.2). The tongue manipulates food for swallowing.
5. Locate the hard and soft palates (Fig. 20.2). The **hard palate** is the ridged roof of the mouth that separates the oral cavity from the nasal passages. The **soft palate** is a smooth region posterior to the hard palate. An extension of the soft palate—the **uvula**—hangs down into the throat in humans. (A pig does not have a uvula.)

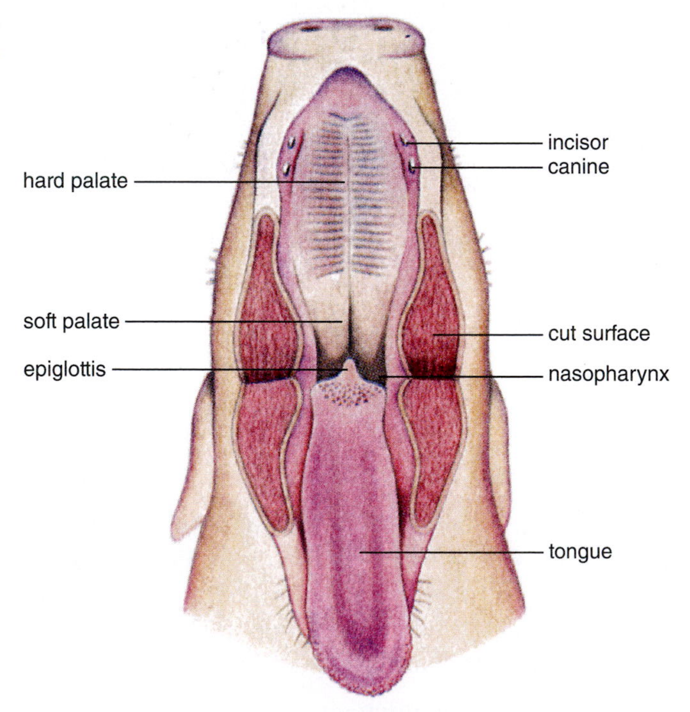

Figure 20.2 Oral cavity of the fetal pig. The roof of the oral cavity contains the hard and soft palates, and the tongue lies above the floor of the oral cavity.

Pharynx

1. Push down on the jaws until they have opened far enough to reveal a slightly pointed flap of tissue that points dorsally (toward the back) (Fig. 20.2). This flap is the **epiglottis,** which covers the glottis. The **glottis** leads to the trachea (Fig. 20.3*a*).
2. Posterior and dorsal to the glottis, find the opening into the **esophagus,** a tube that takes food to the stomach. Note the proximity of the glottis and the opening to the esophagus. Each time the pig—or a human—swallows, the epiglottis instantly closes to keep food and fluids from going into the lungs via the trachea.
3. Insert a blunt probe into the glottis, and note that it enters the trachea. Remove the probe, insert it into the esophagus, and note the position of the esophagus beneath the trachea.
4. Make two lateral cuts at the edge of the hard palate.
5. Posterior to the soft palate, locate the openings to the nasal passages.
6. Explain why it is correct to say that the air and food passages cross in the pharynx.

____

____

____

Figure 20.3 Air and food passages in the fetal pig.
The air and food passages cross in the pharynx. **a.** Drawing. **b.** Dissection of specimen.

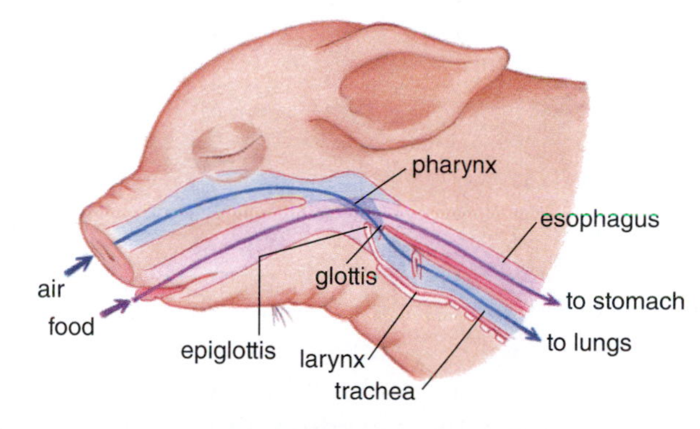

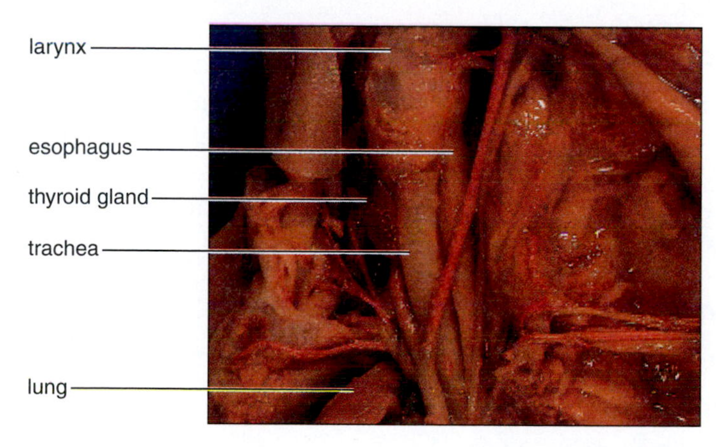

20.3 Thoracic and Abdominal Incisions

First, prepare your pig according to the following directions, and then make thoracic and abdominal incisions so that you will be able to study the internal anatomy of your pig.

Preparation of Pig for Dissection

1. Place the fetal pig on its back in the dissecting pan.
2. Tie a cord around one forelimb, and then bring the cord around underneath the pan to fasten back the other forelimb.
3. Spread the hindlimbs in the same way.
4. With scissors always pointing up (never down), make the following incisions to expose the thoracic and abdominal cavities. The incisions are numbered on Figure 20.4 to correspond with the following steps.

Thoracic Incisions

1. Cut anteriorly up from the **diaphragm,** a structure that separates the thoracic cavity from the abdominal cavity, until you reach the clump of hair below the chin.
2. Make two lateral cuts, one on each side of the midline incision anterior to the forelimbs, taking extra care not to damage the blood vessels around the heart.
3. Make two lateral cuts, one on each side of the midline just posterior to the forelimbs and anterior to the diaphragm, following the ends of the ribs. Pull back the flaps created by these cuts to expose the **thoracic cavity.** You will dissect the thoracic cavity later, but in the meantime list the organs you find in the thoracic cavity. (See Figs. 20.5 and 20.6.) ______________________________

   ____

Abdominal Incisions

4. With scissors pointing up, cut posteriorly from the diaphragm to the umbilical cord.
5. Make a flap containing the umbilical cord by cutting a semicircle around the cord and by cutting posteriorly to the left and right of the cord.
6. Make two cuts, one on each side of the midline incision posterior to the diaphragm. Examine the diaphragm, attached to the chest wall by radially arranged muscles. The central region of the diaphragm, called the **central tendon,** is a membranous area.
7. Make two more cuts, one on each side of the flap containing the umbilical cord and just anterior to the hindlimbs. Pull back the side flaps created by these cuts to expose the **abdominal cavity.**

Complete Preparation of Pig for Dissection

1. Lifting the flap with the umbilical cord requires cutting the **umbilical vein.** Before cutting the umbilical vein, tie a thread on each side of the vein. Cut the vein but keep the threads in place for future reference.
2. As soon as you have opened the abdominal cavity, rinse out the pig. If you have a problem with excess fluid, obtain a disposable plastic pipet to suction off the liquid.

Answer These Questions

1. Name the two cavities separated by the diaphragm. ______________________________
2. You will dissect the abdominal cavity later, but in the meantime list the organs located in the abdominal cavity. (See Figs. 20.5 and 20.6.) ______________________________

   ____

   ____

Figure 20.4 Ventral view of the fetal pig indicating incisions.
These incisions are to be made preparatory to dissecting the internal organs. They are numbered here in the order they should be done.

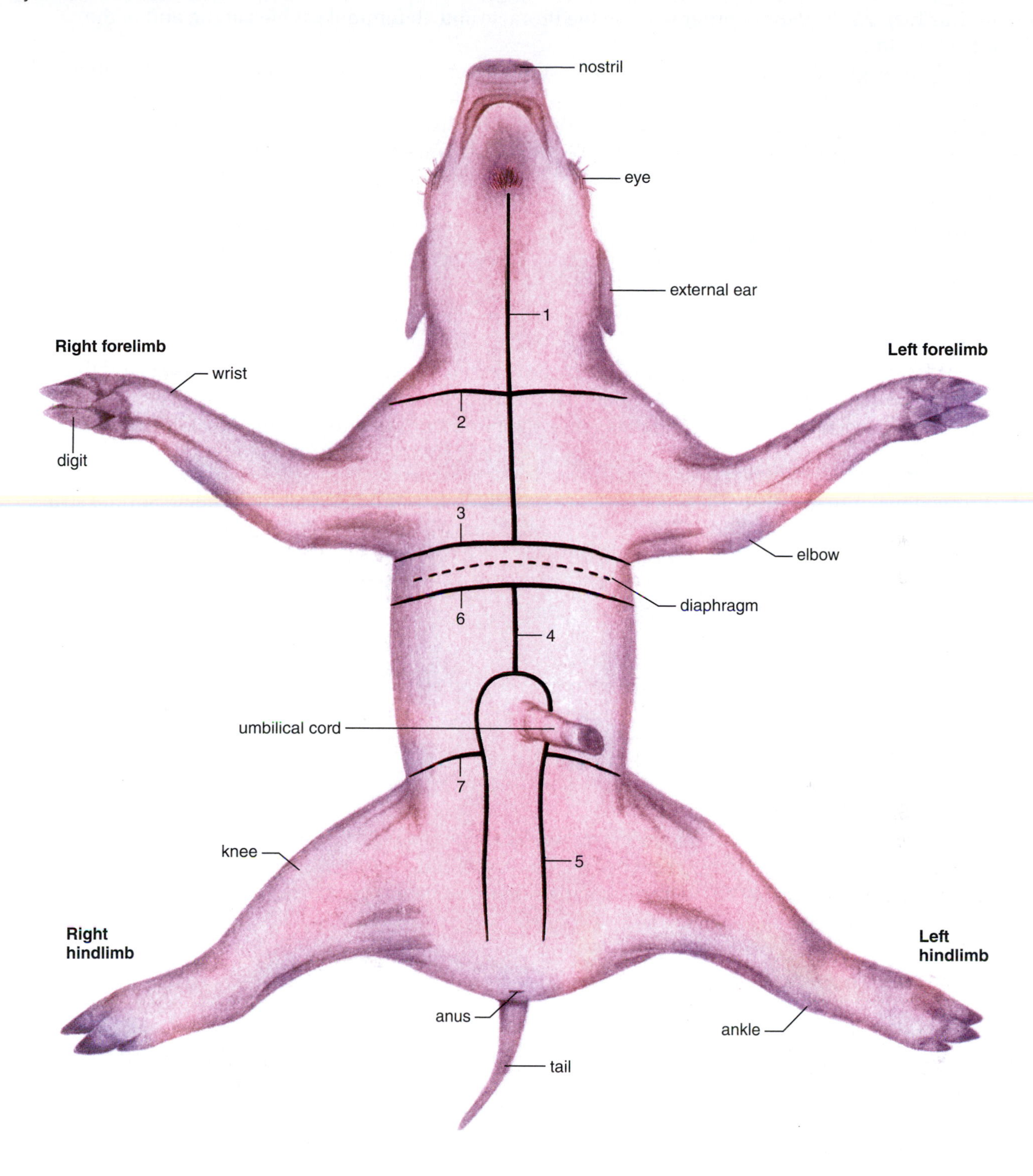

20.4 Neck Region

Several organs in the neck region are of interest. Use Figures 20.3*b* and 20.5 as a guide to locate these organs, but *keep all the flaps* in order to close the thoracic and abdominal cavities at the end of the laboratory session.

The **thymus gland** is a part of the lymphatic system. Certain white blood cells called T (for thymus) lymphocytes mature in the thymus gland and help us fight disease. The **larynx,** or voice box, sits atop the **trachea,** or windpipe. The esophagus is a portion of the digestive tract that leads to the stomach. The **thyroid gland,** a part of the endocrine system, secretes hormones that travel in the blood and act upon other body cells. These hormones (e.g., thyroxine) regulate the rate at which metabolism occurs in cells.

Observation: Neck Region

Thymus Gland

1. Move the skin apart in the neck region just below the hairs mentioned earlier. If necessary, cut the body wall laterally to make flaps.
2. If necessary, *cut through and clear away muscle* to expose the thymus gland, a diffuse gland that lies among the muscles. Later you will notice that the thymus flanks the thyroid and overlies the heart. The thymus is particularly large in fetal pigs, since their immune systems are still developing.

Larynx, Trachea, and Esophagus

1. Probe down into the deeper layers of the neck. Medially (toward the center), beneath several strips of muscle, find the hard-walled larynx and the trachea, which are parts of the respiratory passage. Dorsal to the trachea, find the esophagus.
2. Open the mouth and insert a probe into the glottis and esophagus from the pharynx to better understand the orientation of these two organs.

Thyroid Gland

Locate the thyroid gland just posterior to the larynx, lying ventral to (on top of) the trachea. The thyroid gland secretes hormones that increase metabolism within cells.

20.5 Thoracic Cavity

As previously mentioned, the body cavity of mammals, including humans, is divided by the diaphragm into the thoracic cavity and the abdominal cavity. The heart and lungs are in the thoracic cavity (Figs. 20.5 and 20.6). The **heart** is a pump for the cardiovascular system, and the **lungs** are organs of the respiratory system where gas exchange occurs.

Observation: Thoracic Cavity

Heart and Lungs

1. In order to fold back the chest wall flaps, tear the thin membranes that divide the thoracic cavity into three compartments. The three compartments are the **left pleural cavity** containing the left lung, the **right pleural cavity** containing the right lung, and the **pericardial cavity** containing the heart.
2. Examine the lungs. Locate the four lobes of the right lung and the three lobes of the left lung. The trachea, dorsal to the heart, divides into the **bronchi,** which enter the lungs.
3. Trace the path of air from the nasal passages to the lungs.

____

____

Figure 20.5 Internal anatomy of the fetal pig.
The major organs are featured in this drawing. In the fetal pig, a vessel colored red is an artery, and a vessel colored blue is a vein. (The color does not indicate whether this vessel carries O_2-rich or O_2-poor blood.) Contrary to this drawing, do not cut the flaps, because they can be closed to protect the thoracic and abdominal cavities.

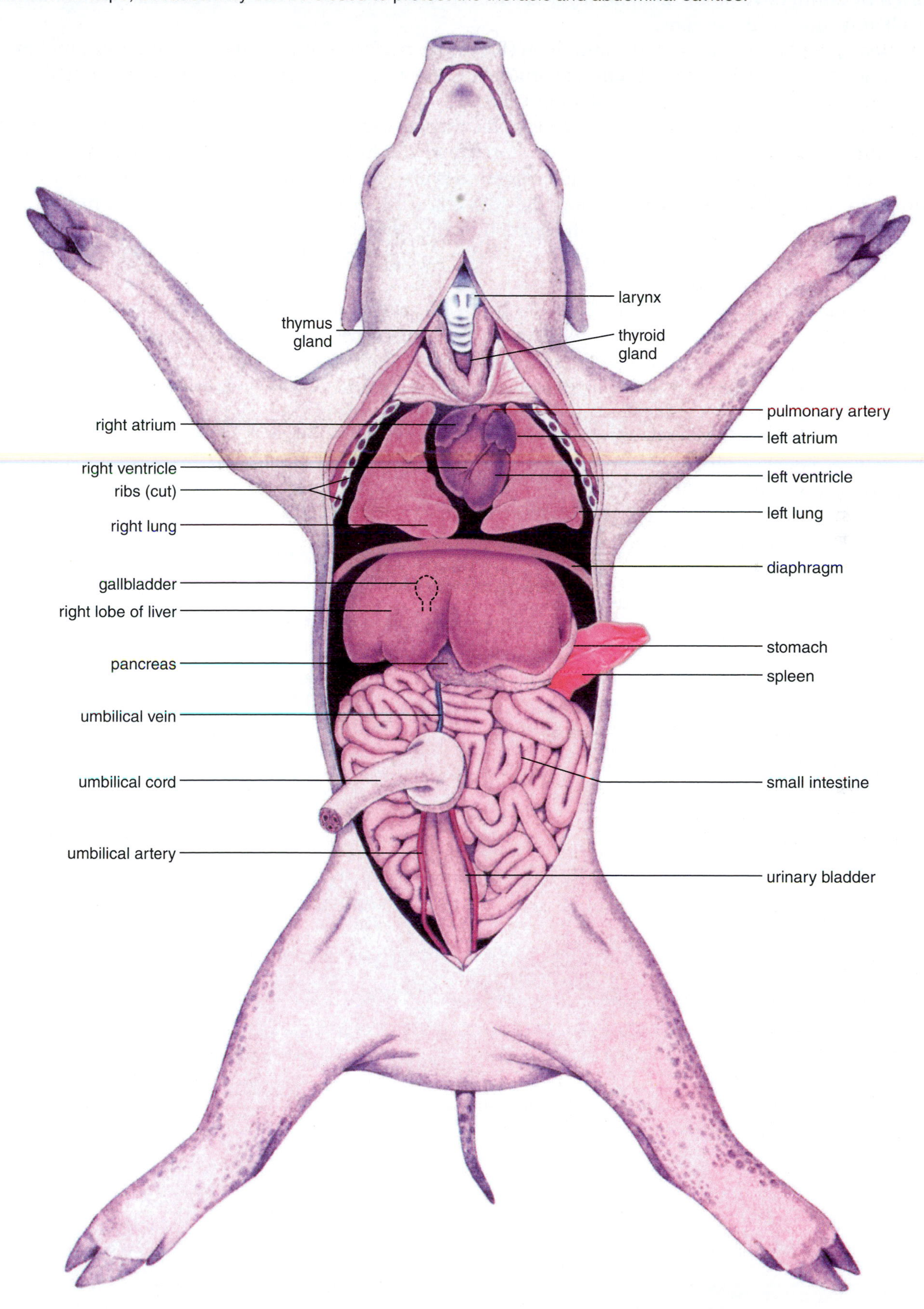

20.6 Abdominal Cavity

The abdominal wall and organs are lined by a membrane called **peritoneum,** consisting of epithelium supported by connective tissue. Double-layered sheets of peritoneum, called **mesenteries,** project from the body wall and support the organs.

The **liver,** the largest organ in the abdomen (Fig. 20.6), performs numerous vital functions, including (1) disposing of worn-out red blood cells, (2) producing bile, (3) storing glycogen, (4) maintaining the blood glucose level, and (5) producing blood proteins.

The abdominal cavity also contains organs of the digestive tract, such as the stomach, small intestine, and large intestine. The **stomach** (see Fig. 20.5) stores food and has numerous gastric glands that secrete gastric juice, which digests protein. The **small intestine** is the part of the digestive tract that receives secretions from the pancreas and gallbladder. Besides being an area for the digestion of all components of food—carbohydrate, protein, and fat—the small intestine absorbs the products of digestion: glucose, amino acids, glycerol, and fatty acids. The **large intestine** is the part of the digestive tract that absorbs water and prepares feces for defecation at the anus.

The **gallbladder** stores and releases bile, which aids the digestion of fat. The **pancreas** (see Fig. 20.5) is both an exocrine and an endocrine gland. As an exocrine gland, it produces and secretes pancreatic juice, which digests all the components of food in the small intestine. Both bile and pancreatic juice enter the duodenum (the first, straight part of the small intestine) by way of ducts. As an endocrine gland, the pancreas secretes the hormones insulin and glucagon into the bloodstream. Insulin and glucagon regulate blood glucose levels.

The **spleen** (see Fig. 20.5) is a lymphoid organ in the lymphatic system that contains both white and red blood cells. It purifies blood and disposes of worn-out red blood cells.

Observation: Abdominal Cavity

Liver

1. If your particular pig is partially filled with dark, brownish material, take your animal to the sink and rinse it out. This material is clotted blood. Consult your instructor before removing any red or blue latex masses, since they may enclose organs you will need to study.
2. Locate the liver, a large, brown organ. Its anterior surface is smoothly convex and fits snugly into the concavity of the diaphragm.
3. Name several functions of the liver. ____________________

____

Stomach and Spleen

1. Push aside and identify the stomach, a large sac dorsal to the liver on the left side.
2. Locate the point near the midline of the body where the **esophagus** penetrates the diaphragm and joins the stomach.
3. Find the spleen, a long, flat, reddish organ attached to the stomach by mesentery.
4. The stomach is a part of the ____________________ system.

   What is its function? ____________________
5. The spleen is a part of the ____________________ system.

   What is its function? ____________________

Figure 20.6 Internal anatomy of the fetal pig.
Most of the major organs are shown in this photograph. The stomach has been removed. The spleen, gallbladder, and pancreas are not visible. *Do not* remove any organs or flaps from your pig.

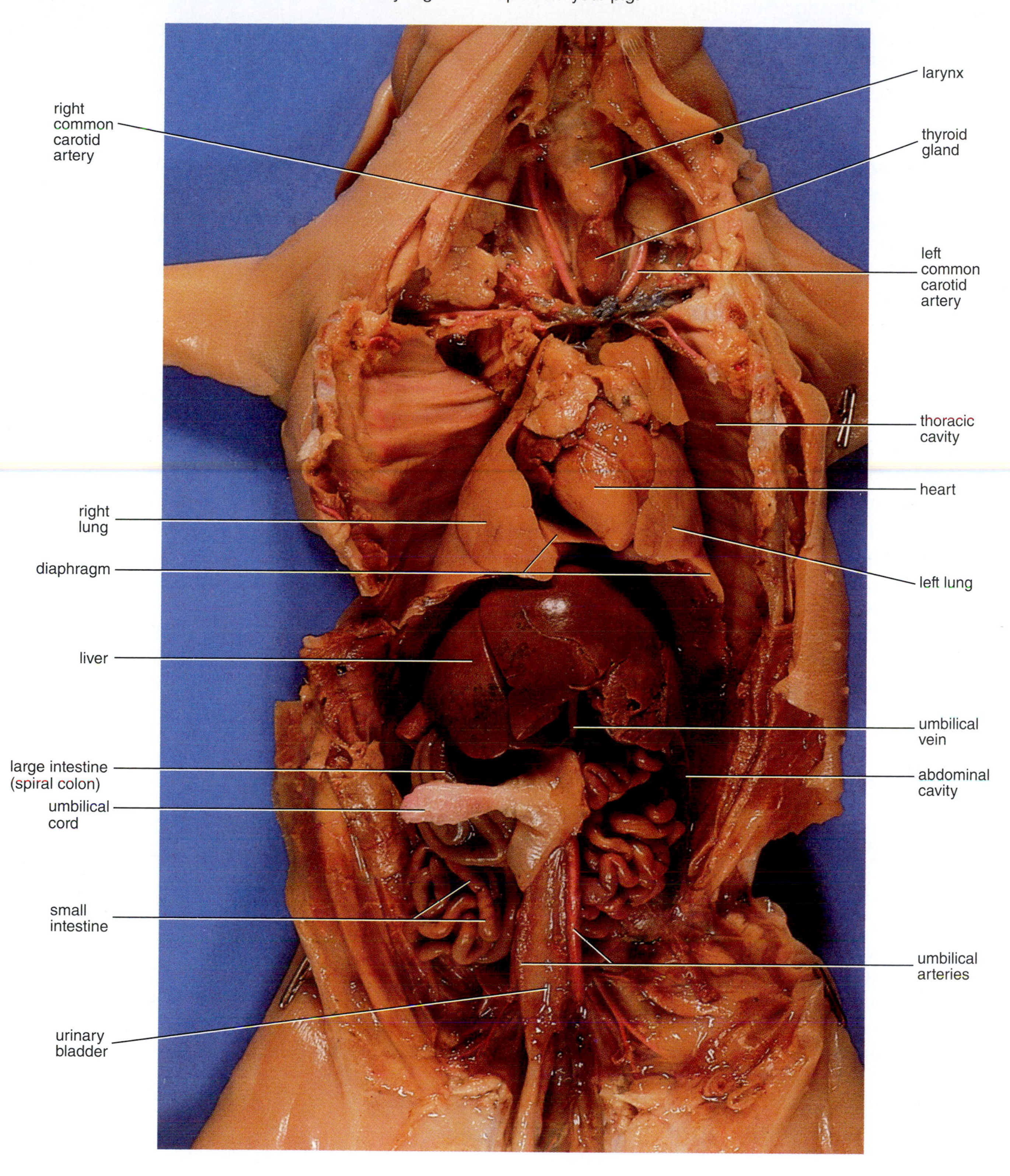

Small Intestine

1. Look posteriorly where the stomach makes a curve to the right and narrows to join the anterior end of the small intestine called the **duodenum.**
2. From the duodenum, the small intestine runs posteriorly for a short distance and is then thrown into an irregular mass of bends and coils held together by a common mesentery.
3. The small intestine is a part of the ____________________ system.

   What is its function? ____

Gallbladder and Pancreas

1. Locate the **bile duct,** which runs in the mesentery stretching between the liver and the duodenum. Find the gallbladder, embedded in the liver on the underside of the right lobe. It is a small, greenish sac.
2. Lift the stomach and locate the pancreas, the light-colored, diffuse gland lying in the mesentery between the stomach and the small intestine. The pancreas has a duct that empties into the duodenum of the small intestine.
3. What is the function of the gallbladder? ____
4. What is the function of the pancreas? ____

   ____

Large Intestine

1. Locate the distal (far) end of the small intestine, which joins the large intestine posteriorly, in the left side of the abdominal cavity (right side in humans). At this junction, note the **cecum,** a blind pouch.
2. Compare the large intestine of a pig to Figure 20.7. The organ does not have the same appearance in humans.
3. Follow the main portion of the large intestine, known as the **colon,** as it runs from the point of juncture with the small intestine into a tight coil (spiral colon), then out of the coil anteriorly, then posteriorly again along the midline of the dorsal wall of the abdominal cavity. In the pelvic region, the **rectum** is the last portion of the large intestine. The rectum leads to the **anus.**
4. The large intestine is a part of the ____________________ system.
5. What is the function of the large intestine? ____
6. Trace the path of food from the mouth to the anus. ____

   ____

Storage of Pigs

1. Before leaving the laboratory, place your pig in the plastic bag provided.
2. Expel excess air from the bag, and tie it shut.
3. Write your name and section on the tag provided, and attach it to the bag. Your instructor will indicate where the bags are to be stored until the next laboratory period.
4. Clean the dissecting tray and tools, and return them to their proper location.
5. Wipe off your goggles.
6. Wash your hands.

20.7 Human Anatomy

Humans and pigs are both mammals, and their organs are similar. A human torso model shows the exact location of the organs in humans (Fig. 20.7). Learn to associate these organs with their particular system. Six systems are color-coded in Figure 20.7.

Figure 20.7 Human internal organs.

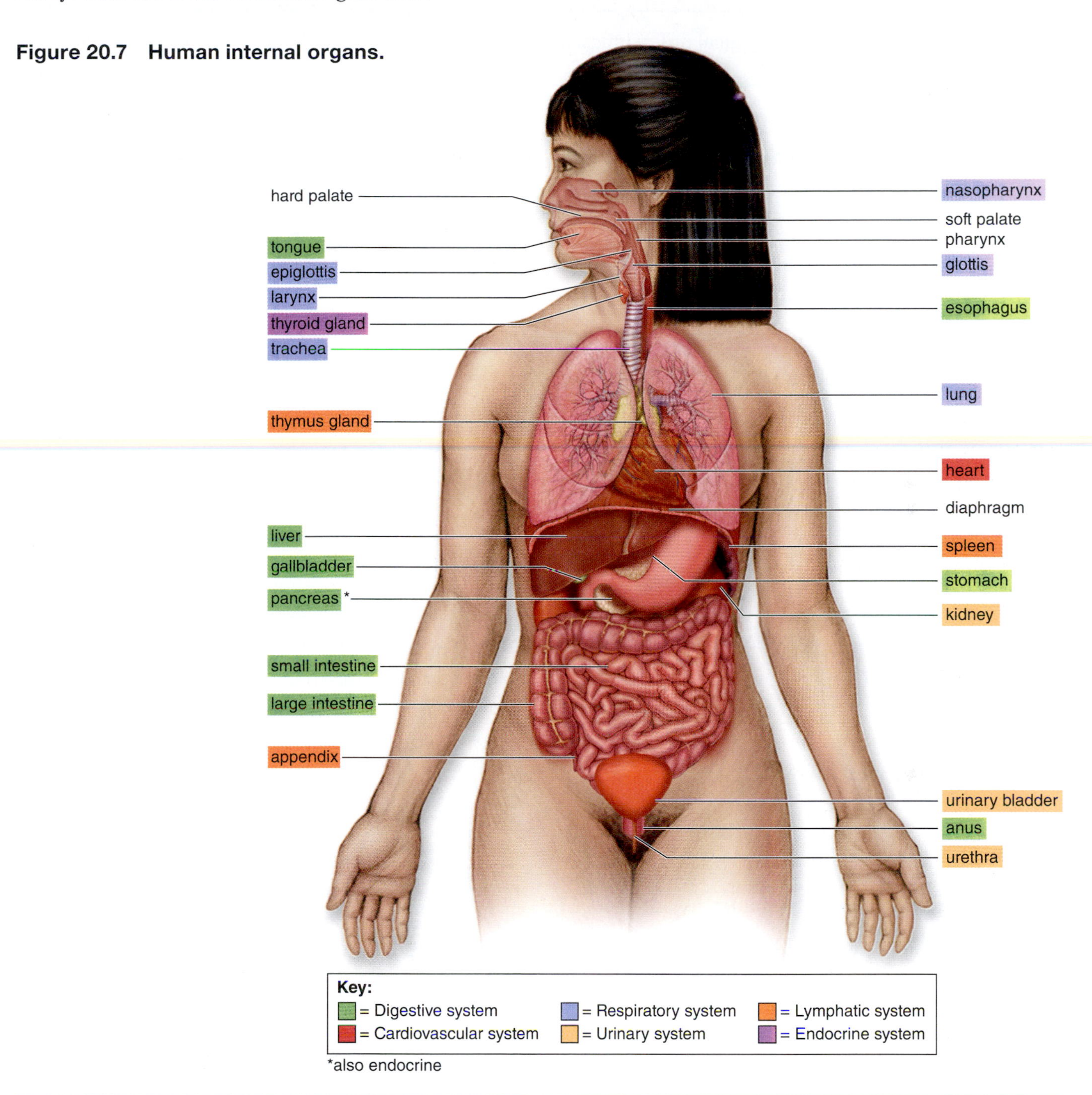

*also endocrine

Observation: Human Torso

1. Examine a human torso model, and using Figure 20.7 as a guide, locate the same organs just dissected in the fetal pig.
2. Name any observed major differences between pig internal anatomy and human internal anatomy.

____

Laboratory Review 20

1. What two features indicate that a pig is a mammal? ______

2. Put the following organs in logical order: lungs, nasal passages, nasopharynx, trachea, bronchi, glottis.

3. What difficulty would probably arise if a person were born without an epiglottis? ______

4. The embryonic coelom may be associated with what two cavities studied in this laboratory? ______

5. Name two principal organs in the thoracic cavity, and give a function for each. ______

6. What difficulty would arise if a person were born without a thymus gland? ______

7. Name the largest organ in the abdominal cavity and list several functions. ______

8. A large portion of the abdominal cavity is taken up with digestive organs. What are they? ______

9. Why is it proper to associate the gallbladder with the liver? ______

10. Where would you find the pancreas? ______

LABORATORY

21

Chemical Aspects of Digestion

Learning Outcomes

Introduction

- Sequence the organs of the digestive tract from the mouth to the anus. 276
- State the contribution of each organ, if any, to the process of chemical digestion. 276

21.1 Protein Digestion by Pepsin

- Associate the enzyme pepsin with the ability of the stomach to digest protein. 277
- Explain why stomach contents are acidic and how a warm body temperature aids digestion. 277–78

21.2 Fat Digestion by Pancreatic Lipase

- Associate the enzyme lipase with the ability of the small intestine to digest fat. 279
- Explain why the emulsification process assists the action of lipase. 279
- Explain why a change in pH indicates that fat digestion has occurred. 279
- Explain the relationship between time and enzyme activity. 280

21.3 Starch Digestion by Pancreatic Amylase

- Associate the enzyme pancreatic amylase with the ability of the small intestine to digest starch. 281

21.4 Requirements for Digestion

- Assuming a specific enzyme, list four factors that can affect the activity of all enzymes. 283
- Explain why the operative procedure that reduces the size of the stomach causes an individual to lose weight. 283

Introduction

In Laboratory 20, you examined the organs of digestion in the fetal pig. Now we wish to further our knowledge of the digestive process by associating certain digestive enyzmes with particular organs, as shown in Figure 21.1. This laboratory will also give us an opportunity to study the action of enzymes, much as William Beaumont did when he removed food samples through a hole in the stomach wall of his patient, Alexis St. Martin. Every few hours, Beaumont would see how well the food had been digested.

In Laboratory 6 we learned that enzymes are very specific and usually participate in only one type of reaction. The active site of an enzyme has a shape that accommodates its substrate, and if an environmental factor such as a boiling temperature or a wrong pH alters this shape, the enzyme loses its ability to function well, if at all. We will have an opportunity to make these observations with controlled experiments. The box on the next page reviews what is meant by a controlled experiment.

Planning Ahead Be advised that protein digestion (page 277) requires 1½ hours and fat digestion (page 279) requires 1 hour. Also a boiling water bath is required for starch digestion (page 281).

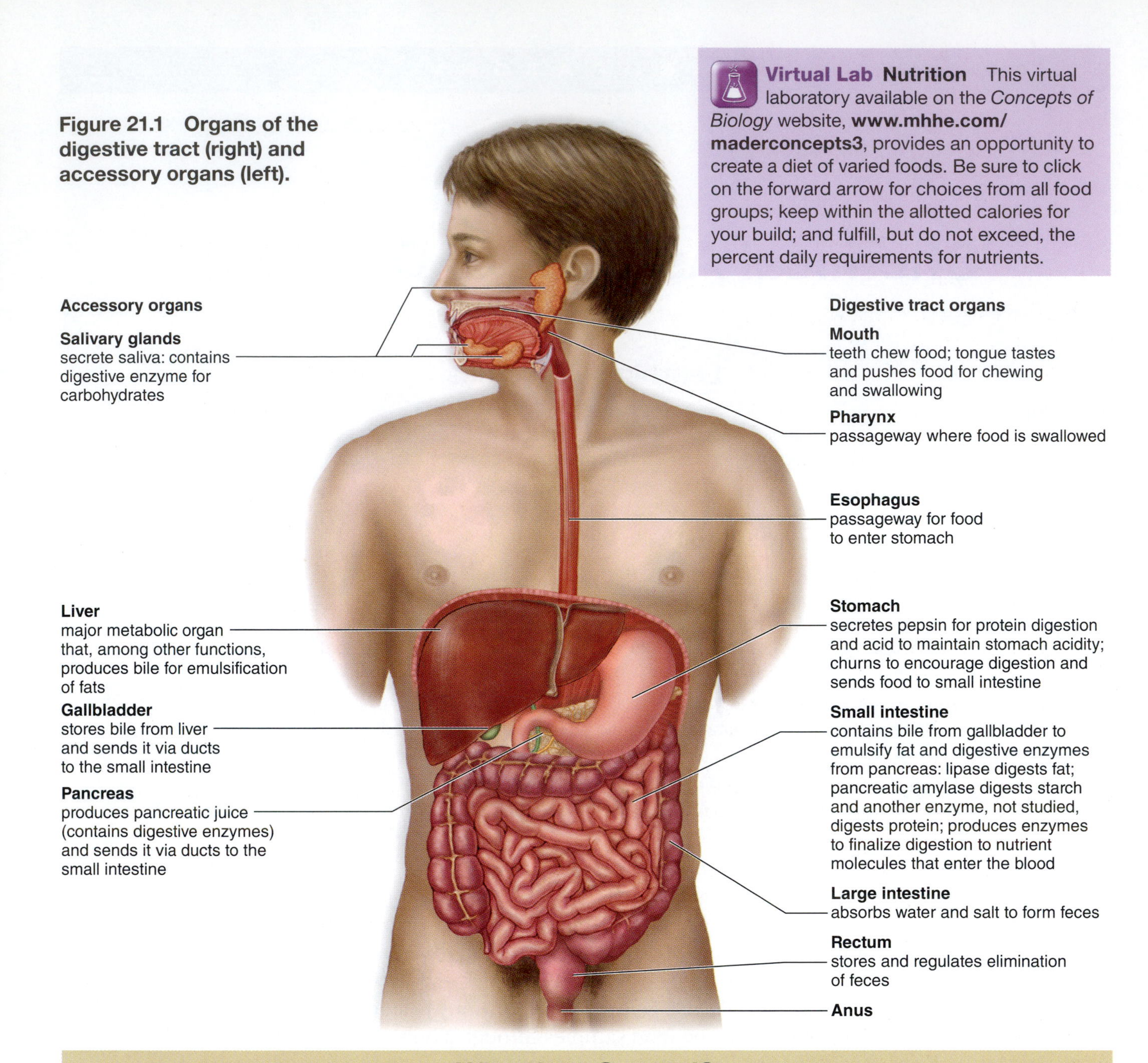

Figure 21.1 Organs of the digestive tract (right) and accessory organs (left).

What Is a Control?

The experiments in today's laboratory have both a positive control and a negative control, *which should be saved for comparison purposes until the experiment is complete*. The **positive control** goes through all the steps of the experiment and does contain the substance being tested. Therefore, positive results are expected. The **negative control** goes through all the steps of the experiment, except it does not contain the substance being tested. Therefore, negative results are expected.

For example, if a test tube contains glucose (the substance being tested) and Benedict's reagent (blue) is added, a red color develops upon heating. This test tube is the positive control; it tests positive for glucose. If a test tube does not contain glucose and Benedict's reagent is added, Benedict's is expected to remain blue. This test tube is the negative control; it tests negative for glucose.

What benefit is a positive control? Positive controls give you a standard by which to tell if the substance being tested is present (or acting properly) in an unknown sample. Negative controls ensure that the experiment is giving reliable results; after all, if a negative control should happen to give a positive result, then the entire experiment may be faulty and unreliable.

21.1 Protein Digestion by Pepsin

Certain foods, such as meat and egg whites, are rich in protein. Egg whites contain albumin, which is the protein used in this Experimental Procedure. Protein is digested by **pepsin** in the stomach (Fig. 21.2), a process described by the following reaction:

$$\text{protein} + \text{water} \xrightarrow{\text{pepsin (enzyme)}} \text{peptides}$$

The stomach has a very low pH. Does this indicate that pepsin works effectively in an acidic or a basic environment? ______________________ This is the pH that allows the enzyme to maintain its normal shape so that it will combine with the substrate. A warm temperature causes molecules to move about more rapidly and increases the encounters between enzyme and substrate. Therefore you would hypothesize that the yield from this enzymatic reaction will be higher if the pH is ____________________ and the temperature is ____________________ (body temperature 37°C).

Test for Protein Digestion

Biuret reagent is used to test for protein digestion. If digestion has not occurred, biuret reagent turns purple, indicating that protein is present. If digestion has occurred, biuret reagent turns pinkish-purple, indicating that peptides are present.

> ⚠ **Biuret reagent** is highly corrosive. Exercise care in using this chemical. If any should spill on your skin, wash the area with mild soap and water. Follow your instructor's directions for its disposal.

Experimental Procedure: Protein Digestion

1. Label four clean test tubes (1 to 4). Using the designated graduated pipet, add 2 ml of the albumin solution to all tubes. Albumin is a protein.
2. Add 2 ml of the pepsin solution to tubes 1 to 3, as listed in Table 21.1.
3. Add 2 ml of 0.2% HCl to tubes 1 and 2. HCl simulates the acidic conditions of the stomach.
4. Add 2 ml of water to tube 3 and 4 ml of water to tube 4, as listed in Table 21.1.
5. Swirl to mix the tubes. Tube 2 remains at room temperature, but the other three are incubated for 1½ hours. Record the temperature for each tube in Table 21.1.
6. Remove the tubes from the incubator and place all four tubes in a tube rack. Add 2 ml of biuret reagent to all tubes and observe. Record your results in Table 21.1 as + or – to indicate digestion or no digestion.

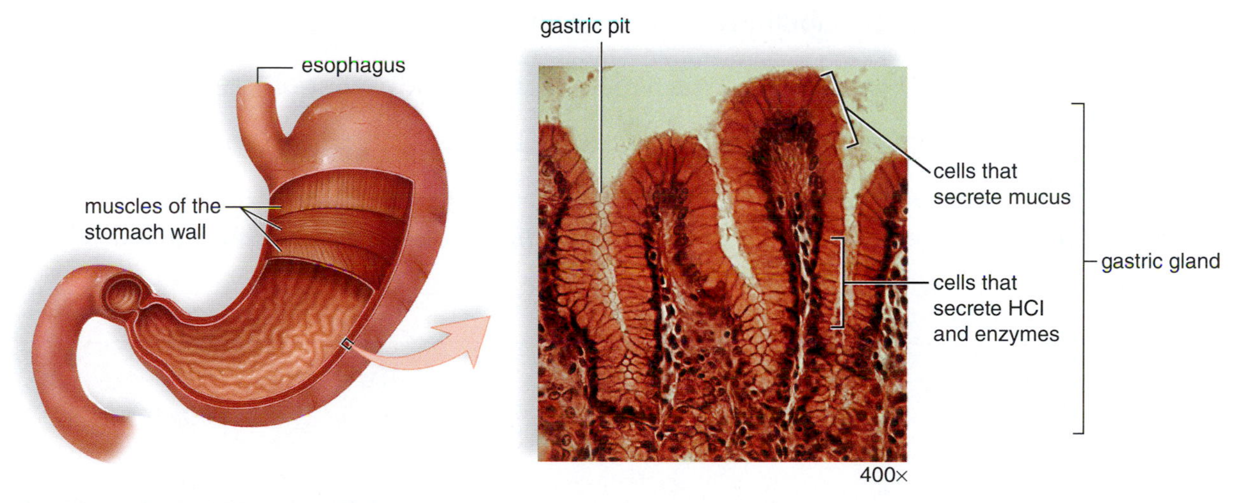

Figure 21.2 Digestion of protein.
Pepsin, produced by the gastric glands of the stomach, helps digest protein.

Table 21.1 Protein Digestion by Pepsin				
Tube	**Contents**	**Temperature**	**Digestion (+ or –)**	**Explanation**
1	Albumin Pepsin HCl Biuret reagent			
2	Albumin Pepsin HCl Biuret reagent			
3	Albumin Pepsin Water Biuret reagent			
4	Albumin Water Biuret reagent			

Conclusions: Protein Digestion

- Explain your results in Table 21.1 by reasoning why digestion did or did not occur.
  To be complete, consider all the requirements for an enzymatic reaction as listed in Table 21.4. Now show here that tube 1 met all the requirements for digestion:

  Pepsin is the correct ______________.

  Albumin is the correct ______________.

  37°C is the optimum ______________.

  HCl provides the optimum ______________.

  1½ hours provides ______________ for the reaction to occur.

- Review "What Is a Control?" on page 276. Which tube was the negative control? ______________

  Explain why it was the negative control. ______________

  ____

- If this control tube had given a positive result for protein digestion, what could you conclude about this experiment? ______________

  ____

21.2 Fat Digestion by Pancreatic Lipase

Lipids include fats (e.g., butterfat) and oils (e.g., sunflower, corn, olive, and canola). Lipids are digested by **pancreatic lipase** in the small intestine (Fig. 21.3).

Figure 21.3 Emulsification and digestion of fat.
Bile from the liver (stored in the gallbladder) enters the small intestine, where lipase in pancreatic juice from the pancreas digests fat.

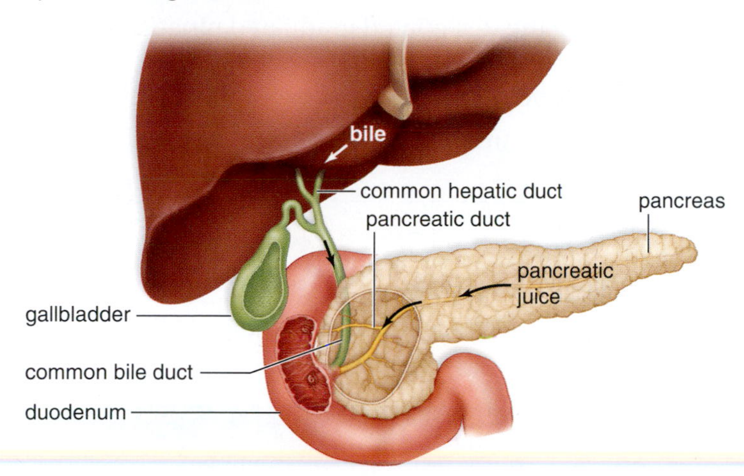

The following two reactions describe fat digestion:

1. fat —bile (emulsifier)→ fat droplets

2. fat droplets + water —lipase (enzyme)→ glycerol + fatty acids

With regard to the first step, consider that fat is not soluble in water; yet, lipase makes use of water when it digests fat. Therefore, bile is needed to emulsify fat—cause it to break up into fat droplets that disperse in water. The reason for dispersal is that bile contains molecules with two ends. One end is soluble in fat, and the other end is soluble in water. Bile can emulsify fat because of this.

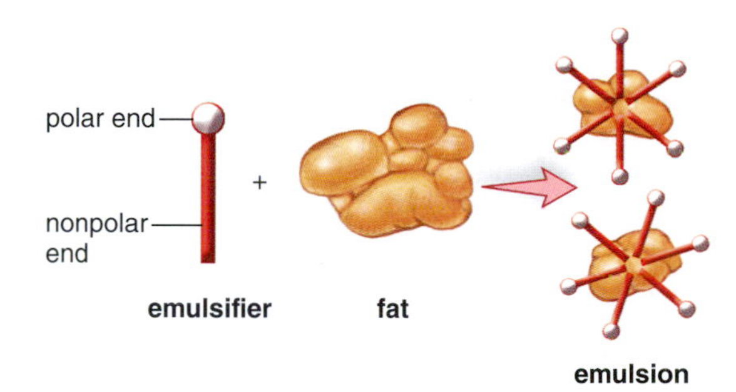

With regard to the second step, would the pH of the solution be lower before or after the enzymatic reaction? (*Hint:* Remember that an acid decreases pH and a base increases pH.) ______________________

____

Test for Fat Digestion

In the test for fat digestion, you will be using a pH indicator, which changes color as the solution in the test tube goes from basic conditions to acidic conditions. Phenol red is a pH indicator that is red in basic solutions and yellow in acidic solutions.

Experimental Procedure: Fat Digestion

1. Label three clean test tubes (1 to 3). Using the designated graduated pipet, add 1 ml of vegetable oil to all tubes.
2. Add 2 ml of phenol red solution to each tube. What role does phenol red play? ______________
3. Add 2 ml of pancreatic lipase (pancreatin) to tubes 1 and 2 and 2 ml of water to tube 3, as listed in Table 21.2. What role does lipase play? ______________
4. Add a pinch of bile salts to tube 1.
5. Record the initial color of all tubes in Table 21.2.
6. Incubate all three tubes at 37°C and check every 20 minutes.
7. Record any color change and how long it took to see this color change in Table 21.2.

Table 21.2 Fat Digestion by Pancreatic Lipase

Tube	Contents	Color		Time Taken	Explanation
		Initial	*Final*		
1	Vegetable oil Phenol red Pancreatin Bile salts				
2	Vegetable oil Phenol red Pancreatin				
3	Vegetable oil Phenol red Water				

Conclusions: Fat Digestion

- Explain your results in Table 21.2 by reasoning why digestion did or did not occur.
- What role did bile salts play in this experiment? ______________

____
- What role did phenol red play in this experiment? ______________
- Review "What Is a Control?" on page 276. Which test tube in this experiment could be considered a negative control? ______________

____

21.3 Starch Digestion by Pancreatic Amylase

Starch is present in bakery products and in potatoes, rice, and corn. Starch is digested by **pancreatic amylase** in the small intestine, a process described by the following reaction:

$$\text{starch} + \text{water} \xrightarrow{\text{amylase (enzyme)}} \text{maltose}$$

1. If digestion *does not* occur, which will be present—starch or maltose? ______________________
2. If digestion *does* occur, which will be present—starch or maltose? ______________________

Tests for Starch Digestion

You will be using two tests for starch digestion:

> ⚠ **Benedict's reagent** is highly corrosive. Use protective eyewear when performing this experiment. Exercise care in using this chemical. If any should spill on your skin, wash the area with mild soap and water. Follow your instructor's directions for disposal of this chemical.

1. If digestion has not taken place, the iodine test for starch will be positive (+). If digestion has occurred, the iodine test for starch will be negative (–).
2. If digestion of starch has taken place, the Benedict's test for sugar (maltose) will be positive (+). If digestion has not taken place, the Benedict's test for sugar will be negative (–).

   To test for sugar, add five drops of Benedict's reagent. Place the tube in a boiling water bath for a few minutes, and note any color changes (see Table 4.5 on page 36). Boiling the test tube is necessary for the Benedict's reagent to react.

Experimental Procedure: Starch Digestion

1. Label six clean test tubes (1 to 6).
2. Using a transfer pipet, add 1 ml of pancreatic-amylase solution to tubes 1 to 4 and 1 ml of water to tubes 5 and 6.
3. Test tubes 1 and 2 immediately.

   Tube 1 Shake the starch solution and add 1 ml of starch solution. Immediately add five drops of iodine to test for starch. Put this tube in a test tube rack and record your results in Table 21.3.

   Tube 2 Shake the starch solution and add 1 ml of starch solution. Immediately add five drops of Benedict's reagent, and place the tube in a boiling water bath to test for sugar. Put this tube in the test tube rack and record your results in Table 21.3.
4. Shake the starch suspension and add 1 ml of starch suspension to tubes 3 to 6. Allow the tubes to stand for 30 minutes.

   Tubes 3 and 5 After 30 minutes, test for starch using the iodine test. Place these tubes in the test tube rack and record your results in Table 21.3.

   Tubes 4 and 6 After 30 minutes, test for sugar using the Benedict's test. Place these tubes in the test tube rack and record your results in Table 21.3.
5. Examine all your tubes in the test rube rack and decide whether digestion occurred (+) or did not occur (–). Complete Table 21.3.

Table 21.3 Starch Digestion by Amylase

Tube	Contents	Time*	Type of Test	Test Resuts (+ or –)	Digestion (+ or –)
1	Pancreatic amylase Starch	0	Iodine	+	
2	Pancreatic amylase Starch				
3	Pancreatic amylase Starch				
4	Pancreatic amylase Starch				
5	Water Starch				
6	Water Starch				

* Enter either 0 for immediately or T for after 30 minutes.

Conclusions: Starch Digestion

- Considering tubes 1 and 2, this experimental procedure showed that __________ must pass for digestion to occur.
- Considering tubes 5 and 6, this experimental procedure showed that an active ______________ must be present for digestion to occur.
- Why would you not recommend doing the test for starch and the test for sugar on the same tube? ____
  ____
- Which test tubes served as a negative control for this experiment? ______________________
  Explain your answer. ____

Absorption of Sugars and Other Nutrients

Figure 21.4 shows that the folded lining of the small intestine has many fingerlike projections called villi. The small intestine not only digests food; it also absorbs the products of digestion, such as sugars from carbohydrate digestion, amino acids from protein digestion, and glycerol and fatty acids from fat digestion at the villi.

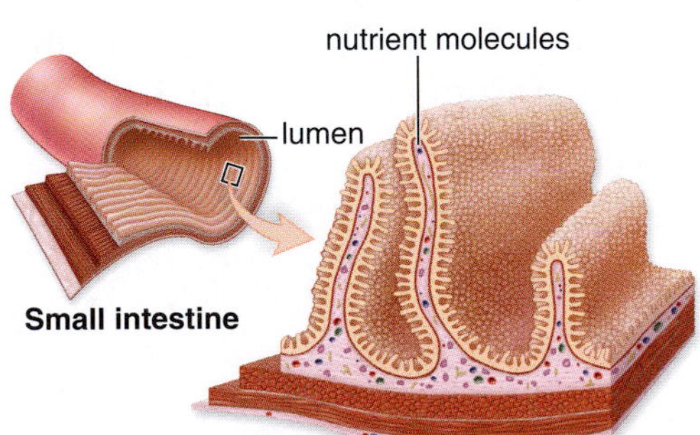

Figure 21.4 Anatomy of the small intestine. Nutrients enter the bloodstream across the much-convoluted walls of the small intestine.

21.4 Requirements for Digestion

Explain in Table 21.4 how each of the requirements listed influences effective digestion.

Table 21.4 Requirements for Digestion

Requirement	Explanation
Specific enzyme	
Specific substrate	
Warm temperature	
Specific pH	
Time	
Fat emulsifier	

To lose weight, some obese individuals undergo an operation in which (1) the stomach is reduced to the size of a golf ball, and (2) food bypasses the duodenum (first 2 feet) of the intestine. Answer these questions to explain how this operation would affect the requirements for digestion.

1. How is the amount of substrate reduced? ______________________
2. How is the amount of digestive enzymes reduced? ______________________
3. How is time reduced? ______________________
4. What makes the pH of the small intestine higher than before? ______________________
5. How is fat emulsification reduced? ______________________
6. How does surgery to reduce obesity sometimes result in malnutrition? ______________________

Laboratory Review 21

1. Where in the body does starch digestion occur? ____________ and ____________ Protein digestion occur? ____________ and ____________ Fat digestion occur? ____________
2. Why would you not expect amylase to digest protein? ____________
3. Relate the expectation of more product per length of time to the fact that enzymes are used over and over. ____________
4. Why do enzymes work better at their optimum pH? ____________
5. Why is an emulsifier needed for the lipase experiment but not for the pepsin and amylase experiments? ____________
6. Which of the following two combinations is most likely to result in digestion?
   - **a.** Pepsin, protein, water, body temperature
   - **b.** Pepsin, protein, hydrochloric acid (HCl), body temperature

   Explain your answer. ____________
7. Which of the following two combinations is most likely to result in digestion?
   - **a.** Amylase, starch, water, body temperature, testing immediately
   - **b.** Amylase, starch, water, body temperature, waiting 30 minutes

   Explain your answer. ____________
8. Relate the composition of fat to the test used for fat digestion. ____________
9. Given that, in this laboratory, you tested for the action of digestive enzymes on their substrates, what substance would be missing from a negative control sample? ____________
10. What substance would be present in a positive control sample? ____________

Concepts of Biology Website

Instructors can find lab prep information and answers to all of the laboratory questions in the Laboratory Resource Guide. _Students_ can practice their knowledge with quizzes, animations, flashcards, and much more.

www.mhhe.com/maderconcepts3

McGraw-Hill Access Science Website

An online encyclopedia of science and technology that provides information, including videos, that can enhance the laboratory experience.

www.accessscience.com

LABORATORY

22

Basic Mammalian Anatomy II

Learning Outcomes

22.1 Cardiovascular System
- Locate and identify the chambers of the heart and the vessels connected to these chambers. 286–87
- Name and locate the valves of the heart, and trace the path of blood through the heart. 286–88
- Name the major vessels of the pulmonary and systemic circuits of the cardiovascular system, and trace the path of blood to various organs. 288–89
- Compare the pulmonary and systemic circuits of vertebrates. 290

22.2 Respiratory, Digestive, and Urinary Systems
- Locate and identify the individual organs of the respiratory system, and state a function for each organ. 291
- Locate and identify the individual organs of the digestive system, and state a function for each organ. 291
- Locate and identify the organs of the urinary system, and state a function for each organ. 293

22.3 Male Reproductive System
- Locate and identify the organs of the male reproductive system. 294–95
- State a function for the organs of the male reproductive system. 294–95
- Compare the pig reproductive system with that of the human male. 296

22.4 Female Reproductive System
- Locate and identify the organs of the female reproductive system. 297–98
- State a function for the organs of the female reproductive system. 297–98
- Compare the pig reproductive system with that of a human female. 299

Introduction

The heart and blood vessels form the cardiovascular system. The heart is a double pump that keeps the blood flowing in one direction—away from and then back to the heart. The blood vessels transport blood and its contents, serve the needs of the body's cells by carrying out exchanges with them, and direct blood flow to those systemic tissues that most require it at the moment. After removing the heart and examining the heart chambers you will have an opportunity to remove and study the lungs and the small intestine. The latter will expose the urinary and reproductive systems. These two systems are so closely associated in mammals that they are often considered together as the urogenital system. In this laboratory, we will focus first on dissecting the urinary and reproductive systems in the fetal pig. We will then compare the anatomy of the reproductive systems in pigs with those in humans.

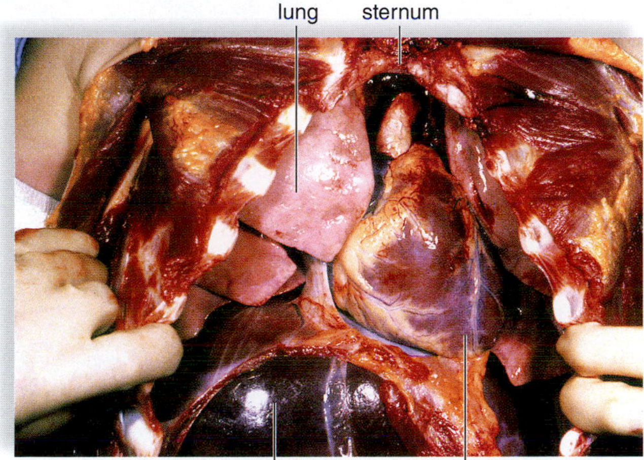

22.1 Cardiovascular System

The **cardiovascular system** includes the heart and two major circular pathways: the **pulmonary circuit** to and from the lungs and the **systemic circuit** to and from the body's organs.

Heart

The mammalian heart has a **right** and **left atrium** and a **right** and **left ventricle.** *To tell the left from the right side, mentally position the heart so that it corresponds to your own body.* Contraction of the heart pumps the blood through the heart and out into the arteries. The right side of the heart sends blood through the smaller pulmonary circuit, and the left side of the heart sends blood through the much larger systemic circuit.

Observation: Heart

Heart Model

1. Study a heart model (Fig. 22.1) or a preserved heart, and identify the four chambers of the heart: right atrium, right ventricle, left atrium, left ventricle.
2. Which ventricle is more muscular? __________ Why is this appropriate? ______________________________

____

Figure 22.1 External view of the mammalian heart.
Externally, notice the coronary arteries and cardiac veins that serve the heart itself.

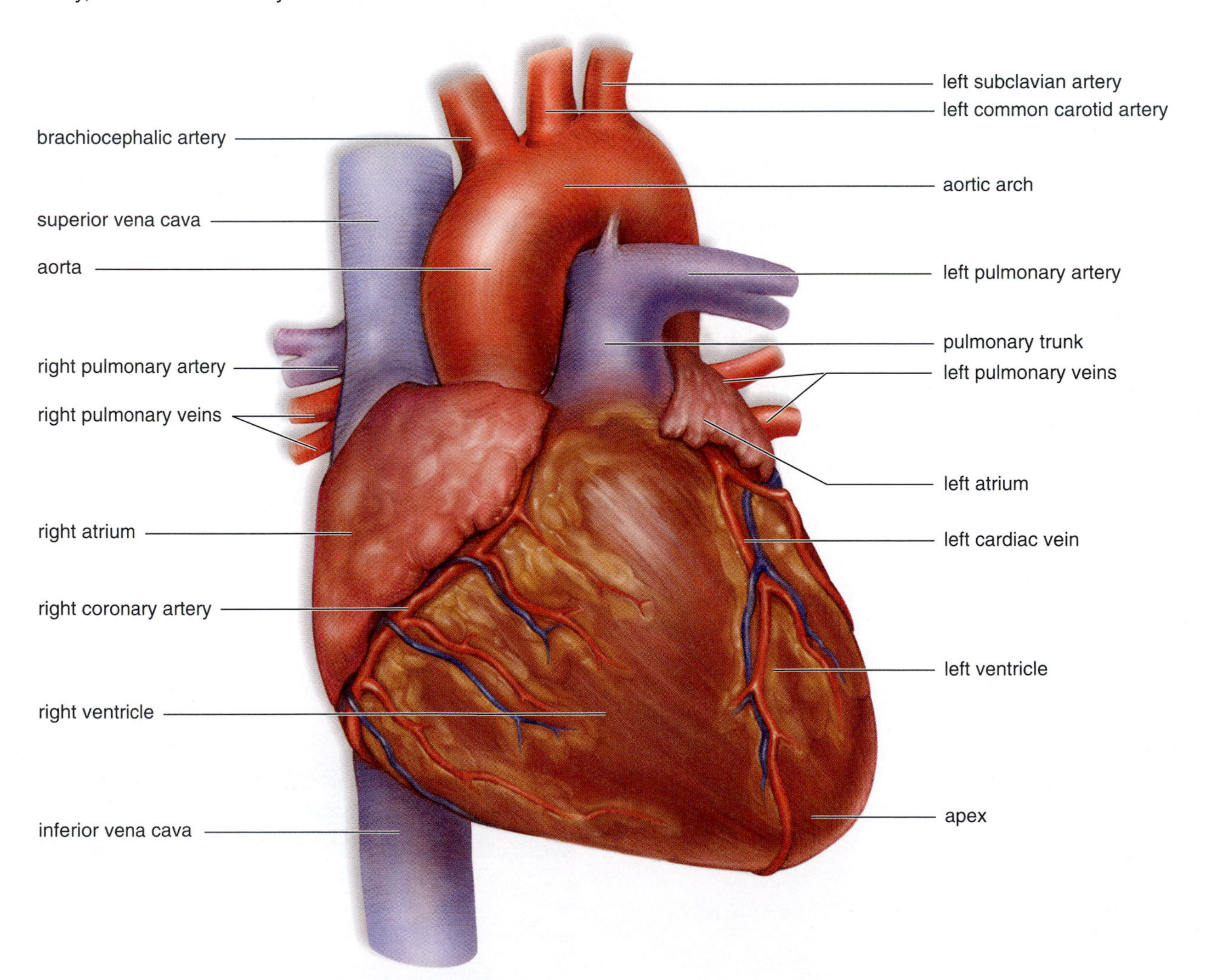

3. Identify the vessels connected to the heart. Locate the following structures:
   a. **Pulmonary trunk:** Leaves the ventral side of the heart from the top of the right ventricle and then passes forward diagonally before branching into the right and left pulmonary arteries.
   b. **Aorta:** Arises from the anterior end of the left ventricle, just dorsal to the origin of the pulmonary trunk. The aorta soon bends to the animal's left as the **aortic arch.**
   c. **Venae cavae:** Anterior (superior) and posterior (inferior) venae cavae enter the right atrium. They bring blood from the head and body, respectively, to the heart.
   d. **Pulmonary veins:** Return blood from the lungs to the left atrium. Why are the pulmonary veins colored red in Figure 22.2? ____

   ____

4. Locate the valves of the heart (Fig. 22.2). Remove the ventral half of the heart model or the ventral half of the heart. Identify the following structures:
   a. **Right atrioventricular (tricuspid) valve** and the **left atrioventricular (bicuspid, mitral) valve.**
   b. **Pulmonary semilunar valve** (in the base of the pulmonary trunk).
   c. **Aortic semilunar valve** (in the base of the aorta).
   d. **Chordae tendineae:** Hold the atrioventricular valves in place while the heart contracts. These extend from the papillary muscles.

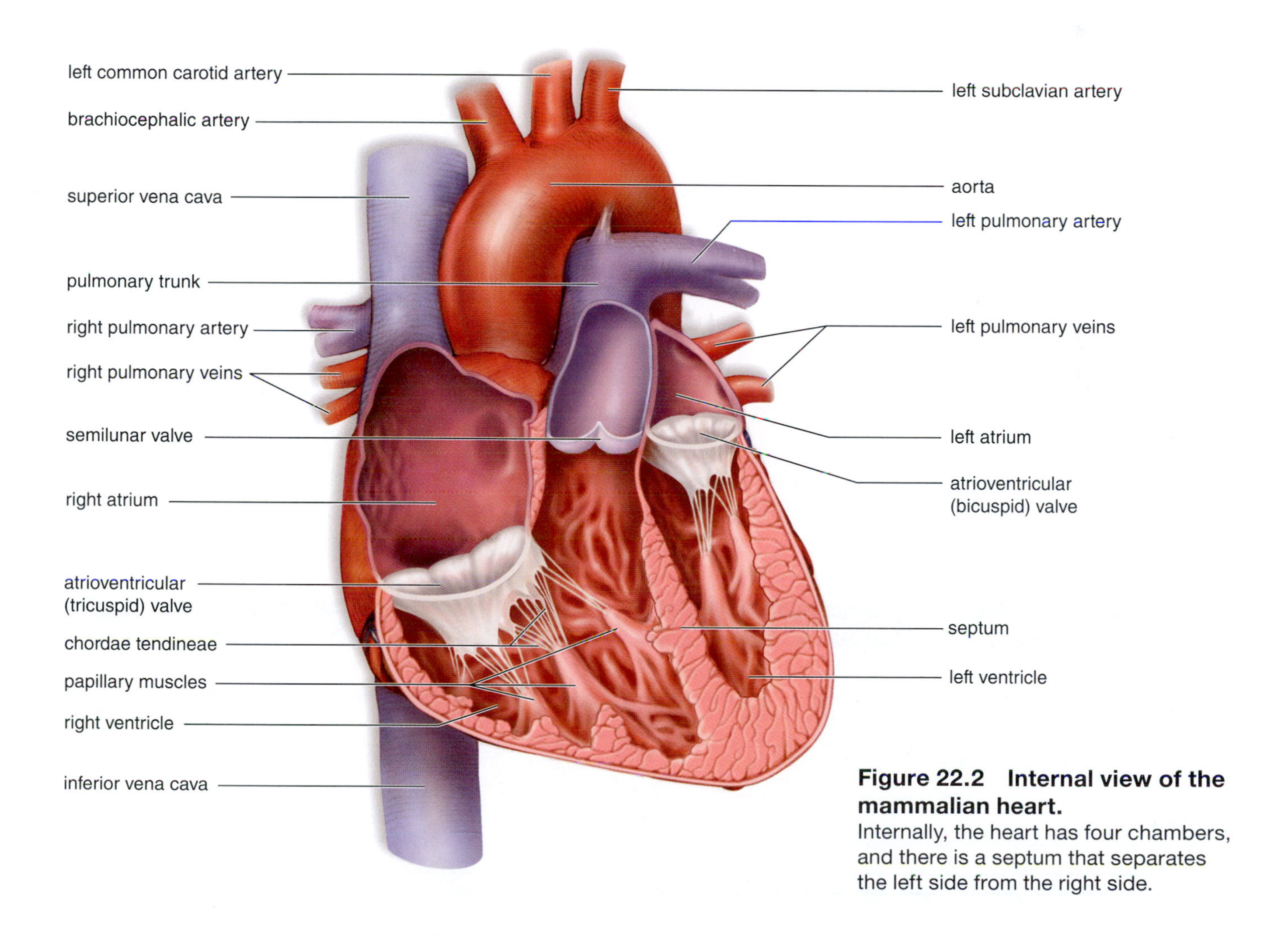

Figure 22.2 Internal view of the mammalian heart. Internally, the heart has four chambers, and there is a septum that separates the left side from the right side.

Tracing the Path of Blood Through the Heart

To demonstrate that O_2-poor blood is kept separate from O_2-rich blood, trace the path of blood from the right side of the heart to the aorta by filling in the blanks with the names of blood vessels and valves. In the adult mammal, note that blood passes through the lungs to go from the right side of the heart to the left side.

From Venae Cavae	From Lungs
______________	______________
______________ valve	______________
______________	______________ valve
______________ valve	______________
______________	______________ valve
______________	
To Lungs	**To Aorta**

Pulmonary Circuit and Systemic Circuit

Blood that leaves the heart enters one of two sets of blood vessels: the **pulmonary circuit,** which takes blood from the heart to the lungs and from the lungs to the heart, and the **systemic circuit,** which takes blood from the heart to the body proper and from the body proper to the heart.

Pulmonary Circuit

1. Sequence the blood vessels in the pulmonary circuit to trace the path of blood from the right ventricle to the left atrium of the heart.

   Right ventricle of the heart

   ____

   ____

   Lungs

   ____

   Left atrium of the heart

2. Which of these blood vessels contains O_2-rich blood? ______________

Systemic Circuit

1. The major blood vessels in the systemic system are the **aorta** (takes blood away from the heart) and the **venae cavae** (take blood to the heart). Otherwise, with the help of Figure 22.3, complete Table 22.1.

Table 22.1 Other Blood Vessels in the Systemic Circuit

Body Part	Artery	Vein
Head		
Front legs in pig (arms in humans)		
Kidney		
Hind legs		

Latex gloves Wear goggles, latex gloves, and protective clothing when handling preserved animal organs.

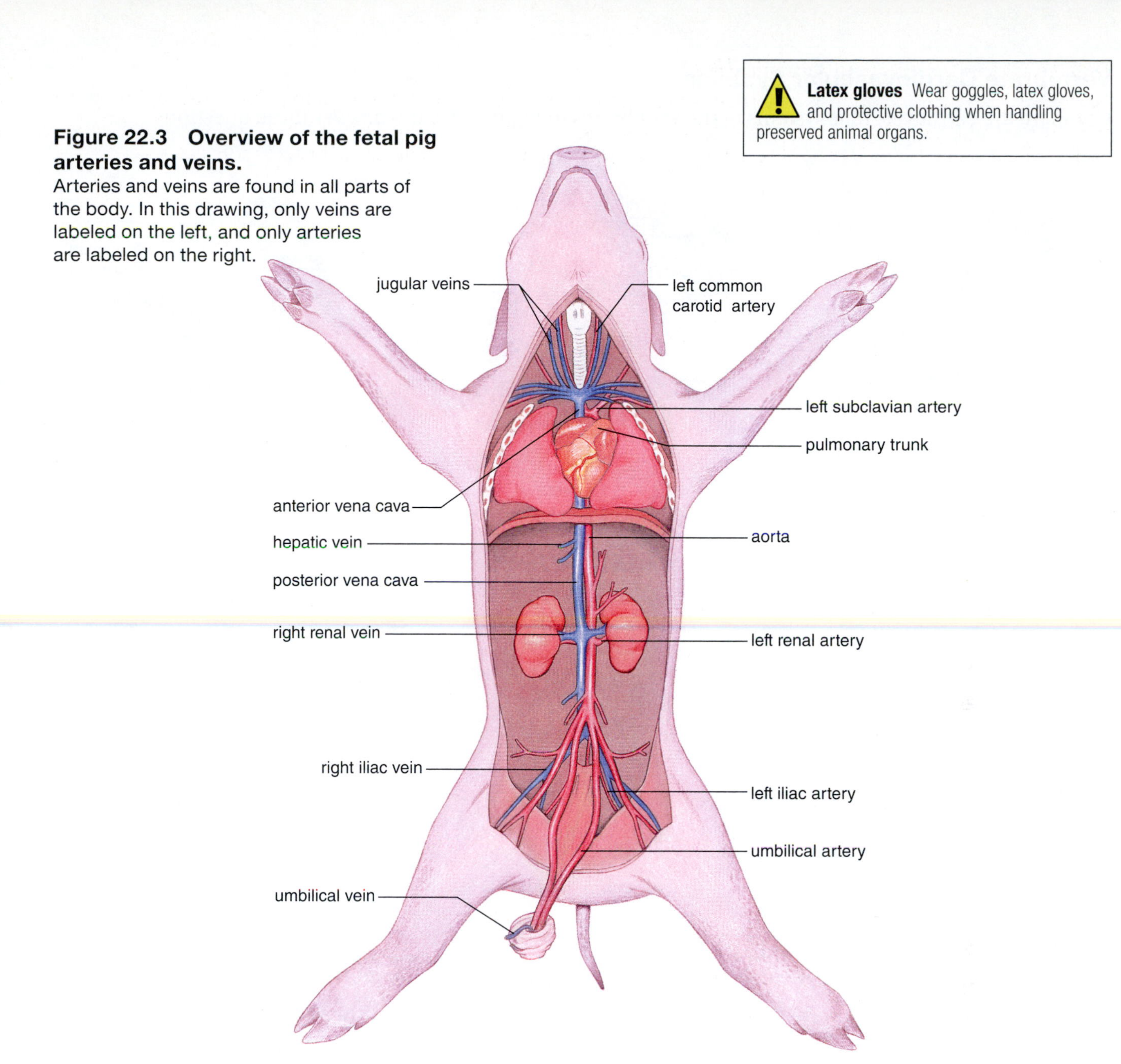

Figure 22.3 Overview of the fetal pig arteries and veins.
Arteries and veins are found in all parts of the body. In this drawing, only veins are labeled on the left, and only arteries are labeled on the right.

2. With the help of Figure 22.3 and Table 22.1, sequence the blood vessels in the systemic circuit to trace the path of blood from the heart to the kidneys and from the kidneys to the heart.

Left ventricle of the heart

____

____

Kidneys

____

____

Right atrium of the heart

Vertebrate Cardiovascular Systems

Compare the cardiovascular systems of the vertebrates in Figure 22.4 and answer these questions.

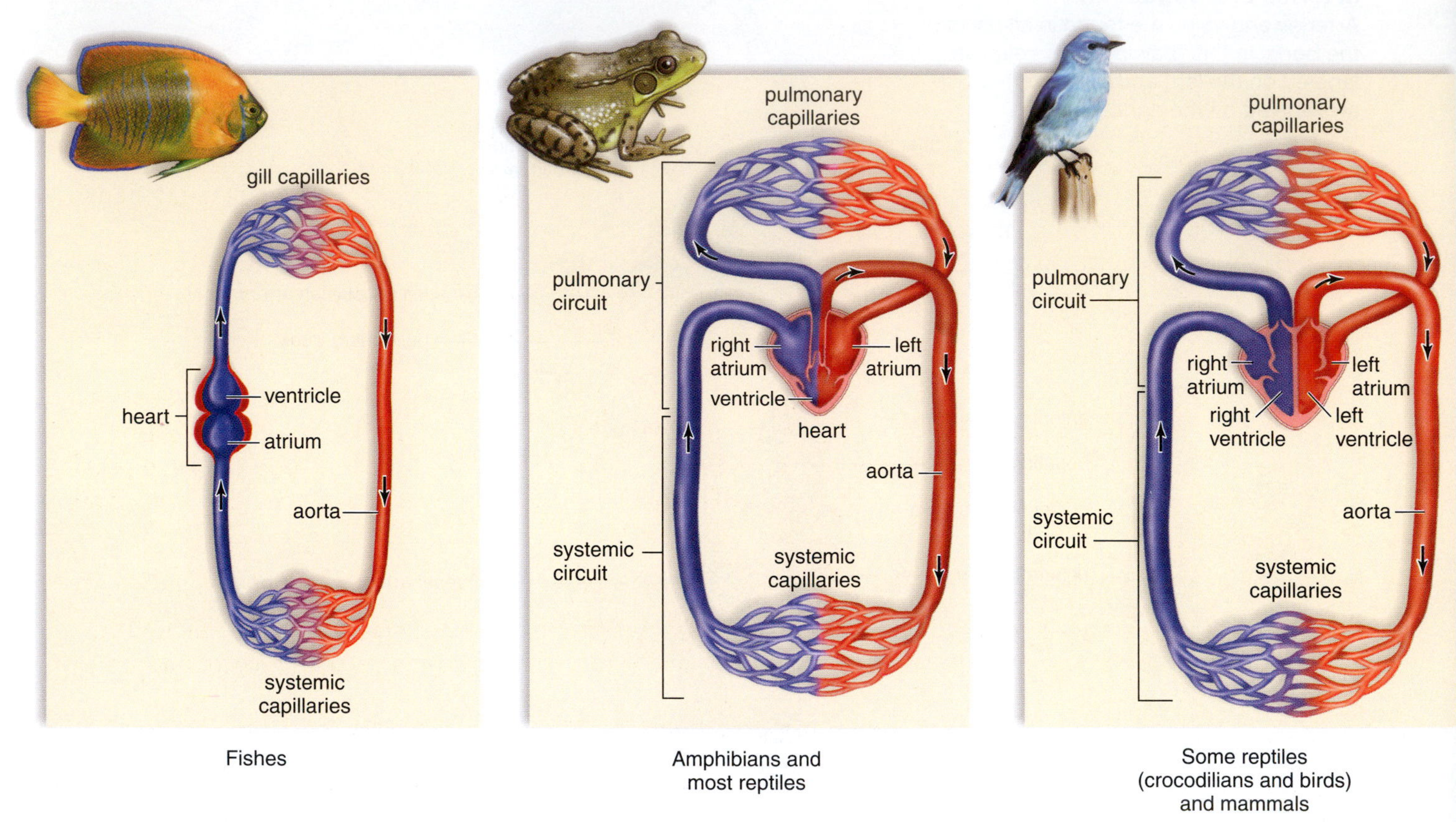

Figure 22.4 Cardiovascular systems in vertebrates.

1. Do fish have a blood vessel that returns blood from the gills to the heart? __________ Would you expect blood pressure to be high or low after blood has moved through the gills? __________
2. What animals studied have pulmonary vessels that take blood from the heart to the respiratory organ and back to the heart? __________ What is the advantage of a pulmonary circuit? __________
3. Which of these animals has a four-chambered heart? __________ What is the advantage of having separate ventricles? __________
4. The circulatory system distributes the heat of muscle contraction in birds and mammals. Is the anatomy of birds and mammals conducive to maintaining a warm internal temperature? __________ Explain your answer. __________

22.2 Respiratory, Digestive, and Urinary Systems

You will now have the opportunity to examine more carefully the organs of the respiratory and digestive systems you merely observed in Laboratory 20. You will also study the urinary system.

Observation: Organs of the Respiratory System and Digestive System

Respiratory System

1. Open a pig's mouth, insert your blunt probe into the **glottis,** and explore the pathway of air in normal breathing by carefully working the probe down through the **larynx** to the level of the bronchi.
2. If you wish, you can slit open the larynx along its midline and observe the small, paired, lateral flaps inside, known as **vocal cords.** These are not yet well developed in this fetal animal.
3. Remove the heart of your pig by cutting the blood vessels that enter and exit heart. Name these blood vessels as you cut them. You will now see the trachea and how it divides into two bronchi. Carefully cut the trachea crosswise just below the larynx. Do not cut the esophagus, which lies just dorsal to the trachea.
4. Lift out the entire portion of the respiratory system you have just freed: trachea, bronchi, bronchioles, and lungs. Place them in a small container of water. Holding the trachea with your forceps, gently but firmly stroke the lung repeatedly with the blunt wooden base of one of your probes. If you work carefully, the alveolar tissue will be fragmented and rubbed away, leaving the branching system of air tubes and blood vessels.

Digestive System

1. With removal of the heart and lungs, you now have a better view of the esophagus. Open the mouth again, and insert a blunt probe into the **esophagus** (see Fig. 20.3). Then trace the esophagus to the stomach.
2. Open one side of the **stomach,** and examine its interior surface. Does it appear smooth or rough?

   ____
3. At the lower end of the stomach, find the **pyloric sphincter,** the muscle that surrounds the entrance to the upper region of the small intestine, the **duodenum.** The pyloric sphincter regulates the entrance of material into the duodenum from the stomach.
4. Find the gallbladder, which is a small, greenish sac dorsal to the liver you also observed in Laboratory 20, page 261. Dissect more carefully now to find the **bile duct system** (see Fig. 20.5) that conducts the bile to the duodenum.
5. Locate the rest of the **small intestine** and the **large intestine** (colon and rectum) first observed in Laboratory 20, page 271. To complete your study, sever the coiled small intestine just below the duodenum, and sever the colon at the point where it joins the **rectum.**
6. Carefully cut the mesenteries holding the coils of the small intestine so that it can be laid out in a straight line. As defined previously, mesenteries are double-layered sheets of membrane that project from the body wall and support the organs.
7. Measure and record in meters the length of the intestinal tract. __________ m
8. Considering that nutrient molecules are absorbed into the blood vessels of the small intestine, why would such a great length be beneficial to the body? ______________________
9. Slit open a short portion of the intestines, and note the corrugated texture of the interior lining.

Figure 22.5 Urinary system of the fetal pig.
In both sexes, urine is made by the kidneys, transported to the bladder by the ureters, stored in the bladder, and then excreted from the body through the urethra.

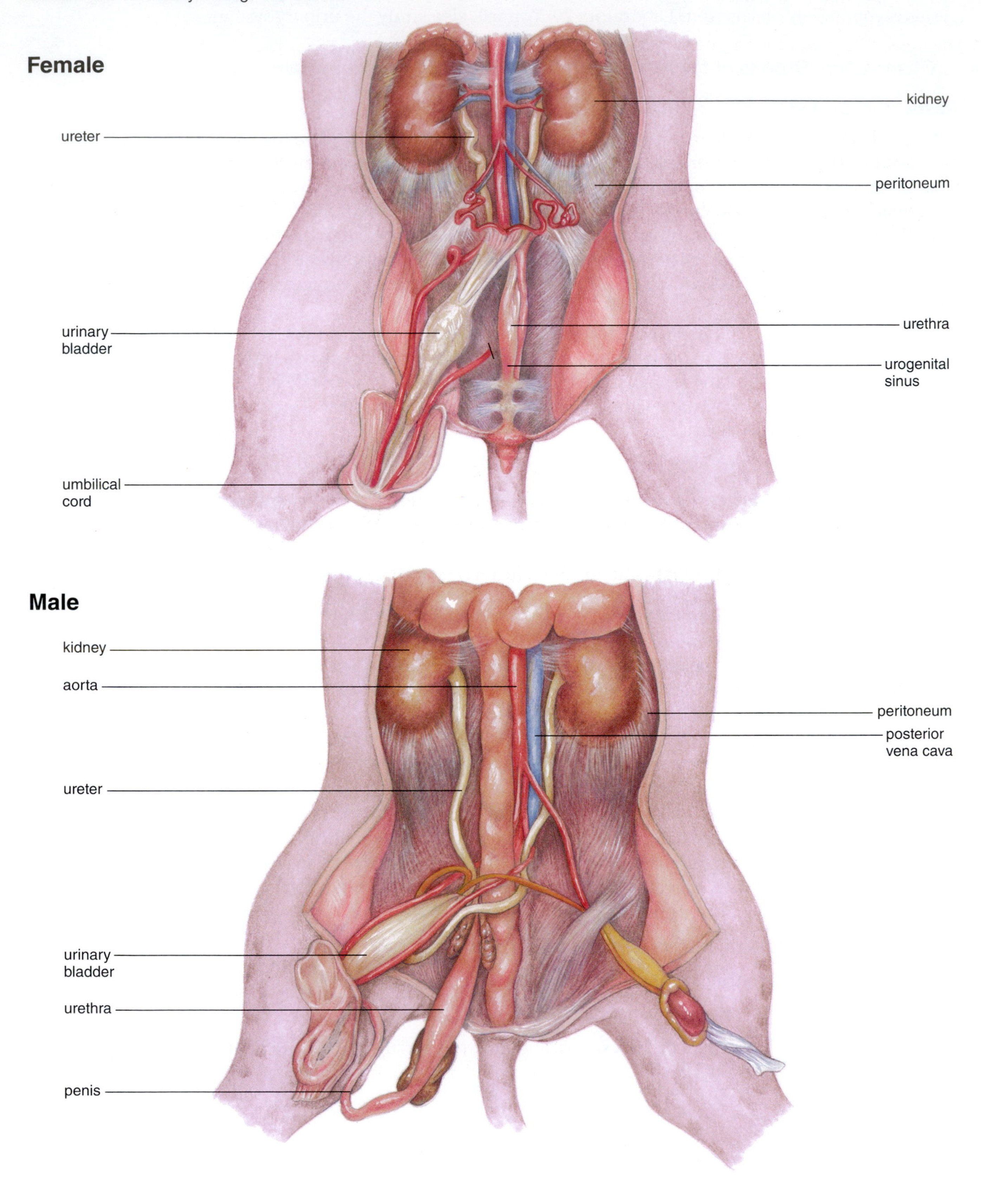

Urinary System in Pigs

The urinary system consists of the **kidneys,** which produce urine; the **ureters,** which transport urine to the **urinary bladder,** where urine is stored; and the **urethra,** which transports urine to the outside. In males, the urethra also transports sperm during copulation.

Observation: Urinary System in Pigs

During this dissection, compare the urinary system structures of male and female fetal pigs. Later in this laboratory period, exchange specimens with a neighboring team for a more thorough inspection.

1. The large, paired kidneys (Fig. 22.5) are reddish organs covered by **peritoneum,** a membrane that anchors them to the dorsal wall of the abdominal cavity, sometimes called the **peritoneal cavity.** Clean the peritoneum away from one of the kidneys, and study it more closely.
2. Locate the **ureters,** which leave the kidneys and run posteriorly under the peritoneum (Fig. 22.5).
3. Clear the peritoneum away, and follow a ureter to the **urinary bladder,** which normally lies in the ventral portion of the abdominal cavity. The urinary bladder is on the inner surface of the flap of tissue to which the umbilical cord was attached.
4. The **urethra** arises from the bladder posteriorly and joins the urogenital sinus. Follow the urethra until it passes from view into the ring formed by the pelvic girdle.
5. Sequence the organs in the urinary system to trace the path of urine from its production to its exit.

____

6. Using a scalpel, section one of the kidneys in place, cutting it lengthwise (Fig. 22.6). At the center of the medial portion of the kidney is an irregular, cavity-like reservoir, the **renal pelvis.** The outermost portion of the kidney (the **renal cortex**) shows many small striations perpendicular to the outer surface. The cortex and the more even-textured **renal medulla** contain **nephrons** (excretory tubules), microscopic organs that produce urine.

Figure 22.6 Anatomy of the kidney.
A kidney has a renal cortex, renal medulla, renal pelvis, and microscopic tubules called nephrons.

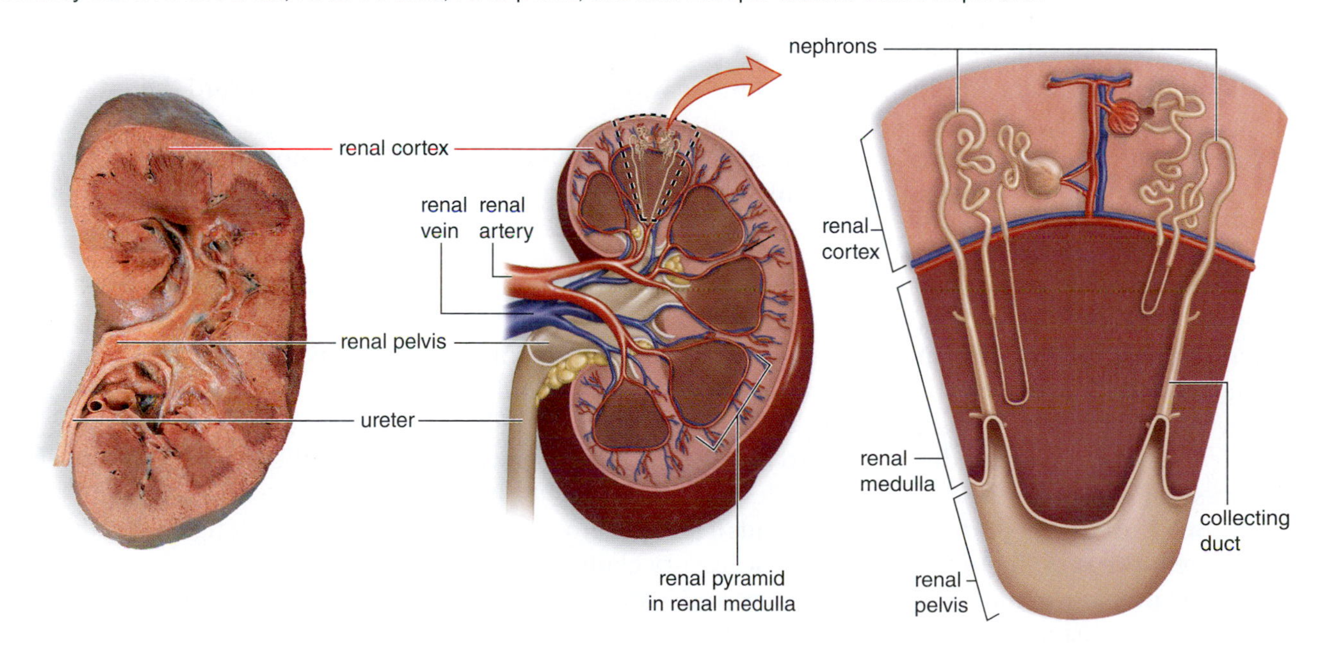

22.3 Male Reproductive System

The **male reproductive system** consists of the **testes** (sing., testis), which produce sperm, and the **epididymides** (sing., epididymis), which store sperm before they enter the **vasa deferentia** (sing., vas deferens). Just prior to ejaculation, sperm leave the vasa deferentia and enter the **urethra,** located in the penis. The **penis** is the male organ of sexual intercourse. **Seminal vesicles,** the **prostate gland,** and the **bulbourethral glands** (Cowper's glands) add fluid to semen (sperm plus fluids) after sperm reach the urethra. Table 22.2 summarizes the male reproductive organs.

The testes begin their development in the abdominal cavity, just anterior and dorsal to the kidneys. Before birth, however, they gradually descend into paired **scrotal sacs** within the scrotum, suspended anterior to the anus. Each scrotal sac is connected to the body cavity by an **inguinal canal,** the opening of which can be found in your pig. The passage of the testes from the body cavity into the scrotal sacs is called the descent of the testes and it occurs in human males. The testes in most of the male fetal pigs being dissected will probably be partially or fully descended.

Observation: Male Reproductive System in Pigs

While doing this dissection, consult Table 22.2 for the function of the male reproductive organs.

Table 22.2 Male Reproductive Organs and Functions

Organ	Function
Testis	Produces sperm and sex hormones
Epididymis	Stores sperm as they mature
Vas deferens	Stores sperm and conducts sperm to urethra
Urethra	Conducts sperm to urogenital opening
Seminal vesicle Prostate gland Bulbourethral glands	Contributes secretions to semen
Penis	Organ of copulation

Inguinal Canal, Testis, Epididymis, and Vas Deferens

1. Locate the opening of the left inguinal canal, which leads to the left scrotal sac (Fig. 22.7).
2. Expose the canal and sac by making an incision through the skin and muscle layers from a point over this opening back to the left scrotal sac.
3. Open the sac, and find the testis. Note the much-coiled tubule—the epididymis—that lies alongside a testis. An epididymis is continuous with a vas deferens, which runs toward the abdominal cavity.
4. Each vas deferens loops over an umbilical artery and ureter and unites with the urethra as it leaves the urinary bladder. We will dissect this juncture below.

Penis, Urethra, and Accessory Glands

1. Cut through the ventral skin surface just posterior to the umbilical cord. This will expose the rather undeveloped penis, which contains a long portion of the urethra.
2. Lay the penis to one side, and then cut down through the ventral midline, laying the legs wide apart in the process (Fig. 22.8). The cut will pass between muscles and through pelvic cartilage (bone has not developed yet). Do not cut any of the ducts or tracts in the region.
3. You will now see the urethra ventral to the rectum. It is somewhat heavier in the male due to certain accessory glands:
   a. Bulbourethral glands (Cowper's glands), about 1 cm in diameter, are further along the urethra and are more prominent than the other accessory glands.
   b. The prostate gland, about 4 mm across and 3 mm thick, is located on the dorsal surface of the urethra, just posterior to the juncture of the bladder with the urethra. It is often difficult to locate and is not shown in Figures 22.7 and 22.8.
   c. Small, paired seminal vesicles may be seen on either side of the prostate gland.

Figure 22.7 Male reproductive system of the fetal pig.
In males, the urinary system and the reproductive system are joined. The vasa deferentia (sing., vas deferens) enter the urethra, which also carries urine.

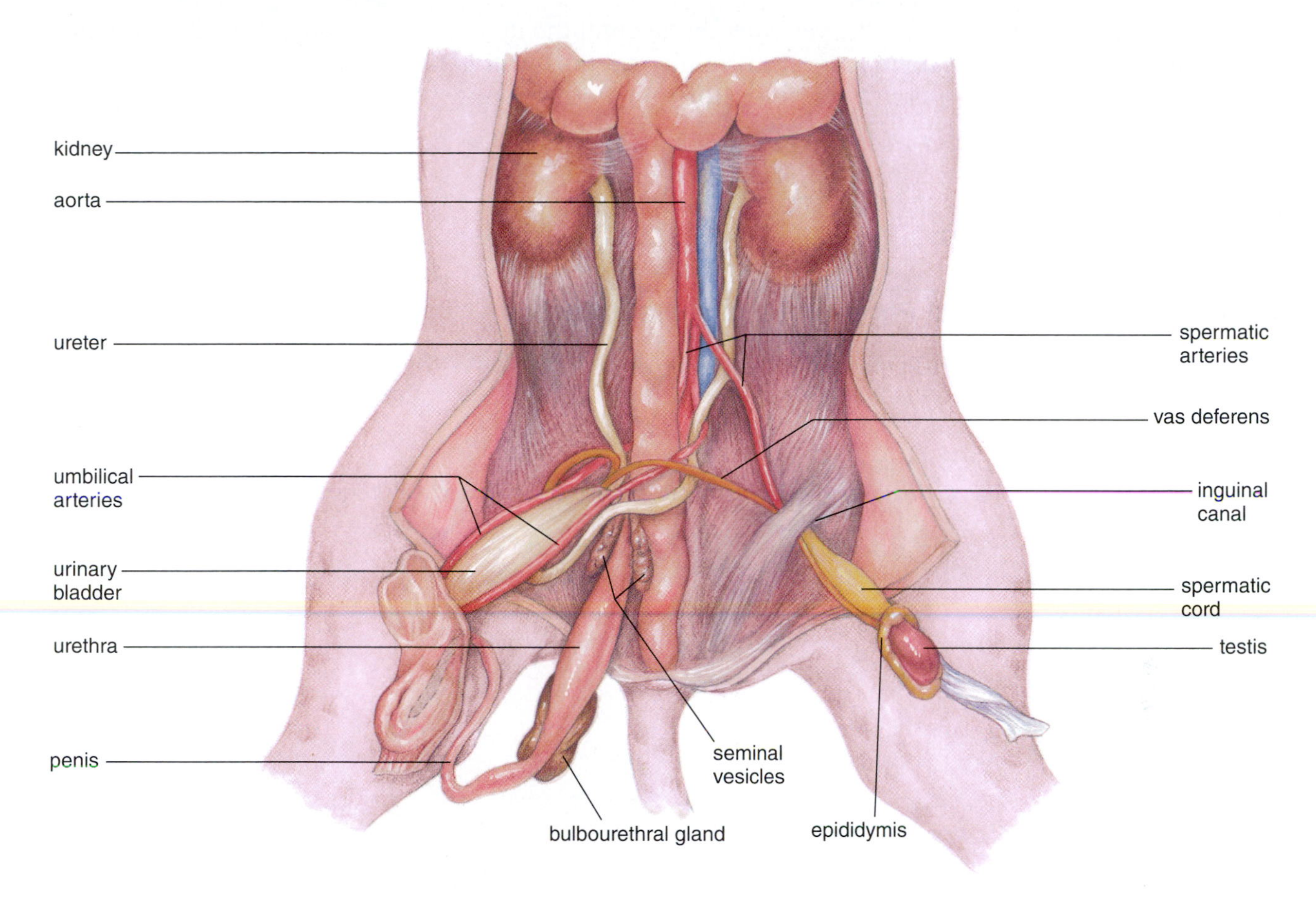

Figure 22.8 Photograph of the male reproductive system of the fetal pig.
Compare the diagram in Figure 22.7 to this photograph to help identify the structures of the male urogenital system.

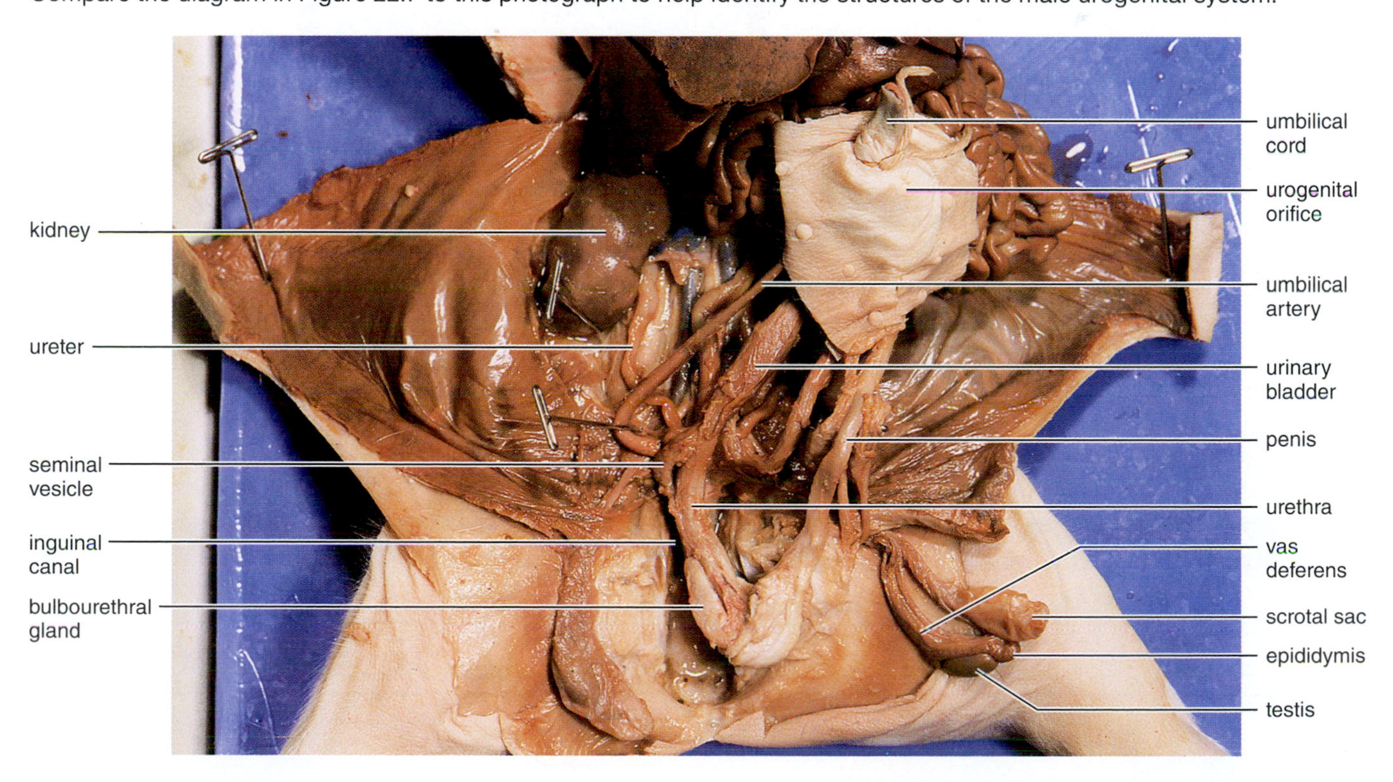

4. Trace the urethra as it leaves the bladder. When it nears the end of the abdominal cavity, it turns rather abruptly and runs anterorally just under the skin where you have just dissected it. This portion of the urethra is within the penis.
5. Now you should be able to see the vasa deferentia enter the urethra. If necessary, free these ducts from surrounding tissue to see them enter the urethra near the location of the prostate gland. In males, the urethra transports both sperm and urine to the urogenital opening (Fig. 20.1*c*).
6. Sequence the organs in the male reproductive system to trace the path of sperm from the organ of production to the penis. ____

____

Comparison of Male Fetal Pig and Human Male

Use Figure 22.9 to help you compare the male pig reproductive system with the human male reproductive system. Complete Table 22.3, which compares the location of the penis in these two mammals.

Figure 22.9 Human male urogenital system.
In the fetal pig, but not in the human male, the penis lies beneath the skin and exits at a urogenital opening.

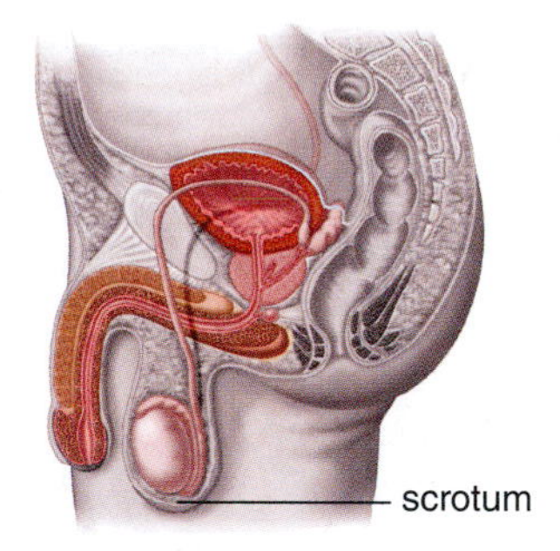

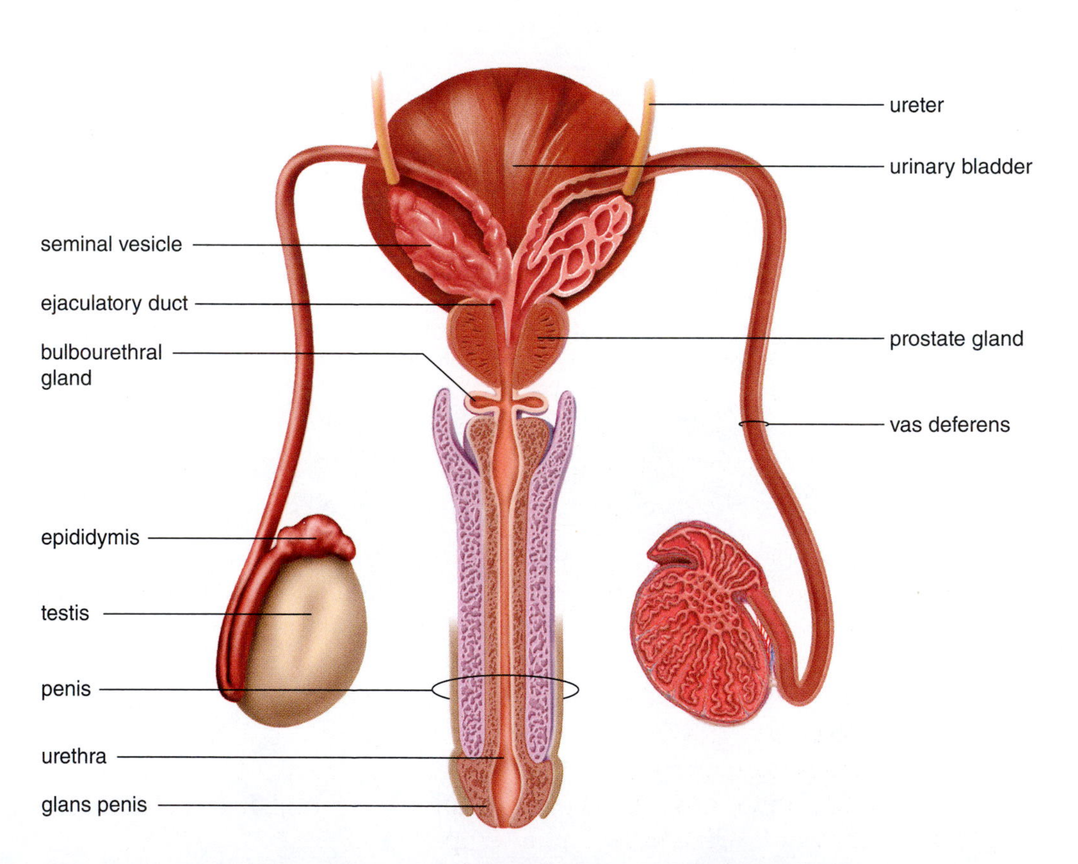

Table 22.3 Location of Penis in Male Fetal Pig and Human Male

	Fetal Pig	Human
Penis		

22.4 Female Reproductive System

The **female reproductive system** (Table 22.4) consists of the **ovaries,** which produce eggs, and the **oviducts,** which transport eggs to the **uterus,** where development occurs. In the fetal pig, the uterus does not form a single organ, as in humans, but is partially divided into external structures called **uterine horns,** which connect with the oviduct. The **vagina** is the birth canal and the female organ of sexual intercourse.

Table 22.4 Female Reproductive Organs and Functions

Organ	Function
Ovary	Produces egg and sex hormones
Oviduct (fallopian tube)	Conducts egg toward uterus
Uterus	Houses developing offspring
Vagina	Receives penis during copulation and serves as birth canal

Observation: Female Reproductive System in Pigs

While doing this dissection, consult Table 22.4 for the function of the female reproductive organs.

Ovaries and Oviducts

1. Locate the paired ovaries, small bodies suspended from the peritoneal wall in mesenteries, posterior to the kidneys (Figs. 22.10 and 22.11).
2. Closely examine one ovary. Note the small, short, coiled **oviduct,** sometimes called the fallopian tube. The oviduct does not attach directly to the ovary but ends in a funnel-shaped structure with fingerlike processes (fimbriae) that partially encloses the ovary. The egg produced by an ovary enters an oviduct where it is fertilized by a sperm, if reproduction will occur. Any resulting embryo passes to the uterus.

Uterine Horns

1. Locate the **uterine horns.** (Do not confuse the uterine horns with the oviducts; the latter are much smaller and are found very close to the ovaries.)
2. Find the body of the uterus located where the uterine horns join.

Vagina

1. Separate the hindlimbs of your specimen, and cut down along the midventral line. The cut will pass through muscle and the cartilaginous pelvic girdle. With your fingers, spread the cut edges apart, and use blunt dissecting instruments to separate connective tissue.
2. Now find the vagina, which passes from the uterus to the **urogenital sinus.** The vagina is dorsal to the urethra, which also enters the urogenital sinus, and ventral to the **rectum,** which exits at the **anus.** The urogenital sinus opens at the urogenital papilla (Figs. 22.10 and 22.11).
3. The vagina is the organ of copulation and is the birth canal. The receptacle for the vagina in a pig, the urogenital sinus, is absent in adult humans and several other adult female mammals in which both the urethra and vagina have their own openings.
4. The vagina plays a critical role in reproduction even though development of the offspring occurs in the uterus. Explain. ____

   ____

Figure 22.10 Female reproductive system of the fetal pig.
In a pig, both the vagina and the urethra enter the urogenital sinus, which opens at the urogenital papilla.

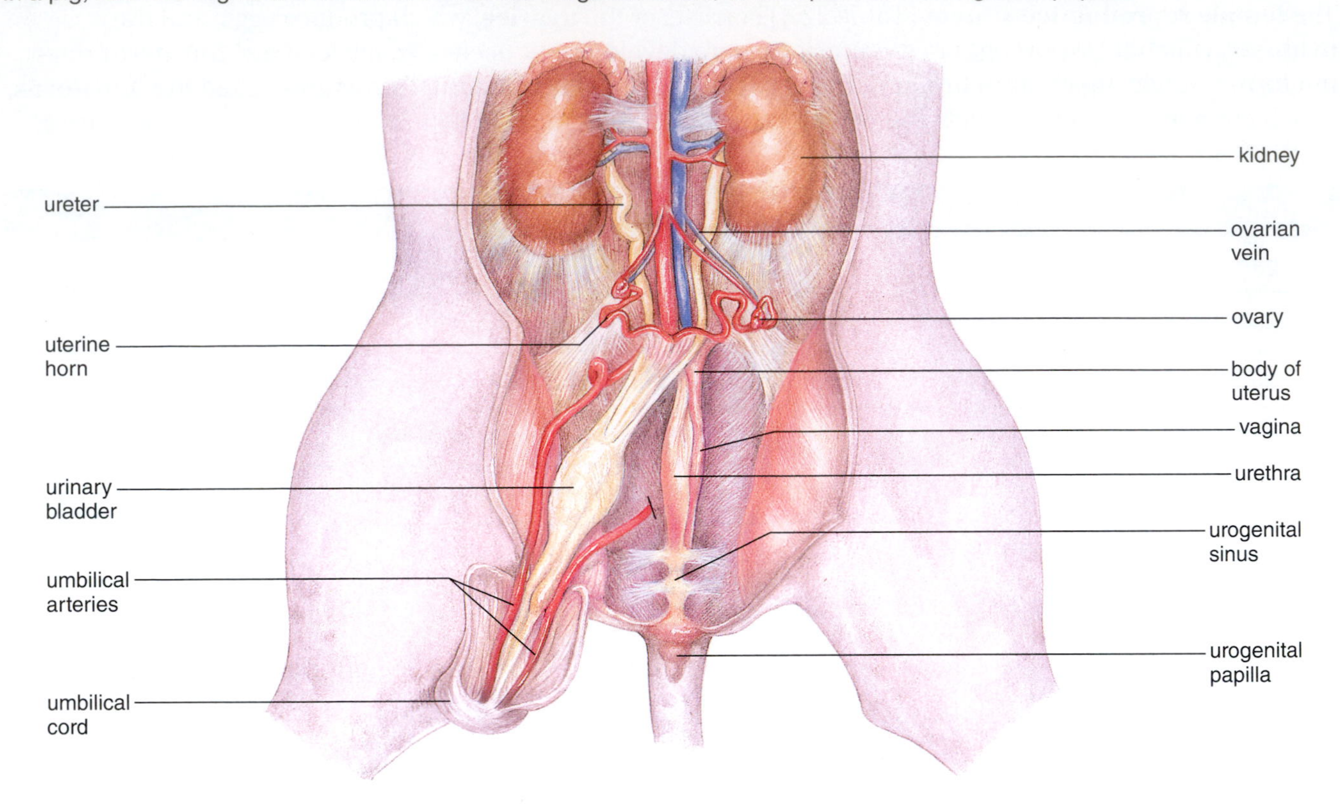

Figure 22.11 Photograph of the female reproductive system of the fetal pig.
Compare the diagram in Figure 22.10 with this photograph to help identify the structures of the female urinary and reproductive systems.

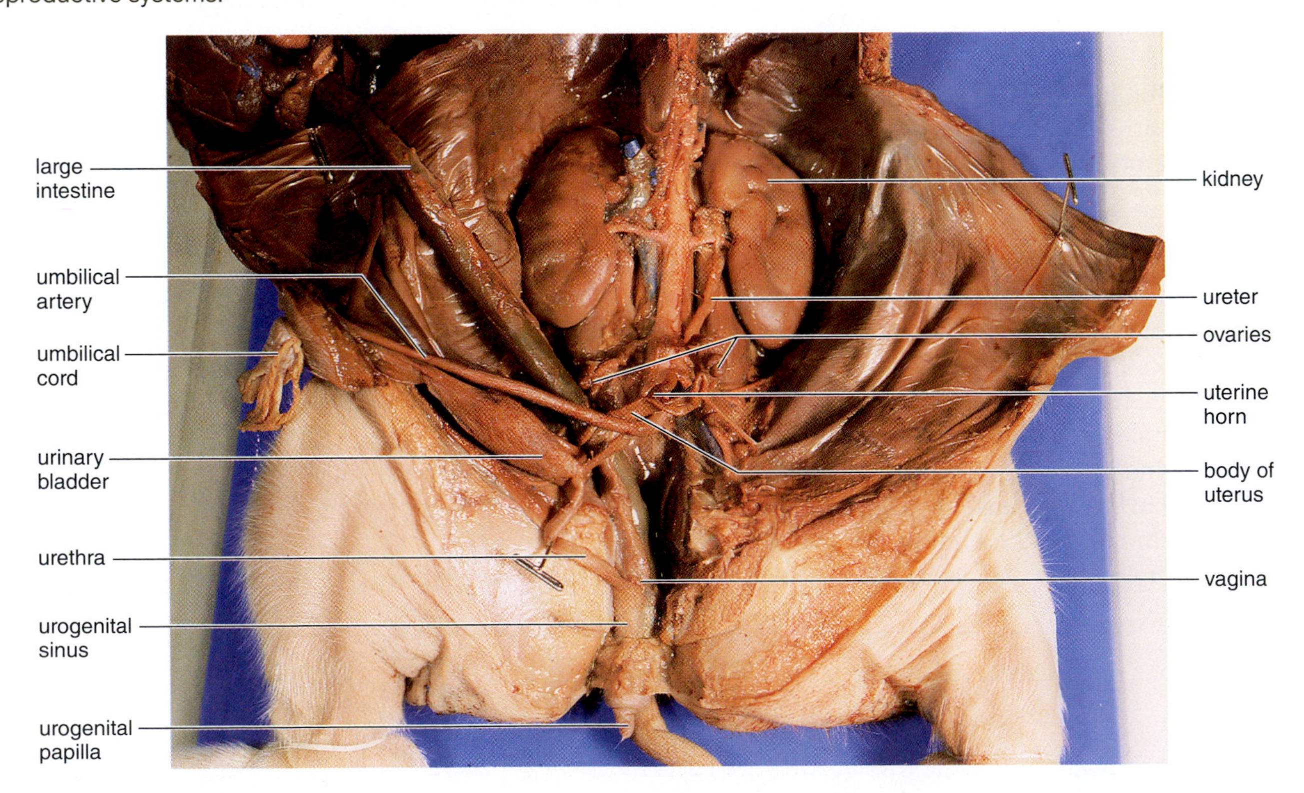

Comparison of Female Fetal Pig with Human Female

Use Figure 22.12 to compare the female pig reproductive system with the human female reproductive system. Complete Table 22.5, which compares the appearance of the oviducts and the uterus, as well as the presence or absence of a urogenital sinus in these two mammals.

Figure 22.12 Human female reproductive system.
Especially compare the anatomy of the oviducts in humans with that of the uterine horns in a pig. In a pig, the fetuses develop in the uterine horns; in a human female, the fetus develops in the body of the uterus.

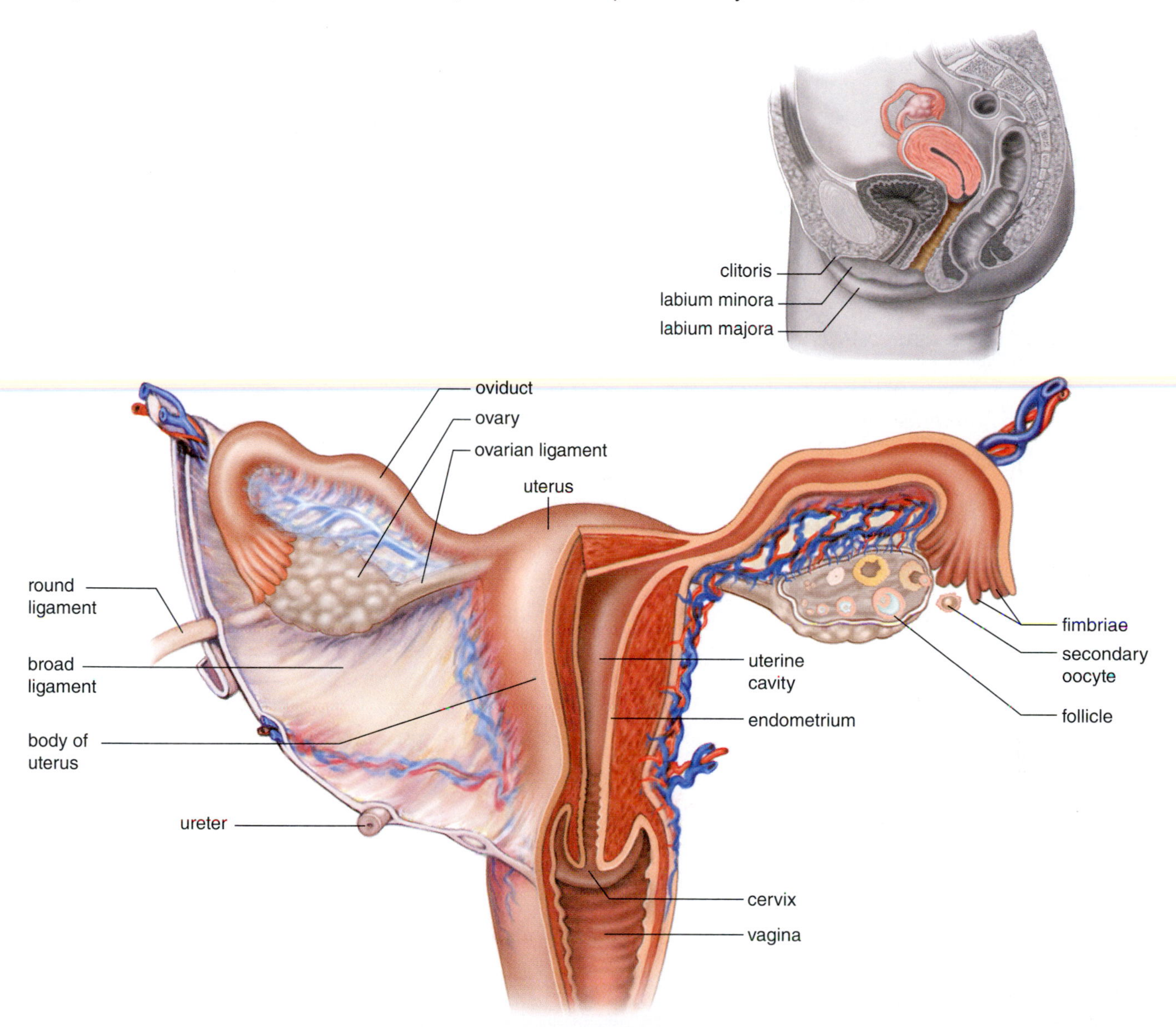

Table 22.5 Comparison of Female Fetal Pig with Human Female

	Fetal Pig	Human
Oviducts		
Uterus		
Urogenital sinus		

Storage of Pigs

Before leaving the laboratory, dispose of or store your pig according to the directions of your instructor. Clean the dissecting tray and tools, and return them to their proper location. Wipe off your goggles and wash your hands.

Laboratory Review 22

1. What are the four chambers of the mammalian heart? ______________________________

2. Contrast the pumping function of the right and left sides of the heart. ______________________________

3. Sequence the blood vessels in the systemic circuit to trace the path of blood from the left ventricle to the kidneys and back to the right atrium. ______________________________

4. Sequence the organs of the respiratory system from the glottis to the lungs. ______________________________

5. What's the difference between the ureters and the urethra in the urinary system? ______________________________

6. Sequence the following organs: stomach, large intestine, small intestine, pharynx, mouth, esophagus, and anus. ______________________________

7. Sequence the path of sperm from the testes to the urogenital opening. ______________________________

8. What organs enter the urogenital sinus in female pigs? ______________________________

9. Which organ in males produces sperm, and which organ in females produces eggs? ______________________________

10. How and when do sperm acquire access to an egg in mammals? ______________________________

Concepts of Biology Website

Instructors can find lab prep information and answers to all of the laboratory questions in the Laboratory Resource Guide. *Students* can practice their knowledge with quizzes, animations, flashcards, and much more.

www.mhhe.com/maderconcepts3

McGraw-Hill Access Science Website

An online encyclopedia of science and technology that provides information, including videos, that can enhance the laboratory experience.

www.accessscience.com

LABORATORY

23

Homeostasis

Learning Outcomes

Introduction

Homeostasis refers to the dynamic equilibrium of the body's internal environment. The **internal environment** consists of blood and tissue fluid. The body's cells take nutrients from tissue fluid and return their waste molecules to it. Tissue fluid, in turn, exchanges molecules with the blood. This is called capillary exchange. All internal organs contribute to homeostasis, but this laboratory specifically examines the contributions of the blood, lungs, and kidneys (Fig. 23.1).

Figure 23.1 Contributions of organs to homeostasis.
The lungs exchange gases with blood; the kidneys remove nitrogenous wastes from blood; and the intestinal tract adds nutrients as regulated by the liver to blood.

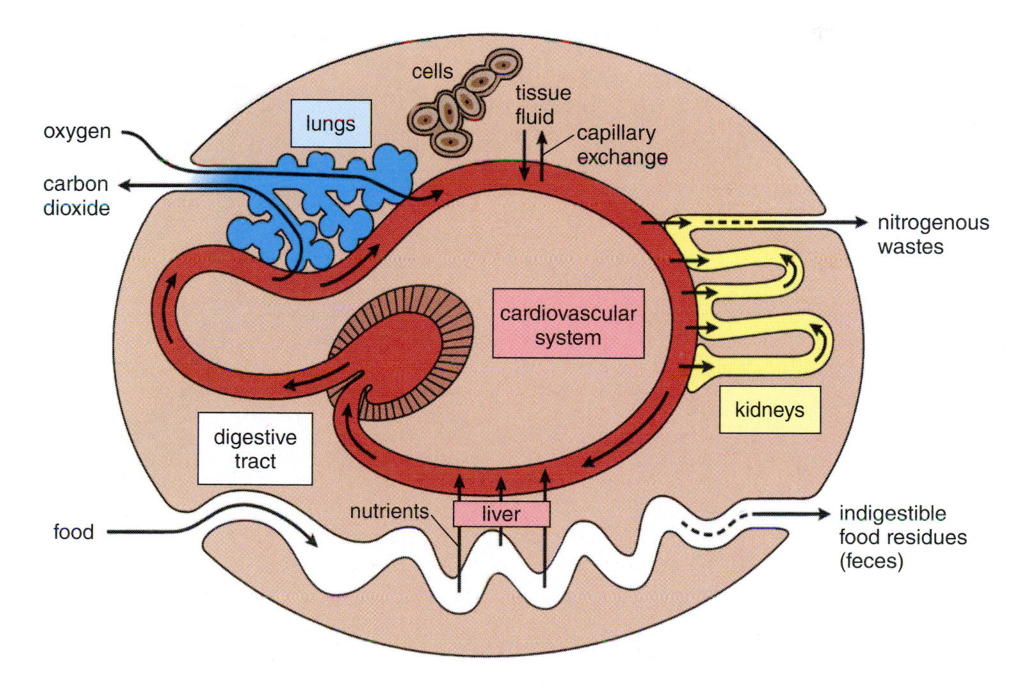

23.1 Heartbeat and Blood Flow

Recall that the cardiovascular system consists of the heart, blood vessels, and blood (Fig. 23.2). Arteries carry blood away from the heart while veins transport blood toward the heart. Arteries branch into smaller vessels called arterioles that enter capillary beds. Capillary beds are present throughout the organs and tissues of the body. An exchange of gases takes place across the thin walls of **pulmonary capillaries.** In the lungs, CO_2 leaves the blood and O_2 enters the blood. An exchange of gases and nutrients for metabolic wastes takes place across the thin walls of **systemic capillaries.** In the body tissues, O_2 and nutrients exit the blood, while CO_2 and metabolic wastes enter the blood.

We will see that the heart is vital to homeostasis because its contraction (called the **heartbeat**) keeps the blood moving in the arteries and arterioles which take blood to the capillaries. The exchanges that take place across capillaries help maintain homeostasis.

Liver

The hepatic portal vein lies between the

_________________________________ and

the _________________________________.

This placement allows the liver to regulate what molecules enter the blood from the digestive tract. For example, if the hormone insulin is present, the liver removes excess glucose and stores it as glycogen. Later, the liver breaks down glycogen to glucose to keep the blood glucose concentration constant.

Figure 23.2 The circulatory system.
The heart provides the pumping action that transports the blood through the arteries, capillary beds, and veins.

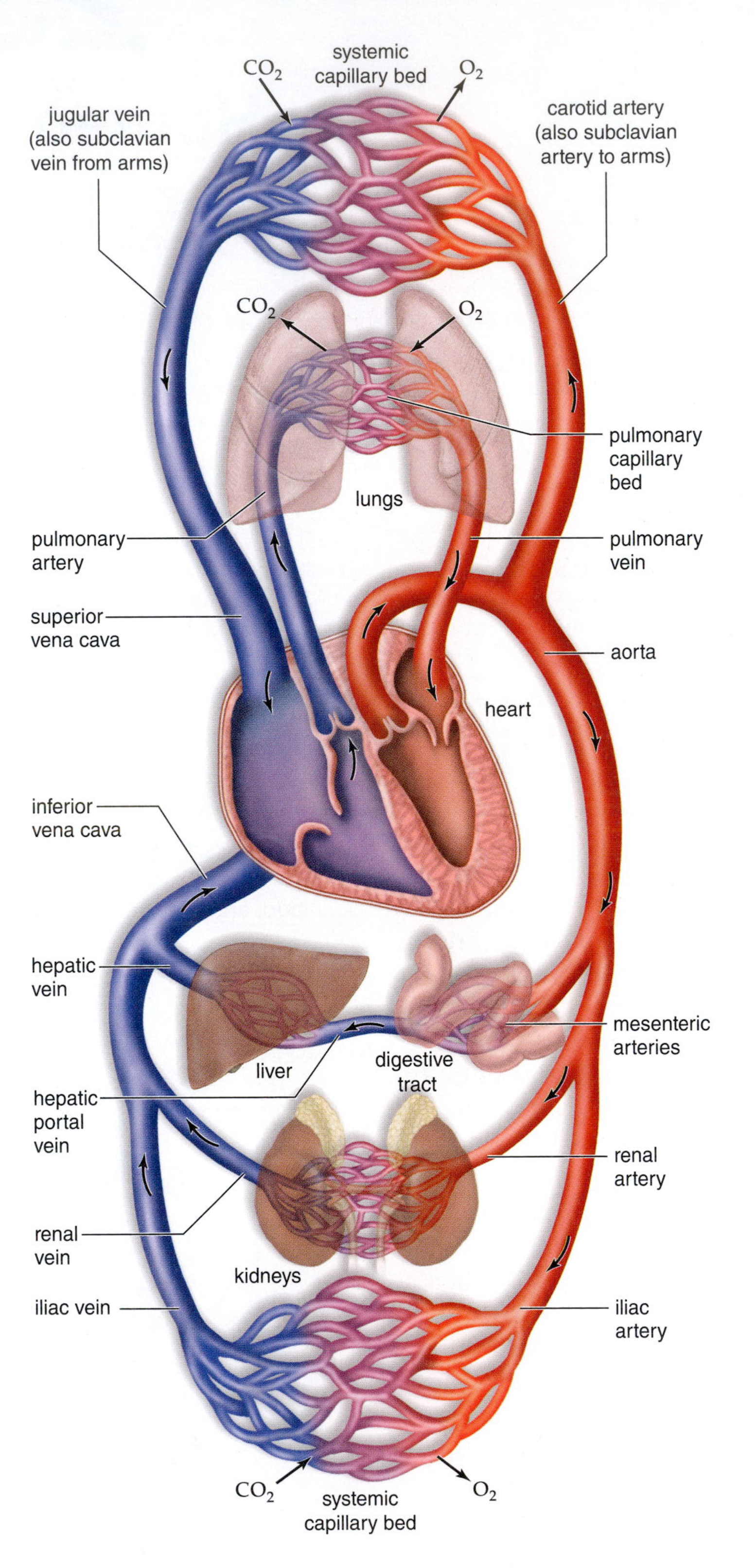

Heartbeat

During a heartbeat, first the atria contract and then the ventricles contract. When a chamber contracts, it is called **systole,** and when a chamber relaxes, it is called **diastole.** The atria and ventricles take turns being in systole.

Time	*Atria*	*Ventricles*
0.15 sec	Systole	Diastole
0.30 sec	Diastole	Systole
0.40 sec	Diastole	Diastole

Usually, there are two heart sounds with each heartbeat (Fig. 23.3). The first sound *(lub)* is low and dull and lasts longer than the second sound. It is caused by the closure of valves following atrial systole. The second sound *(dub)* follows the first sound after a brief pause. The sound has a snapping quality of higher pitch and shorter duration. The *dub* sound is caused by the closure of valves following ventricle systole.

Figure 23.3 The heartbeat sounds.
a. When the atria contract (are in systole), the ventricles fill with blood. **b.** Closure of the valves between atria and ventricles results in a **lub** sound. When the ventricles contract, blood enters the attached arteries (aorta and pulmonary trunk). **c.** Closure of valves between ventricles and arteries results in a **dub** sound. Blood enters the heart from the attached veins (vena cavae and pulmonary veins) once more.

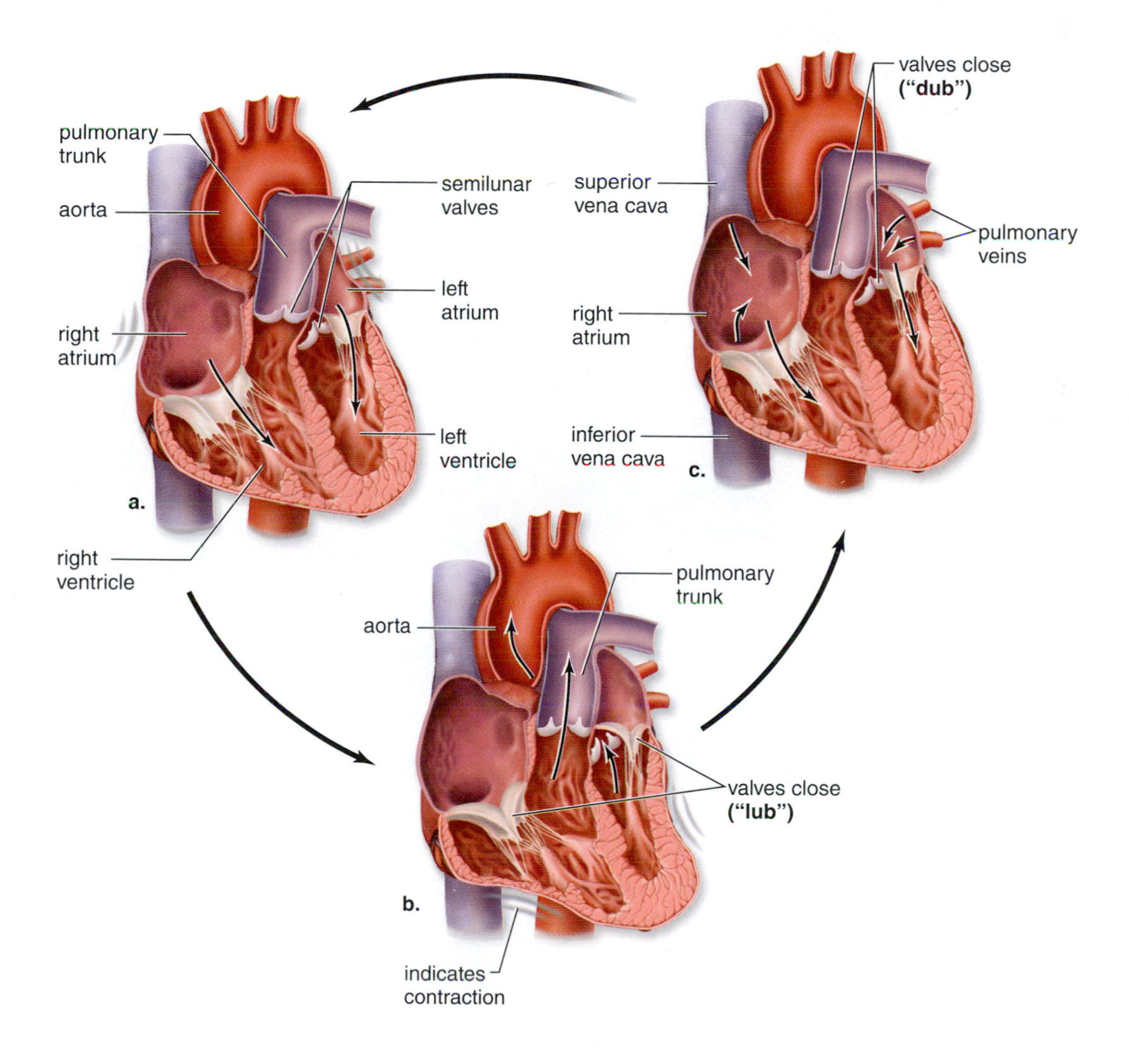

Experimental Procedure: Heartbeat at Rest

In the following procedure, you will work with a partner and use a stethoscope to listen to the heartbeat. It will not be necessary for you to count the number of beats per minute.

1. Obtain a stethoscope, and properly position the earpieces. They should point forward. Place the bell of the stethoscope on the left side of your partner's chest between the fourth and fifth ribs. This is where the apex (tip) of the heart is closest to the body wall.
2. Which of the two sounds (lub or dub) is louder? ______________________________
3. Now switch, and your partner will determine your heartbeat.

Blood Pressure

Blood pressure is highest just after ventricular systole (contraction) and it is lowest during ventricular diastole (relaxation). Why? ______________________________

We would expect a person to have lower blood pressure readings at rest than after exercise. Why?

____

Experimental Procedure: Blood Pressure and Pulse at Rest and After Exercise

A number of different types of digital blood pressure monitors are available, and your instructor will instruct you on how to use the type you will be using for this Experimental Procedure. The resting blood pressure readings for an individual are displayed on the monitor shown in Figure 23.4. A blood pressure reading at or below 120/80 (systolic/diastolic) is considered normal.

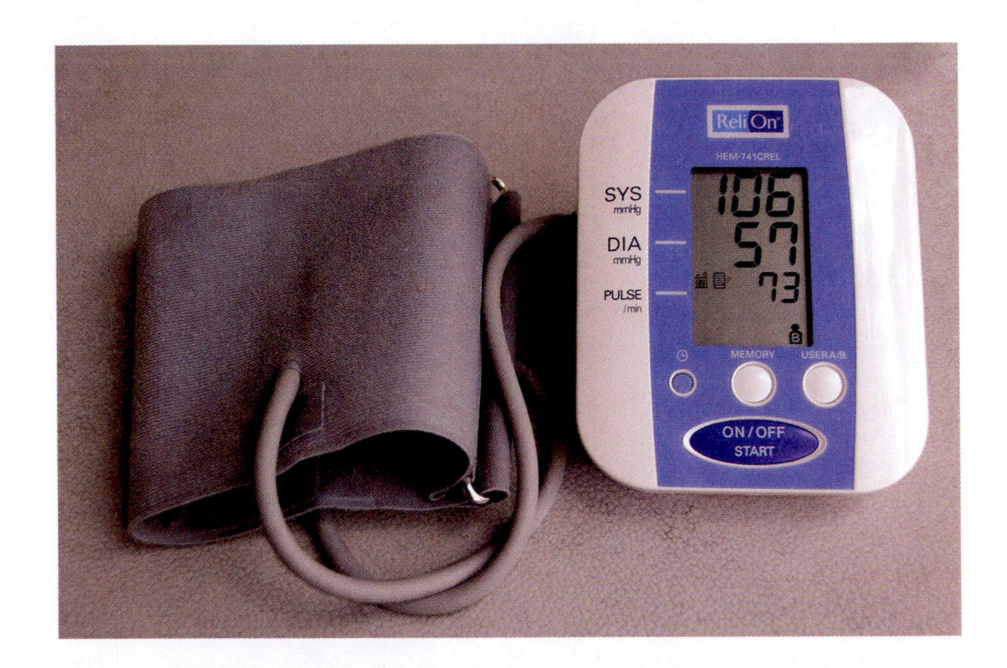

Figure 23.4 Measurement of blood pressure and pulse. There are many different types of digital blood pressure/pulse monitors now available. The one shown here uses a cuff to be placed on the arm. Others use a cuff for the wrist.

During this Experimental Procedure, you may work with a partner or by yourself. If working with a partner, each of you will assist the other in taking blood pressure readings. When you note the blood pressure readings, also note the pulse reading.

Blood Pressure and Pulse at Rest

1. Reduce your activity as much as possible.
2. Use the blood pressure monitor to obtain several blood pressure readings, average them, and record your results in Table 23.1. Also note the pulse rates and average. Record in Table 23.1.

Blood Pressure and Pulse After Exercise

1. Run in place for 1 minute.
2. Immediately use the blood pressure monitor to obtain a blood pressure reading, and record it in Table 23.1. Also note the pulse rate and record in Table 23.1.

Table 23.1 Blood Pressure

	Rates at Rest		Rates After Exercise	
	Blood Pressure	Pulse	Blood Pressure	Pulse
Partner				
Yourself				

Conclusions: Blood Pressure

- Knowing that exercise increases the heart rate, offer an explanation for your results. ______________________
- Under what conditions in everyday life would you expect the heart rate and the blood pressure to increase, even though you were not exercising? ______________________

  When might this be an advantage? ______________________

  A disadvantage? ______________________

23.2 Blood Flow and Systemic Capillary Exchange

We associate death with lack of a heartbeat, but the real problem is lack of blood flow to the capillaries.

Blood Flow

The beat of the heart moves blood into the aorta, which divides into arterioles, and then arterioles divide into capillaries. Venules, which receive blood from capillaries, combine to form veins, which take blood back to the heart.

Experimental Procedure: Blood Flow

1. Observe blood flow through arterioles, capillaries, and venules, either in the tail of a goldfish or in the webbed skin between the toes of a frog, as prepared by your instructor.
2. Examine under low and high power of the microscope.
3. Watch the pulse and the swiftly moving blood in the arterioles.
4. Contrast this with the more slowly moving blood that circulates in the opposite direction in the venules. Many criss-crossing capillaries are visible.
5. Look for blood cells floating in the bloodstream. Don't confuse blood cells with chromatophores, irregular black patches of pigment that may be visible in the skin.

Systemic Capillary Exchange

The beat of the heart creates blood pressure, and blood pressure is necessary to capillary exchange. Blood pressure acts to move water out of a capillary, while osmotic pressure (created by the presence of proteins in the blood) acts to move water into a capillary (Fig. 23.5). Blood pressure is higher than osmotic pressure at the arteriole end of a capillary and water moves out of a capillary. But blood pressure lessens as the blood moves through a capillary bed. This means that osmotic pressure is higher than blood pressure at the venule end of a capillary and water moves back into a capillary.

Figure 23.5 Systemic capillary exchange.
At a systemic capillary, an exchange takes place across the capillary wall. Between the arterial end and the venule end, molecules follow their concentration gradient. Oxygen and nutrients move out of a capillary, while carbon dioxide and wastes move into the capillary.

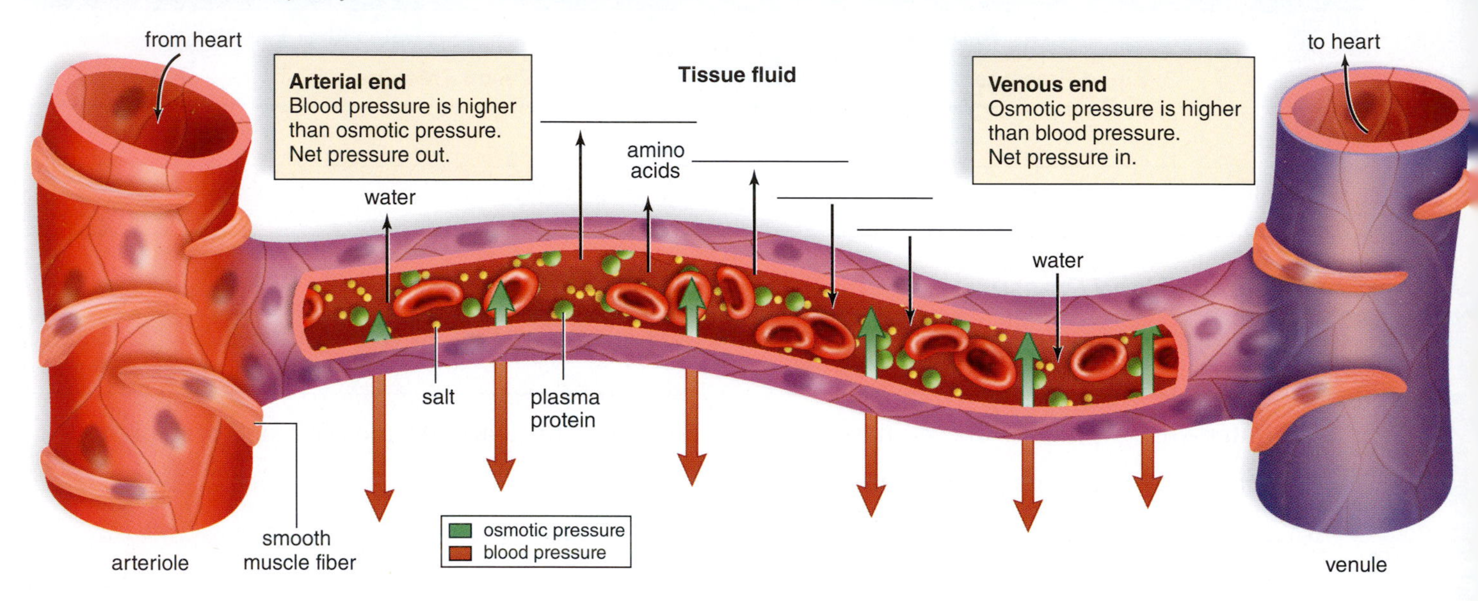

Conclusions: Systemic Capillary Exchange

- What generates blood pressure? ______________________________
- Why are tissue cells always in need of glucose and oxygen? ______________________________

  Add glucose at the end of an appropriate arrow in Figure 23.5. Do the same for oxygen.
- Why are tissue cells always producing carbon dioxide? ______________________________

  Add carbon dioxide at the start of an appropriate arrow in Figure 23.5. Do the same for metabolic wastes.

23.3 Lung Structure and Human Respiratory Volumes

The right and left lungs lie in the thoracic cavity on either side of the heart. Air moves from the nasal passages to the trachea, bronchi, bronchioles, and finally, lungs.

Lung Structure

A **lung** is a spongy organ consisting of irregularly shaped air spaces called **alveoli** (sing., alveolus) (Fig. 23.6*a*). The alveoli are surrounded by a rich network of tiny blood vessels called pulmonary capillaries. *In Fig. 23.6*a*, use a labeled arrow to show oxygen entering blood from an alveoli and another labeled arrow to show carbon dioxide entering alveoli from the blood.*

Observation: Lung Structure

1. Observe a prepared slide of a stained section of a lung (Fig. 23.6*b*). In stained slides, the nuclei of the cells forming the thin alveolar walls appear purple or dark blue.
2. Look for areas that show red or orange disc-shaped **erythrocytes.** These are the red blood cells that contain hemoglobin, which takes up oxygen and transports it to the tissues. When these appear in strings, you are looking at capillary vessels in side view.
3. In some part of the slide, you may even observe an artery. Thicker, circular or oval structures with a lumen (cavity) are cross sections of **bronchioles,** tubular pathways through which air reaches the air spaces.
4. In Figure 23.6*c*, note that in emphysema, alveoli have burst. In smokers, small bronchioles collapse and trapped air in alveoli causes them to burst. Now gas exchange is minimal.

Figure 23.6 Healthy lung tissue versus emphysema.
a. The lungs normally contain many air sacs called alveoli where gas exchange occurs. **b.** Micrograph of normal lung tissue. **c.** In smokers, emphysema can occur; the alveoli burst and gas exchange is inadequate.

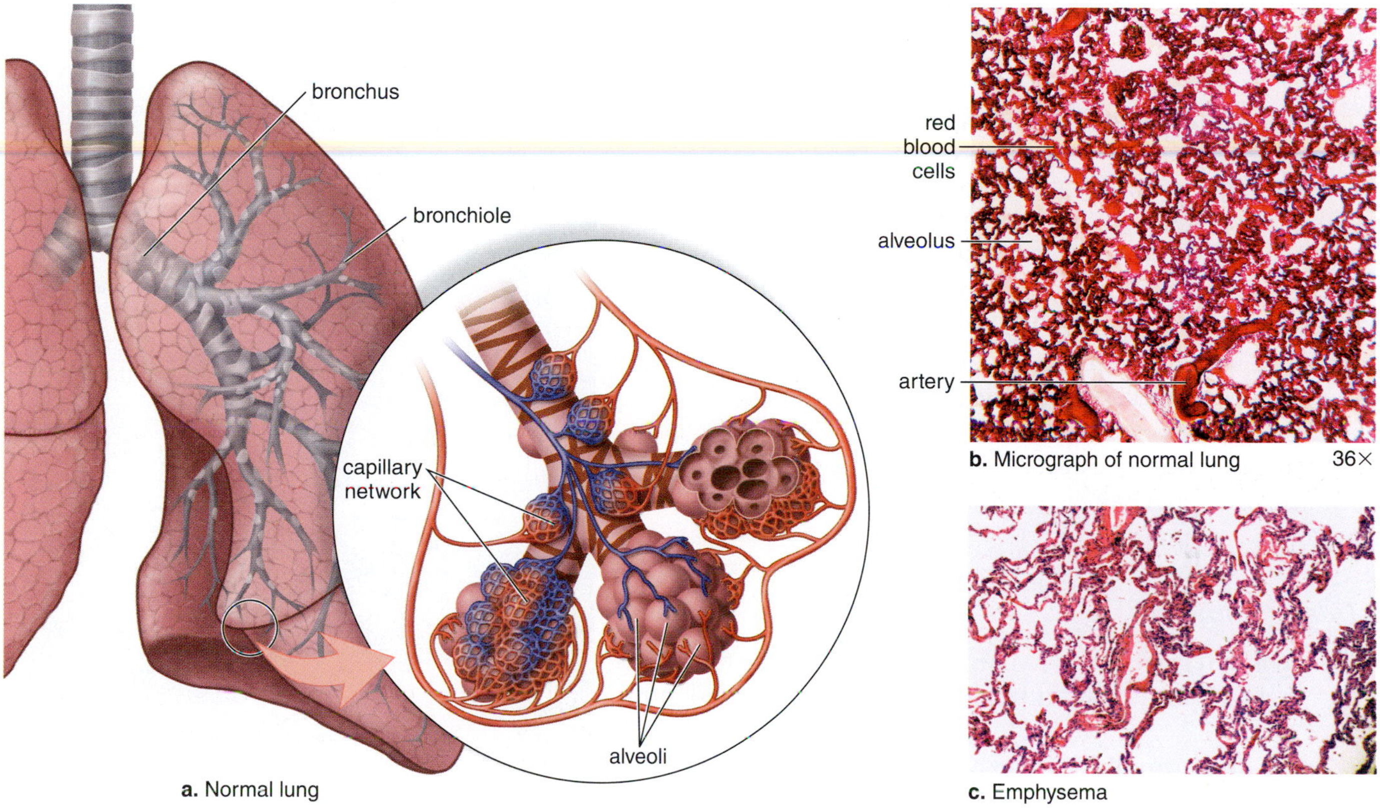

a. Normal lung

b. Micrograph of normal lung 36×

c. Emphysema

Human Respiratory Volumes

Breathing in, called **inspiration** or inhalation, is the active part of breathing because that's when contraction of rib cage muscles causes the rib cage to move up and out, and contraction of the diaphragm causes the diaphragm to lower. Due to an enlarged thoracic cavity, the lungs expand and air is drawn into them. Breathing out, called **expiration,** or exhalation, occurs when relaxation of these same muscles causes the thoracic cavity to resume its original capacity. Now air is pushed out of the lungs (Fig. 23.7).

Figure 23.7 Inspiration and expiration.
a. Inspiration occurs after the rib cage moves up and out and the diaphragm moves down. Air rushes into lungs because they expand as the thoracic cavity expands. **b.** Expiration occurs as the rib cage moves down and in and the diaphragm moves up. As the thoracic cavity and lungs get smaller, air is pushed out.

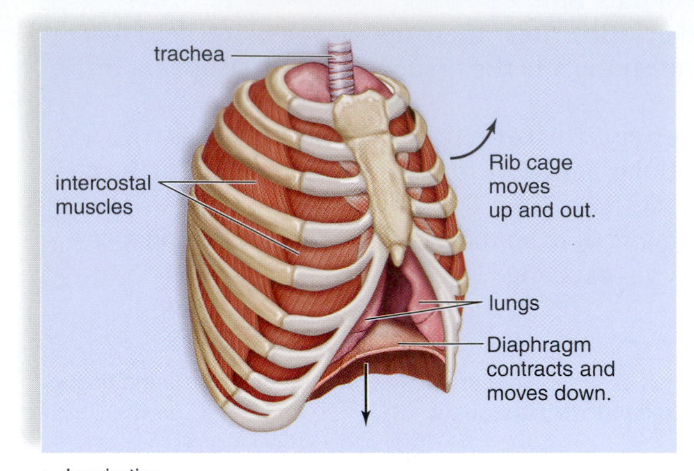

a. Inspiration

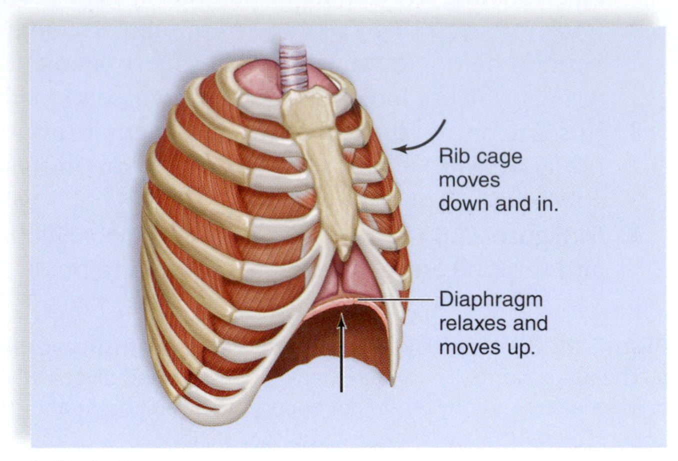

b. Expiration

Experimental Procedure: Human Respiratory Volumes

During this Experimental Procedure, you will be working with a spirometer, an instrument that measures the amount of exhaled air (Fig. 23.8). Normally, about 500 to 600 ml of air move into and out of the lungs with each breath. This is called the **tidal volume (TV).** You can inhale deeply after a normal breath and more air will enter the lungs; this is the **inspiratory reserve volume (IRV).** You can also force more air out of your lungs after a normal breath; this is the **expiratory reserve volume (ERV). Vital capacity** is the volume of air that can be forcibly exhaled after forcibly inhaling.

Figure 23.8 Nine-liter student wet spirometer.

Tidal Volume (TV)

1. When it's your turn to use the spirometer, install a new disposable mouthpiece and set the spirometer to zero.
2. Inhale normally, then exhale normally (with *no* extra effort) through the mouthpiece of the spirometer. Record your measurement in Table 23.2.
3. Three readings are needed, so twice more set the spirometer to zero and repeat the same procedure. Record your measurements in Table 23.2.
4. Later, if necessary, change your readings to milliliters (ml), and calculate your average TV in ml.

   In your own words, what is tidal volume? ____

   ____

Expiratory Reserve Volume (ERV)

1. Make sure the spirometer is set to zero.
2. Inhale and exhale normally and then force as much air out as possible into the spirometer. Record your measurement in Table 23.2.

3. Three readings are needed, so twice more, set the spirometer to zero and repeat the same procedure. Record your measurements in Table 23.2.
4. Later, if necessary, change your readings to ml, and calculate your average ERV.

   In your own words, what is expiratory reserve volume? ____

Vital Capacity (VC)

1. Make sure the spirometer is set to zero.
2. Inhale as much as possible and then exhale as much as possible into the spirometer.
3. Three readings are needed, so twice more, set the spirometer to zero and repeat the same procedure. Record your measurements in Table 23.2.
4. Later, if necessary, change your readings to ml, and calculate your average VC.

   In your own words, what is vital capacity? ____

Inspiratory Reserve Volume (IRV)

It will be necessary for us to calculate IRV because a spirometer measures only exhaled air, not inhaled air. Explain. ____

From having measured vital capacity (VC) you can see that VC = TV + IRV + ERV. To calculate IRV, simply subtract the average TV + the average ERV from the value you recorded for the average VC:

IRV = VC − (TV + ERV) = __________ ml. Record your IRV in Table 23.2.

Table 23.2 Measurements of Lung Volumes

Tidal Volume (TV)	Expiratory Reserve Volume (ERV)	Vital Capacity (VC)	Inspiratory Reserve Volume (IRV)
1st	1st	1st	____
2nd	2nd	2nd	____
3rd	3rd	3rd	____
Average ml	Average ml	Average ml	Calculated value = ml

Conclusions: Human Respiratory Volumes

- Vital capacity varies with age, sex, and height; however, typically for men, vital capacity is about 5,200 ml, and for women it is about 4,000 ml. How does your vital capacity compare to the typical values for your gender? ____________________ If smaller than normal, are you a smoker or is there any health reason why it would be smaller? If larger than normal, are you a sports enthusiast or do you play a musical instrument that involves inhaling and exhaling deeply? ____________________
- Diffusion alone accounts for pulmonary gas exchange. Therefore, how does good lung ventilation assist gas exchange? ____

23.4 Kidneys

The **kidneys** are bean-shaped organs that lie along the dorsal wall of the abdominal cavity.

Kidney Structure

Figure 23.9 shows the structure of a kidney, macroscopic and microscopic. The macroscopic structure of a kidney is due to the placement of over 1 million **nephrons.** Nephrons are tubules that do the work of producing urine.

Figure 23.9 Longitudinal section of a kidney.
a. The kidneys are served by the renal artery and renal vein. **b.** Macroscopically, a kidney has three parts: renal cortex, renal medulla, and renal pelvis. **c.** Microscopically, each kidney contains over a million nephrons.

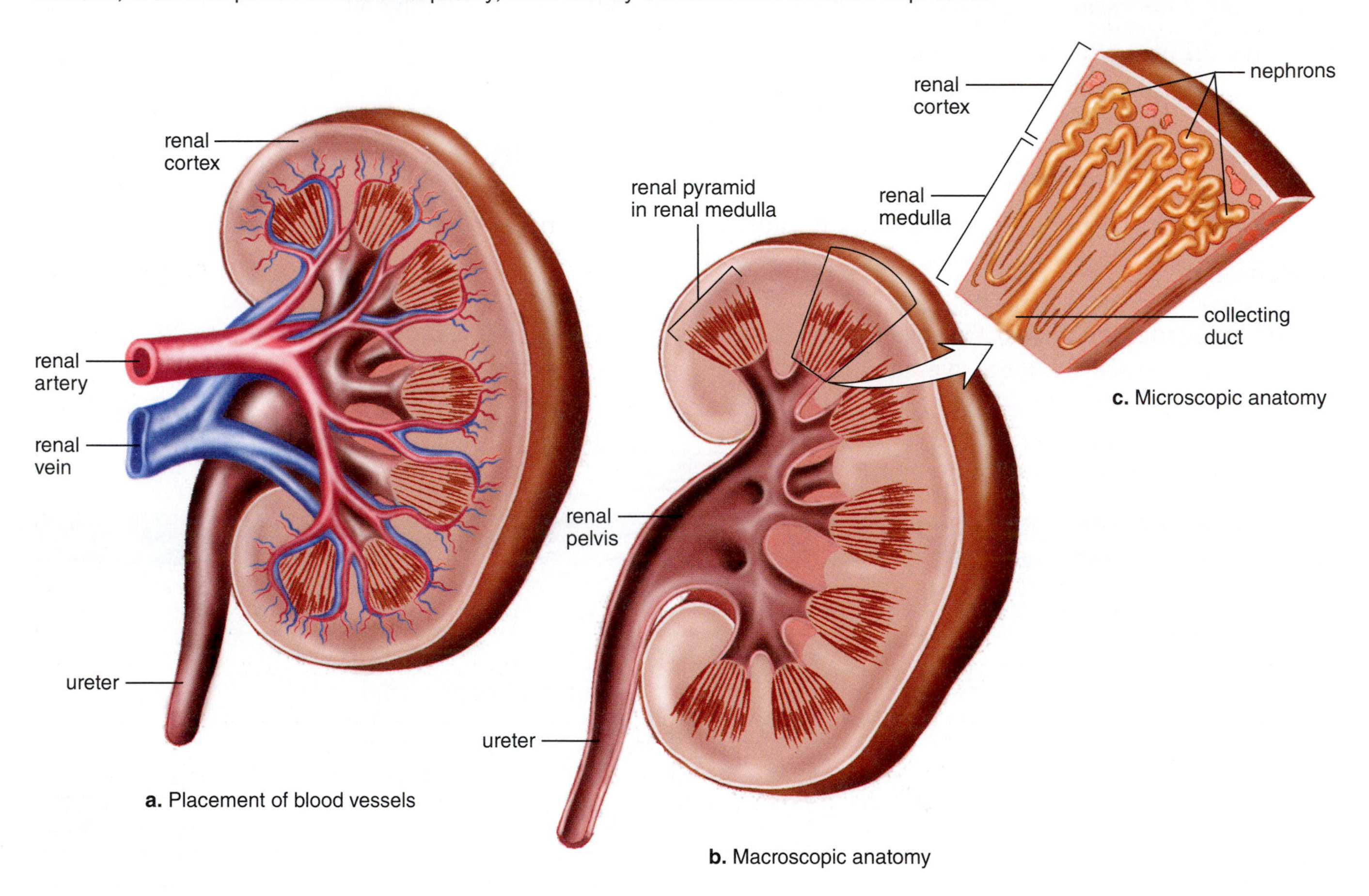

Observation: Kidney Model

Study a model of a kidney, and with the help of Figure 23.9, locate the following:

1. **Renal cortex:** A granular region.
2. **Renal medulla:** Contains the renal pyramids.
3. **Renal pelvis:** Where urine collects.

Observation: Nephron Structure

Study a nephron model and, with the help of Figure 23.10, identify the following parts of a nephron:

1. **Glomerular capsule** (Bowman's capsule): Closed end of the nephron pushed in on itself to form a cuplike structure; the inner layer has pores that allow **glomerular filtration** to occur; substances move from the blood to inside the nephron.
2. **Proximal convoluted tubule:** The inner layer of this region has many microvilli that allow **tubular reabsorption** to occur; substances move from inside the nephron to the blood.
3. **Loop of the nephron:** Nephron narrows to form a U-shaped portion that functions in water reabsorption.
4. **Distal convoluted tubule:** Second convoluted section that lacks microvilli and functions in **tubular secretion;** substances move from blood to inside nephron.

Several nephrons enter one collecting duct. The **collecting ducts** also function in water reabsorption, and they conduct urine to the pelvis of a kidney.

Observation: Circulation About a Nephron

Study a nephron model and, with the help of Figure 23.10 and Table 23.3, trace the path of blood from the renal artery to the renal vein:

1. **Afferent arteriole:** Small vessel that conducts blood from the renal artery to a nephron.
2. **Glomerulus:** Capillary network that exists inside the glomerular capsule; small molecules move from inside the capillary to the inside of the glomerulus during glomerular filtration.
3. **Efferent arteriole:** Small vessel that conducts blood from the glomerulus to the peritubular capillary network.
4. **Peritubular capillary network:** Surrounds the proximal convoluted tubule, the loop of the nephron, and the distal convoluted tubule.
5. **Venule:** Takes blood from the peritubular capillary network to the renal vein.

Table 23.3 Blood Vessels Serving the Nephron

Name of Structure	Significance
Afferent arteriole	Brings arteriolar blood to the glomerulus
Glomerulus	Capillary tuft enveloped by glomerular capsule
Efferent arteriole	Takes arteriolar blood away from the glomerulus
Peritubular capillary network	Capillary bed that envelops the rest of the nephron
Venule	Takes venous blood away from the peritubular capillary network

Kidney Function

The kidneys produce urine and in doing so help maintain homeostasis in several ways. Urine formation requires three steps: **glomerular filtration, tubular reabsorption,** and **tubular secretion** (see Fig. 23.10).

Figure 23.10 Nephron structure and blood supply.
The three main processes in urine formation are described in boxes and color coded to arrows that show the movement of molecules out of or into the nephron at specific locations. In the end, urine is composed of the substances within the collecting duct (see brown arrow).

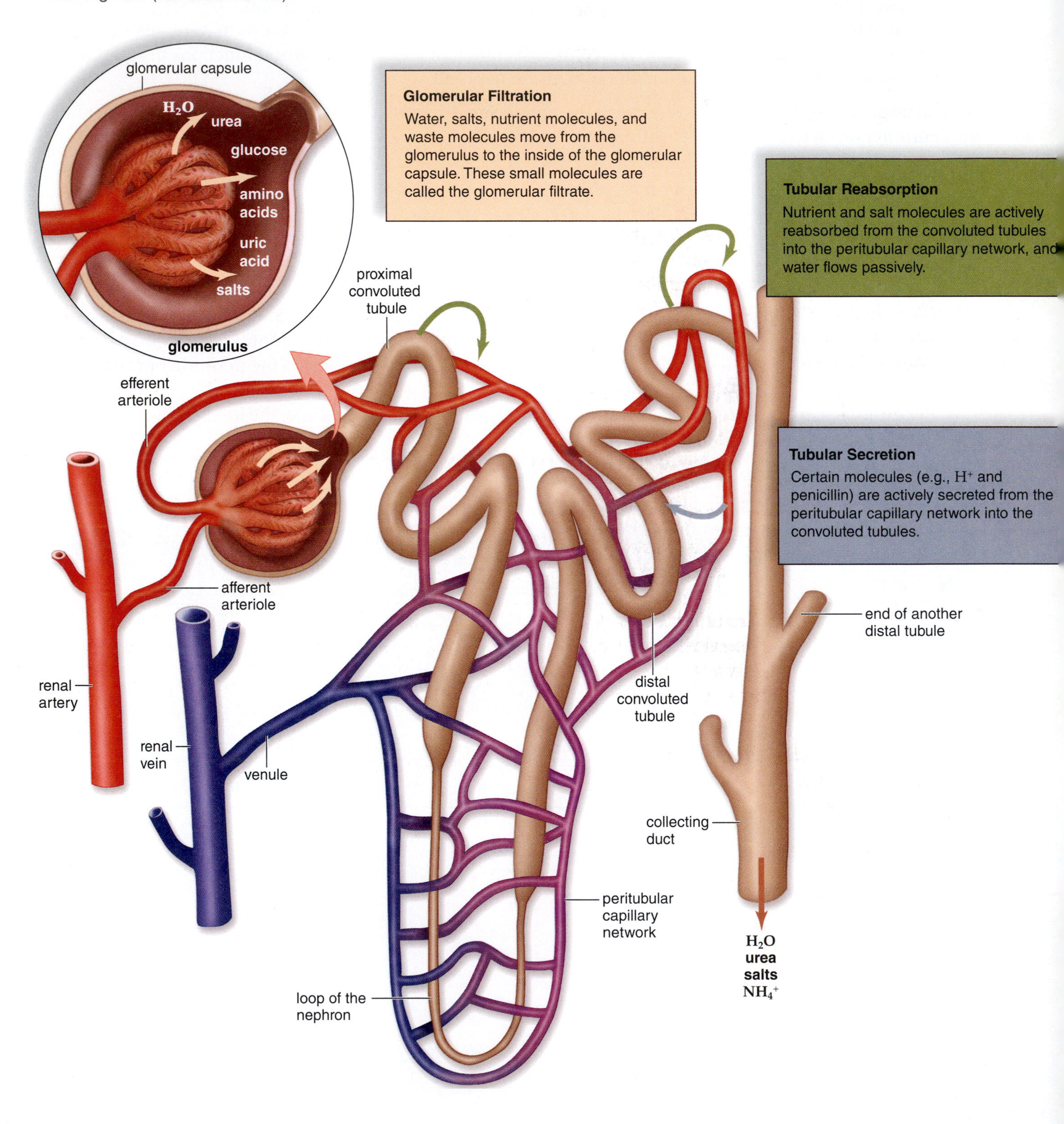

Glomerular Filtration

1. Blood entering the glomerulus contains blood cells, proteins, glucose, amino acids, salts, urea, and water. Blood cells and proteins are too large to pass through the glomerular wall and enter the filtrate.
2. Blood pressure causes small molecules of glucose, amino acids, salts, urea, and water to exit the blood and enter the glomerular capsule. The fluid in the glomerular capsule is called the **filtrate.**
3. In the list that follows, draw an arrow from left to right for the small molecules that leave the glomerulus and become part of the filtrate.

Glomerulus	**Glomerular (Filtrate)**
Cells	
Proteins	
Glucose	
Amino acids	
Urea	
Water and salts	

4. Complete the second column in Table 23.4, page 314. Use an X to indicate that the substance is at the locations noted.

Tubular Reabsorption

1. When the filtrate enters the proximal convoluted tubule, it contains glucose, amino acids, urea, water, and salts. Some water and salts remain in the nephron, but enough are *passively* reabsorbed into the peritubular capillary to maintain blood volume and blood pressure. Use this information to state a way kidneys help maintain homeostasis. ______________________________

____

2. The cells that line the proximal convoluted tubule are also engaged in active transport and usually completely reabsorb nutrients (glucose and amino acids) into the peritubular capillary. What would happen to cells if the body lost all its nutrients by way of the kidneys? ______________________________

____

3. Which of the filtrate substances is reabsorbed the least and will become a part of urine? ______________

Urea is a nitrogenous waste. State here another way kidneys contribute to homeostasis. ______________

____

4. In the list that follows, draw an arrow from left to right for all those molecules passively reabsorbed into the blood of the peritubular capillary. Use darker arrows for those that are reabsorbed completely by active transport.

Proximal Convoluted Tubule	**Peritubular Capillary**
Water and salts	
Glucose	
Amino acids	
Urea	

Tubular Secretion

1. During tubular secretion, certain substances—for example, penicillin and histamine—are actively secreted from the peritubular capillary into the fluid of the tubule. Also, hydrogen ions (H^+) and ammonia (NH_3) are secreted as NH_4^+ as necessary. Complete the last column of Table 23.4. Check your entries against Figure 23.10.
2. The blood is buffered, but only the kidneys can excrete H^+. The excretion of H^+ by the kidneys raises the pH of the blood. Use this information to state a third way the kidneys contribute to homeostasis. ______

______

Table 23.4 Urine Constituents

In Glomerulus	In Filtrate	In Urine
Blood cells		
Proteins		
Glucose		
Amino acids		
Urea		
Water and salts		
NH_4^+		

Urinalysis: A Diagnostic Tool

Urinalysis can indicate whether the kidneys are functioning properly or whether an illness such as diabetes mellitus is present. The procedure is easily performed with a Chemstrip test strip, which has indicator spots that produce specific color reactions when certain substances are present in urine.

Experimental Procedure: Urinalysis

A urinalysis has been ordered, and you are to test the urine for a possible illness. (In this laboratory, you will be testing simulated urine.)

Assemble Supplies

1. Obtain three Chemstrip urine test strips each of which tests for leukocytes, pH, protein, glucose, ketones, and blood, as noted in Figure 23.11.
2. The color key on the diagnostic color chart or on the Chemstrip vial label will explain what any color changes mean in terms of the pH level and amount of each substance present in the urine sample. You will use these color blocks to read the results of your test.
3. Obtain three "specimen containers of urine" marked 1 through 3. Among them are a normal specimen and two that indicate the patient has an illness. Have a piece of absorbent paper ready to use.

Test the Specimen

1. Briefly (no longer than 1 second) dip a test strip into the first specimen of urine. Be sure the chemically treated patches on the test strip are totally immersed.
2. Draw the edge of the strip along the rim of the specimen container to remove excess urine.
3. Turn the test strip on its side, and tap once on a piece of absorbent paper to remove any remaining urine and to prevent the possible mixing of chemicals.
4. After 60 seconds, read the results as follows: Hold the strip close to the color blocks on the diagnostic color chart (Fig. 23.11) or vial label, and match carefully, ensuring that the strip is properly oriented to the color chart. Enter the test results in Figure 23.11. Use a negative symbol (−) for items that are not present in the urine, a plus symbol (+) for those that are present, and a number for the pH.
5. Test the other two specimens.

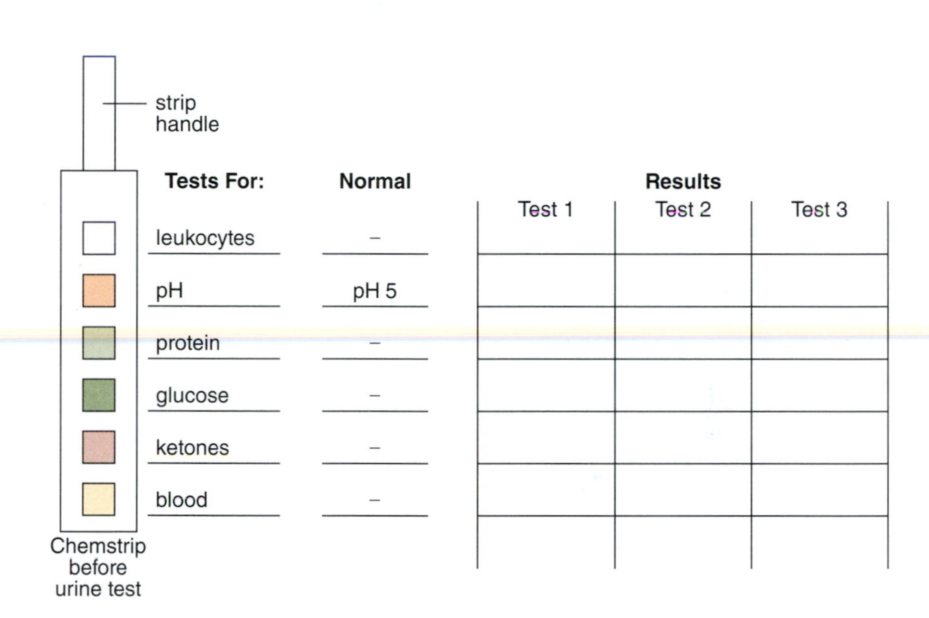

Tests For:	Normal	Results Test 1	Test 2	Test 3
leukocytes	–			
pH	pH 5			
protein	–			
glucose	–			
ketones	–			
blood	–			

Figure 23.11 Urinalysis test. A Chemstrip test strip can help determine illness in a patient by detecting substances in the urine. If leukocytes (white blood cells), protein, or blood are in the urine, the kidneys are not functioning properly. If glucose and ketones are in the urine, the patient has type 1 or type 2 diabetes.

Conclusion: Urinalysis

- State below if the urinalysis is normal or indicates a urinary tract infection (leukocytes, blood, and possibly protein in the urine) or that the patient has diabetes.

  Test strip 1 ______________________________

  Test strip 2 ______________________________

  Test strip 3 ______________________________

- The hormone insulin promotes the uptake of glucose by cells. When glucose is in the urine, either the pancreas is not producing insulin (type 1 diabetes) or cells are resistant to insulin (type 2 diabetes). Ketones (acids) are also in the urine because the cells are metabolizing fat instead of glucose. Explain why. ______________________________

  ____

  Why is the pH of urine lower than normal? ______________________________

  ____

- If urinalysis shows that proteins are excreted instead of retained in the blood, would capillary exchange in the tissues (see Fig. 23.5) be normal? __________ Why or why not? ______________________________

  ____

Laboratory Review 23

1. In your own words, what is homeostasis? ______
2. Explain how the systemic capillaries help maintain homeostasis. ______
______
3. Relate a blood pressure of 120/80 to systole and diastole of the ventricles during a heartbeat. When would blood pressure be 120? ______ When would blood pressure be 80? ______
______
4. If a smoker has a low tidal volume, why might he feel tired and run down? ______
______
5. How would you measure the effects of exercise on vital capacity? ______
______
6. What role do the kidneys have in maintaining blood pressure and volume? ______
______
7. List the three steps in urine formation and define.
   a. ______
   b. ______
   c. ______
8. With regard to urine formation, name a substance found in both the filtrate and the urine. ______
   Explain why the substance is in both places. ______
   ______
9. With regard to urine formation, name a substance found in the filtrate and not in the urine. ______
   Explain why the substance is only in the filtrate and not in the urine. ______
   ______
10. After a urinalysis test, what medical condition would be indicated by a positive test for glucose in the urine? ______
    ______

***Concepts of Biology* Website**

Instructors can find lab prep information and answers to all laboratory questions in the Laboratory Resource Guide. *Students* can practice their knowledge with quizzes, animations, flashcards, and much more.

www.mhhe.com/maderconcepts3

McGraw-Hill Access Science Website

An online encyclopedia of science and technology that provides information, including videos, that can enhance the laboratory experience.

www.accessscience.com

LABORATORY

24

The Nervous System and Senses

Learning Outcomes

Introduction

The vertebrate nervous system has two major divisions: the central nervous system (CNS), consisting of the brain and spinal cord, and the peripheral nervous system (PNS), which contains cranial nerves and spinal nerves (Fig. 24.1). Sensory receptors detect changes in environmental stimuli, and nerve impulses move along sensory nerve fibers to the brain and the spinal cord. The brain and spinal cord sum up the data before sending impulses via motor nerve fibers to effectors (muscles and glands) so a response to stimuli is possible. Nervous tissue consists of neurons; whereas the brain and spinal cord contain all parts of neurons, nerves contain only axons.

Figure 24.1 The nervous system.
The central nervous system (CNS) is in the midline of the body, and the peripheral nervous system (PNS) is outside the CNS.

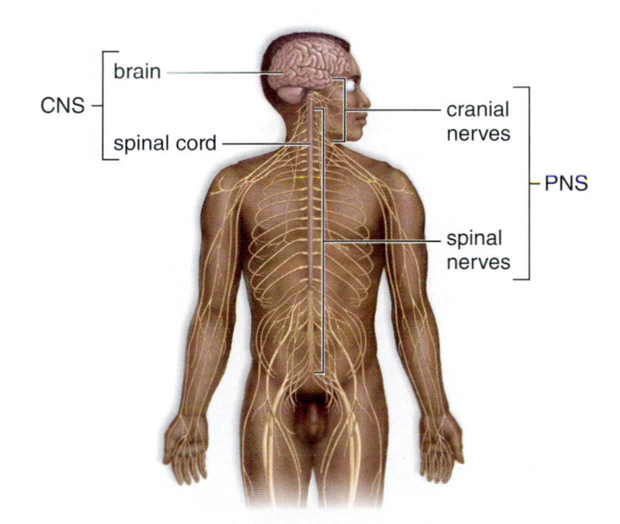

24.1 Central Nervous System

The brain is the enlarged anterior end of the spinal cord; it contains centers that receive input from and can command other regions of the nervous system.

Latex gloves Wear protective latex gloves when handling preserved animal organs. Use protective eyewear and exercise caution when using sharp instruments during this laboratory. Wash hands thoroughly upon completion of this laboratory.

Preserved Sheep Brain

The sheep brain (Fig. 24.2) is often used to study the brain. It is easily available and large enough that individual parts can be identified.

Observation: Preserved Sheep Brain

Examine the exterior and a midsagittal (longitudinal) section of a preserved sheep brain or a model of the human brain, and with the help of Figure 24.1, identify the following:

1. **Ventricles:** Interconnecting spaces that produce and serve as a reservoir for cerebrospinal fluid, which cushions the brain. Toward the anterior, note the lateral ventricle (on one longitudinal section) and similarly a lateral ventricle (on the other longitudinal section). Trace the second ventricle to the third and then the fourth ventricles.
2. **Cerebrum:** Most developed area of the brain; responsible for higher mental capabilities. The cerebrum is divided into the right and left **cerebral hemispheres,** joined by the **corpus callosum,** a broad sheet of white matter. The outer portion of the cerebrum is highly convoluted and divided into the following surface lobes (see Fig. 24.2):
   a. **Frontal lobe:** Controls motor functions and permits voluntary muscle control; it also is responsible for abilities to think, problem solve, speak, and smell.
   b. **Parietal lobe:** Receives information from sensory receptors located in the skin and also the taste receptors in the mouth. A groove called the **central sulcus** separates the frontal lobe from the parietal lobe.
   c. **Occipital lobe:** Interprets visual input and combines visual images with other sensory experiences. The optic nerves split and enter opposite sides of the brain at the optic chiasma, located in the diencephalon.
   d. **Temporal lobe:** Has sensory areas for hearing and smelling. The olfactory bulb contains nerve fibers that communicate with the olfactory cells in the nasal passages and take nerve impulses to the temporal lobe.
3. **Diencephalon:** Portion of the brain where the third ventricle is located. The hypothalamus and thalamus are also located here.
   a. **Thalamus:** Two connected lobes located in the roof of the third ventricle. The thalamus is the highest portion of the brain to receive sensory impulses before the cerebrum. It is believed to control which received impulses are passed on to the cerebrum. For this reason, the thalamus sometimes is called the "gatekeeper to the cerebrum."
   b. **Hypothalamus:** Forms the floor of the third ventricle and contains control centers for appetite, body temperature, blood pressure, and water balance. Its primary function is homeostasis. The hypothalamus also has centers for pleasure, reproductive behavior, hostility, and pain.
4. **Cerebellum:** Located just posterior to the cerebrum as you observe the brain dorsally, the cerebellum's two lobes make it appear rather like a butterfly. In cross section, the cerebellum has an internal pattern that looks like a tree. The cerebellum coordinates equilibrium and motor activity to produce smooth movements.

Figure 24.2 The sheep brain.

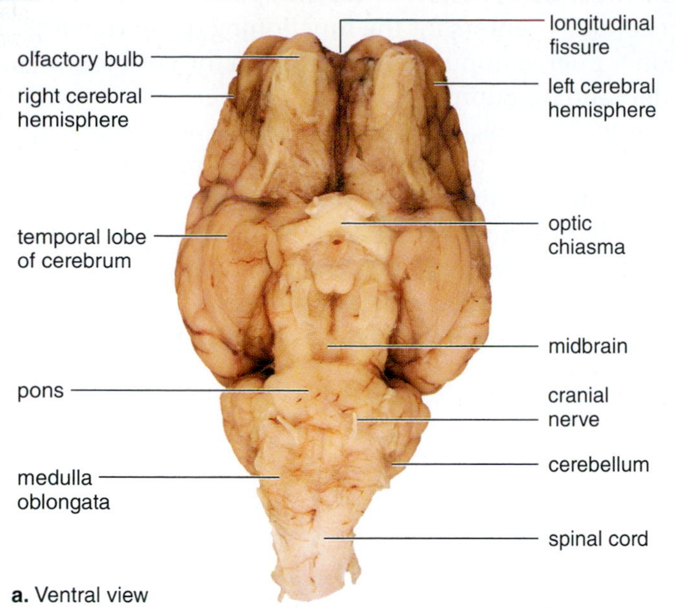

a. Ventral view

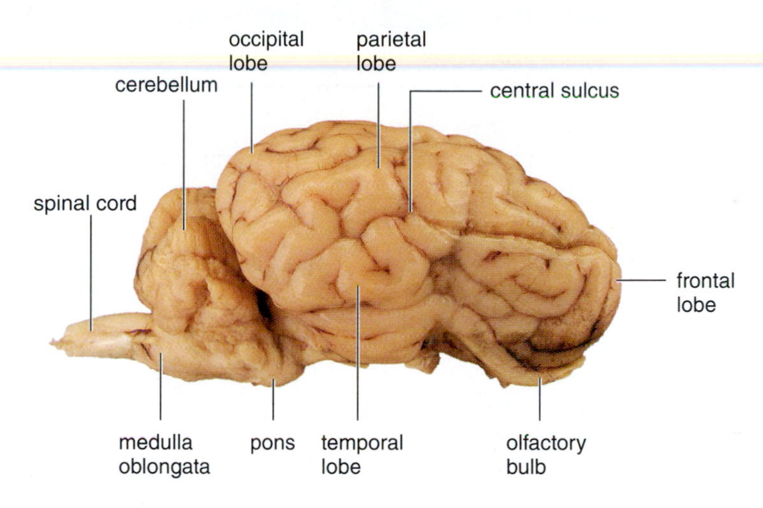

b. Lateral view

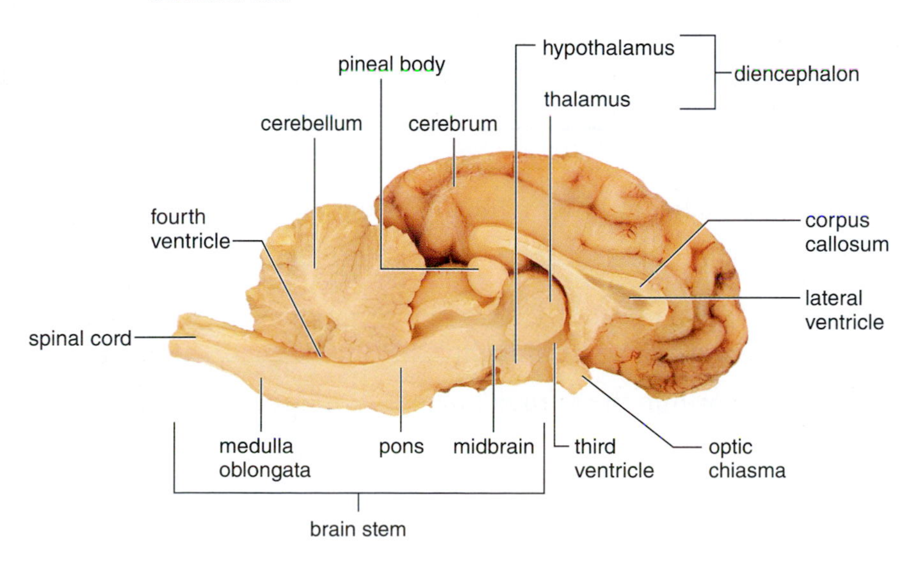

c. Longitudinal section

5. **Brain stem:** Part of the brain that connects with the spinal cord. Because it includes the pons and medulla oblongata, it contains centers for the functioning of internal organs; because of its location, it serves as a relay station for nerve impulses passing from the cord to the brain. Therefore, it helps keep the rest of the brain alert and functioning.
   a. **Midbrain:** Anterior to the pons, the midbrain serves as a relay station for sensory input and motor output. It also contains a reflex center for eye muscles.
   b. **Pons:** The ventral, bulblike enlargement on the brain stem. It serves as a passageway for nerve impulses running between the medulla and the higher brain regions.
   c. **Medulla oblongata** (or simply **medulla**): The most posterior portion of the brain stem. It controls internal organs; for example, blood pressure, cardiac, and breathing control centers are present in the medulla. Nerve impulses pass from the spinal cord through the medulla to and from higher brain regions.

The Human Brain

Based on your knowledge of the sheep brain, complete Table 24.1 by stating the major functions of each part of the brain listed. *Also label Figure 24.3.*

Table 24.1 Summary of Brain Functions

Part	Major Functions
Cerebrum	
Cerebellum	
Diencephalon Thalamus	
Hypothalamus	
Brain stem Midbrain	
Pons	
Medulla oblongata	

Which parts of the brain work together to achieve the following?

1. Good eye-hand coordination ______
2. Concentrating on homework when TV is playing ______
3. Avoiding dark alleys while walking home at night ______
4. Keeping the blood pressure within the normal range ______

Figure 24.3 The human brain (longitudinal section).
The cerebrum is larger in humans than in sheep. Label where indicated.

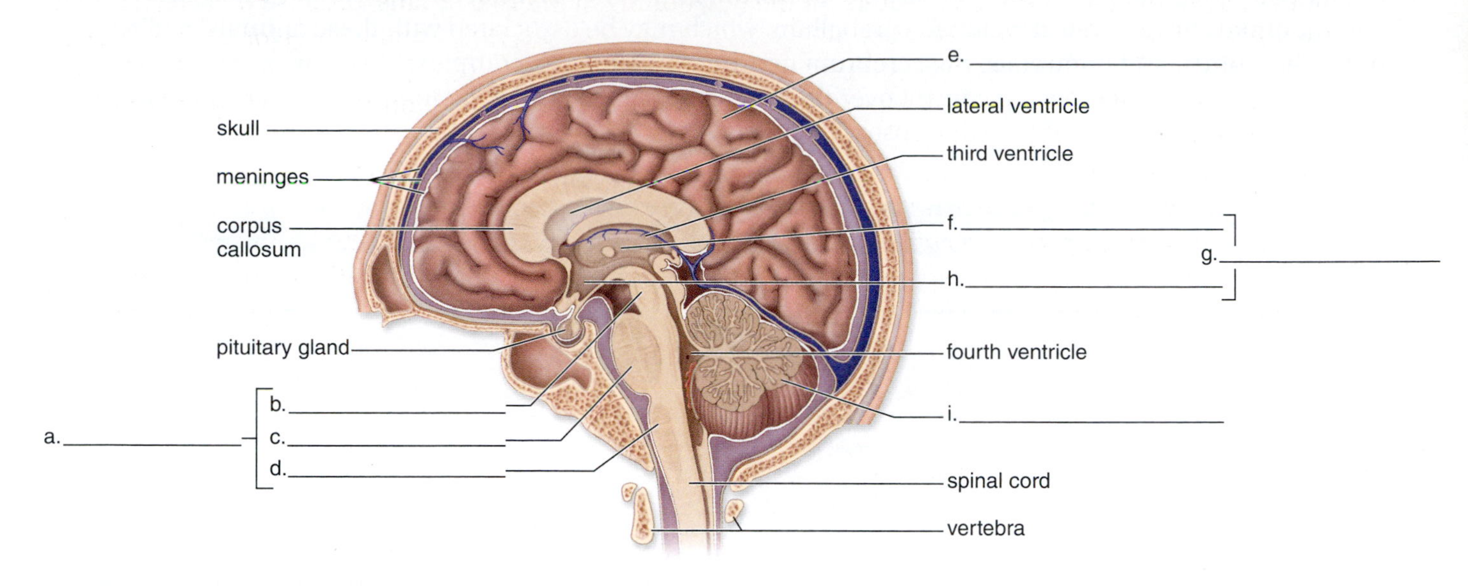

Comparison of Vertebrate Brains

The vertebrate brain has a forebrain, midbrain, and hindbrain. In the earliest vertebrates, the forebrain was largely a center for sense of smell, the midbrain was a center for the sense of vision, and the hindbrain was a center for the sense of hearing and balance. How the functions of these parts changed to accommodate the lifestyles of different vertebrates can be traced.

Observation: Comparison of Vertebrate Brains

1. Examine the brain of a fish, amphibian, bird, and mammal (Fig. 24.4).
2. Compare the sizes of the following three areas:
   a. **Forebrain:** Contains olfactory bulb and cerebrum.
   b. **Midbrain:** Contains optic lobe (and other structures).
   c. **Hindbrain:** Contains cerebellum (and other structures).

Figure 24.4 Comparative vertebrates brains.
Vertebrate brains differ in particular by the comparative sizes of the cerebrum, optic lobe, and cerebellum.

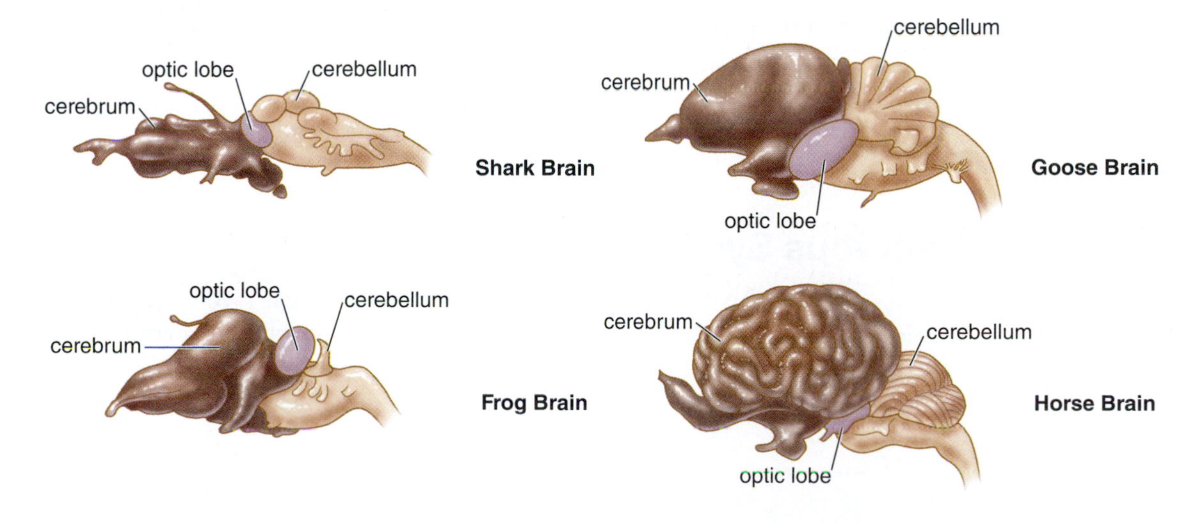

In fishes and amphibians, the midbrain is often the most prominent region of the brain because it contains higher control centers. In these animals, the cerebrum largely has an olfactory function. Fishes, birds, and mammals have a well-developed cerebellum, which may be associated with these animals' agility. In reptiles, birds, and mammals, the cerebrum becomes increasingly complex. This may be associated with the cerebrum's increasing control over the rest of the brain and the evolution of areas responsible for thought and reasoning. Record your observations in Table 24.2.

Table 24.2 Comparison of Vertebrate Brains

Vertebrate	Observations
Fish (shark)	
Amphibian (frog)	
Bird (goose)	
Mammal (horse)	

The Spinal Cord

The spinal cord is a part of the central nervous system. It lies in the middorsal region of the body and is protected by the vertebral column.

Observation: The Spinal Cord

1. Examine a prepared slide of a cross section of the spinal cord under the lowest magnification possible. For example, some microscopes are equipped with a short scanning objective that enlarges about 3.5×, with a total magnification of 35×. If a scanning objective is not available, observe the slide against a white background with the naked eye.
2. Identify the following with the help of Figure 24.5:
   a. **Gray matter:** A central, butterfly-shaped area composed of masses of short nerve fibers, interneurons, and motor neuron cell bodies.
   b. **White matter:** Masses of long fibers that lie outside the gray matter and carry impulses up and down the spinal cord. In animals, white matter appears white because an insulating myelin sheath surrounds long fibers.

Figure 24.5 The spinal cord.
Photomicrograph of spinal cord cross section.

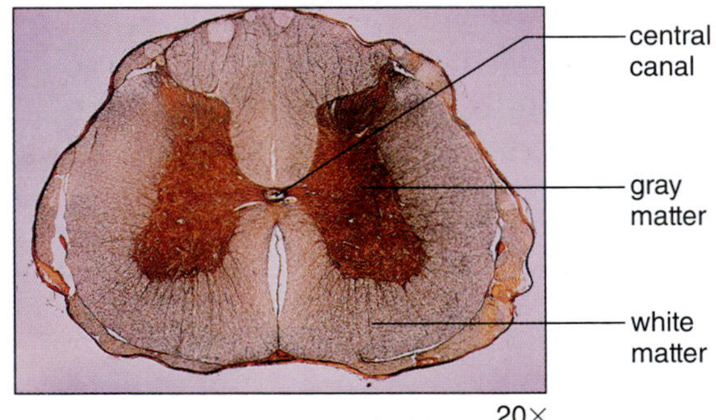

24.2 Peripheral Nervous System

The peripheral nervous system contains the cranial nerves and the spinal nerves. Twelve pairs of cranial nerves project from the inferior surface of the brain. The cranial nerves are largely concerned with nervous communication between the head, neck, and facial regions of the body and the brain. The 31 pairs of spinal nerves emerge from either side of the spinal cord (Fig. 24.6).

Spinal Nerves

Each spinal nerve contains long fibers of sensory neurons and long fibers of motor neurons. In Figure 24.6, identify the following:

1. **Sensory neuron:** Takes nerve impulses from a sensory receptor to the spinal cord. The cell body of a sensory neuron is in the dorsal root ganglion.
2. **Interneuron:** Lies completely within the spinal cord. Some interneurons have long fibers and take nerve impulses to and from the brain. The interneuron in Figure 24.6 transmits nerve impulses from the sensory neuron to the motor neuron.
3. **Motor neuron:** Takes nerve impulses from the spinal cord to an effector—in this case, a muscle. Muscle contraction is one type of response to stimuli.

Suppose you were walking barefoot and stepped on a prickly sandbur. Describe the pathway of information, starting with the pain receptor in your foot, that would allow you to both feel and respond to this unwelcome stimulus. ____

____

____

Spinal Reflexes

A **reflex** is an involuntary and predictable response to a given stimulus that allows a quick response to environmental stimuli without communicating with the brain. When you touch a sharp tack, you immediately withdraw your hand (see Fig. 24.6). When a spinal reflex occurs, a sensory receptor is

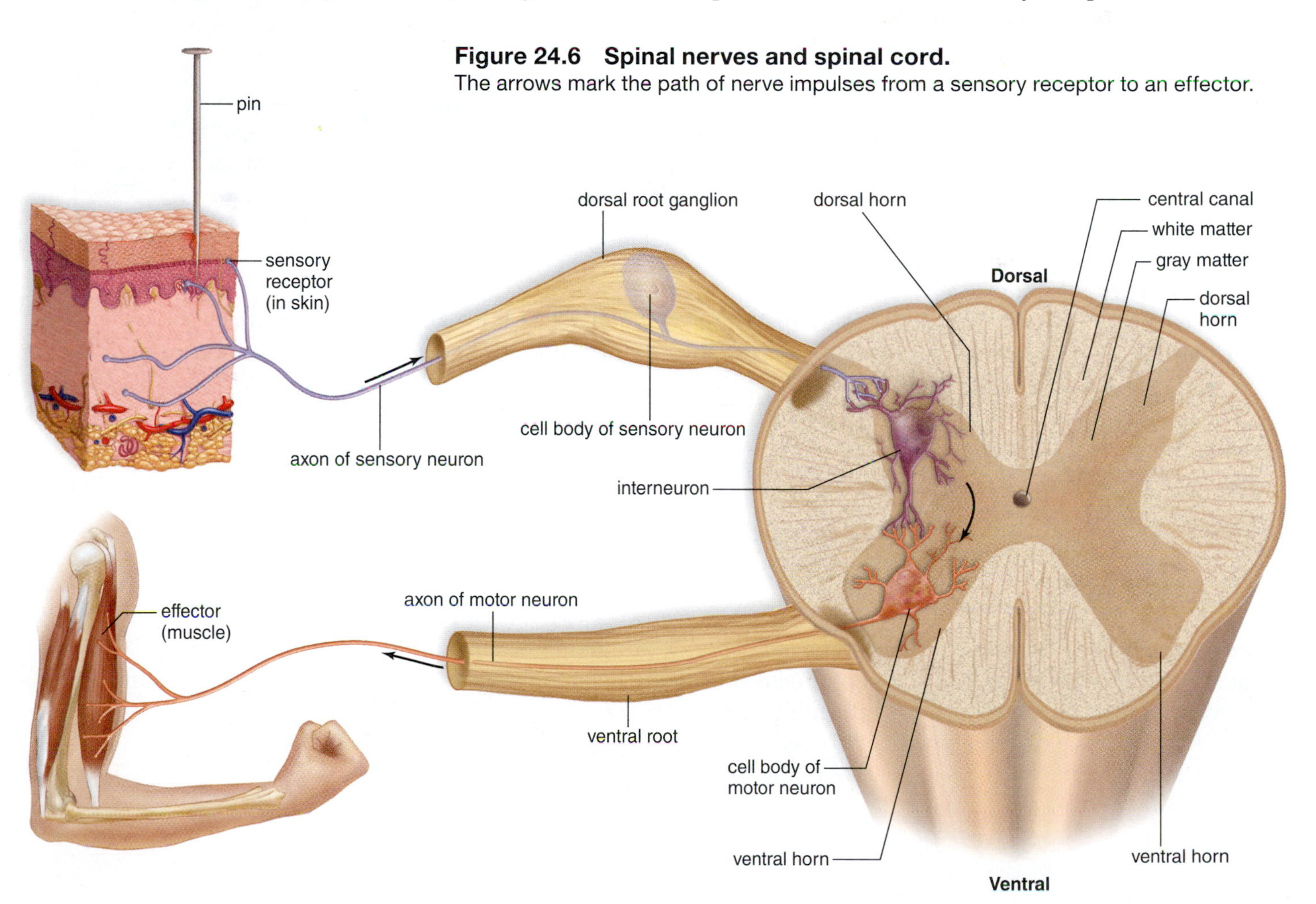

Figure 24.6 Spinal nerves and spinal cord.
The arrows mark the path of nerve impulses from a sensory receptor to an effector.

stimulated and generates nerve impulses that pass along the three neurons mentioned earlier—the sensory neuron, interneuron, and motor neuron—until the effector responds. In the spinal reflexes that follow, a receptor detects the tap, and sensory neurons conduct nerve impulses to interneurons in the spinal cord. The interneurons send a message via motor neurons to the effectors, muscles in the leg or foot. These reflexes are involuntary because the brain is not involved in formulating the response. *Consciousness* of the stimulus lags behind the response because information must be sent up the spinal cord to the brain before you can become aware of the tap.

Experimental Procedure: Spinal Reflex

Although many reflexes occur in the body, only a tendon reflex is investigated in this Experimental Procedure. One easily tested tendon reflex involves the **patellar tendon.** When this tendon is tapped with a reflex hammer (Fig. 24.7) or, in this experiment, with a meterstick, the attached muscle is stretched. This causes a receptor to generate nerve impulses, which are transmitted along sensory neurons to the spinal cord. Nerve impulses from the cord then pass along motor neurons and stimulate the muscle, causing it to contract. As the muscle contracts, it tugs on the tendon, causing movement of a bone opposite the joint. Receptors in other tendons, such as the Achilles tendon, respond similarly. Such reflexes help the body automatically maintain balance and posture.

Figure 24.7 Knee-jerk reflex. The quick response when the patellar tendon is stimulated by tapping with a reflex hammer indicates that a reflex has occurred.

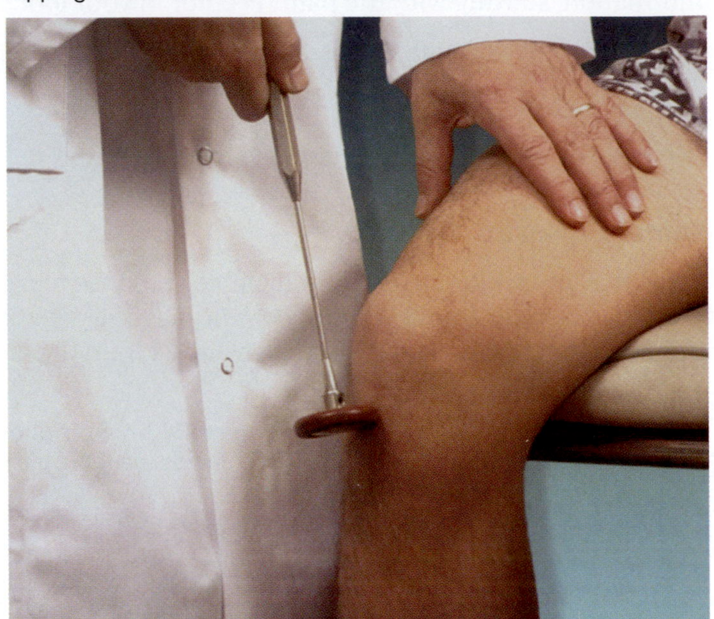

Knee-jerk (patellar) reflex

Knee-Jerk (Patellar) Reflex

1. Have the subject sit on a table so that his or her legs hang freely.
2. Sharply tap one of the patellar tendons just below the patella (kneecap) with a meterstick.
3. In this relaxed state, does the leg flex (move toward the buttocks) or extend (move away from the buttocks)? ______________________________

24.3 Animal Eyes

The eye is a special sense organ for detecting light rays in the environment.

Anatomy of Invertebrate Eyes

Arthropods have **compound eyes** composed of many independent visual units, called ommatidia, each of which has its own photoreceptor cells (Fig. 24.8). Each unit "sees" a separate portion of the object. How well the brain combines this information is not known.

A squid has a camera type of eye. A single lens focuses an image of the visual field on the photoreceptors, packed closely together (Fig. 24.9).

Figure 24.8 Compound eye of a fly.
Each visual unit of a compound eye has a cornea and lens that focus light onto photoreceptor cells. These cells generate nerve impulses transmitted to the brain, where interpretation produces a mosaic image.

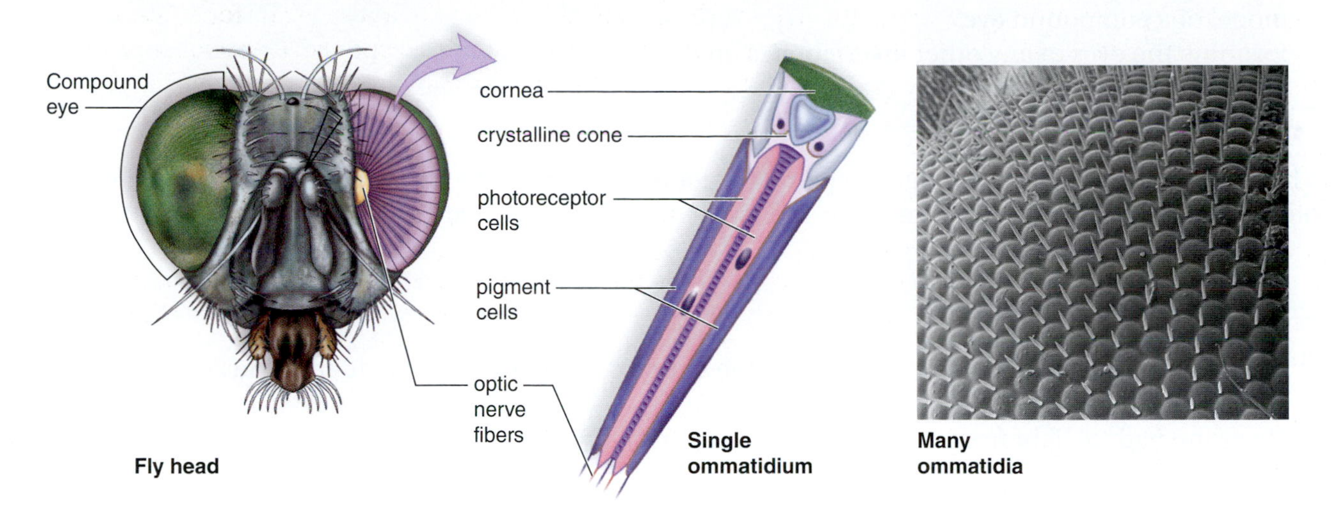

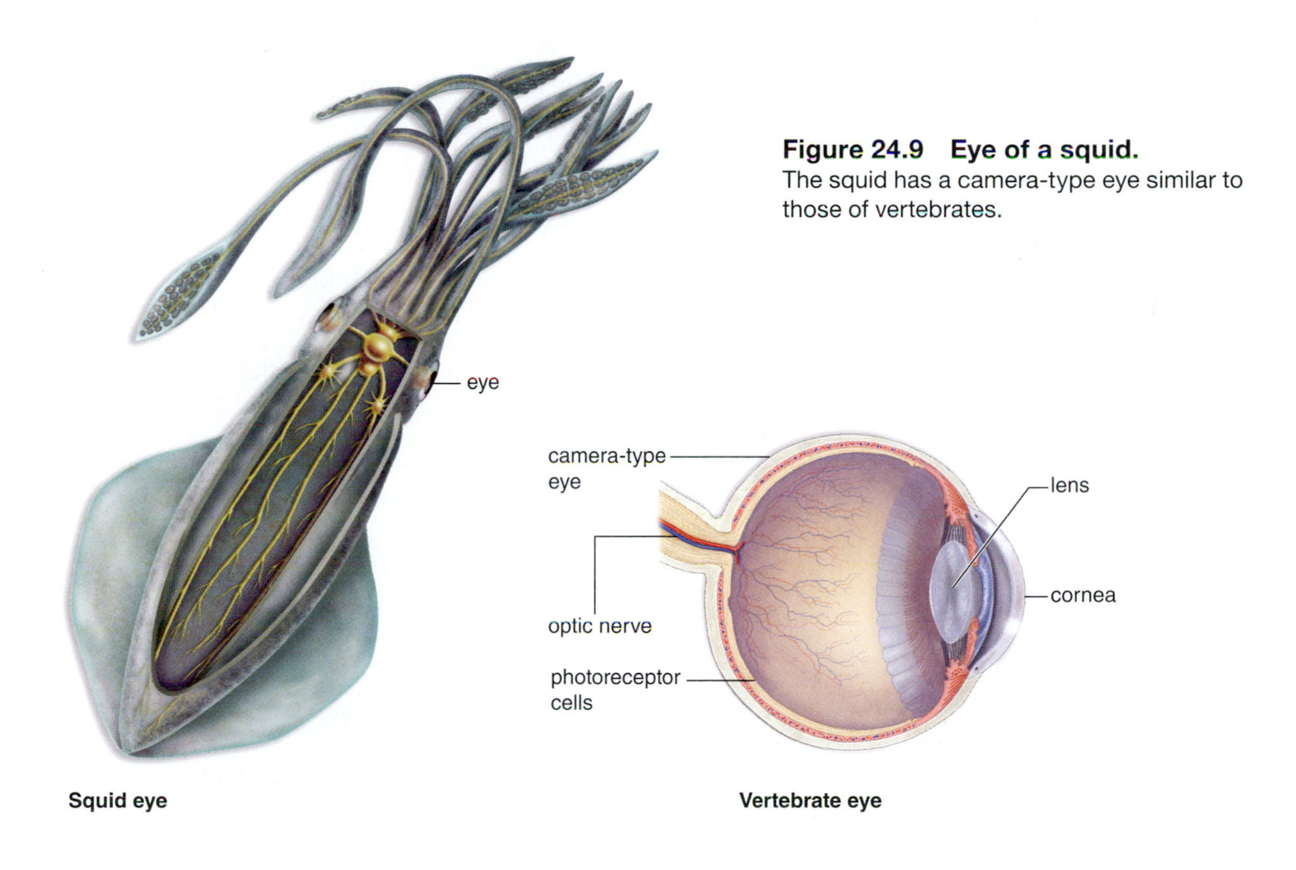

Figure 24.9 Eye of a squid.
The squid has a camera-type eye similar to those of vertebrates.

Observation: Invertebrate Eyes

1. Examine the demonstration slide of a compound eye set up under a stereomicroscope or examine a model of a compound eye.
2. Examine the eyes of any other invertebrates on display.

Anatomy of the Human Eye

The human eye is responsible for sight. Light rays enter the eye and strike the **rod cells** and **cone cells,** the photoreceptors for sight. The rods and cones generate nerve impulses that go to the brain via the optic nerve.

Observation: The Human Eye

1. Examine a human eye model, and identify the structures listed and depicted in Table 24.3 and Figure 24.10.
2. Trace the path of light from outside the eye to the retina.

   ____

   ____

   ____

3. During **accommodation,** the lens rounds up to aid in viewing near objects or flattens to aid in viewing distant objects. Which structure holds the lens and is involved in accommodation?

   ____

   ____

4. **Refraction** is the bending of light rays so that they can be brought to a single focus. Which of the structures listed in Table 24.3 aid in refracting and focusing light rays?

   ____

5. Specifically, what are the sensory receptors for sight, and where are they located in the eye?

   ____

   ____

6. What structure takes nerve impulses to the brain from the rod cells and cone cells?

   ____

7. Which cerebral lobe processes nerve impulses from an eye? ______________________

Table 24.3 Parts of the Human Eye

Part	Location	Function
Sclera	Outer layer of eye	Protects and supports eyeball
Cornea	Transparent portion of sclera	Refracts light rays
Choroid	Middle layer of eye	Absorbs stray light rays
Retina	Inner layer of eye	Contains receptors for sight
Rod cells	In retina	Make black-and-white vision possible
Cone cells	Concentrated in fovea centralis	Make color vision possible
Fovea centralis	Special region of retina	Makes acute vision possible
Lens	Interior of eye between cavities	Refracts and focuses light rays
Ciliary body	Extension from choroid	Holds lens in place; functions in accommodation
Iris	More anterior extension of choroid	Regulates light entrance
Pupil	Opening in middle of iris	Admits light
Humors (aqueous and vitreous)	Fluid media in anterior and posterior compartments, respectively, of eye	Transmit and refract light rays; support eyeball
Optic nerve	Extension from posterior of eye	Transmits impulses to occipital lobe of brain

Figure 24.10 Anatomy of the human eye.
The sensory receptors for vision are the rod cells and cone cells present in the retina of the eye.

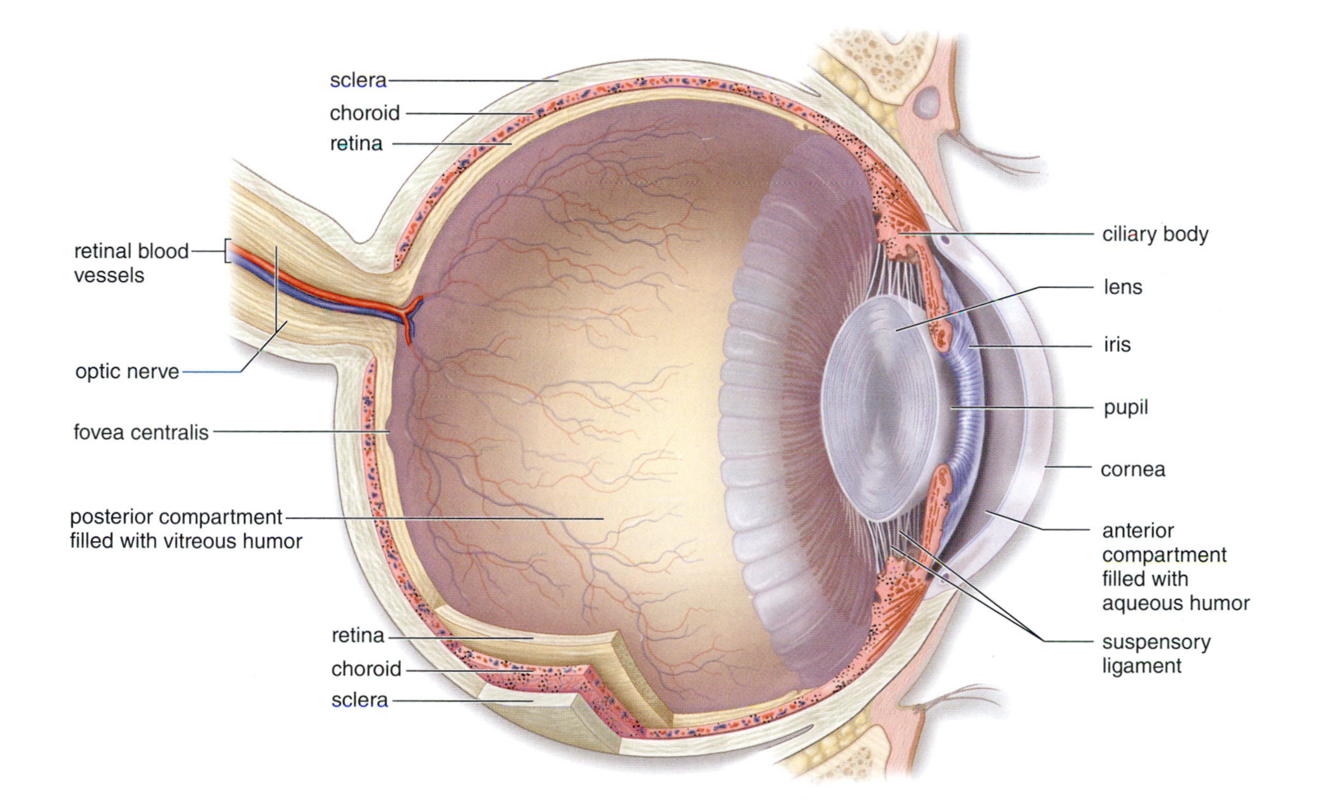

The Blind Spot of the Eye

The **blind spot** occurs where the optic nerve fibers exit the retina. No vision is possible at this location because of the absence of rod cells and cone cells.

Experimental Procedure: Blind Spot of the Eye

This Experimental Procedure requires a laboratory partner. Figure 24.11 shows a small circle and a cross several centimeters apart.

Figure 24.11 Blind spot.
This dark circle (or cross) will disappear at one location because there are no rod cells or cone cells at each eye's blind spot, where vision does not occur.

Left Eye

1. Hold Figure 24.11 approximately 30 cm from your eyes. The cross should be directly in front of your left eye. If you wear glasses, keep them on.
2. Close your right eye.
3. Stare only at the cross with your left eye. You should also be able to see the circle in the same field of vision. Slowly move the paper toward you until the circle disappears.
4. Repeat the procedure as many times as needed to find the blind spot.
5. Then slowly move the paper closer to your eyes until the circle reappears. Because only your left eye is open, you have found the blind spot of your left eye.
6. With your partner's help, measure the distance from your eye to the paper when the circle first disappeared. Left eye: __________ cm

Right Eye

1. Hold Figure 24.11 approximately 30 cm from your eyes. The circle should be directly in front of your right eye. If you wear glasses, keep them on.
2. Close your left eye.
3. Stare only at the circle with your right eye. You should also be able to see the cross in the same field of vision. Slowly move the paper toward you until the cross disappears.
4. Repeat the procedure as many times as needed to find the blind spot.
5. Then slowly move the paper closer to your eyes until the cross reappears. Because only your right eye is open, you have found the blind spot of your right eye.
6. With your partner's help, measure the distance from your eye to the paper when the cross first disappeared. Right eye: __________ cm

Why are you unaware of a blind spot under normal conditions? Although the eye detects patterns of light and color, it is the brain that determines what we visually perceive. The brain interprets the visual input based in part on past experiences. In this exercise, you created an artificial situation in which you became aware of how your perception of the world is constrained by the eye's anatomy.

Accommodation of the Eye

When the eye accommodates to see objects at different distances, the shape of the lens changes. The lens shape is controlled by the ciliary muscles attached to it. When you are looking at a distant object, the lens is in a flattened state. When you are looking at a closer object, the lens becomes more rounded. The elasticity of the lens determines how well the eye can accommodate. Lens elasticity decreases with increasing age, a condition called **presbyopia.** Presbyopia is the reason many older people need bifocals to see near objects.

Experimental Procedure: Accommodation of the Eye

This Experimental Procedure requires a laboratory partner. It tests accommodation of either your left or your right eye.

Figure 24.12 Accommodation.
When testing the ability of your eyes to accommodate to see a near object, always keep the pencil in this position.

1. Hold a pencil upright by the eraser and at arm's length in front of whichever of your eyes you are testing (Fig. 24.12).
2. Close the opposite eye.
3. Move the pencil from arm's length toward your eye.
4. Focus on the end of the pencil.
5. Move the pencil toward you until the end is out of focus. Measure the distance (in centimeters) between the pencil and your eye: __________ cm
6. At what distance can your eye no longer accommodate for distance? __________ cm
7. If you wear glasses, repeat this experiment without your glasses, and note the accommodation distance of your eye without glasses: ______________ cm. (Contact lens wearers need not make these determinations, and they should write the words "contact lens" in this blank.)
8. The "younger" lens can easily accommodate for closer distances. The nearest point at which the end of the pencil can be clearly seen is called the **near point.** The more elastic the lens, the "younger" the eye (Table 24.4). How "old" is the eye you tested? __________

Table 24.4 Near Point and Age Correlation

Age (years)	10	20	30	40	50	60
Near point (cm)	9	10	13	18	50	83

24.4 Animal Ears

Ears contain specialized receptors for detecting sound waves in the environment. They also often function as organs of balance.

Anatomy of Invertebrate Ears

Among invertebrates, only certain arthropod groups—crustaceans, spiders, and insects—have receptors for detecting sound waves. The invertebrate ear usually has a simple design: a pair of air pockets enclosed by a tympanum that passes sound waves to sensory neurons.

Observation: An Invertebrate Ear

Examine the preserved grasshopper on display, and with the help of Figure 24.13*a*, locate the tympanum. The tympanum covers an internal air sac that allows the tympanum to vibrate when struck by sound waves. Sensory neurons attached to the tympanum are stimulated directly by the vibration.

Figure 24.13 Evolution of the human ear.
a. A few invertebrates have ears such as that of the grasshopper. **b.** The lateral line of fishes contains hair cells with cilia embedded in a gelatinous cupula. **c.** Hair cells in the human ear have stereocilia embedded in the gelatinous tectorial membrane. When the basal membrane vibrates, bending of the stereocilia generates nerve impulses.

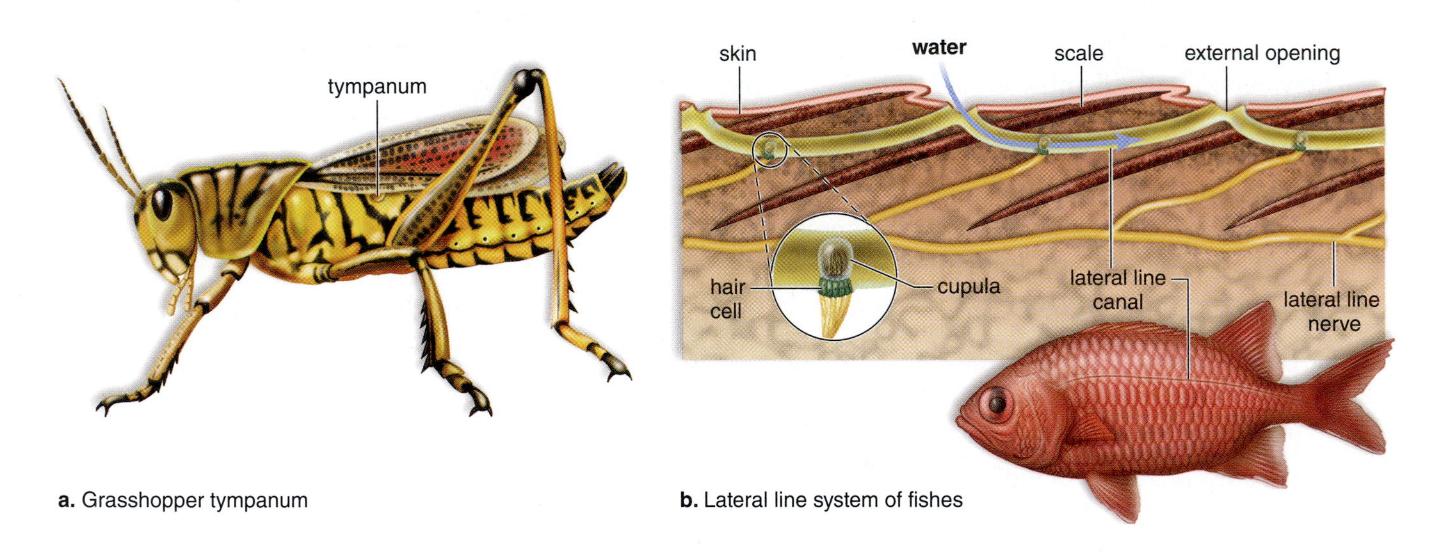

a. Grasshopper tympanum

b. Lateral line system of fishes

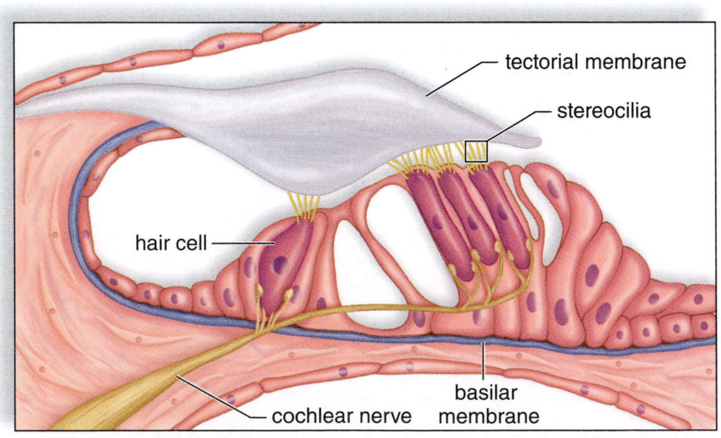

c. Cochlear cross section of the human ear

Anatomy of the Human Ear

The human ear, whose parts are listed and depicted in Table 24.5 and Figure 24.14, serves two functions: hearing and balance.

Observation: The Human Ear

Examine a human ear model, and find the structures depicted in Figure 24.14 and listed in Table 24.5.

Table 24.5 Parts of the Human Ear

Part	Medium	Function	Mechanoreceptor
Outer ear	Air		
Pinna		Collects sound waves	—
Auditory canal		Filters air	—
Middle ear	Air		
Tympanic membrane and ossicles		Amplify sound waves	—
Auditory tube		Equalizes air pressure	—
Inner ear	Fluid		
Semicircular canals		Rotational equilibrium	Stereocilia embedded in cupula
Vestibule (contains utricle and saccule)		Gravitational equilibrium	Stereocilia embedded in otolithic membrane
Cochlea (spiral organ)		Hearing	Stereocilia embedded in tectorial membrane

Figure 24.14 Anatomy of the human ear.
The outer ear extends from the pinna to the tympanic membrane. The middle ear extends from the tympanic membrane to the oval window. The inner ear encompasses the semicircular canals, the vestibule, and the cochlea.

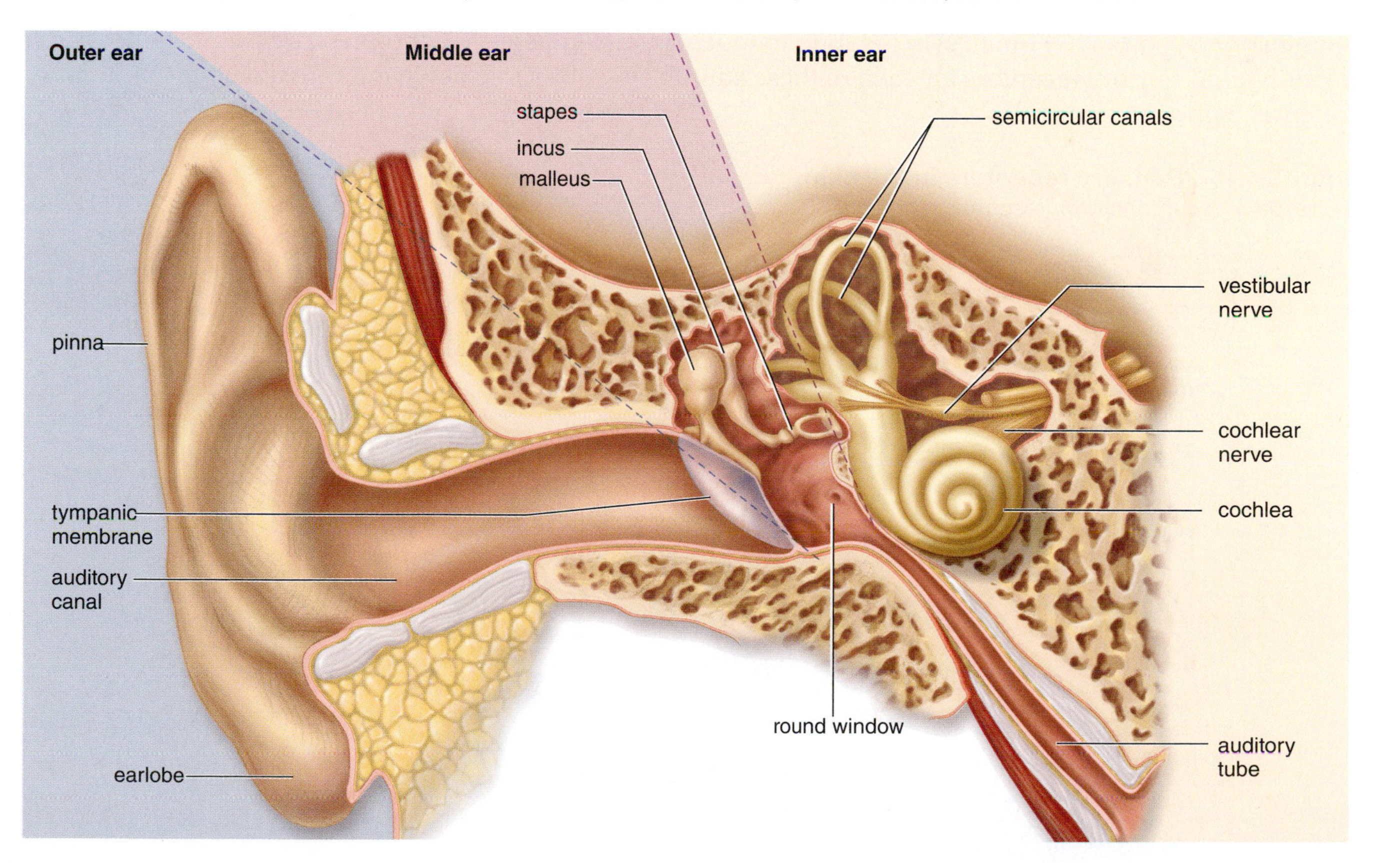

Physiology of the Human Ear

When you hear, sound waves are picked up by the **tympanic membrane** and amplified by the **malleus, incus,** and **stapes.** This creates pressure waves in the canals of the **cochlea** that lead to stimulation of **hair cells,** the receptors for hearing. Hair cells in the utricle and saccule of the vestibule and in semicircular canals are receptors for equilibrium (i.e., balance). Nerve impulses from the cochlea travel by way of the cochlear nerve and the vestibular nerve to the brain and eventually are interpreted by the brain as sound.

Experimental Procedure: Locating Sound

Humans locate the direction of sound according to how well it is detected by either or both ears. A difference in the hearing ability of the two ears can lead to a mistaken judgment about the direction of sound. You and a laboratory partner should perform this Experimental Procedure on each other.

1. Ask the subject to be seated, with eyes closed. Then strike a tuning fork or rap two spoons together at the five locations listed in number 2. Use a random order.
2. Ask the subject to give the exact location of the sound in relation to his or her head. Record the subject's perceptions when the sound is
   a. directly below and behind the head. ____________________
   b. directly behind the head. ____________________
   c. directly above the head. ____________________
   d. directly in front of the face. ____________________
   e. to the side of the head. ____________________
3. Is there an apparent difference in hearing between the subject's two ears? ____________________

24.5 Sensory Receptors in Human Skin

The sensory receptors in human skin respond to touch, pain, temperature, and pressure (Fig. 24.15). There are individual sensory receptors for each of these stimuli, as well as free nerve endings able to respond to pressure, pain, and temperature.

Figure 24.15 Sensory receptors in the skin.
Each type of receptor shown responds primarily to a particular stimulus.

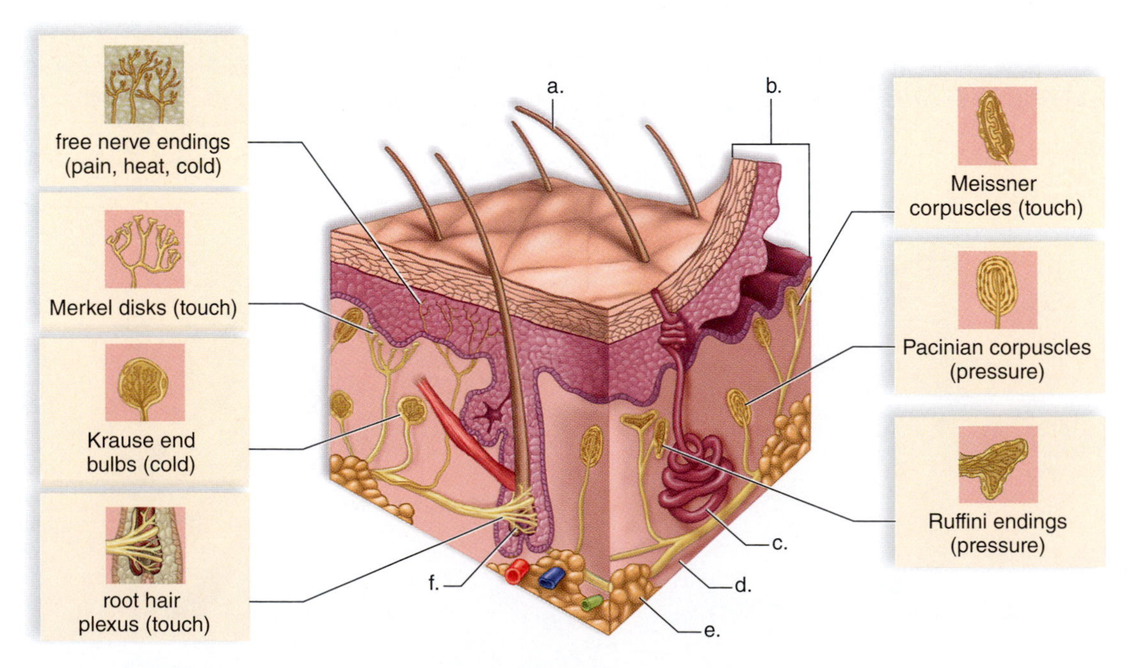

Experimental Procedure: Sensory Receptors in Human Skin

The dermis of the skin contains touch and temperature receptors.

Sense of Touch

You will need a laboratory partner to perform this Experimental Procedure. Enter *your* data, not the data of your partner, in the spaces provided.

1. Ask the subject to be seated, with eyes closed. You are going to test the subject's ability to discriminate between the two points of a hairpin or a pair of scissors at the four locations noted below.
2. Hold the points of the hairpin or scissors on the given skin area, with both points gently touching the subject. Ask the subject whether the experience involves one or two touch sensations.
3. Record the shortest distance between the hairpin or scissor points for a two-point discrimination.

   a. Forearm: ________ mm

   b. Back of the neck: ________ mm

   c. Index finger: ________ mm

   d. Back of the hand: ________ mm
4. Which of these areas apparently contains the greatest density of touch receptors? ________________

   Why is this useful? ____
5. Do you have a sense of touch at every point in your skin? ________ Explain. ________________

   ____
6. What specific part of the brain processes nerve impulses from touch and pain receptors?

   ____

Sense of Heat and Cold

1. Obtain three 1,000 ml beakers, and fill one with *ice water,* one with *tap water* at room temperature, and one with *warm water* (45°–50°C).
2. Immerse your left hand in the ice-water beaker and your right hand in the warm-water beaker for 30 seconds.
3. Then place both hands in the beaker with room-temperature tap water.
4. Record the sensation in the right and left hands.

   a. Right hand ____

   b. Left hand ____
5. Explain your results. ____

   ____

24.6 Human Chemoreceptors

The taste receptors, called ______________________, located in the mouth, and the smell receptors, called ______________________, located in the nasal cavities, are the chemoreceptors that respond to molecules in the air and water. Nerve impulses from taste receptors go to the ______________________ lobe of the brain, while those from smell receptors go to the ______________________ lobe of the brain.

Experimental Procedure: Sense of Taste and Smell

You will need a laboratory partner to perform the following procedures. It will not be necessary for all tests to be performed on both partners. You should take turns being either the subject or the experimenter. Dispose of used cotton swabs in a hazardous waste container or as directed by your instructor.

Taste and Smell

1. Students work in groups. Each group has one experimenter and several subjects.
2. The experimenter should obtain a LifeSavers candy from the various flavors available, without letting the subject know what flavor it is.
3. The subject closes both eyes and holds his or her nose.
4. The experimenter gives the LifeSavers candy to the subject, who places it on his or her tongue.
5. The subject, while still holding his or her nose, guesses the flavor of the candy. The experimenter records the guess in Table 24.6.
6. The subject releases his or her nose and guesses the flavor again. The experimenter records the guess and the actual flavor in Table 24.6.

Table 24.6 Taste and Smell Experiment

Subject	Actual Flavor	Flavor While Holding Nose	Flavor After Releasing Nose
1			
2			
3			

Conclusions: Sense of Taste and Smell

- From your results, how would you say that smell affects the taste of LifeSavers candy?

  ____

- What do you conclude about the effect of smell on your sense of taste?

  ____

Laboratory Review 24

1. Describe the cerebrum of the human brain, and state a function. ___

2. The brain stem includes the medulla oblongata, the pons, and the midbrain. Explain the expression *brain stem* as an anatomical term. ___

3. Describe the location of the gray/white matter of the spinal cord, and give a function for each. ___

4. State, in order from receptor to effector, the neurons associated with a spinal reflex. ___

5. Trace the path of light in the human eye—from the exterior to the retina and then from retinal nerve impulses to the brain. ___

6. Contrast the eye of an arthropod with the eye of a squid and human. ___

7. If you move an illustration that contains a dark circle and a dark cross toward an eye, one or the other may disappear. Give an explanation for this. ___

8. Trace the path of sound waves in the human ear—from the tympanic membrane to the receptors for hearing. ___

9. Compare the manner in which a grasshopper "hears" to the way a human hears. ___

10. Name four structures located in the dermis of the skin. ___

Concepts of Biology Website

Instructors can find lab prep information and answers to all laboratory questions in the Laboratory Resource Guide. *Students* can practice their knowledge with quizzes, animations, flashcards, and much more.

www.mhhe.com/maderconcepts3

McGraw-Hill Access Science Website

An online encyclopedia of science and technology that provides information, including videos, that can enhance the laboratory experience.

www.accessscience.com

LABORATORY

25

Animal Development

Learning Outcomes

25.1 The First Stages of Development

- Identify the cellular stages of development in sea stars, frogs, and mammals. 338
- Identify the tissue stages of development in frogs and mammals. 340–41

25.2 The Organ Stages of Development

- Associate the germ layers with the development of various organs. 341
- Describe the significance of the process of induction during development. 342
- Identify which organs develop first in a vertebrate embryo (e.g., frog, chick, and mammal). 342–44
- Compare the later development of humans to that of chicks. 346–47

25.3 Extraembryonic Membranes, the Placenta, and the Umbilical Cord

- Distinguish between, and give a function for, the extraembryonic membranes, the placenta, and the umbilical cord. 348–50

25.4 Human Fetal Development

- Trace the main events of human fetal development. 351–52

Introduction

The early development of animals is quite similar, regardless of the species. The fertilized egg, or zygote, undergoes successive divisions by cleavage, forming a mulberry-shaped ball of cells called a morula and then a hollow ball of cells called a blastula. The fluid-filled cavity of the blastula is the blastocoel. Later, some of the surface cells fold inward, or invaginate, eventually forming a double-walled structure. The outer layer is called the ectoderm, and the inner layer is the endoderm. Between these layers, a middle layer, or mesoderm, arises. The embryo is now called a gastrula. In particular, the presence of yolk (nutrient material) influences how the gastrula comes about. All later development can be associated with the three **germ layers** (ectoderm, endoderm, and mesoderm) that give rise to different tissues and organ systems.

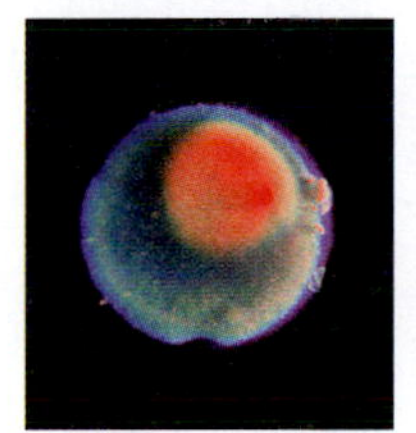

zygote

embryo at one week; implants in uterine wall

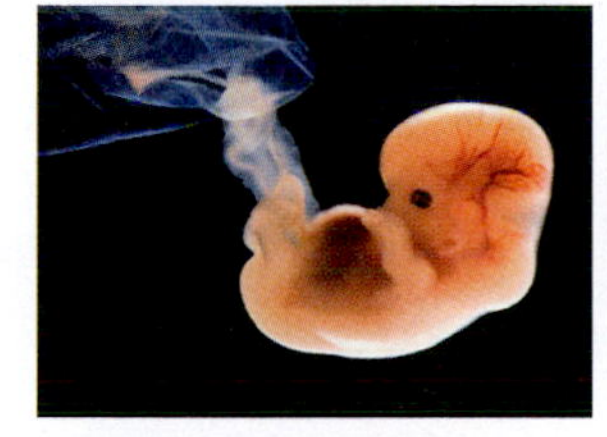

embryo at eight weeks

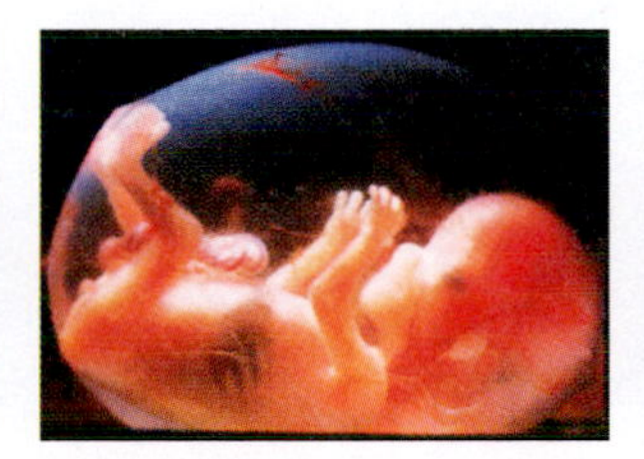

fetus at three months

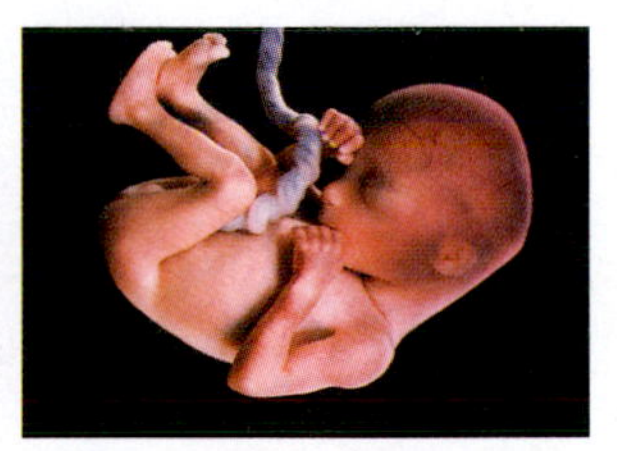

fetus at five months

Development occurs in stages.

25.1 The First Stages of Development

In this section, we will consider the cellular and tissue layer stages of development.

Cellular Stages of Development

The cellular stages of development include the following:

- **Zygote formation:** A single sperm fertilizes an egg and the result is a zygote, the first cell of the new individual.
- **Morula formation:** Zygote divides into a number of smaller cells until there is a cluster of 16 to 32 cells called a morula.
- **Blastula formation:** The morula becomes a blastula, a hollow ball of cells.

Observation: Cellular Stages of Development in the Sea Star

The cellular stages of development are remarkably similar in all animals. Therefore, we can view slides of sea star development and know that all animals develop in this same way (Fig. 25.1). A sea star is an invertebrate that develops in the ocean and, therefore, will develop easily in the laboratory where it can be observed.

Obtain slides or view a model of sea star development and note the following:

1. **Zygote:** Both plants and animals begin life as a single cell, a zygote. A zygote contains chromosomes from each parent. Explain. ______________________________

   ____
2. **Cleavage:** View slides showing various numbers of cells due to the process of cleavage, cell division without growth until the morula stage. Is the morula about the same size as the zygote? ______________

   Explain. ______________________________

   ____
3. **Blastula:** The cavity of a blastula is called the blastocoel. *Label the blastocoel in Figure 25.1.* The formation of a hollow cavity is important to the next stage of development.

Figure 25.1 Sea star development.
All animals, including sea stars and mammals, go through the same cellular stages from cleavage to blastula. (Magnification 75×)

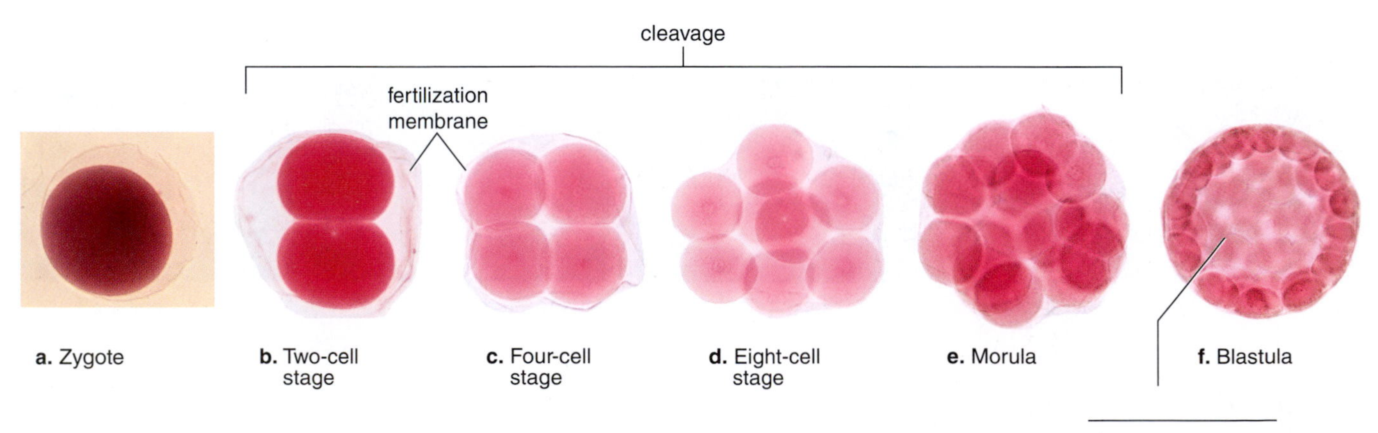

Observation: Cellular Stages of Development in Other Animals

Observe models or film loops that show the cellular stages of development in a mammal (Fig. 25.2). Note that fertilization in mammals occurs in an oviduct following ovulation. As the embryo undergoes cleavage, it travels in the oviduct to the uterus.

The blastula in mammals is called a blastocyst. The blastocyst contains an **inner cell mass** that becomes the embryo, and the outer group of cells (the trophoblast) will become membranes that nourish and protect it. When the blastocyst reaches the uterus and implants into the uterine wall, it will receive nourishment from the mother's bloodstream.

What's the main difference between the cellular stages in a sea star and in a mammal? ______________

____

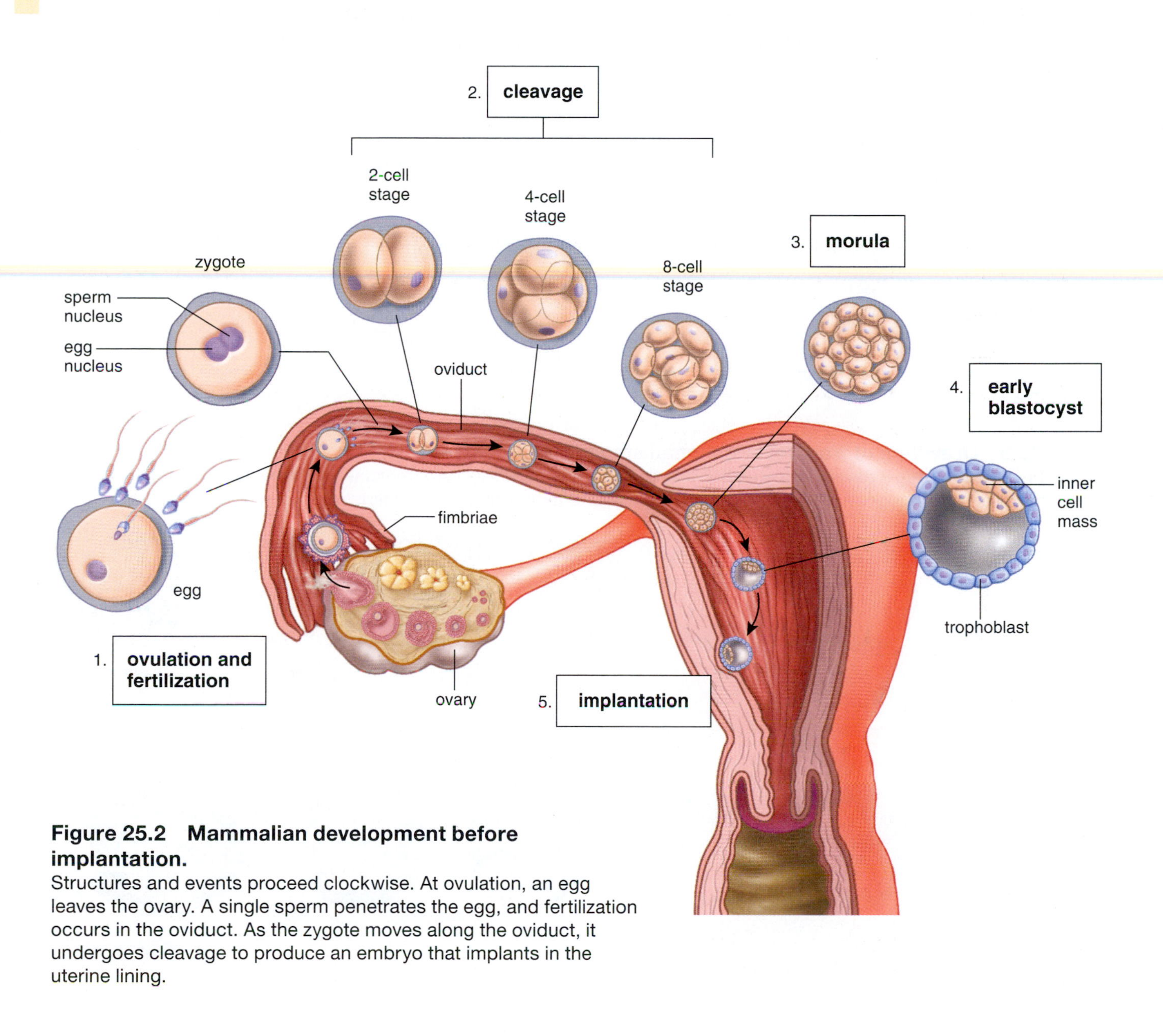

Figure 25.2 Mammalian development before implantation.
Structures and events proceed clockwise. At ovulation, an egg leaves the ovary. A single sperm penetrates the egg, and fertilization occurs in the oviduct. As the zygote moves along the oviduct, it undergoes cleavage to produce an embryo that implants in the uterine lining.

Tissue Stages of Development

The tissue stages of development include the following:

- **Early gastrula stage:** This stage begins when certain cells begin to push or invaginate into the blastocoel at the **blastopore,** creating a double layer of cells. The outer layer is called the **ectoderm,** and the inner layer is called the endoderm.
- **Late gastrula stage:** Gastrulation is not complete until there are three layers of cells. The third layer called mesoderm occurs between the other two layers already mentioned.

Observation: Tissue Stages of Development in Frogs

It is traditional to view gastrulation in a frog. A frog is a vertebrate, and so its development is expected to be closer to that of a mammal than is a sea star. Figure 25.3 shows the development of the frog including tissue development.

1. **Early gastrula stage:** Obtain a cross section of a frog gastrula. Most likely, your slide is the equivalent of Figure 25.3*b*, number 3, in which case you will see two cavities, the old blastocoel and newly forming *archenteron,* which forms once cells have invaginated. The archenteron will become the digestive tract, and the blastopore will become the anus.
2. **Late gastrula stage:** Invagination of cells occurs at the lateral and ventral lips of the blastopore only because cells heavy laden with yolk (yellow cells) do not invaginate.

   Name the germ (tissue) layers that are now present in the embryo. ______________________

   ____

   ____

Figure 25.3 Drawings of frog developmental stages.
a. Cellular stages. During cleavage, the number of cells increases but overall size remains the same. **b.** Tissue stages. During gastrulation, three germ (tissue) layers form. Blue = ectoderm; yellow = endoderm; red = mesoderm.

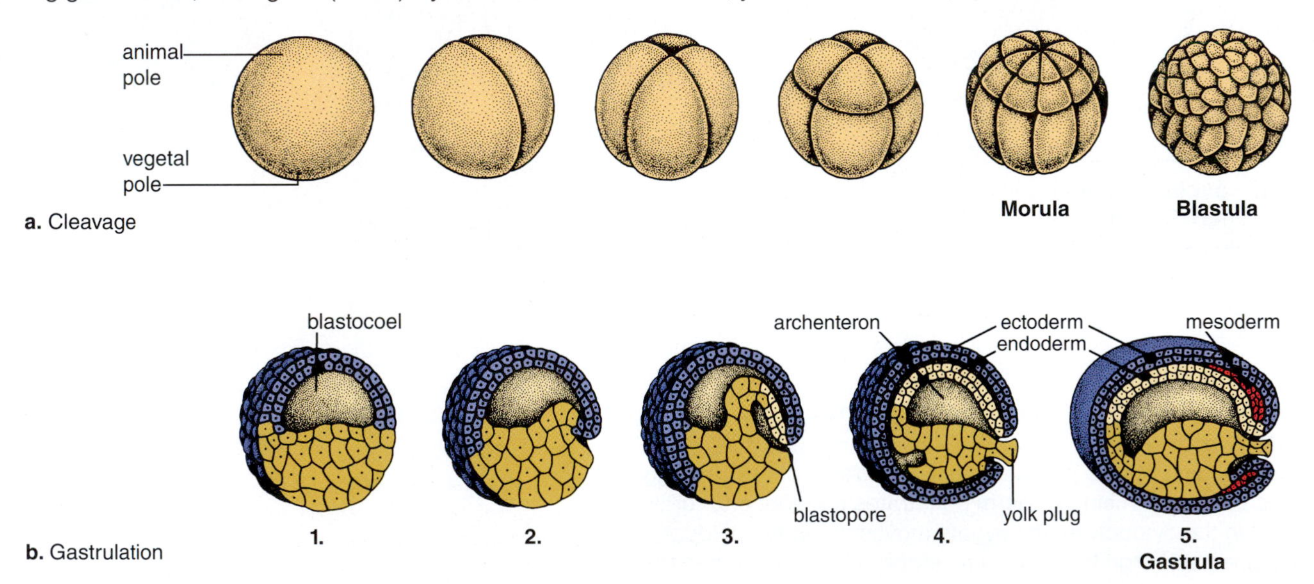

Observation: Tissue Stages of Development in Mammals

Figure 25.4 allows you to observe the results of gastrulation in a mammal. After implantation, gastrulation in mammals turns the inner cell mass into the **embryonic disk.** Figure 25.4 shows the embryonic disk, which has the three layers of cells we have been discussing: the ectoderm, mesoderm, and endoderm. Figure 25.4 also shows the significance of these layers, often called the **germ layers.** The future organs of an individual can be traced back to the germ layers. Each layer gives rise to specific organs.

Figure 25.4 Embryonic disk.
The embryonic disk has three germ layers called ectoderm, mesoderm, and endoderm. Organs and tissues can be traced back to a particular germ layer as indicated in this illustration.

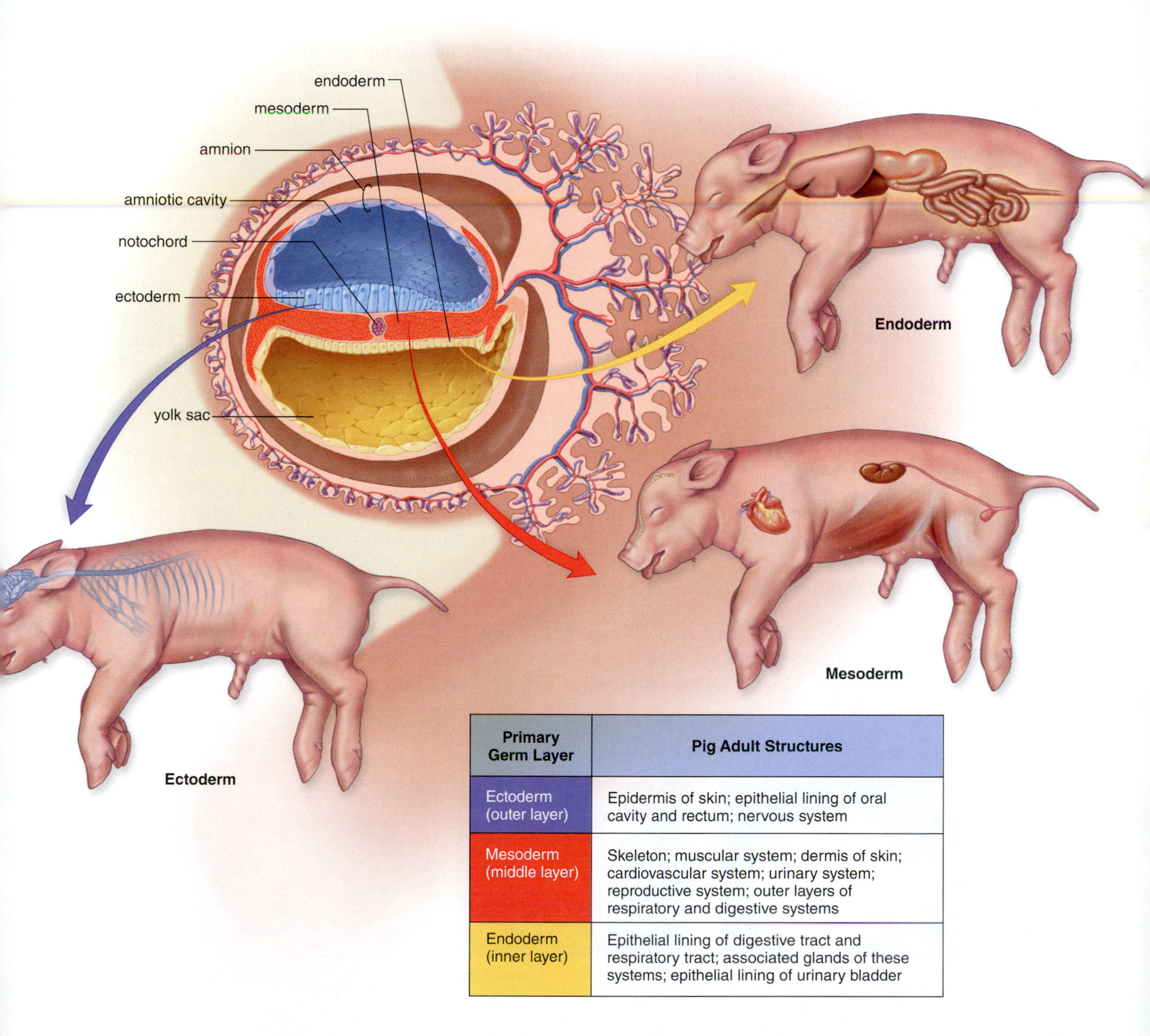

Primary Germ Layer	Pig Adult Structures
Ectoderm (outer layer)	Epidermis of skin; epithelial lining of oral cavity and rectum; nervous system
Mesoderm (middle layer)	Skeleton; muscular system; dermis of skin; cardiovascular system; urinary system; reproductive system; outer layers of respiratory and digestive systems
Endoderm (inner layer)	Epithelial lining of digestive tract and respiratory tract; associated glands of these systems; epithelial lining of urinary bladder

25.2 The Organ Stages of Development

As soon as all three embryonic tissue layers (ectoderm, endoderm, and mesoderm) are established, the organ level of development begins. It continues until all organs have formed. Figure 25.4 pertains to the organ stages of development because it shows which germ layers give rise to which organs.

The first organs to develop are the (1) digestive tract (you have already observed the start of the archenteron during gastrulation); (2) neural tube and brain; and (3) heart.

Development of the Neural Tube and Brain

One of the first systems to form is the nervous system. Why might it be beneficial for the nervous system to begin development first? ______________________________

____

During nervous system development in the frog, two folds of ectoderm grow upward as the neural folds with a groove between them. The flat layer of ectoderm between them is the **neural plate.** The tube resulting from closure of the folds is the **neural tube.** An examination of the neurula in cross section shows that the neural tube develops directly above the **notochord,** a structure that arises from mesoderm in the middorsal region. Later, the notochord is replaced by vertebrae and the neural tube is then called the nerve cord. The anterior portion of the neural tube becomes the brain.

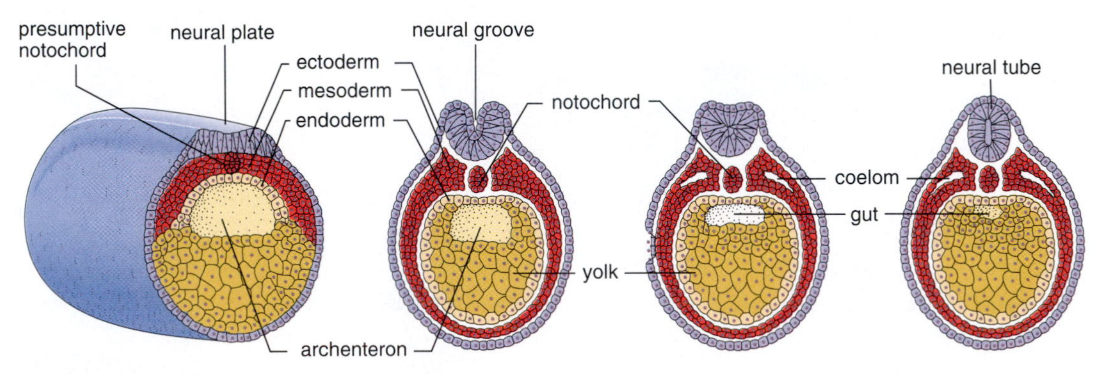

The notochord is said to **induce** the formation of the nervous system. Experiments have shown that if contact with notochord tissue is prevented, no neural plate is formed. **Induction** is believed to be one means by which development is usually orderly. The part of the embryo that induces the formation of an adjacent organ is said to be an **organizer** and is believed to carry out its function by releasing one or more chemical substances.

Observation: Development of Neural Tube and Brain

1. Obtain a slide showing a cross section of a frog neurula stage, and match it to one of the drawings above. Which drawing seems to best match your slide? Your instructor will confirm your match for you.
2. Obtain and examine frog embryos for a three-dimensional view of neurulation in the frog (Fig. 25.5).

Figure 25.5 Photographs of frog during neurulation.

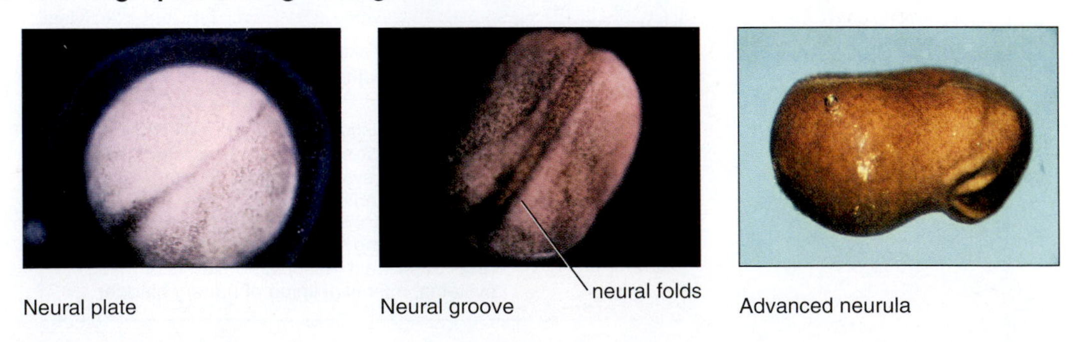

Development of the Heart

A chick embryo offers an opportunity to view a beating heart in an embryo. Your instructor may show you various stages. In particular you will want to observe the chick embryo from the 48-hour stage up to the 96-hour stage.

Observation: Development of Heart

Forty-Eight-Hour Chick Embryo

1. Follow the standard procedure (see box on this page) for selecting and opening an egg containing a 48-hour chick embryo.
2. The embryo has turned so that the head region is lying on its side. Refer to Figure 25.6, and identify the following:
   a. **Shape of the embryo,** which has started to bend. The head is now almost touching the heart.
   b. **Heart,** contracting and circulating blood. Can you make out a ventricle, an atrium, and the aortic arches in the region below the head? Later, only one aortic arch will remain.
   c. **Vitelline arteries** and **veins,** which extend over the yolk. The vitelline veins carry nutrients from the yolk sac to the embryo.
   d. **Brain** with several distinct regions.
   e. **Eye,** which has a developing lens.
   f. **Margin (edge) of the amnion,** which can be seen above the vitelline arteries (see next section for amnion).
   g. **Somites,** blocks of developing muscle tissue that differentiate from mesoderm, which now number 24 pairs.
   h. **Caudal fold** of the amnion. The embryo will be completely enveloped when the head fold and caudal fold meet the margin of the amnion.

Observing Live Chick Embryos

Use the following procedure for selecting and opening the eggs of live chick embryos:

1. Choose an egg of the proper age to remove from the incubator, and put a penciled × on the uppermost side. The embryo is just below the shell.
2. Add warmed chicken Ringer solution to a finger bowl until the bowl is about half full. (Chicken Ringer solution is an isotonic salt solution for chick tissue that maintains the living state.) The chicken Ringer solution should not cover the yolk of the egg.
3. On the edge of the dish, gently crack the egg on the side opposite the ×.
4. With your thumbs placed over the ×, hold the egg in the chicken Ringer solution while you pry it open from below and allow its contents to enter the solution. If you open the egg too slowly or too quickly, the shell may damage the delicate membranes surrounding the embryo.

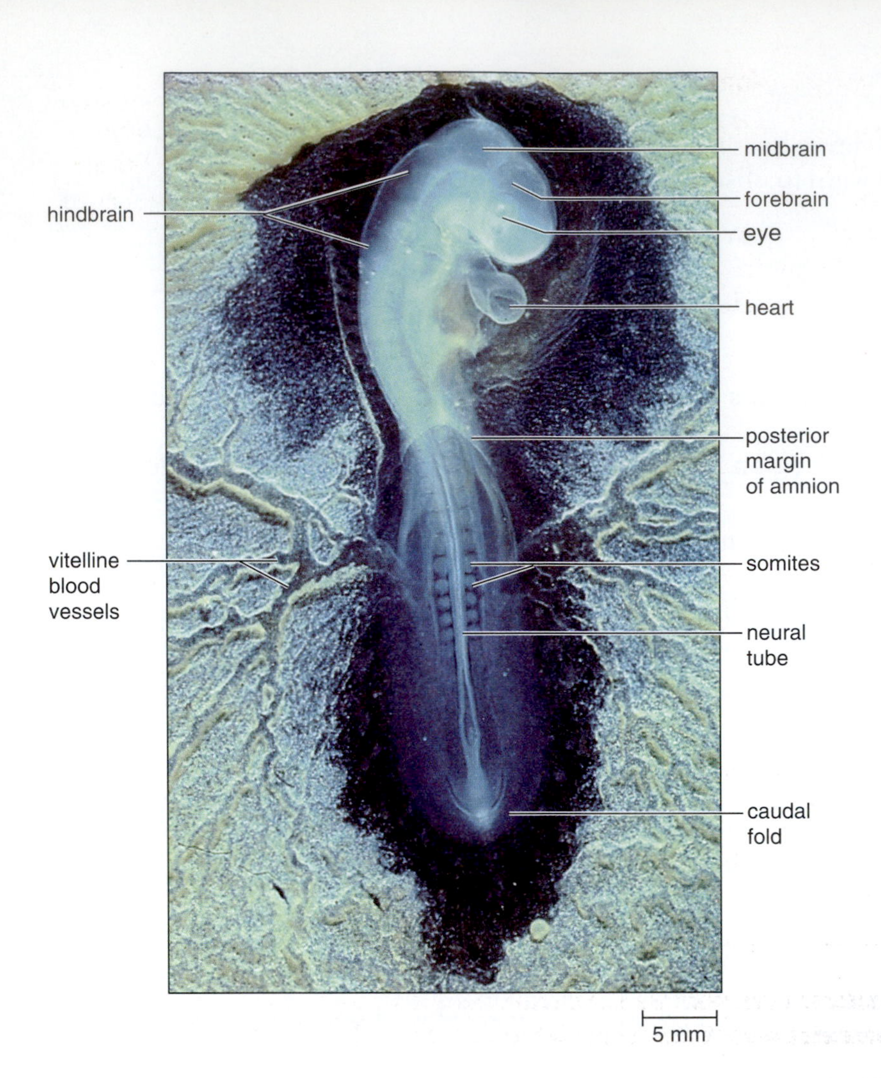

Figure 25.6 Forty-eight-hour chick embryo.
The most prominent organs are labeled.

Older Chick Embryos

As a chick embryo continues to grow, various organs differentiate further (Fig. 25.7). The neural tube closes along the entire length of the body and is now called the spinal cord. The allantois, an extraembryonic membrane, is seen as a sac extending from the ventral surface of the hindgut near the tail bud. The digestive system forms specialized regions, and there are both a mouth and an anus. The yolk sac, the extraembryonic membrane that encloses the yolk, is attached to the ventral wall, but when the yolk is used up, the ventral wall closes.

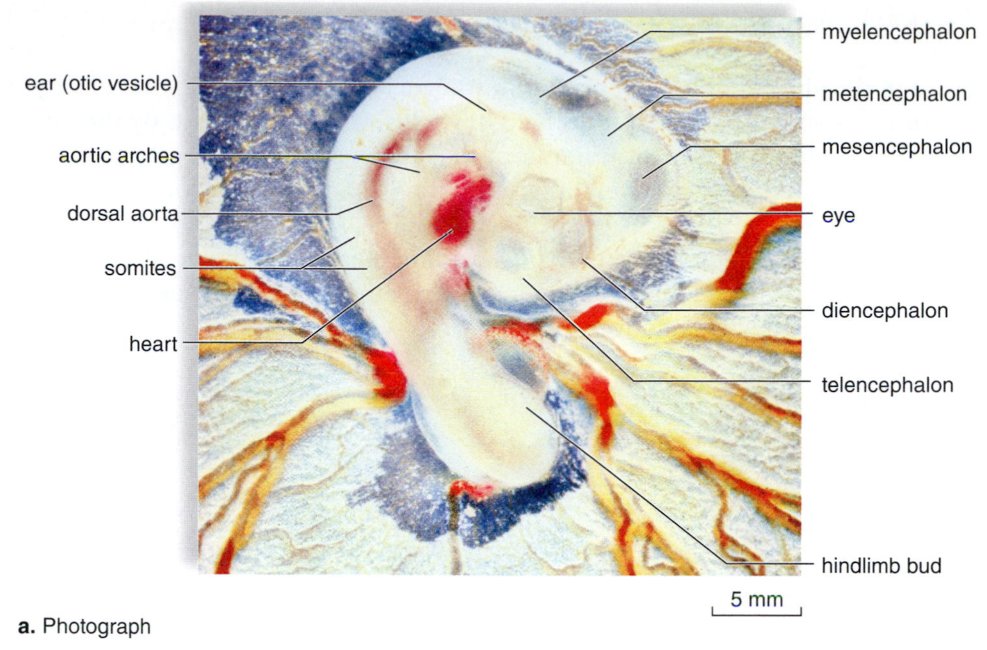

a. Photograph

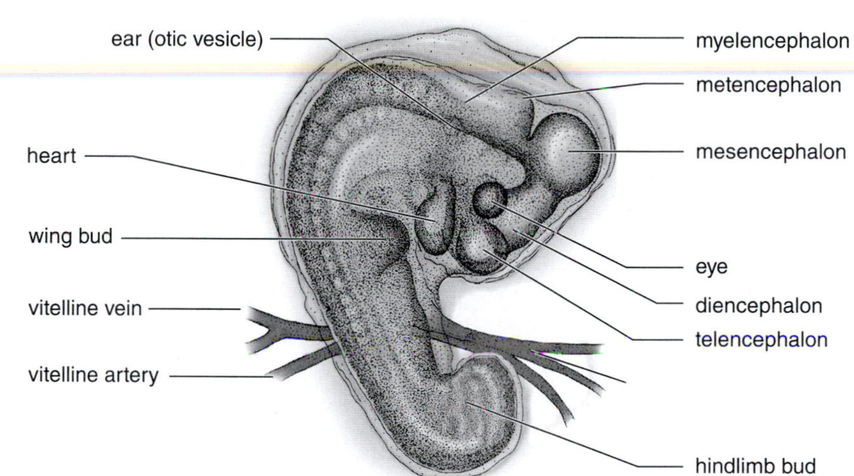

b. Drawing

Key:

Brain region	**Becomes:**
telencephalon	cerebrum
diencephalon	thalamus, hypothalamus, posterior pituitary gland, pineal gland
mesencephalon	midbrain
metencephalon	cerebellum, pons
myelencephalon	medulla oblongata

Figure 25.7 Ninety-six-hour chick embryo.
Brain regions listed in the key can now be seen.

Observation: Comparison of Human to Chick Development

Human development requires nine months of development. Human embryonic development encompasses the first two months (eight weeks) of development. By the end of embryonic development, all organs have formed, and a human form has taken shape. During human fetal development (the next seven months), refinements and weight gain occur. See Section 25.3 for human fetal development.

1. Study models or other study aids available that show the development of human embryos at different ages. Also view Figure 25.8, which depicts the external appearance of the embryo from the fourth to the seventh week of development.

Figure 25.8 External appearance of the human embryo.
a. Weeks 4 to 5 and (**b**) weeks 6 to 7.

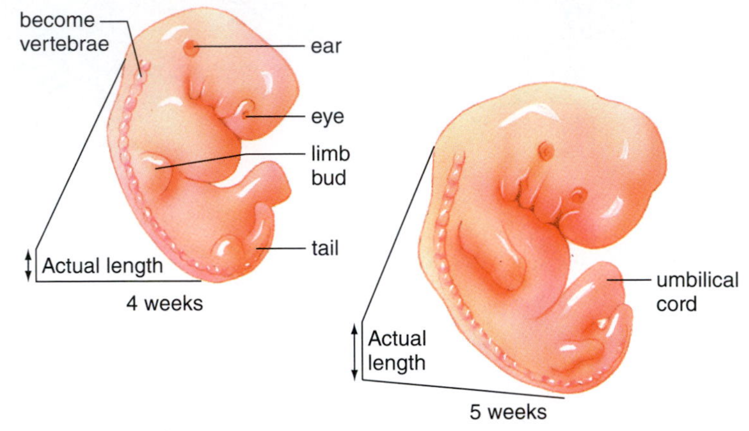

a. Weeks 4 and 5
- Head dominant, but body getting longer.
- Limb buds are visible.
- Eyes and ears begin to form.
- Tissue for vertebrae extend into tail.

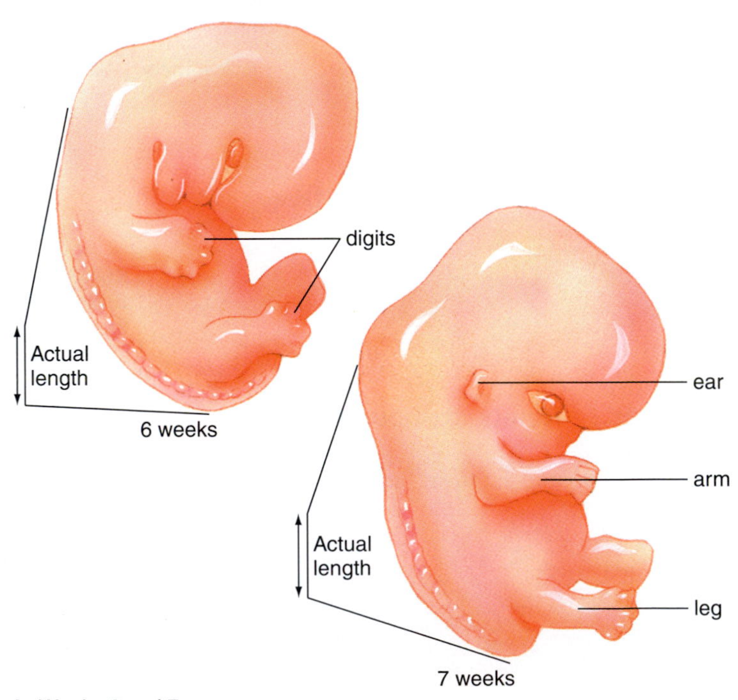

b. Weeks 6 and 7
- Head still dominant, but tail has disappeared.
- Facial features continue to develop.
- Hands and feet have digits.
- All organs are more developed.

2. As illustrated in Figure 25.9, the early stages of human development are quite similar to those of the chick. Differences become marked only as development proceeds.

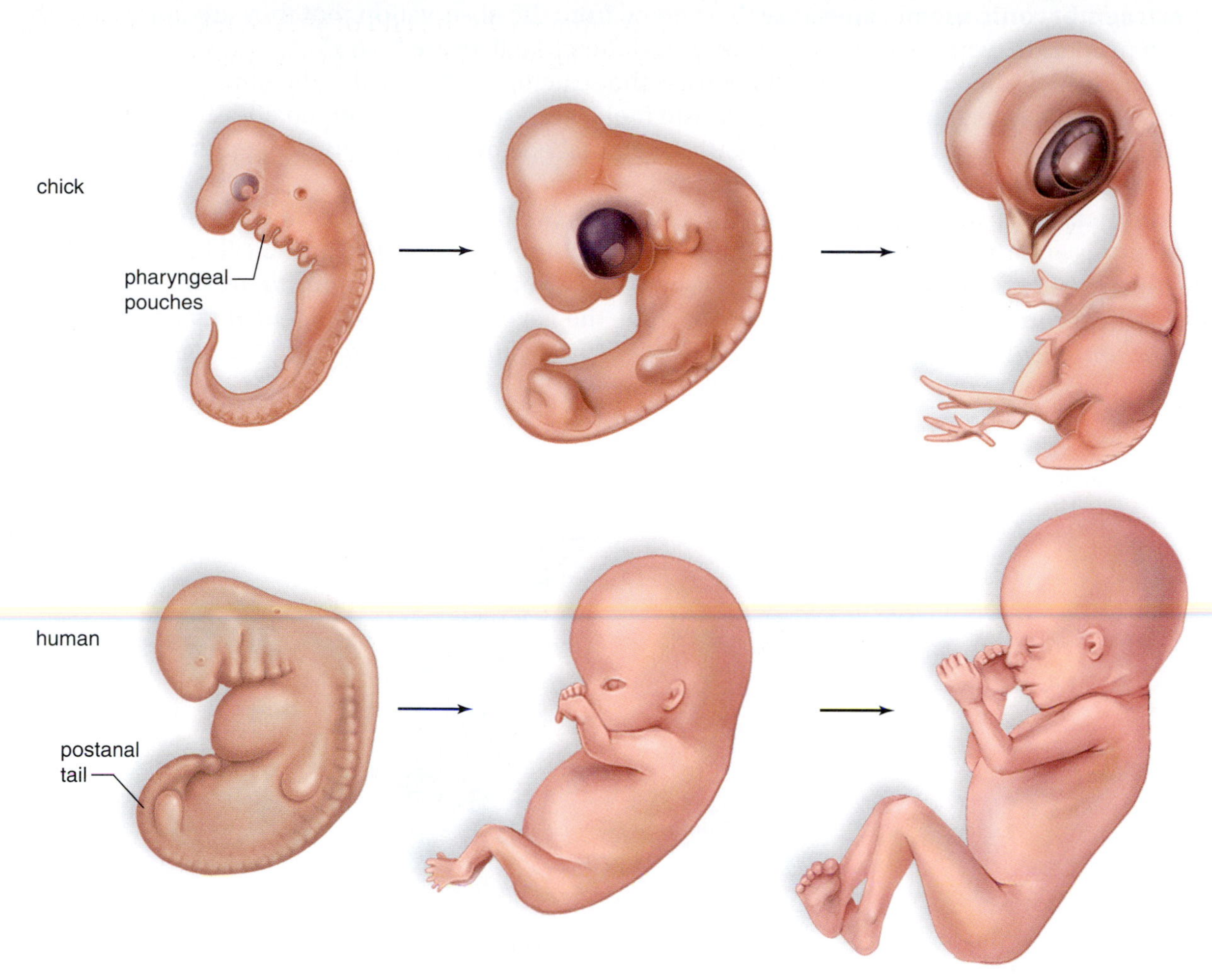

Figure 25.9 Comparison of chick and human embryos. Successive stages in the development of chick and human. Early stages (far left) are similar; differences become apparent as development continues.

25.3 Extraembryonic Membranes, the Placenta, and the Umbilical Cord

- The **extraembryonic membranes** take their name from the observation that they are not part of the embryo proper. They are outside the embryo, and therefore they are "extra."
- In mammals including humans, the **placenta** is the structure that provides the embryo with nutrient molecules and oxygen and takes away its waste molecules, such as carbon dioxide. The fetal half of the placenta contains the fetal capillaries. The maternal half of the placenta is the uterine wall where maternal blood vessels meet the fetal capillaries.
- The **umbilical cord** is a tubular structure that contains two of the extraembryonic membranes (the allantois and the yolk sac) and also the **umbilical blood vessels.** The umbilical blood vessels bring fetal blood to and from the placenta. When the baby is born and begins to breathe on its own, the umbilical cord is cut and the remnants become the navel. *In this drawing, identify the umbilical cord, the umbilical blood vessels, and the placenta, which contains the maternal blood vessels.*

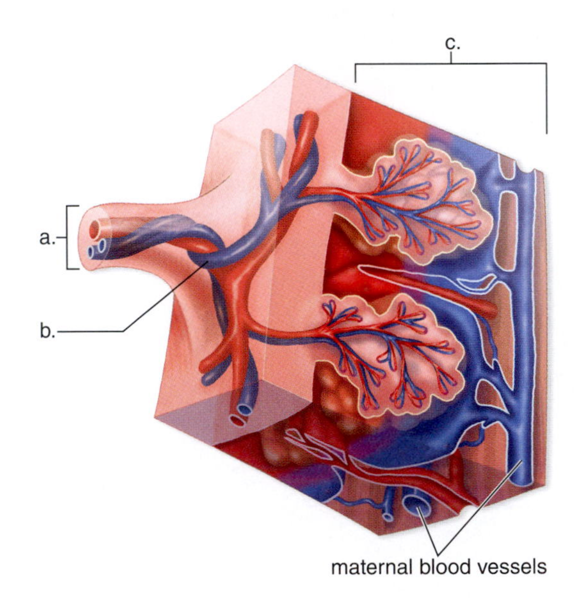

Observation: The Extraembryonic Membranes

In a model, and in Figure 25.10, trace the development of the extraembryonic membranes. Also, note the development of the placenta and the umbilical cord. The extraembryonic membranes are as follows:

- **Chorion:** The outermost membrane, and in chicks it lies just below the porous shell, where it functions in gas exchange. In mammals, an outer layer of cells surrounding the inner cell mass at the blastocyst stage becomes the chorion. Notice in Figure 25.10 that the treelike **chorionic villi** are a part of the chorion.
- **Amnion:** Forms the amniotic cavity, which envelops the embryo and contains the amniotic fluid that cushions and protects the developing offspring (Fig. 25.11). All animals, whether the sea star, the frog, the chick, or a human, develop in an aqueous environment. Birth of a human is imminent when "the water breaks," and the amniotic fluid is lost.
- **Allantois:** Serves as a storage area for metabolic waste in the chick. In mammals, the allantois extends into the umbilical cord. It accumulates the small amount of urine produced by the fetal kidneys and later contributes to urinary bladder formation. Its blood vessels become the umbilical blood vessels.
- **Yolk sac:** The first embryonic membrane to appear. In the chick, the yolk sac does contain yolk, food for the developing embryo. In mammals, the yolk sac contains plentiful blood vessels and is the first site of blood cell formation.

Figure 25.10 Development of extraembryonic membranes in mammals including humans.
a. At first, no organs are present in the embryo, only tissues. The amniotic cavity is above the embryonic disk, and the yolk sac is below. The chorionic villi are present. **b, c.** The allantois and yolk sac, two more extraembryonic membranes, are positioned inside the body stalk as it becomes the umbilical cord. **d.** At 35+ days, all membranes are present, and the umbilical cord takes blood vessels between the embryo and the chorion (placenta).

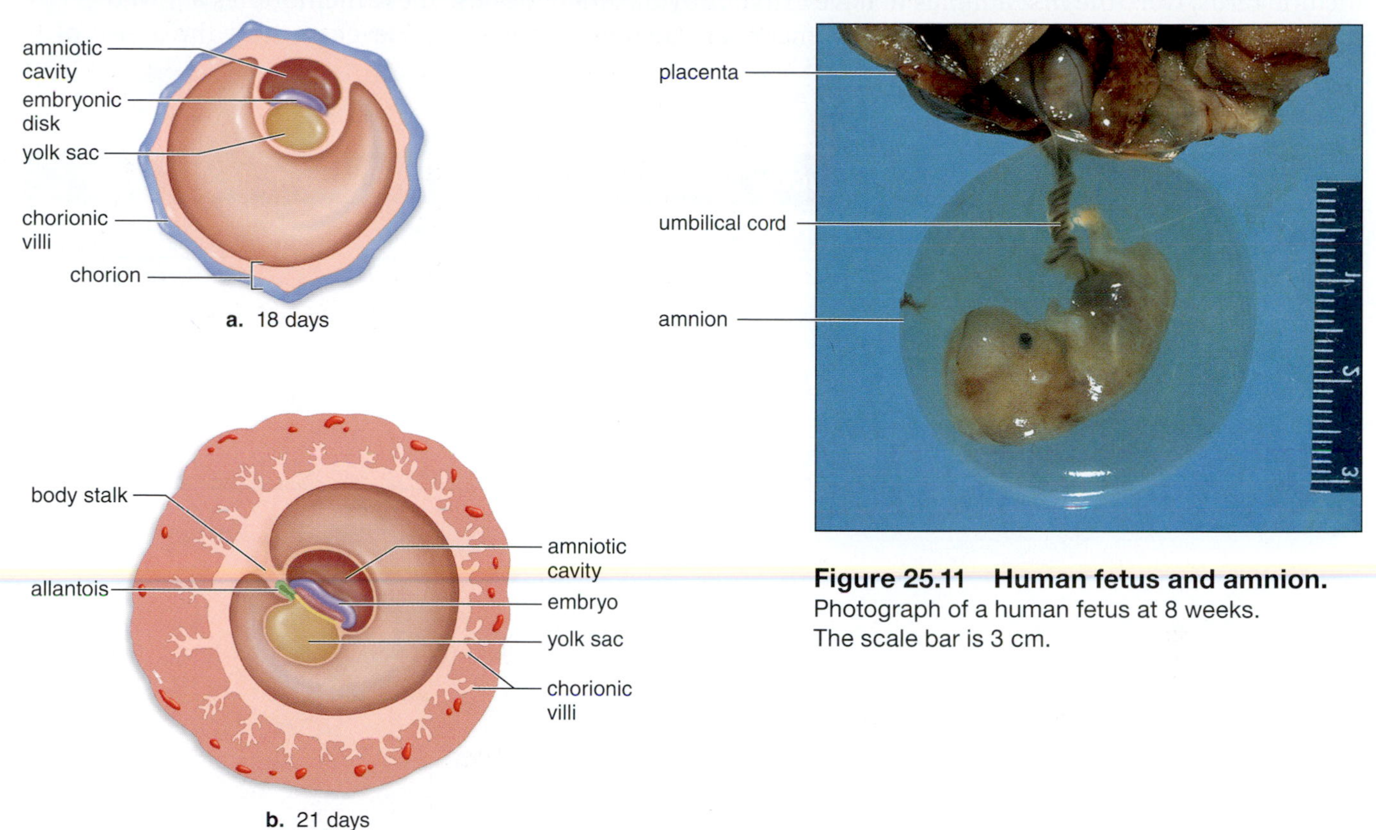

Figure 25.11 Human fetus and amnion.
Photograph of a human fetus at 8 weeks. The scale bar is 3 cm.

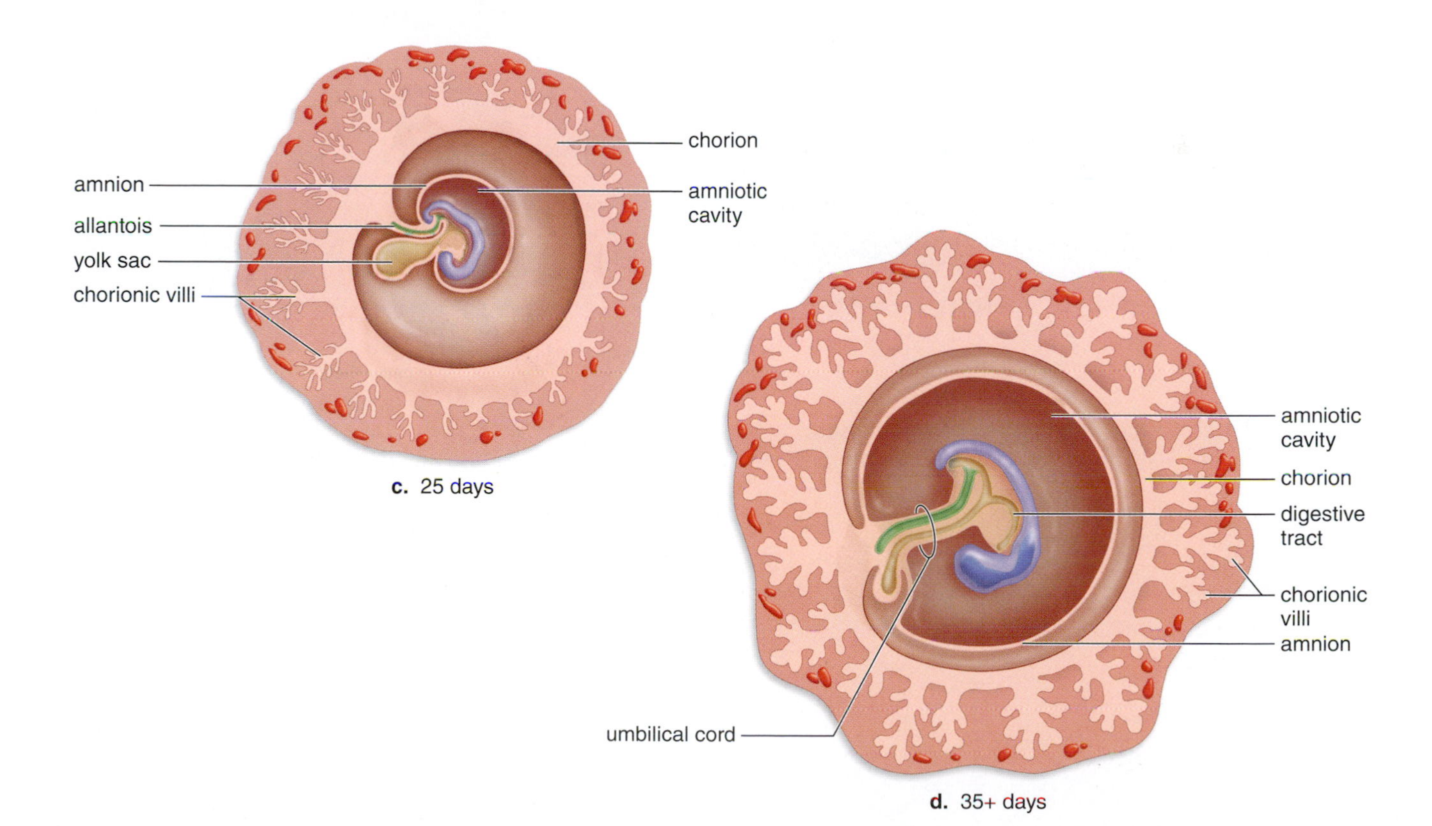

Comparison of Extraembryonic Membranes in Chicks and Mammals

Consult Figure 25.12 and use the information on page 350 to complete Table 25.1, which compares the function of the extraembryonic membranes in chicks and in mammals. Reptiles, which we now know include birds, were the first animals to have extraembryonic membranes. These membranes allowed reptiles to develop on land. They also allow mammals, including humans, to develop inside the uterus of the mother.

Table 25.1 Functions of Extraembryonic Membranes in Chicks and Mammals

Membrane	Chick	Mammal
Amnion		
Allantois		
Chorion		
Yolk sac		

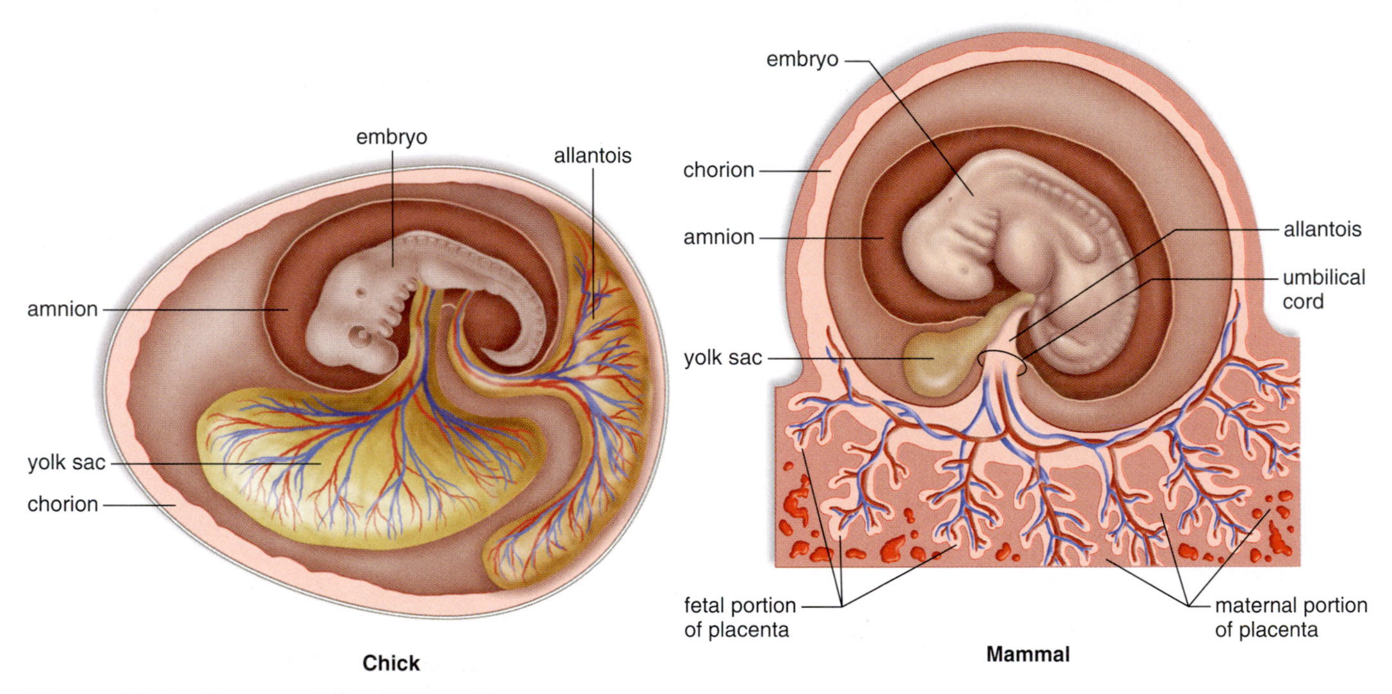

Figure 25.12 Extraembryonic membranes.
Chicks and mammals have the same extraembryonic membranes but, except for the amnion, they have different functions.

25.4 Human Fetal Development

During fetal development (last seven months in humans), the skeleton becomes ossified (bony), reproductive organs form, arms and legs develop fully, and the fetus enlarges in size and gains weight.

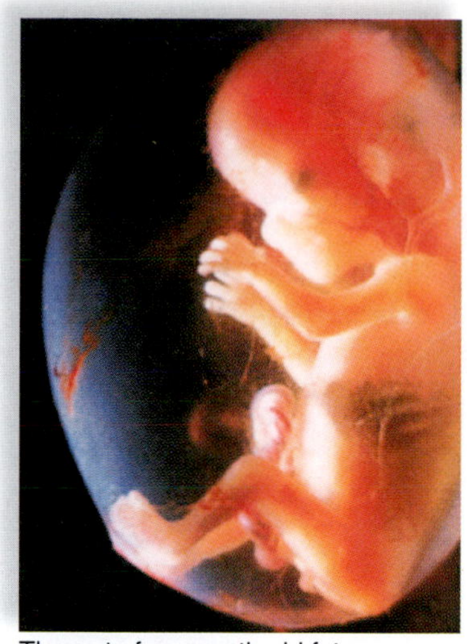

Three- to four-month-old fetus

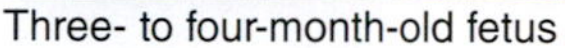

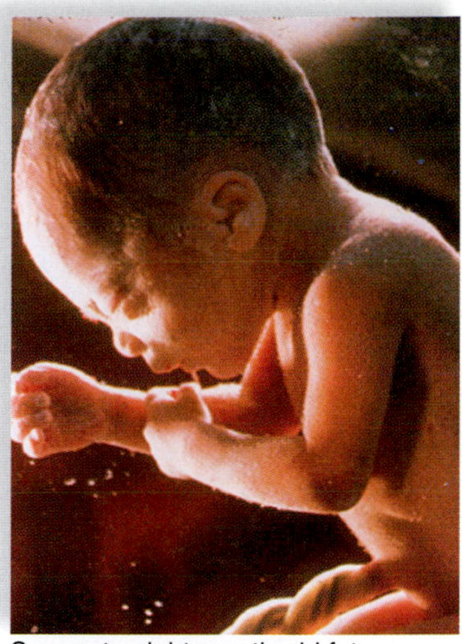

Seven- to eight-month-old fetus

Observation: Fetal Development

1. Using Table 25.2 and Figure 25.13 to assist you, examine models of fetal development.
2. In Table 25.2, note the following.
   a. **External genitals:** About the third month, it is possible to tell male from female if an ultrasound is done.
   b. **Quickening:** Fetal movement is felt during the fourth or fifth months.
   c. **Vernix caseosa:** Beginning with the fifth month, the skin is covered with a cheesy coating called vernix caseosa.
   d. **Lanugo:** During the sixth and seventh months, the body is covered with fine, downy hair termed lanugo.

Table 25.2 Human Fetal Development

Month	Events for Mother	Events for Baby
Third month	Uterus is the size of a grapefruit.	Possible to distinguish sex. Fingernails appear.
Fourth month	Fetal movement is felt by those who have been previously pregnant. Heartbeat is heard by stethoscope.	Bony skeleton visible. Hair begins to appear. 150 mm (6 in.), 170 g (6 oz.).
Fifth month	Fetal movement is felt by those who have not been previously pregnant. Uterus reaches up to level of umbilicus and pregnancy is obvious.	Protective cheesy coating, called vernix caseosa, begins to be deposited. Heartbeat can be heard.
Sixth month	Doctor can tell where baby's head, back, and limbs are. Breasts have enlarged, nipples and areolae are darkly pigmented, and colostrum is produced.	Body is covered with fine hair called lanugo. Skin is wrinkled and reddish.
Seventh month	Uterus reaches halfway between umbilicus and rib cage.	Testes descend into scrotum. Eyes are open. 300 mm (12 in.), 1,350 g (3 lb.).
Eighth month	Weight gain is averaging about a pound a week. Difficulty in standing and walking because center of gravity is thrown forward.	Body hair begins to disappear. Subcutaneous fat begins to be deposited.
Ninth month	Uterus is up to rib cage, causing shortness of breath and heartburn. Sleeping becomes difficult.	Ready for birth. 530 mm (20½ in.), 3,400 g (7½ lb.).

Figure 25.13 Human development.
Changes occurring from the fifth week to the eighth month.

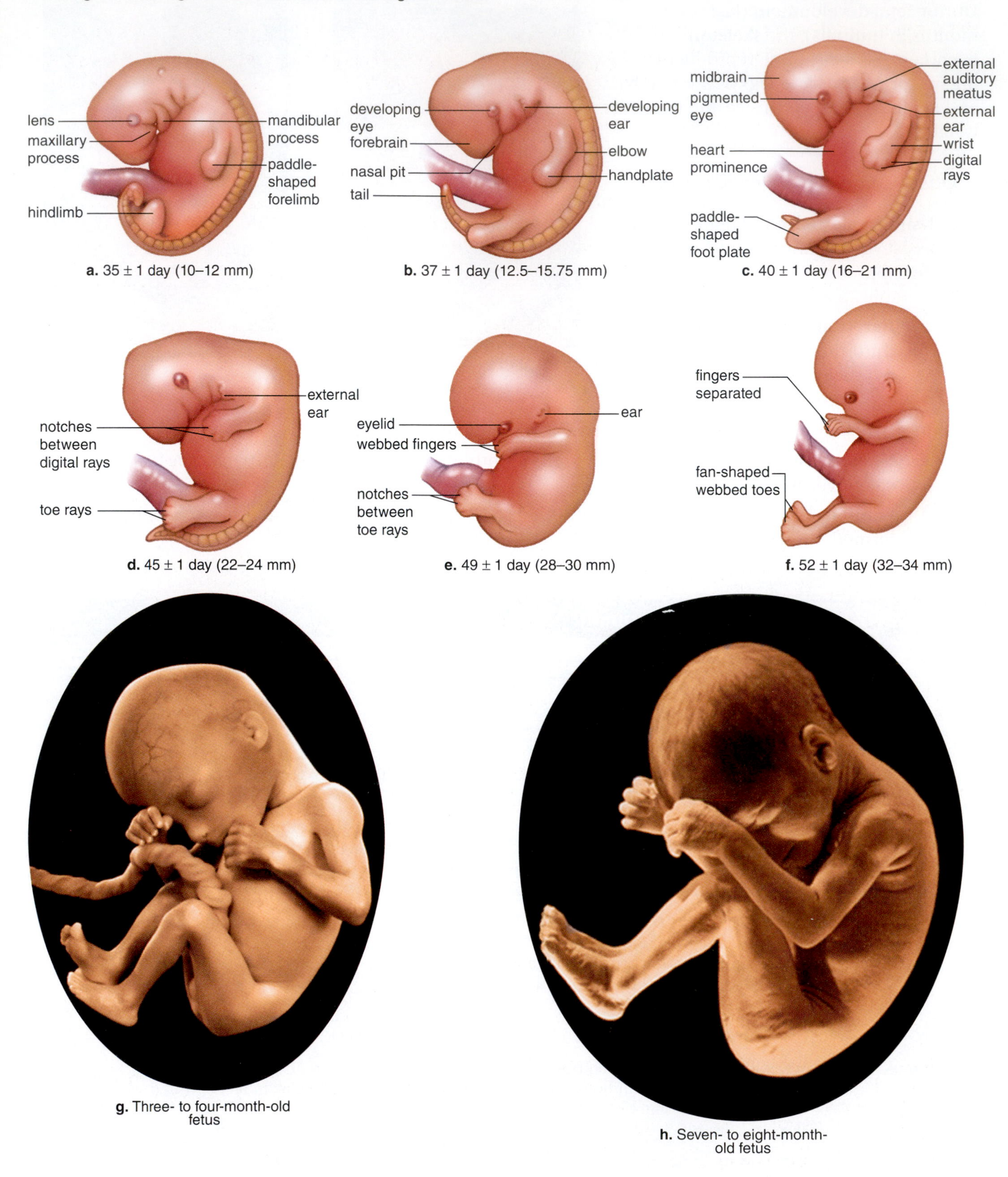

a. 35 ± 1 day (10–12 mm)

b. 37 ± 1 day (12.5–15.75 mm)

c. 40 ± 1 day (16–21 mm)

d. 45 ± 1 day (22–24 mm)

e. 49 ± 1 day (28–30 mm)

f. 52 ± 1 day (32–34 mm)

g. Three- to four-month-old fetus

h. Seven- to eight-month-old fetus

Laboratory Review 25

Complete this table by placing a checkmark in the appropriate square if the feature pertains to the organism.

Comparison of Embryonic Features of a Developing Sea Star, Frog, and Chick

Feature	Sea Star	Frog	Mammal
1. Blastula is a circular cavity.			
2. Germ layers are present.			
3. Notochord is present.			
4. Waste is deposited in water.			

5. Describe how an animal embryo becomes a morula, blastula, and gastrula. ___

6. What type of cells influence how gastrulation occurs in a frog? ___

7. Describe how induction may control development. ___

8. Name the four extraembryonic membranes, and state the function of each in birds and mammals.

	Extraembryonic Membrane	Function in Birds	Function in Mammals
a.			
b.			
c.			
d.			

9. Name three features that are quite noticeable in a 48-hour chick embryo. ___

10. List the two stages of human development, and state a reason for dividing human development into these two stages. ___

LABORATORY

26

Effects of Pollution on Ecosystems

Learning Outcomes

Introduction

This laboratory will consider three causes of aquatic pollution: thermal pollution, acid pollution, and cultural eutrophication. **Thermal pollution** occurs when water temperature rises above normal. As water temperature rises, the amount of oxygen dissolved in water decreases, possibly depriving organisms and their cells of an adequate supply of oxygen. Deforestation, soil erosion, and the burning of fossil fuels contribute to thermal pollution, but the chief cause is use of water from a lake or the ocean as a coolant for the waste heat of a power plant.

When sulfur dioxide and nitrogen oxides enter the atmosphere, usually from the burning of fossil fuels, they are converted to acids, which return to Earth as **acid deposition** (acid rain or snow). Acid deposition kills plants, aquatic invertebrates, and also decomposers, threatening the entire ecosystem.

Figure 26.1 Cultural eutrophication. Eutrophic lakes tend to have large populations of algae and rooted plants.

Cultural eutrophication, or overenrichment, is due to runoff from agricultural fields, wastewater from sewage treatment plants, and even excess detergents. These sources of excess nutrients cause an algal bloom seen as a green scum on a lake (Fig. 26.1). When algae overgrow and die, decomposition robs the lake of oxygen, causing a fish die-off.

26.1 Studying the Effects of Pollutants

We are going to study the effects of pollution by observing its effects on hay infusion organisms, on seed germination, and on an animal called *Gammarus.*

Study of Hay Infusion Culture

A hay infusion culture (hay soaked in water) contains various microscopic organisms, but we will be concentrating on how the pollutants in our study affect the protozoan populations in the culture. We will consider both of these aspects:

1. **Species composition:** Number of different types of protozoans.
2. **Species diversity:** Diversity increases as the relative abundance of each type protozoan.

Experimental Procedure: Effect of Pollutants on a Hay Infusion Culture

During this Experimental Procedure you will examine, by preparing a wet mount, hay infusion cultures that have been treated in the following manner.

1. **Control culture:** This culture simulates the species composition and diversity of an untreated culture. Prepare a wet mount, and with the assistance of Figure 26.2 and any guides available in the laboratory, identify as many different types of protozoans as possible in the hay infusion culture. State whether species composition is high, medium, or low. Record your estimation in the second column of Table 26.1. Do you judge species diversity to be high, medium, or low? Record your estimation in the third column of Table 26.1.
2. **Oxygen-deprived culture:** Thermal pollution causes water to be oxygen deprived; therefore, when we study the effects of low oxygen on a hay infusion culture, we are studying an effect of thermal pollution. Prepare a wet mount of this culture, and determine if there is a change in species composition and diversity. Again record the species composition and species diversity as high, medium, or low in Table 26.1.

Figure 26.2 Microorganisms in hay infusion cultures.
Organisms are not to size.

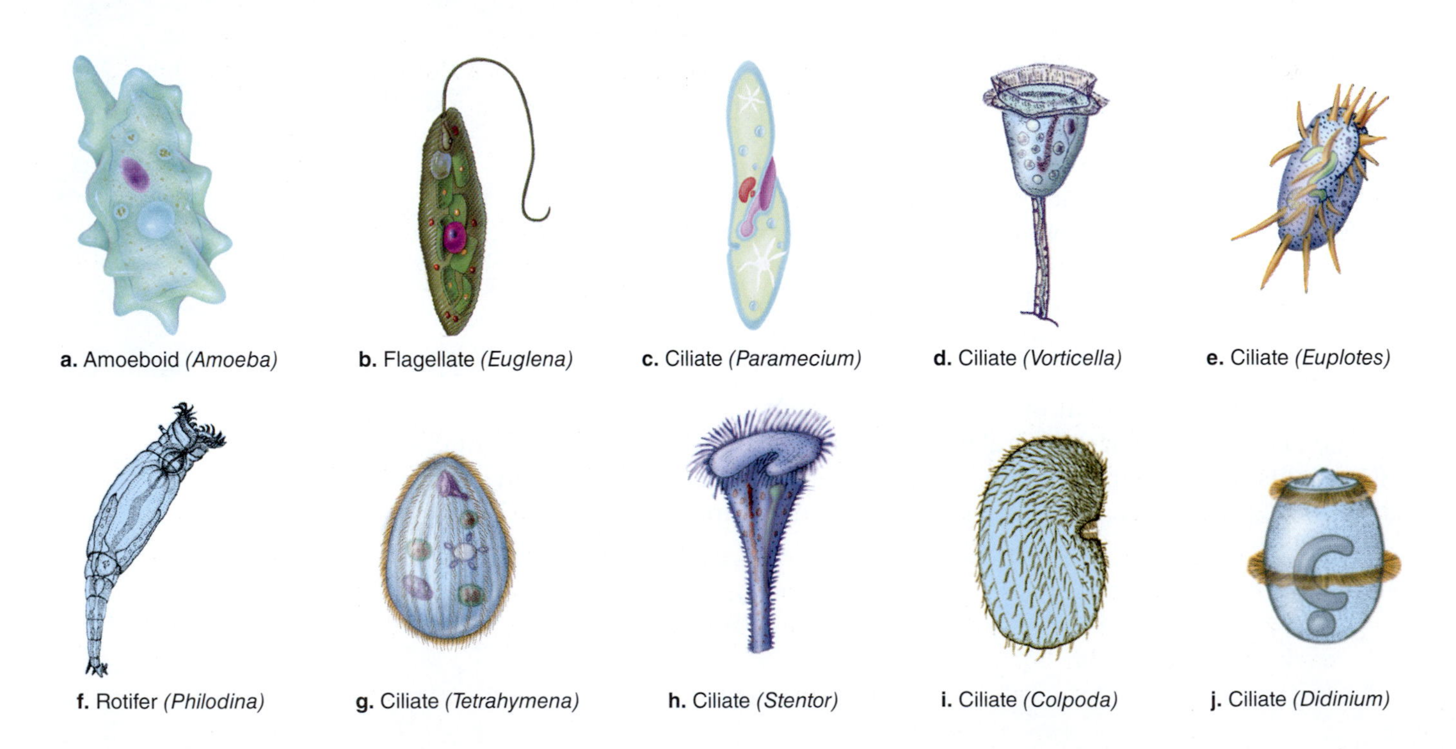

Table 26.1 Effect of Pollution on a Hay Infusion Culture			
Type of Culture	**Species Composition (High, Medium, or Low)**	**Species Diversity (High, Medium, or Low)**	**Explanation**
Control			
Oxygen deprived			
Acidic			
Enriched			

3. **Acidic culture:** In this culture, the pH has been adjusted to 4 with sulfuric acid (H_2SO_4). This simulates the effect of acid rain on a hay infusion culture. Prepare a wet mount of this culture, and determine if there is a change in species composition and diversity. Again record the species composition and species diversity as high, medium, or low in Table 26.1.
4. **Enriched culture:** More organic nutrients have been added to this culture. These nutrients will cause the algae population, which is food for most protozoans, to increase. In the short term, their species composition should increase. Eventually, as the algae die off decomposition will rob the water of oxygen, and the protozoans may start to die off. Prepare a wet mount of this culture, and determine if there is a change in species composition and diversity. Again record the species composition and species diversity as high, medium, or low in Table 26.1.

Conclusions: Effect of Pollution on a Hay Infusion Culture

- What could be a physiological reason for the adverse effects of oxygen deprivation on a hay infusion culture? If consistent with your results, enter this explanation in the last column of Table 26.1.
- What could be a physiological reason for the adverse effects of a low pH on a hay infusion culture? If consistent with your results, enter this explanation in the last column of Table 26.1.
- What could be an environmental reason for the adverse affects of an enriched culture? If consistent with your results, enter this explanation in the last column of Table 26.1.

Effect of Acid Rain on Seed Germination

Seeds depend on favorable environmental conditions of temperature, light, and moisture to germinate, grow, and reproduce. Like any other biological process, germination requires enzymatic reactions that can be adversely affected by an unfavorable pH.

Experimental Procedure: Effect of Acid Rain on Seed Germination

In this Experimental Procedure, we will test whether there is a negative correlation between acid concentration and germination. In other words, it is hypothesized that as acidity increases, the more likely seeds will ______________________________.

Your instructor has placed 20 sunflower seeds in each of five containers with water of increasing acidity: 0% vinegar (tap water), 1% vinegar, 5% vinegar, 20% vinegar, and 100% vinegar.

1. Test and record the pH of solutions having the vinegar concentrations noted above. Record the pH of each solution in Table 26.2.
2. Count the number of germinated sunflower seeds in each container, and complete Table 26.2.

Table 26.2 Effect of Increasing Acidity on Germination of Sunflower Seeds			
Concentration of Vinegar	**pH**	**Number of Seeds that Germinated**	**Percent Germination**
0%			
1%			
5%			
20%			
100%			

Conclusions: Effect of Increasing Acidity on Germination of Sunflower Seeds

- As you know, each enzyme has an optimum pH. Explain why acid rain is expected to inhibit metabolism, and therefore, seedling development. ______________________________
- Do the data support or falsify your hypothesis? ______________________________

Study of *Gammarus*

A small crustacean called *Gammarus* lives in ponds and streams (Fig. 26.3) where it feeds on debris, algae, or anything smaller than itself, such as some of the protozoans in Figure 26.2. In turn, fish like to feed on *Gammarus*.

Experimental Procedure: Gammarus

- Add 25 ml of spring water to a beaker and record the pH of the water. ______________ pH
- Add four *Gammarus* to the container. Do they all use their legs in swimming? ______________
- Which legs are used in jumping and climbing? ______________________________
- What do *Gammarus* do when they "bump" into each other? ______________________________

Control Sample

After observing *Gammarus,* decide what behaviors are most often observed. During a 5-minute time span, total the amount of time spent doing each of these behaviors.

Behaviors	**Amount of Time**	**Total Time**
1. ______________	______________	________
2. ______________	______________	________
3. ______________	______________	________

Test Sample

If so directed by your instructor, put a *Gammarus* in a beaker of spring water adjusted to pH 4 by adding vinegar. During a 5-minute time span, total the amount of time spent doing each of these behaviors.

Behaviors	**Amount of Time**	**Total Time**
1. ______________	______________	________
2. ______________	______________	________
3. ______________	______________	________

Figure 26.3 ***Gammarus.*** *Gammarus* is a type of crustacean classified in a subphylum that also includes shrimp.

***Conclusions:* Gammarus**

- Draw a conclusion from this study. ____________________
  ____
- Create a food chain that shows who eats whom when the food chain includes algae, protozoans, *Gammarus,* fish, and humans. ____________________
  What would happen to this food chain if the water was oxygen deprived? ____________________
  ____
  Acidic? ____________________
  Enriched with inorganic nutrients (short term and long term)? ____________________
  ____

Conclusions: Studying the Effects of Pollutants

- Give an example to show that the hay infusion study pertains to real ecosystems. ____________________
  ____
- What are the potential consequences of acid rain on crops that reproduce by seeds? ____________________
  ____
  On the food chains of the ocean? ____________________
- How does the addition of nutrients affect species composition and species diversity of an ecosystem over time? ____________________

26.2 Studying the Effects of Cultural Eutrophication

Chlorella, the green alga used in this study, is considered to be representative of algae in bodies of fresh water. *Daphnia* (Fig. 26.4) feeds on green algae such as *Chlorella.* First, you will observe how *Daphnia* feeds, and then you will determine the extent to which *Daphnia* could keep the effects of cultural eutrophication from occurring in a hypothetical example. Keep in mind that this case study is an oversimplification of a generally complex problem.

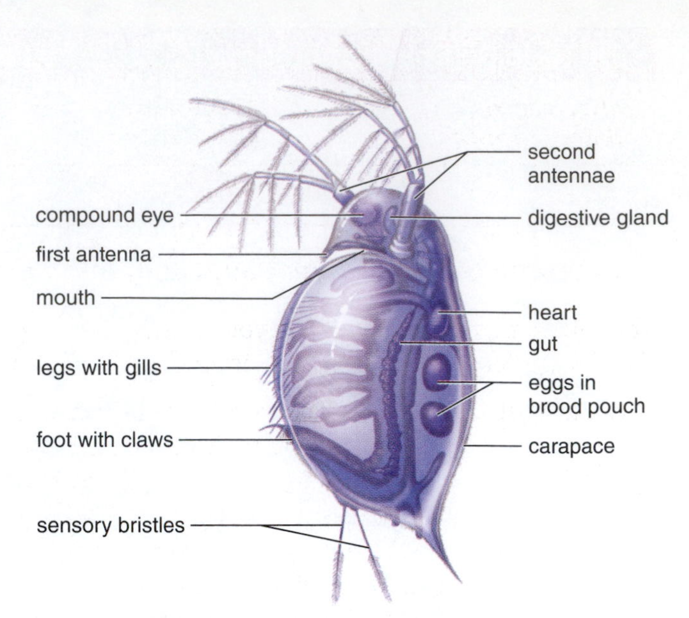

Figure 26.4 Anatomy of *Daphnia,* an arthropod.

Observation: Daphnia *Feeding*

1. Place a small pool of petroleum jelly in the center of a small petri dish.
2. Use a dropper to take a *Daphnia* from the stock culture, place it on its back (covered by water) in the petroleum jelly, and observe it under the stereomicroscope (see Fig. 26.4).
3. Note the clamlike carapace and the legs waving rapidly as the *Daphnia* filters the water.
4. Add a drop of *carmine solution,* and observe how the *Daphnia* filters the "food" from the water and passes it through the gut. The gut is more visible if you push the animal onto its side. In this position, you may also observe the heart beating in the region above the gut and just behind the head.
5. Allow the *Daphnia* to filter-feed for up to 30 minutes, and observe the progress of the carmine particles through the gut. Does the carmine travel completely through the gut in 30 minutes?

____

Experimental Procedure: Daphnia *Feeding on* Chlorella

This exercise requires the use of a spectrophotometer. Absorbance will be a measure of the algal population level; the greater the number of algal cells, the greater the absorbance. The higher the absorbance, the greater the amount of light absorbed and *not* passed through the solution.

1. Obtain two spectrophotometer tubes (cuvettes) and a Pasteur pipet.
2. Fill one of the cuvettes with distilled water, and use it to zero the spectrophotometer. Save this tube for step 6.
3. Use the Pasteur pipet to fill the second cuvette with *Chlorella.* Gently aspirate and expel the sample several times (without creating bubbles) to give a uniform dispersion of the algae.
4. Add ten hungry *Daphnia,* and following your instructor's directions, immediately measure the absorbance with the spectrophotometer. If a *Daphnia* swims through the beam of light, a strong deflection should occur; do not use any such higher readings—instead, use the lower figure for the absorbance. Record your reading in the first column of Table 26.3.
5. Remove the cuvette with the *Daphnia* to a safe place in a test tube rack. Allow the *Daphnia* to feed for 30 minutes.
6. Rezero the spectrophotometer with the distilled water cuvette.
7. Measure the absorbance of the experimental cuvette again. Record your data in the second column of Table 26.3, and explain your results in the third column.

Table 26.3 Spectrophotometer Data/*Daphnia* Feeding on *Chlorella*

Absorbance Before Feeding	Absorbance After Feeding	Explanation

Experimental Procedure: Case Study in Cultural Eutrophication

The following problem will test your understanding of the ecological value of a single species—in this case, *Daphnia.* Please realize that this is an oversimplification of a generally complex problem.

1. Assume that developers want to build condominium units on the shores of Silver Lake. Homeowners in the area have asked the regional council to determine how many units can be built without altering the nature of the lake. As a member of the council, you have been given the following information:

   The current population of *Daphnia,* 10 animals/liter, presently filters 24% of the lake per day, meaning that it removes this percentage of the algal population per day. This is sufficient to keep the lake essentially clear. Predation—the eating of the algae—will allow the *Daphnia* population to increase to no more than 50 animals/liter. Therefore, 50 *Daphnia*/liter will be available for feeding on the increased number of algae that would result from building the condominiums.

   Using this information, complete Table 26.4.

Table 26.4 *Daphnia* Filtering

Number of *Daphnia*/liter	Percent of Lake Filtered
10	24%
50	

2. The sewage system of the condominiums will add nutrients to the lake. Phosphorus output will be 1 kg per day for every 10 condominiums. This will cause a 30% increase in the algal population. Using this information, complete Table 26.5.

Table 26.5 Cultural Eutrophication

Number of Condominiums	Phosphorus Added	Increase in Algal Population
10	1kg	30%
20		
30		
40		
50		

Conclusion: Cultural Eutrophication

- Assume that phosphorus is the only nutrient that will cause an increase in the algal population and that *Daphnia* is the only type of zooplankton available to feed on the algae. How many condominiums would you allow the developer to build? ____________________

- What other possible impacts could condominium construction have on the condition of the lake?

  ____

Laboratory Review 26

1. What type of population would you expect to be the largest in most ecosystems? ______

   Explain. ______

2. What causes acid rain? ______
3. Why is acid deposition harmful to organisms? ______
4. Name the type of pollution that results when water from rivers and ponds is used for cooling power plants, and explain why it has detrimental effects. ______
5. Give an example to show that the pollutants studied in this laboratory can have an effect on the human population. ______
6. When excess nutrients enter an aquatic ecosystem, long-term effects can result. Why? ______
7. Describe how the cultural eutrophication study supports the hypothesis that a balance of population sizes in ecosystems is beneficial. ______
8. When pollutants enter an ecosystem, they have far-ranging effects. Use acid rain and a food chain to support this statement. ______
9. Contrast species composition with species diversity of ecosystems. ______
10. Suppose among sunflower seeds, a particular variety can germinate despite acidic conditions. What do you predict about the survival of that sunflower variety in today's acidic environment compared to the rest of the population? ______

    What do we call a change in a population's phenotype composition due to the presence of an environmental agent? ______

Credits

Image Researcher: Evelyn Jo Johnson

Laboratory 1

Page 1 (pillbugs): © James Robinson/Animals Animals; **1.1:** © National Cancer Institute/Science Source; **1.2:** Courtesy Leica Microsystems, Inc.

Laboratory 2

Figure 2.1: © Hill Street Studios/Harmik Nazarian/Getty RF; **2.2:** © Bruce M. Johnson; **2.5:** © B.A.E. Inc./Alamy RF; **2.6** (both): © Richard Hutchings; **2.8:** © moodboard/Alamy RF.

Laboratory 3

Figure 3.2a: © Michael Ross/Science Source; **3.2b:** © CNRI/SPL/Science Source; **3.2c:** © Steve Gschmeissner/Science Source; 3.3–3.4: Courtesy Leica Microsystems, Inc. **3.6:** © Dr. Gopal Murti/Science Source; **3.7:** © Kevin and Betty Collins/Visuals Unlimited; **3.8:** © Tom Adams/Visuals Unlimited; **p. 33** (biuret test): © David S. Moyer.

Laboratory 4

Figure 4.3a: © Jeremy Burgess/SPL/Science Source; **p. 35** (iodine test): © Martin Shields/Science Source; **p. 37** (Benedict's test): © David S. Moyer; **4.6–4.8:** © McGraw-Hill Education. John Thoeming, photographer.

Laboratory 5

Figure 5.3 (both): Courtesy Ray F. Evert, University of Wisconsin; **5.7** (all): © David Phillips/Visuals Unlimited; **p. 52** (elodea cells): © Alfred Owczarzak/Biological Photo Service.

Laboratory 6

Figure 6.2: © McGraw-Hill Education. Jill Braaten, photographer and Anthony Arena, Chemical Consultant.

Laboratory 8

Figure 8.1: © Dwight Kuhn; **8.3** (shoot tip): © Steven P. Lynch; **8.3** (root tip), **8.4:** Courtesy Ray F. Evert, University of Wisconsin; **8.6b:** © Biodisc/Visuals Unlimited; **8.7b:** © Carolina Biological Supply Company/Phototake; **8.8a:** © John D. Cunningham/Visuals Unlimited; **8.8b:** Courtesy George Ellmore, Tufts University; **8.8a:** © Dr. Robert Calentine/Visuals Unlimited; **8.8b:** © McGraw-Hill Education. Evelyn Jo Johnson, photographer; **8.8c:** © Brad Mogen/Visuals Unlimited; **8.8d:** © Tim Laman/National Geographic/Getty RF; **8.10a:** © Ed Reschke; **8.10b:** Courtesy Ray F. Evert, University of Wisconsin; **8.11** (left): © Carolina Biological Supply Company/Phototake; **8.11** (right): © Kingsley Stern; **8.13b:** © Carolina Biological Supply Company/Phototake.

Laboratory 9

Figure 9.1: © Science Photo Library/Getty Images; **9.3:** © Andrew Syred/Science Source; **9.5** (early prophase, anaphase, telophase): © Ed Reschke; **9.5** (metaphase): © Ed Reschke/Getty Images; **9.6** (top): © Thomas Deerinck/Visuals Unlimited; **9.6** (bottom): © Steve Gschmeissner/SPL/Getty RF; **9.7** (interphase, prophase, metaphase, anaphase): © Carolina Biological Supply Company/Visuals Unlimited; **9.7** (telophase): © Robert Calentine/Visuals Unlimited; **9.8:** © Biophoto Associates/Science Source.

Laboratory 10

Figure 10.1: © Thierry Berrod, Mona Lisa Production/Science Source; **10.3–10.4** (all): © Ed Reschke; **p. 106** (dogs): © American Images, Inc/Getty Images.

Laboratory 11

Figure 11.4: © Evelyn Jo Johnson; **11.5:** © Carolina Biological Supply Company/Phototake.

Laboratory 12

Figure 12.2a: © Superstock; **12.2b:** © HFPA, 63rd Golden Globe Awards; **12.2c–f, h:** © McGraw-Hill Education. Bob Coyle, photographer; **12.2g:** © Corbis RF; **p. 134** (Turner): Courtesy UNC Medical Illustration and Photography; **p. 134** (Poly X): Courtesy The McElligott Family; **p. 134** (Klinefelter): Courtesy Stefan D. Schwarz, http://klinefeltersyndrome.org; **p. 134** (Jacob): Courtesy The Giles Family; **12.6a–d:** © CNRI/SPL/Science Source.

Laboratory 13

Figure 13.9 (both): © Bill Longcore/Science Source; **153** (fossil): © Annie Griffiths Belt/Corbis.

Laboratory 14

Figure 14.2 (trilobites): © Danita Delimont/Getty Images; **14.2** (pseudoscorpion, mantis): © John Cancalosi/Getty Images; **14.2** (snails): © Ed Reschke/Getty Images; **14.2** (echinoderms): © DEA/G. Nimatallah/Getty Images; **14.2** (ammonite): © Carl Pendle/Getty Images; **14.3** (frog): © DEA/G. Cigolini/Getty Images; **14.3** (snake): © John Cancalosi/Getty Images; **14.3** (duckbill dinosaur): © Kevin Schafer/Getty Images; **14.3** (bird): © WaterFrame/Alamy; **14.3** (deerlike mammal): © Gary Ombler/Getty Images; **14.3** (fish): © John Cancalosi/Getty Images; **14.4** (sassafras leaf): © Jonathan Blair/Getty Images; **14.4** (ferns): © John Cancalosi/Getty Images; **14.4** (seed plant leaves): © Sinclair Stammers/Science Source; **14.4** (maple leaf): © Biophoto Associates/Science Source; **14.4** (poplar leaf): © James L. Amos/Science Source; **14.4** (flower): © Barbara Strnadova/Science Source; **14.5** (bird): © McGraw-Hill Education; **14.5** (man): © McGraw-Hill Education. Eric Wise, photographer; **14.5** (bat): © Jack Milchanowski/Getty Images; **14.5** (cat): © Marc Henrie/Getty Images; **14.5** (lizard): © Mauricio Handler/Getty Images; **14.9** *(A. ramidus)*: © Richard T. Nowitz/Science Source; **14.9** *(A. afarensis)*: © Scott Camazine/Alamy; **14.9** *(A. africanus)*: © Philippe Plailly/Science Source; **14.9** *(H. habilis)*: © Kike Calvo VWPics/Superstock; **14.9** *(H. sapiens)*: © Kenneth Garrett/Getty Images.

Laboratory 15

Figure 15.1 (bacteria): © Dr. Dennis Kunkel/Phototake; **15.1** (paramecium): © Michael Abbey/Visuals Unlimited; **15.1** (morel): © Corbis RF; **15.1** (sunflower): © Photodisc Green/Getty RF; **15.1** (snow goose): © Winfried Wisniewski/Getty RF; **p. 172** (Gram-stained bacteria): © Science Source; **15.2:** © Henry Aldrich/Visuals Unlimited; **15.3** (both): © Kathy Park Talaro; **15.4a:** © Dr. Richard Kessel & Dr. Gene Shih/Visuals Unlimited; **15.4b:** © Gary Gaugler/Visuals Unlimited; **15.4c:** © SciMAT/Science Source; **15.5a:** Courtesy Steven R. Spilatro, Marietta College, Marietta, OH; **15.5b:** © Sherman Thomas/Visuals Unlimited; **15.6:** © M.I. Walker/Science Source; **15.7:** © R. Knauft/Science Source; **15.9:** © M.I. Walker/Science Source; **15.10** (left): © John D. Cunningham/Visuals Unlimited; **15.10** (right): © Carolina Biological Supply Company/Visuals Unlimited; **15.11** (rockweed): © D.P Wilson/Eric & David Hosking/Science Source; **15.12a:** © Steven P. Lynch; **15.12b:** © Gary R. Robinson/Visuals Unlimited; **15.13:** © Dennis Kunkel Microscopy, Inc./Phototake; **15.14:** © Biophoto Associates/Science Source; **15.15c:** © Eye of Science/Science Source; **15.15d** (left): © London School of Hygiene and Tropical Medicine/Science Source; **15.15d** (right): Courtesy CDC/Dr. Mae Melvin; **15.16b:** © Tom E. Adams/Visuals

Unlimited; **15.18a:** © Biophoto Associates/Science Source; **15.18b:** © Bill Keogh/Visuals Unlimited; **15.18c:** © Gary R. Robinson/Visuals Unlimited; **15.18d:** © Carol Wolfe/photographer; **15.19a:** © Gary T. Cole/Biological Photo Service; **15.20:** © Garry DeLong/Getty Images; **15.21a-b:** © Carolina Biological Supply Company/Visuals Unlimited; **15.23a:** © Everett S. Beneke/Visuals Unlimited; **15.23b:** © John Hadfield/SPL/Science Source; **15.23c:** © P. Marazzi/SPL/Science Source.

Laboratory 16

Figure 16.1: © Steven P. Lynch; 16.3 (right): © Kingsley Stern; **16.3** (left): © Dr. John D. Cunningham/Visuals Unlimited; **16.5** (top): © Peter Lilja/Getty Images; **16.5** (bottom): © Steven P. Lynch; **16.6:** © Steve Solum/Photoshot; **16.7:** © Matt Meadows/Getty Images; **16.8:** © McGraw-Hill Education. Carlyn Iverson, photographer; **16.9** (cycad): © D. Cavagnaro/Visuals Unlimited; **16.9** (ginkgo): © Kingsley Stern; **16.9** (conifer): © D. Giannechini/Science Source; **16.9** (gnetophyte): © Virigina Weinland/Science Source; **16.10** (pollen grains): © Carolina Biological Supply Company/Phototake; **16.12a:** © J.R. Waaland/Biological Photo Service; **16.12b:** © Ed Reschke; **16.13:** © Carolina Biological Supply Company/Phototake.

Laboratory 17

Figure 17.3a: © Amar and Isabelle Guillen, Guillen Photography/Alamy; **17.3b:** © Andrew J. Martinez/Science Source; **17.3c:** © Kenneth M. Highfill/Science Source; **17.5:** © Carolina Biological Supply Company/Visuals Unlimited; **17.6** (both): © Kim Taylor/npl/Minden Pictures; **17.7a:** © Azure Computer & Photo Services/Animals Animals; **17.7b:** © Ron & Valerie Taylor/Photoshot; **17.7c:** © NHPA/Charles Hood/Photoshot RF; **17.7d:** © Amos Nachoum/Corbis; **17.9:** © Carolina Biological Supply Company/Phototake; **17.10** (left): Photography by Marc C. Perkins, Orange Coast College, Costa Mesa, CA; image blending by Heather Bartell, Huntington Beach, CA; **17.10** (right): © Tom E. Adams/Visuals Unlimited; **17.12a:** © James Webb/Phototake; **17.13a, c:** © Lauritz Jensen/Visuals Unlimited; **17.14:** © Carolina Biological Supply Company/Phototake; **p. 222** (elephantiasis): © Vanessa Vick/The New York Times/Redux; **17.15a:** © Wim van Egmond/Visuals Unlimited.

Laboratory 18

Figure 18.1a: © Fred Bavendam/Minden; **18.1b:** © Andrew J. Martinez/Science Source; **18.1c:** © IT Stock Free/Alamy RF; **18.1d:** © Douglas Faulkner/Science Source; **18.3b:** © Ken Taylor/Wildlife Images; **18.5a:** © Roger K. Burnard/Biological Photo Service; **18.5b:** © R. DeGoursey/Visuals Unlimited; **18.5c:** © Diane R. Nelson; **18.5d:** © C.P. Hickman/Visuals Unlimited; **18.6:** © John Cunningham/Visuals Unlimited; **18.10a** (honeybee): © R. Williamson/Visuals Unlimited; **18.10a** (millipede): © Bill Beatty/Visuals Unlimited; **18.10a** (centipede): © Adrian Wenner/Visuals Unlimited; **18.10b** (spider): © Dr. William Weber/Visuals Unlimited; **18.10b** (scorpion): © David M. Dennis; **18.10b** (horseshoe crab): © E.R. Degginger/Science Source; **18.10c** (crab): © Tom McHugh/Science Source; **18.10c** (shrimp): © Alex Kerstitch/Visuals Unlimited; **18.10c** (barnacles): © Kjell Sandved/Visuals Unlimited; **18.13a:** © Daniel Gotshall/Visuals Unlimited; **18.13b:** © Hal Beral/Visuals Unlimited; **18.13c:** © Robert Dunne/Science Source; **18.13d:** © Neil McDan/Science Source; **18.13e:** © Robert Clay/Visuals Unlimited; **18.13f:** © Alex Kerstitch/Visuals Unlimited; **18.14** (both): © BiologyImaging.com; **18.16** (shark): © Hal Beral/Visuals Unlimited; **18.16** (fish): © Patrice/Visuals Unlimited; **18.16** (frog): © Rod Planck/Science Source; **18.16** (turtle): © Suzanne and Joseph Collins/Science Source; **18.16** (bird): © Bill Horn/Acclaim Images RF; **18.16** (fox): © Craig Lorenz/Science Source.

Laboratory 19

Figure 19.1 (simple squamous, pseudostratified, cuboidal, columnar, cardiac, skeletal, nervous, cartilage): © Ed Reschke; **19.1** (smooth, bone, dense): © McGraw-Hill Education. Dennis Strete, photographer; **19.1** (blood, adipose): © McGraw-Hill Education. Al Telser, photographer; **p. 248** (simple squamous): © Ed Reschke; **p. 249** (simple cuboidal, columnar), **p. 250** (pseudostratified), **p. 251** (loose): © Ed Reschke; **p. 251** (dense): © McGraw-Hill Education. Dennis Strete, photographer; **p. 252** (adipose): © McGraw-Hill Education. Al Telser, photographer; **p. 252** (bone): © Ed Reschke/Getty Images; **p. 253** (hyaline), **19.2a-d:** © Ed Reschke; **19.2e:** © R. Kessel/Visuals Unlimited; **p. 255** (skeletal, cardiac): © Ed Reschke; **p. 256** (smooth): © McGraw-Hill Education. Dennis Strete, photographer; **19.3b–19.4:** © Ed Reschke; **19.5:** © Garry DeLong/Science Source.

Laboratory 20

Figure 20.3b, 20.6: © Ken Taylor/Wildlife Images.

Laboratory 21

Figure 21.2: © Ed Reschke.

Laboratory 22

Page 285 (heart and lungs): © SIU/Visuals Unlimited; **22.6:** © Ralph T. Hutchings/Visuals Unlimited; **22.8, 22.11:** © McGraw-Hill Education. Carlyn Iverson, photographer.

Laboratory 23

Figure 23.4: © Ilene MacDonald/Alamy RF; **23.6b:** © Dr. Keith Wheeler/Science Source; **23.6c:** © CMSP/Getty Images; **23.8:** Courtesy Phipps & Bird, Inc., Richmond, VA.

Laboratory 24

Figure 24.2 (all): Courtesy Dr. J. Timothy Cannon; **24.5:** © Kage Mikrofotografie/Phototake; **24.7:** © P.H. Gerbier/SPL/Science Source; **24.8:** © S.L. Flegler/Visuals Unlimited.

Laboratory 25

Page 337 (zygote): © Anatomical Travelogue/Science Source; **p. 337** (embryo, 1 week): © Bettman/Corbis; **p. 337** (embryo, 8 weeks): © Neil Harding/Getty Images; **p. 337** (fetus, 3 months): © Petit Format/Science Source; **p. 337** (fetus, 5 months): © Ralph T. Hutchings/Visuals Unlimited; **25.1a:** © Dr. Robert Calentine/Visuals Unlimited; **25.1b–f, 25.5** (neural plate, neural groove): © Carolina Biological Supply Company/Phototake; **25.5** (advanced neurula): © Alfred Owczarzak/Biological Photo Service; **25.6–25.7a:** © Carolina Biological Supply Company/Phototake; **25.11:** © Martin Rotker/Phototake; **p. 351** (3-4 & 7-8 month fetus): © Petit format/Science Source; **25.12g–h:** © Ralph Hutchings/Visuals Unlimited.

Laboratory 26

Figure 26.1: © McGraw-Hill Education. Pat Watson, photographer; **26.3:** © NOAA/Visuals Unlimited.

Credit Lines

Lab 1: Laboratory adapted from Kathy Liu, "Eye to Eye with Garden Snails." Reprinted by permission of Kathy Liu, Port Townsend WA. http://www.accessexcellence.org/AE/AEC/AEF/1994/liu_snails.html

Figure 18.11: After Carolina Biological Supply.

Figure 18.12: After Carolina Biological Supply.

Figure 24.14: From *The Complementarity of Structure and Function: A Laboratory Block* by A. Glenn Richards. Copyright © 1965 by D.C. Heath, a division of Houghton Mifflin Harcourt Publishing Company. All rights reserved. Reprinted by permission of Holt McDougal, a division of Houghton Mifflin Harcourt Publishing Company.

Figure 26.3: Drawing by Kristine A. Kohn. From Harriett Stubbs' "Acid Precipitation Awareness Curriculum Materials in the Life Sciences." *The American Biology Teacher* (1983), 45(4), 221. With permission from the National Association of Biology Teachers.

Index

Note: Page numbers followed by *f* refer to figures; page numbers followed by *t* refer to tables.

A

B

D

E

G

H

M

Q

R